I0752409

TEXAS BRIGADIER TO THE FALL OF ATLANTA

JOHN BELL HOOD

STEPHEN DAVIS

MERCER UNIVERSITY PRESS
Macon, Georgia
1979–2019
40 Years of Publishing Excellence

MUP/ H980

© 2019 by Mercer University Press
Published by Mercer University Press
1501 Mercer University Drive
Macon, Georgia 31207
All rights reserved

9 8 7 6 5 4 3 2 1

Books published by Mercer University Press are printed on acid-free paper that meets the requirements of the American National Standard for Information Sciences—Permanence of Paper for Printed Library Materials.

Printed and bound in the United States.

This book is set in Adobe Caslon Pro (text) and Georgia (display).

Cover/jacket design by Burt&Burt

ISBN 978-0-88146-720-8
Cataloging-in-Publication Data is available from the Library of Congress

To my most perfectest wife, Billie

Contents

List of Maps

Cartography by Hal Jesperson

Preface

It has been more than three decades since the publication of Richard M. McMurry's seminal biography, *John Bell Hood and the War for Southern Independence* (1982). Since that time, scholars have added to the literature about the Confederacy's youngest full general. Brian Craig Miller's *John Bell Hood and the Fight for Civil War Memory* (2010), while not a biography as such, has added details to Hood's life-story. Stephen M. Hood's *John Bell Hood: The Rise, Fall, and Resurrection of a Confederate General* (2013) presents essentially a bibliographical polemic defending the historical reputation of his distant ancestor. The same author offers more value in his collecting and editing of a trove of General Hood's personal papers that he discovered in the Philadelphia attic of a descendant in 2012 (Stephen M. Hood, ed., *The Lost Papers of Confederate General John Bell Hood* [2015]).

These works suggest—happily—that a century and a half after the Civil War, there is still previously unknown material about John Bell Hood to be found and discussed.

Here's an example. In one of the previously unpublished letters in *Lost Papers,* the widow of Bishop Polk wrote Hood from Nashville in June 1874, thanking him for sending her a copy of his "strictures" against Johnston's *Narrative,* the book that had just been published. Hood's article, Mrs. Polk wrote, had been printed in "the Times." I then had to figure out which *Times*? I believed New York and Philadelphia were out; but what about the New Orleans *Times,* which was published in the 1870s before merging with the *Picayune*? Fortunately, I chanced upon a monograph from 1912—on Federal blockhouses, of all things—which confirmed that General Hood had written articles for the New Orleans *Times* in the spring of 1874.

In further research, the excellent staff of the Louisiana Research Collection at Tulane University's Howard-Tilton Library told me that they had the New Orleans *Times* of the 1870s, but only in originals, not on microfilm or online. They invited me to visit the library, which I did a few years ago, thanks to an invitation to speak to the New Orleans Civil War Round Table. Going through the pristine original copies of the *Times,* sure enough, I found on the front page of its May 10, 1874 issue, "Gen. J. B. Hood's Reply to Certain Statements Contained in Gen. Joseph E. Johnston's Book." It was followed by another "Reply" in the paper's issue of May 31, which was continued in the *Times* of June 7. Reading the articles, I realized their similarity to the six chapters that form the central section of Hood's memoir, *Advance and Retreat.* With closer look I saw that in long sections, the text of the *Times* articles and Hood's book are identical.

Previous writers, even McMurry, have been unaware that portions of Hood's Chapters IV-IX were printed in New Orleans newspapers five years before his death in 1879. My discovery explains what Hood meant when, in *A & R,* he states, "as I have remarked in my reply to General Johnston." It also explains why the chapters in *Advance and Retreat* read like a somewhat disjointed midsection, kind of plopped into the book. When Stephen Hood complains that the "Reply" chapters in *A & R* "turned the book into a polemic," I want to say, Sam, they *read* like a polemic because Hood originally *wrote* them as a polemic.

There's more. Students of the war's western theater are aware that since the Civil War Centennial of the 1960s, historians' assessment of John Bell Hood's generalship has been evolving. Thomas Lawrence Connelly led the way in *Autumn of Glory: The Army of Tennessee, 1862-1865* (1971) with interpretations that cast Hood's leadership in new light. A decade later, McMurry's biography brought a trained historian's perspective to the subject. I began writing about the Atlanta Campaign in the late 1980s, as Albert Castel was finishing his definitive *Decision in the West: The Atlanta Campaign of 1864* (1992), which elevated Hood's performance even as it diminished Sherman's.

More recent monographs have treated battles fought during Hood's service as corps commander in the Army of Tennessee (February-July 1864) and his defense of Atlanta (July-September): Gary Ecelbarger, *The Day Dixie Died: The Battle of Atlanta* (2010); Earl J. Hess, *Kennesaw Mountain: Sherman, Johnston, and the Atlanta Campaign* (2013); and Robert D. Jenkins, *The Battle of Peach Tree Creek: Hood's First Sortie, 20 July 1864* (2013).

Clearly, there is much happening in Hoodiana. Several years ago I came upon the belief that the time was right for a long view: an analysis of Hood in 1864, the crucial year of his military career, both his six months as corps commander, and his half-year as leader of the Army of Tennessee (July 1864-January 1865). Recent writing on Hood's Tennessee Campaign—James Lee McDonough's *Nashville: The Western Confederacy's Final Gamble* (2004), Eric A. Jacobson's *For Cause & For Country: A Study of the Affair at Spring Hill and the Battle of Franklin* (2008) and Derek Smith's *In the Lion's Mouth: Hood's Tragic Retreat from Nashville, 1864* (2011)—encouraged me to engage in an in-depth discussion of the events of Spring Hill, Franklin and Nashville in November-December 1864. At the same time, I have kept an eye on the standard sources, especially the two early biographies, Richard O'Connor, *Hood: Cavalier General* (1949) and John P. Dyer, *The Gallant Hood* (1950), not to mention the general's memoir, *Advance and Retreat* (1880).

Reflecting on this literature is one of the purposes of my work. As a consequence, this book is a blend of historical narrative and historiographical discourse. Here I wish to thank Dr. Marc A. Jolley, Director of Mercer University Press, for the latitude to pursue my subject in this way.
Mercer has also generously not imposed limits on the length of my manuscript. When finished, Marc and I agreed that it was better to publish it in two volumes. The present one, *Texas Brigadier to the Fall of Atlanta,* includes Hood's rise from lieutenant to major general in Virginia. Volume two, *Into Tennessee and Failure,* will follow Hood and his troops through their legendary charge at Franklin and their tragic rout at Nashville, to Hood's resignation in January 1865. An Epilogue will carry Bell Hood after the war to marriage in New Orleans, fatherhood, business ventures, memoir writing and eventually to death in August 1879, when yellow fever struck the city. Volume two should be available in late 2020.

We will include a bibliography of works cited in each volume. I also wish to thank the talented cartographer Hal Jesperson of Redwood City, California, for the maps featured in our work.

With this explanation, I will close with an author's oft-expressed hope that my book will have advanced Civil War scholarship.

Introduction

Late in life, as he was completing his memoirs, John Bell Hood wrote, "no man is justly entitled to be considered a great General, unless he has won his spurs."

To be sure, Hood did not explain how an officer "earned his spurs," but he really didn't have to. One may assume that such an accomplishment came about when a soldier conscientiously performed his duty and gave his all in carrying out his country's expectations of him.[1]

In summer 1864, Hood had become the Confederacy's youngest full general and had been given command of one of the South's two largest armies, with instructions to save the city of Atlanta from enemy capture. As we shall see in this work—the first of two volumes exploring Hood's career as Confederate general—Hood failed to hold Atlanta, which Sherman's troops captured on September 2, 1864.

I will attempt to show that even if Hood failed to save Atlanta—and thereby achieve greatness as a general—he undeniably had "earned his spurs" in doing everything he could to save the city.

In the end, of course, Hood had failed his first, dramatic test as an officer with full Confederate general's rank. As a result, writers of history have concluded that Hood was a nineteenth-century embodiment of the "Peter Principle"—the American sociological notion that in hierarchical organizations, individuals will rise to their levels of incompetence.[2] Historians have judged that Hood's ambition took him to levels of military rank for which he was not qualified, and that the Confederate government mistakenly appointed him to those positions.[3]

Logical as it is, and so easily demonstrable, such a view casts Hood in a tragic, foredoomed, even predestined light. History, and historiography, should demand a closer look. How did John Bell Hood come to be the soldier

---

[1] Hood, *Advance and Retreat*, 316. Cited hereafter as *A & R*.

[2] Peter and Hull, *The Peter Principle*.

[3] Patterson, "John Bell Hood," 12 ("the blunder of advancing him far beyond his abilities"); McMurry, "John Bell Hood," 3:127 ("he is often cited as an example of the 'Peter Principle'—one promoted beyond the level of his or her competency"); Hattaway, "The General Whom the President Elevated Too High," 84–103, 187–88; Bohannon, "'A Bold Fighter,'" 249–87.

that he was? How did he rise through the Confederate army from first lieutenant to full general? And having achieved that rank, how did he perform his duty?

These are the questions we are about to explore.

Prologue: Hood before 1864

In his reminiscences, written in the late 1870s, John Bell Hood devotes just eleven pages to his antebellum life and military career before the Civil War—his first thirty years. But from his first written line, Hood's life was all about soldiering. His grandfathers had fought under Washington, so he believed he had it in his blood. His father, a physician, lost out when he suggested a career in medicine; no, young Hood wanted to go to West Point. An uncle in Congress, Richard French, of Kentucky's Ninth District, helped get him an appointment to the military academy, which Hood entered in July 1849 at the age of eighteen. He does not mention his courses and studies in his memoir, probably because he did not do well in them. Hood graduated in 1853, forty-fourth in his class of fifty-two cadets. Nor was his behavior as cadet exemplary; he accrued 374 demerits during his four years at West Point. Yet a recent student, Brian Craig Miller, has noted that Hood's record was similar to a lot of other cadets', and that Superintendent Robert E. Lee never had to write Hood's parents about their son's academic performance. Moreover, Miller notes that Hood's class standing, forty-fourth out of fifty-two, was actually a lot better than a number of other classmates who would later rise to fame in the Civil War. Jefferson Davis, William J. Hardee, and Edmund Kirby Smith all graduated in the bottom half of their respective classes. In the class of 1842, Lafayette McLaws was forty-eighth out of fifty-six; Earl Van Dorn fifty-second out of fifty-six; and James Longstreet, Lee's famed "War Horse," fifty-fourth out of fifty-six. Confederate cavalry leader Joseph Wheeler was nineteenth out of twenty-two, and Fitzhugh Lee, Robert E. Lee's nephew, was forty-fifth out of his class of forty-six. George Pickett and Harry Heth, division commanders in the Army of Northern Virginia, were both dead-last in their classes.[1]

---

[1] Hood, *A & R*, 1880 ed., 5; S. Davis, "John Bell Hood," 152; McMurry, *John Bell Hood*, 6; Miller, *John Bell Hood*, 22–23. Angelina V. Winkler of Corsicana, Texas, in an early biographical narrative of Hood, mentions that "his father was a physician…and was desirous his son should adopt the same profession," to the point of offering a European medical education (Winkler, *The Life and Character of Gen. John B. Hood*, 6). Winkler was the wife of Col. Clinton M. Winkler of the 4th Texas (N. A. Davis, *Chaplain Davis and Hood's Texas Brigade*, 17). My sincerest thanks go to Sean Benjamin, public services librarian, Louisiana Research Collection

Upon graduation, cadets were given the rank of second lieutenant in the army; the branch of it was determined by one's class ranking. The elite units were the engineers, then came the cavalry, finally the infantry. Hood was obviously destined for the latter, but he hoped to be assigned to the dragoons. Again hc called on his uncle for assistance. On May 24, 1853, French wrote Adjutant General Samuel Cooper, "A nephew of mine, John Hood, is about to graduate at West Point, and writes me he wishes to be attached to the Draggoon Service." This time the uncle with connections failed to help; Richard French had been out of Congress four years. Hood received his commission in July 1853 and was assigned to the 4th Infantry. The regiment was posted at Fort Jones in northern California, where Hood served as post adjutant during 1854–1855. The post was isolated and service there was dull and lonely. Hood was not alone in wanting transfer elsewhere, especially to the cavalry, which promised more excitement. The massacre of U.S. soldiers by Indians in Utah Territory in August 1854 caused Secretary of War Jefferson Davis to call for several new army regiments to protect the frontier. Congress was initially unresponsive, but word of Davis's proposal spread through the army. Hood determined to apply for transfer to the cavalry should the politicians expand the military. Uncle French was no longer able to help; he had died three months before. For assistance this time, Hood wrote to another Kentuckian and family friend, John C. Breckinridge. Breckinridge had retired from Congress to private life in Lexington, but he was still well connected in Washington. Hood asked Breckinridge "to use your influence towards my getting transferred to one of the Cavalry Regiments, in case an increase of the army should be brought about." Then, on the last day of the 33rd Congress, March 3, 1855, the legislators finally acted on Secretary Davis's proposal, authorizing four new army regiments—two of infantry, two cavalry. The War Department acted quickly: announcement of the new cavalry regiments went out on March 12, and two weeks later officers of the 2nd Cavalry were announced. When some of those appointees declined, the vacancies gave Hood another chance to apply for transfer. This time he went straight to the top, in a letter of April 25, 1855, to the secretary of war himself. Hood knew he was breaking the rules in so corresponding and apologized to Davis. But he feared going through channels would delay his application and knew also that any openings would be filled quickly. Three weeks later, on May 12, Hood wrote another letter from Fort Jones, this one again to former Congressman

---

at the Howard-Tilton Memorial Library of Tulane University for locating and providing me a copy of Winkler's seldom-cited booklet.

Breckinridge. Hood reminded him of his earlier letter concerning transfer, adding now that Congress had created the new cavalry regiments:

> I have forwarded an application to the Secretary of War to be laid before the President, for such a transfer, in case someone should refuse to accept their transfer from an older Reg't, and as I know of *one* that has *refused*, I would be more than obliged if you would use your influence so soon as you get this, to have me transferred to that arm of the Service.

Hood urged Breckinridge's speedy attention, as he worried that forthcoming graduates at West Point, getting their commissions on July 1, might get one of the cavalry openings. Whether Breckinridge acted is not known; Hood's letter to the secretary of war alone may have done the job. On August 4, while out in the field, Hood learned of his transfer to the 2nd Cavalry as second lieutenant. He was leading a detachment escorting engineers as they looked for possible route for a railroad through the western wilderness—Lt. Phil Sheridan brought the news.

Three months later, in October 1855, Hood reported to his new regiment at Jefferson Barracks near St. Louis. With the 2nd Hood marched through Arkansas and the Indian Territory to Texas. Second-in-command of the regiment was Lt. Col. Robert E. Lee, who had been superintendent of West Point during Hood's last year there. Years later, Hood recalled the fatherly advice Lee gave him during one of their rides over the prairie: "Never marry unless you can do so into a family which will enable your children to feel proud of both sides of the house." Lee's sage words may have been one of the reasons Hood married so late in life.[2]

In July 1857, while leading a mounted patrol near the Rio Grande, Lieutenant Hood got his first taste of combat when some Comanches attacked. The skirmish was brief but bloody, and Hood was wounded by an arrow through his left hand. But the cavalrymen drove off the Indians; General David Twiggs, in charge of the Department of Texas, wrote his superiors, "Lieut. Hood's affair was a most gallant one." The young officer was proud of his battle-baptism, to which he devoted a full six pages in his memoir, based on the copy of the after-action report he kept in his personal papers. Biographers too see importance in Hood's first fight. Writing in 1950, John P. Dyer judged that the battle showed Hood to be "a superb combat officer whom battle transformed from a shy young dandy into a fierce, almost terrible, fighter." Indeed, Dyer asserted, "nothing revealed this metamorphosis more

---

[2] McMurry, *John Bell Hood*, 10–11, 13–15, 207 n. 7; Bohannon, "'A Bold Fighter,'" 250–51; Dyer, *Gallant Hood*, 37; Hood, *A & R*, 8.

than his eyes. Normally sad and faunlike, they blazed in battle with a light that almost frightened his own men."[3]

Hood's only promotion in the old army, to first lieutenant, occurred in August 1858. But he had evidently caught the attention of his superiors as an officer of talent. In November 1860, Hood received an order to report to West Point and assume a faculty position as instructor of cavalry tactics. By then, however, Abraham Lincoln had been elected president. The secession of Southern states loomed and with it the possibility of civil war. In Washington, Hood appeared before the army's adjutant general, Col. Samuel Cooper, and asked to be relieved from the order. He recorded Cooper's response: "Lieutenant, you surprise me; this is a post and position sought by almost every soldier." Hood, though, had made up his mind, explaining that if war came, he wanted to be able to follow his conscience. Besides, duty at West Point was bound to be duller than service in the cavalry out west had been. Cooper granted his request and Hood spent the next several months on furlough. On January 15, 1861, with four Southern state conventions having passed ordinances of secession, Lieutenant Hood wrote Gov. Beriah Magoffin of Kentucky. "I see that dissolution is now regarded as a fixed fact," he began, "and that Kentucky will have an important part to perform in this great movement." The "great movement" to establish Southern independence would likely lead to war, so Hood informed the governor, "I thereby have the honor to offer my sword & services to my native state." A few days later he explained his situation to the governor in a second letter. He was still a U. S. army officer and would remain so while Kentucky stayed in the Union. "But whenever Kentucky leaves the Union," he affirmed, "I shall, with the greatest pleasure obey any call from the governor." Later, in the company of brother officers, Hood announced his "intention to resign if Kentucky should secede." But the Bluegrass State did not. Hood was in Kentucky at the time of the firing on Fort Sumter, talking with political leaders about what was likely to happen. On April 16, the day after Lincoln called for 75,000 volunteers to suppress the "rebellion," Hood wrote his letter of resignation from the army.[4]

Within two weeks of his resignation, Hood traveled to Montgomery, the capital of the newly formed Confederate States of America. He offered his

---

[3] Hood, *A & R*, 8–15; Twiggs to Lt. Col. Lorenzo Thomas, August 5, 1857, and Hood to Lt. Charles Phifer, July 27, 1857, in S. Hood, *Lost Papers*, 6–8; H. B. Simpson, *Cry Comanche*, 86–93; Dyer, *Gallant Hood*, 45.

[4] McMurry, *John Bell Hood*, 17, 21, 25; Hood, *A & R*, 15; S. Hood, *Lost Papers*, 9–10, 41 (to his death Hood kept his first lieutenant's promotion certificate, of which he was evidently proud).

services to the C. S. army and was made first lieutenant of cavalry on April 20. Soon Hood was ordered to report to Governor Magoffin at Frankfort; Confederate adjutant general Samuel Cooper recorded that Hood would assist in recruiting activity at Louisville. But he was back in a month. On May 24, Hood, in Nashville, was reimbursed for the costs of his travel from Montgomery to Frankfort.

The War Department's plans for Hood soon changed. By orders dated May 16, Hood was instructed to proceed to Richmond—which would soon become the Confederate capital—and there report to Robert E. Lee, who was serving both as major general of Virginia state forces and as brigadier in the Confederate army.[5] Lieutenant Hood held Lee in great respect, both from his time at West Point and out west. As Richard McMurry, Hood's most thorough biographer, writes, "Hood's admiration for Lee seems to have been especially intense." For this reason, in his memoir Hood detailed his meeting in Richmond with Lee, which probably occurred in late May. Writing more than a decade later, Hood recalled that the general's office was on the third or fourth floor of Mechanic's Hall, across the street from the state capitol. "He extended me a most cordial greeting, and, taking me by the hand, said, 'I am glad to see you. I want you to help me.'" Lee elaborated: "I wish you to go to Yorktown and report to Colonel Magruder." Hood asked when and described Lee's answer. "He turned his head, looked at the clock, and, with a smile, said he would like me to go before I dined." It was 11 A.M.; Lee wanted Hood to leave immediately.[6]

It is likely that Hood had been promoted to captain by this time, as General Lee so referred to him in his letter to Colonel Magruder of May 27 ("Captains Cosby and Hood, of the Confederate Army, have been ordered to report to you for the purpose of instructing the cavalry troop"). Within a few days Hood was at Yorktown, where Magruder announced that Hood was now "in charge of all the cavalry on the York River": the Old Dominion Company, Charles City Troop, and New Kent Troop. By June 10, he had been promoted to the rank of major. During these first months of the war, as new recruits were doing everywhere, Hood and his men busied themselves with drills and

---

[5] Hood, *A & R*, 16; Register of Appointments, Cooper order no. 31, and payment voucher, May 24, 1861, Hood Service Record; S. Cooper endorsement, April 30, 1861, U. S. War Department, *The War of the Rebellion: A Compilation of the Official Records* (cited hereafter as afterward *OR*), vol. 52, pt. 2, 68; Dyer, *Gallant Hood*, 49. Dyer cites Special Order No. 51, C. S. Adjutant and Inspector General's Office with its date of May 16. Hood's recollection in *A & R* that he arrived in Richmond "about the 5th of May" is therefore off.

[6] McMurry, *John Bell Hood*, 23; Hood, *A & R*, 16–17.

training exercises. These latter included scouting expeditions beyond Magruder's lines at Yorktown, looking for any Federals venturing from their base at Newport News. On July 12, Hood ran into some of them. Four miles from Newport News a detachment of the 7th New York Infantry was found to be lying in ambush. Hood ordered some of his troopers to dismount and advance. They flushed the Yankees into the open whereupon Hood led his command in a charge that routed the enemy. Hood reported twelve New Yorkers captured and two killed, buried on the field, and fourteen muskets gathered up. His own loss was just one horse.

Even in small actions like this, officers can demonstrate their mettle, and the young cavalry leader did so here. After Hood wrote his report on the skirmish, Magruder sent it on to Richmond with a very complimentary endorsement. The fight of the 12th was "a brilliant little affair"; "too much praise cannot be bestowed upon Major Hood." Hood's report found its way to General Lee, who was so pleased that he forwarded it to President Davis. "Will you express to Major Hood," Lee wrote back to Magruder, "and the gallant men who were engaged in the affair the pleasure which their conduct has given both myself and the President." Within a month and a half, when Magruder's cavalry companies were organized as a regiment, Hood was promoted to lieutenant colonel and became second in command.[7]

Hood, the cavalryman in Texas and on the Peninsula, would not long remain so. Arrival in Richmond of several infantry companies from Texas allowed the formation of two regiments, the 4th and 5th Texas. Hood was promoted to colonel of the 4th on September 30, took command the next day, and set to drilling and training his new recruits. As they were encamped a few miles from Richmond, Colonel Hood had opportunity to spend much time in the capital city. He did not make much of an impression. Col. J. C. Haskell, a staff officer for Maj. Gen. Gustavus W. Smith, was introduced to Hood by Smith. Haskell remembered Hood as looking like "a raw backwoodsman, dressed up in an ill-fitting uniform." But it was this very appearance, plus his other attributes, that impressed the rough Texans. In the words of one writer, Hood's "commanding appearance, manly deportment, quick perception, courteous manners and decision of character readily impressed the officers

---

[7] Lee to Magruder, May 27, 1861, *OR*, vol. 2, 883; McMurry, *John Bell Hood*, 25–26 (skirmish of July 12), 208 n.1 (including McMurry's doubt that Hood held a captaincy); *OR*, vol. 51, pt. 2, 122 (Hood as major, May 31); Douglas, "Hood's Messmates at Yorktown," 78; Hood report, *OR*, vol. 2, 297–98; Johnston to Crosby, July 27, OR, vol. 2, 1003 ("Lieutenant-Colonel Hood"); Hood, *A & R*, 18. Hattaway mentions the firefight of July 12 in "The General Whom the President Elevated Too High," 84.

and men that he was the man to govern them in the camp and command them on the field."[8]

On October 22, the 1st, 4th and 5th Texas were brigaded together and came to be called the "Texas Brigade." Its commander, Brig. Gen. Louis T. Wigfall, had been a U.S. senator from Texas before the war. Within a few weeks the brigade was ordered to the front—or at least close to it—where they would become part of Gen. Joseph E. Johnston's Confederate Army of the Potomac. On November 8, Hood and the 4th Texas entrained at Richmond for Fredericksburg, then marched twenty-five miles farther north to Dumfries, which they reached on the 13th. Bushrod W. Frobel, serving in Hood's command, recalled after the war an exchange about that time between opposing soldiers on the picket line.

> YANKEE.—Halloo, Mr. Secesh! What regiment is that?
>
> CONFEDERATE.—First Texas.
>
> YANKEE.—That's a lie. If you'd a been Texicans, you'd 'a come over here long ago and scalped us."

Then the "Texicans" went into winter quarters, building log-and-canvas shacks that were photographed by a brigade member, Solomon T. ("Tom") Blessing. The 4th Texas called their area "Camp Hood," but the other two regiments named their camps from nearby streams. "We built rude huts," remembered a Texan, "and then, with the exception of having to cut green pine poles and coax them to burn to make our huts comfortable, there was not much to do." One winter night Hood and some officers whiled time by opening the Bible to random passages and selecting one for reading aloud. Hood's choice made an impression upon him: "I shall not die, but live" (Ps 118:17).

About November 20, the 18th Georgia regiment arrived at Dumfries and was added to Wigfall's Texas Brigade. During this time, as Susannah Ural has recently written, the Texans observed of Hood "how much time he spent caring for them, listening to them, and visiting with them in camp." The Georgians also took a shine to the new commander. "I am mutch better pleas with gen Hood than I was at first," wrote Milton Barrett of the 18th. Barrett also saw how disciplining the Texans required special handling; "the Texas regiments is the most disapated foot soldiers in servis and it takes the tites dissipland to keep them rite."

---

[8] McMurry, *John Bell Hood*, 27–28; Hood Service Record; Dyer, *Gallant Hood*, 331 n.29 (date of colonelcy); O'Connor, *Cavalier General*, 41; "Gen. Hood," *Mobile Telegraph*, July 26, 1864, reprinted in *New York Times*, August 7, 1864.

Hood also turned his attention to the army's practice of officer election. On January 16, 1862, he wrote General Wigfall, "For Gods sake and our bleeding Country, get Congress to stop this thing of election of Company officers, after they are once organized, let all vacancies be filled by appointment upon the recommendation of the Captains & Colonels. And for the field officers anything to prevent an election." John B. Hood, West Pointer and army veteran, recognized the need for strong leadership, even at the company level.[9]

An aspect of that leadership meant setting high standards for the men and instilling discipline while sometimes good-naturedly winking at minor infractions. Stealing, of course, was a punishable offense, but mere carousing was another matter. Before the regiment had left Richmond, one soldier had so tellingly succumbed to "a decoction of brown sugar and corn juice" that he was brought before the city mayor for an upbraiding. In these cases, as Richard O'Connor has written, "according to all regimental accounts, Hood often laughed off their minor examples of talent for hell-raising." Pvt. Joseph B. Polley (Co. F, 4th Texas) credited Hood's successful leadership as rooted in his "understanding the peculiarities of Texans character," especially how volunteer soldiers "would not submit to the same restriction that would be imposed on regulars." Whenever he got the chance, though, he reminded his men that they were responsible for one another's behavior, especially when they got into town: if one soldier got out of line, comrades should "take him in charge, and not allow his misconduct to bring discredit upon the regiment." Meanwhile, Hood kept the men in exercises of training. "Marksmanship was one subject in which they required little instruction," according to O'Connor, "but drill, parade, tactical maneuvers were drummed into them from sunup to sundown."

Hood also understood that his troops, like many recruits, would respond better to orders whose purpose the officers explained. For instance, the command for lights out and no noise in the soldiers' shanties might have seemed unduly restrictive, but Hood in his memoirs recalled how he explained the rationale: in active operations troops needed their rest, and one loud man at night among six or seven could ruin it for his messmates the next day. Then, too, came the commander's responsibility for exhortation to bravery. "I lost

[9] McMurry, *John Bell Hood*, 29–33; [Frobel], "Field and Camp," (January 1867): 6; Ural and Eiserman, "The Winter That Made the Texas Brigade," 39; David Fuller map of the Dumfries area, in Ural and Eiserman, 43; Rick Eiserman, "'We Have Had a Picture Taken,'" in Ural and Eiserman, 46–47; Lemon, *Feed Them the Steel!*, 18; E. B. Williams, *Hood's Texas Brigade*, 35; Hunter, "A Thirty Days' Scout," 457; Ural, *Hood's Texas Brigade*, 79–80.

no opportunity whenever the officers or men came to my quarters, or whenever I chanced to be in conversations with them," Hood remembered, "to arouse their pride, to impress upon them that no regiment in that Army should ever be allowed to go forth upon the battle-field and return with more trophies of war than the Fourth Texas;—that the number of colors and guns captured, and prisoners taken, constituted the true test of the work done by any command in an engagement." Colors, guns, prisoners captured; Hood was telling his troops that he was ready for battle, and that they were too.[10]

Before that came, Hood was promoted. General Wigfall, the brigade commander, resigned from the army effective February 20; he had been elected to the C. S. Senate from his native Texas. Col. James J. Archer of the 5th Texas was at first given command of the brigade—Archer was senior colonel—but on March 6 Hood was promoted to brigadier general, to rank from the 3rd, and placed in charge. Hood had entered the Confederate service from Texas. This official tie, plus Hood's service on the Texas plains and the Texans' own complaints about being led by any officer not one of their own, justified the War Department's decision (Archer was a native of Maryland). Besides, Archer was not a West Pointer, a factor which always mattered to President Davis. Archer took his oversloughing in gentlemanly manner, visiting Hood at his tent, congratulating him, and professing willingness to serve under him.[11]

Within several days word circulated that Confederate forces would retreat from their position near Manassas to south of the Rappahannock. Some of the men grumbled, but General Hood consoled them in a statement he delivered to his troops: "I had hoped that when we left our winter-quarters, it would be to move forward; but those who have better opportunities of judging than we have, order otherwise. You must not regard it as a disgrace—it is never a disgrace to retreat when the welfare of your country requires such a movement."

The brigade broke camp on April 8 and set out on its march to Fredericksburg. At one point, an icy creek halted the column. Hood led the way by dismounting and wading in: "Come on men, right through. It's not deep." The troops, delighted by their leader's example, followed him across, although

---

[10] McMurry, *John Bell Hood*, 31; Ural, *Hood's Texas Brigade*, 80 (Polley, "Texans character"); O'Connor, *Cavalier General*, 47, 49; Hood, *A & R*, 19.

[11] Hood Service Record; Dyer, *Gallant Hood*, 62, 332 n.34 (date of brigadier's commission); Hood, *A & R*, 16, 20; Hattaway, "The General Whom the President Elevated Too High," 8.

more than one noticed that Hood was wearing waterproof boots and India rubber pants.[12]

The Texas Brigade went into camp a few miles west of Fredericksburg and spent much of a month there. Meanwhile the strategic picture in Virginia was changing. Union Maj. Gen. George B. McClellan, commanding the Army of the Potomac near Washington, was charged with capturing Richmond. Rather than advancing overland, McClellan persuaded the Lincoln administration that it would be better to shift the front from northern Virginia to the peninsula formed by the James and York Rivers, southeast of Richmond, and advance from there. The troop movement began on March 17; by the end of the first week in April, McClellan's entire army of 100,000 men had landed at Fort Monroe. Facing them were only 11,000 Confederates under now-Maj. Gen. John B. Magruder, manning a thirteen-mile line of fortifications stretching from river to river. But the theatrical Magruder marched his troops repeatedly back and forth in full view of the Yankees, such that McClellan believed he faced a larger force and resolved to bring up heavy cannon for a slow siege of Magruder's lines. Meanwhile General Johnston in northern Virginia was directed to send reinforcements, and Hood's troops were among them. On April 7, the brigade began its march to Richmond, taking the train part of the way; on the 14th it set out in march to Yorktown, arriving five days later. Hood's troops took their place in the Confederates' fortified lines. "Our men had an enjoyable time while there," remembered a Texan, as the Federals stayed rather quiet. McClellan was preparing his bombardment of the Rebel works when on May 3 Magruder began a quiet withdrawal, joining forces with Maj. Gen. James Longstreet at Williamsburg and continuing a slow retreat up the peninsula toward Richmond. Everyone knew that a battle loomed ahead. One of Hood's staff officers, J. T. Hunter, remembered a conversation with him, when Hood asked, "Lieutenant, have you any presentiment of being hurt in this battle?" Hunter replied, "No, General. I have just made up my mind to do my duty as I understand it without reference to results, as far as I am individually concerned." Hood replied, "You are just right; that is the way every soldier ought to go into battle."

After Longstreet held off pursuing Federals at Williamsburg on May 5, Hood's brigade, as part of Brig. Gen. William H. C. Whiting's division, became rear guard for the army. Charged with protecting their wagon train, the Confederates' challenge came not so much from aggressive pursuit by McClellan as from enemy forces landing upriver and threatening to cut off the route of retreat. Indeed, at noon of the 6th, Whiting received news that

---

[12] Dyer, *Gallant Hood*, 61; McMurry, *John Bell Hood*, 34; Ural, *Hood's Texas Brigade*, 82.

the Federals had landed at Eltham's Landing on the York River, some eighteen miles northwest of Williamsburg. Maj. Gen. Gustavus Smith, commanding corps and Whiting's superior, rode up, confirmed the report, and ordered Whiting to attack the enemy forces the next morning, driving them back to the cover of their gunboats. Whiting assigned Hood and his brigade the task, and, at 7 A.M. on the 7th, Colonel Archer and the 5th Texas drove in the enemy skirmishers. Hood followed with his other three regiments, supported on his right by Col. Wade Hampton's brigade. The men were marching with unloaded muskets (lest an accidental discharge alert the enemy) as Hood led them into position. Suddenly, a Federal skirmish line "met me face to face," as Hood relates in *Advance and Retreat*; "I did not discover the Federals till we were almost close enough to shake hands." Hood jumped from his horse and ordered his men to load. Just then, a Union corporal raised his rifle to fire at Hood. Pvt. John Deal, Co. A, 4th Texas, had disobeyed orders; "he instantly killed the corporal, who fell within a few feet of me." The literature accepts this as possibly Hood's closest scrape with death during the war. "Fortunately for him," Sears writes, Deal had disobeyed orders and "shot down the Yankee corporal before he could fire."

Then came Hood's charge upon the enemy, in which the Confederates drove the Northerners back. "We were overjoyed to see the enemy fleeing across an open field to a skirt of timber half a mile away," one of them wrote. The Southerners repelled a counterattack before retiring around 2:30 P.M., "having driven the enemy a distance of 1½ miles," Hood reported, "through a most difficult forest, forcing him under the protection of his gunboats." In this little brush at Eltham's, the Texas Brigade lost 15 killed and 25 wounded. It had driven the enemy from the field and, according to Hood, had inflicted no less than 300 casualties and taken 40 prisoners (the Federals' actual loss, as reported by Brig. Gen. William B. Franklin, was 48 killed, 110 wounded, and 28 captured or missing). More important for Hood himself was the praise he received for his battlefield bravery. The compliments became more effusive up the chain of command, beginning with Whiting's commendation of Hood's "conspicuous gallantry." Smith credited the Texans with "the largest share of the honors of the day." Johnston made note to General Lee of Hood's "additional evidence of...high merit." By the time word of Eltham's Landing reached Richmond, President Davis told a Texan that Hood and his men "saved the rear of the army and the whole of our baggage train."[13]

[13] McMurry, *John Bell Hood*, 34, 38–39; Sears, *To the Gates of Richmond*, 19–20, 22, 36–38, 86; Magruder report, May 3, *OR*, vol. 11, pt. 1, 405; Casdorph, *Prince John Magruder*, 152–53; Hunter, "At Yorktown in 1862 and What Followed," (February 1918): 66, and (March

Hood and his troops were developing a reputation for courage. After the war, Maj. J. H. Littlefield, assistant quartermaster for the 5th Texas, wrote of overhearing Hood's exchange with Johnston after the fight at Eltham's, in which the commanding general mildly chided Hood for exceeding his orders to "feel the enemy gently and fall back": "What would your Texans have done, sir, if I had ordered them to charge and drive back the enemy?" "I suppose, General," Hood answered, "they would have driven them into the river, and tried to swim out and capture the gunboats."[14]

Johnston's withdrawal up the Peninsula ended when his army crossed the Chickahominy River on May 15; Hood's troops were the last Confederates to do so. The passage led to another anecdote about the Texans' moxie. General Whiting, supervising his troops' crossing, had become impatient, "exhorting us to 'close up,'" as one Lone Star Stater later wrote, "and for God's sake to hurry." This vexed the men who were having to slog though some heavy mud ("a perfect mortar had been formed from one to three feet deep"). "Hurry up, men, hurry up! Don't mind a little mud," Whiting shouted, which of course provoked one of Hood's men to call out, "Do'ye call this a *little mud?* S'pose you git down and try it, stranger; I'll hold your horse." General Whiting took offense and asked the soldier whom he thought he was addressing. "General," the soldier answered, "don't you reckon I know a general from a long-tongued courier?" Repeated a couple of different ways, the Texan's retort caused General Whiting to ride away, seeking situations he could more favorably control.

Johnston seems not to have informed the government that he had ordered his army across the Chickahominy, the last major river barrier between McClellan and the capital. Postmaster General John H. Reagan, of Texas, used the arrival of Hood's Texans to pay them a visit and spent the night of May 16–17 in their camps. The next day he rode with them when Johnston shifted his position to within three miles of Richmond. Afterward Reagan was riding back into the city when he came upon President Davis, who invited

---

1918): 112; Whiting report, May 8, *OR*, vol. 11, pt. 1, 629–30; Polley, "To Charming Nellie—Not So Funny," 44; Andrew, Jr., *Wade Hampton*, 93 ("elements of Hampton's brigade, including the 19th Georgia, the infantry battalion of the Hampton Legion, and the legion artillery"); Hood, *A & R*, 21 (Private Deal); Hood report, May 7, *OR*, vol. 11, pt. 1, 631; Franklin report, May 17, *OR*, vol. 11, pt. 1, 618; Smith report, May 12, *OR*, vol. 11, pt. 1, 627; Johnston to Lee, May 8, *OR*, vol. 11, pt. 3, 500.

[14] [Littlefield], "Hood—'Feeling the Enemy,'" 2:276; [McCabe], *The Grayjackets*, 237.

The author of "'Feeling the Enemy'" is not identified in Johnson and Buel, eds., *Battles and Leaders*; my assumption is that he was Major Littlefield. His death in 1901 is observed in "Maj. J. H. Littlefield," 33.

him to ride out to the Chickahominy to visit General Johnston. "I said to him that he would not have to go to the Chickahominy to see the General," Reagan recalled in his memoirs. "From what occurred afterward it was apparent that he had not caught my words." Sure enough, as they approached an expanse of soldiers' tents, the president asked what troops were there. Reagan answered that they were Hood's: "'No!' [Davis] exclaimed; 'Hood's brigade is down on the Chickahominy.'" Reagan explained that he had just spent the night with them, adding, "If you want to see General Johnston, he is in the brick house off to our right." Reagan continued, "Again the president objected, not seeming to be able to realize the situation: 'No, General Johnston is down on the Chickahominy.' To which I answered that I had seen him and his staff go to that house that day. The look of surprise which swept over his face showed a trace of pain."

Reagan explained that "Mr. Davis and one of his staff officers, I believe it was Colonel Ives, turned off to the General's headquarters; and I rode on to Hood's camp." Reagan learned later from a staff officer about Davis's and Johnston's conversation. "The President never told me what occurred between him and General Johnston; but his staff officer did. He said the President inquired of the General why he was in the suburbs of Richmond, and had not contested the crossing of the Chickahominy with General McClellan," Reagan recalled. "General Johnston's answer was that the army was out of provisions, that the ground near the Chickahominy was low and marshy and the water bad; and that he had brought the army near Richmond where the ground was dry, the water good, and to be that much nearer needed supplies." This led the president to ask Johnston if he intended to give up Richmond without a battle. "[N]ot getting a satisfactory answer," Reagan wrote, Davis then told Johnston that "if he was not going to give battle, he would appoint some one to the command who would."

The president was justified in wondering if Johnston would give up the capital without a fight. His inactivity allowed McClellan to slowly bring his army by May 25 to within seven miles of the city. Then it was McClellan's inactivity, while he awaited reinforcements (Maj. Gen. Irvin McDowell's corps, which would never arrive) that allowed Johnston to plan an attacking battle for May 29, aimed against McClellan's right wing north of the Chickahominy. But then Johnston changed his objective to the enemy corps south of the river and planned his assault for May 31.

The attack primarily involved Longstreet and Maj. Gen. Daniel H. Hill, with G. W. Smith and Whiting in supporting roles. But miscommunication—Longstreet received only verbal orders from Johnston—lack of coordination, and foul-ups in their approach routes led the Confederates on the afternoon of May 31 to launch largely unsuccessful attacks that cost over 6,000 casualties. In this engagement Hood saw no combat as his brigade was ordered from this point to that during the day. Hood later recounted that "this marching and counter-marching ended the part taken by my troops in the battle of Seven Pines or Fair Oaks."

The most important consequence of Seven Pines was the wounding of Joseph E. Johnston, struck by shell fragments in the chest and thigh late in the afternoon as the battle was ebbing. Johnston was borne from the field; Gustavus Smith took command and tried to renew the fight the next day but with no success. On the afternoon of June 1, Robert E. Lee took command of the Army of Northern Virginia.[15]

In the next several weeks Lee reordered his army's command structure, tactfully exiting generals such as G. W. Smith (who had been found wanting at Seven Pines). Among the changes, the infantry of Hampton's Legion was attached to Hood's brigade (Colonel Hampton had been wounded in the battle). While the Confederates strengthened their fortifications outside the capital, McClellan seemed paralyzed, promising his superiors a big, decisive battle as soon as more reinforcements were sent. Meanwhile he shifted the bulk of his army to the south bank of the Chickahominy. This gave Lee his opportunity to seize the initiative, especially after a recon ride by Brig. Gen. J. E. B. Stuart's cavalry fixed the position of the Federal forces' right flank. Lee

---

[15] Sears, *To the Gates of Richmond*, 95, 110; McMurry, *John Bell Hood*, 40–41; Davis, *Hood's Texas Brigade*, 64 (Whiting at the river); Govan and Livingood, *A Different Valor*, 128 (crossed May 15; moved again May 17); Reagan, *Memoirs with Special Reference to Secession and the Civil War*, 138–39; W. C. Davis, *Jefferson Davis*, 421–22 (accepting Reagan's account); Dowdey, *The Seven Days*, 69 ("Johnston did not inform the War Department even of his whereabouts"); Davis, *The Rise and Fall of the Confederate Government*, 2:103 ("surprised to learn that the whole army had crossed the Chickahominy"); Symonds, *Joseph E. Johnston*, 158, 162–63, 172; Freeman, *Lee's Lieutenants*, 1:225.

Although Johnston had commanded the Department of Northern Virginia since October 1861, he liked to call his forces the "Army of the Potomac," despite the Federals' use of that name for McClellan's army. Lee had begun terming Johnston's force the Army of Northern Virginia by March 28, 1862 (*OR*, vol. 11, pt. 3, 408). Freeman notes that this now famous name for Lee's army came about more "on usage, not on formal, authorized adoption" (Freeman, *R.E. Lee*, 2:78n).

Thomas has commented scathingly on Johnston's retreat up the Peninsula to Richmond. "Would the Confederate Army retreat forever? ...Davis all but begged Johnston for some assurance, some plan, and Johnston responded by asking how his army would be supplied if Richmond were lost" (E. M. Thomas, "The Peninsula Campaign: Part I," 45).

ordered Maj. Gen. Thomas J. ("Stonewall") Jackson and his troops, who had bedeviled and whipped the Yankees in the Shenandoah Valley, to join the main army for an attack on McClellan's right north of the river. "To be efficacious the movement must be secret," Lee advised Jackson. To distract the enemy, Lee called for Whiting's division to march toward the Valley as if to reinforce Jackson. On June 11, Hood's brigade and others loudly paraded through Richmond (for any Northern spies to see) before boarding trains to the Valley. To preserve secrecy, all soldiers were kept in the dark and given the same orders, which Hood recited to each of his regiments. As a member of the 4th Texas recalled, the general called out,

> Soldiers of the Fourth: I know as little of our destination as you do. If, however, any of you learn or suspect it, keep it a secret. To every one who asks questions, answer, 'I don't know.' We are now under the orders of Gen. Jackson and I repeat them to you. I can only tell you further, that those of you who stay with the command on this march will witness and participate in grand events.

The grand events would take a while. Joining Jackson's forces at Staunton, the Texans on the 20th about-faced and proceeded with them by road and rail back to Richmond and Lee's army.

General Lee with his lieutenants—Jackson, Longstreet, Harvey Hill, and Maj. Gen. Ambrose Powell Hill—met on the 23rd to discuss Lee's plan for them to attack the lone Union infantry corps (Brig. Gen. Fitz John Porter's) north of the Chickahominy. The assault, to be led by Jackson's troops, could take place on June 25 if Jackson, who had the longest approach route, could get into position in time. On Wednesday the 25th, however, Jackson's column, with Hood's brigade in the lead, was miles behind schedule. The fighting that day was launched not by Lee but by McClellan, who had finally decided to act. South of the river, at Oak Grove, Maj. Gen. Joe Hooker's and Brig. Gen. Phil Kearny's divisions attacked Maj. Gen. Benjamin Huger's part of the Confederate line, but by the end of the day the Federals had merely advanced their pickets.[16]

Hood and his troops prepared for action the next morning; rations and ammunition were issued near Ashland, more than a dozen miles north of where Jackson planned to launch his attack. But the march on the morning

---

[16] Freeman, *Lee's Lieutenants*, 1:262–63; Coxe, "In the Battle of Seven Pines," 25; Sears, *To the Gates of Richmond*, 155–56, 174–77, 189; Lemon, *Feed Them the Steel!*, 21; Sears, *George B. McClellan*, 200; O'Connor, *Cavalier General*, 69–71; McMurry, *John Bell Hood*, 44–45; Polley, "Another 'Charming Nellie' Letter," 118.

of June 26 started late and made its sluggish way with a slowness uncharacteristic of Jackson's performance in the Valley. Hood and his brigade were at the head of Jackson's column and at about noon ran into enemy cavalry outposts, which were driven in. Several hours later, at a creek Hood found the enemy burning the bridge and chopping trees to block the road. He called up a battery and pushed his skirmishers forward. The Yankees retreated in such a hurry, Hood recalled, "that they left their axes in the trees." Late in the afternoon the Confederates reached a place called Hundley's Corner, several miles northeast of Mechanicsville. There A. P. Hill around 3 P.M. launched a reconnaissance in force that Lee turned into an attack, even though Jackson had not reached the battlefield (Hood and his men could hear the sound of the musketry). In this battle of Mechanicsville the Confederates suffered a costly repulse.[17]

June 27, 1862, would be one of the most illustrious days of Hood's career though it did not begin auspiciously for Lee. Having tipped his hand that he would be moving against Porter's corps north of the river, he had to maintain the offensive. In actuality there was no threat that McClellan, the misnamed "Young Napoleon," would seize the initiative himself. Even before the gunfire at Mechanicsville had ended, Mac had decided to retreat from before Richmond—or, as he chose to put it, to change his base southward to the James River. Thus on the afternoon of the 26th he had his engineers choose a position in rear of Porter's Mechanicsville lines so that his corps could move closer to the Chickahominy. Early in the morning of the 27th, Porter's troops withdrew to this new line, nearly two miles long, drawn up on high ground east of Boatswain's Swamp, a marshy creek bed that would impede any infantry attack. Though the enemy position was formidable, Lee resolved to attack it.

Powell Hill's division had followed Porter's rearguard from Mechanicsville and across Powhite Creek, where Dr. William Gaines had built a grist mill. With Hill at the center, Lee's line took shape with Longstreet on the right and Jackson, then D. H. Hill to A. P. Hill's left. Whiting's division marched at the rear of Jackson's column as it moved past Dr. Gaines's millpond and into position. The day's march was slow and marred by a wrong road-turn. Hood, impatient, asked Jackson what he could do to help things. "Press on sir—press on," was Stonewall's customary reply.

Powell Hill's troops led off the attack at 2:30. The Confederates charged bravely, and at a point managed to break the enemy's first line before being driven back. Yet the Federals' hot artillery fire kept most of Hill's troops from

[17] Hood, *A & R*, 25; McMurry, *John Bell Hood*, p. 45; Lemon, *Feed Them the Steel!*, 21; Sears, *To the Gates of Richmond*, 199–203: Gruber, "'Texas need not feel ashamed,'" 54.

getting much past Boatswain's Creek. Around 3:30 Lee sent in Maj. Gen. Richard S. Ewell's division, which was also repulsed. Then it was Longstreet's turn; his attack fizzled as well. After deploying as the left of Lee's line, Jackson and Harvey Hill ordered their assaults. Hill's men overran a Union battery before being pushed back by a counterattack. Late in the afternoon, approaching dusk, Lee ordered one last, desperate assault along the whole line. Hood and the Texas Brigade were approaching the battlefield when General Lee encountered Hood. Lt. James Lemon of the 18th Georgia saw the two generals talking, then watched Hood ride off to examine the ground. Upon returning, Hood ordered his troops a couple of hundred yards forward, then rode to Lee again. "I could see that Lee was directing Hood's attention to a far ridge," Lemon later wrote, as Lee seemed to explain to Hood the challenge before him: eight hundred yards of ground to cover, three enemy lines of infantry, strong artillery support. "This must be done. Can you break this line?" Lee asked. Hood promised, "I will try."[18]

A variant of the Hood-Lee conversation before the final charge has Hood riding up to and saluting Lee, "who said that the efforts to break the enemy's lines in front of us had been unsuccessful and that it was of the utmost importance to do so. Gen. Hood replied: 'We will do it.' As Gen. Lee turned his horse to ride away, he lifted his hat and said, 'May God be with you!'"[19]

Another Gaines's Mill story has Hood conversing with General Whiting:

> Whiting said: "That battery ought to be taken, Hood," "Then why has it not been taken?" asked Hood. "Because," replied Whiting, "the position is too strong. My brigade is composed of veteran troops, but they can do nothing with it." "I have a regiment that can take it," declared Hood, and scarcely waiting for Whiting's assent to his undertaking the task, galloped in the direction of the Fourth Texas

The challenge ahead was to advance several hundred yards of open ground sloping down to Boatswain's Creek, then to cross the swampy bottom before charging up the hill, defended by several lines of enemy infantry. In the attack formation Hood's brigade was split, half to the left of Col. Evander M. Law's brigade, and two regiments (thc 4th Texas and 18th Georgia), between Law's right and Brigadier George Pickett's brigade. Sgt. Val C. Giles

---

[18] Sears, *To the Gates of Richmond*, 211–17, 225–28; McMurry, *John Bell Hood*, 45–52; Cullen, "Gaines's Mill," 12–16; Dowdey, *Seven Days*, 221–36; R. E. L. Krick, "The Men Who Carried This Position Were Soldiers Indeed," 187, 204; Lemon, *Feed Them the Steel!*, 22.

[19] Hamby, "Fourth Texas," 183 ("May God be with you!"); Polley, *Hood's Texas Brigade*, 46 ("I have a regiment").

remembered Hood addressing the men of the 4th Texas. "Soldiers, I have come to fulfill a promise I made to you while colonel of your regiment. I promised to lead you personally in the first great battle. The time has come, and I am here." Another Texan also recalled the scene: "Hood always looked grand in battle. Now he looked sublime." Giving the order to advance, Hood led the Georgians and Texans down the slope; the men were hunkered forward in what was called the "hunter's slouch." "Gen. Hood was in our front until we were within one hundred yards of the creek," wrote a Texan, "when he wheeled his horse to our right and ordered us to fix bayonets and charge at double-quick."[20]

Then, having dismounted on foot he led his men through the swampy creek bottom, all at a quick pace, heading for the center of the Union line. They took losses, but the enemy had been weakened by the Confederates' previous attacks. Thus, with the 4th Texas in the lead—his old regiment—Hood led a charge that overran the Yankees' first line. Federals ran up the slope and into their second line with Hood's men close behind—so close that this second enemy line could not get off successful volleys before it too was broken. "We flew towards the breastworks, cleared them, and slaughtered the retreating devils as they scampered up the hill towards their battery," wrote Lt. Decimus et Ultimus Barziza of the 4th Texas. At one point, Hood aligned his ranks in an orchard for the final push as they came under Northern shellfire. His nonchalance impressed the men. "I tell you what, Joe," said one after the fight. "I got mighty nervous and shaky while we were forming in the apple orchard to make that last desperate charge on the batteries. But when I looked behind me and saw old Hood resting on one foot, his arm raised above his head, his hand grasping the limb of a tree, looking as unconcerned as if we were on dress parade, I just determined that if he could stand it, I would." Hood then led his men with commands they could hear above the battle din. "Forward! Steady!" "Don't fire here!" he ordered, when soldiers stopped to fire. "Charge right down on them, and drive them out with the bayonet!" They did; Hood's brigade broke the enemy line. The 5th Texas captured an entire New Jersey regiment while the 4th and 18th overran and took fourteen Northern cannon. The Confederates chased the fleeing Yankees to near the river's edge when darkness put an end to the fighting.

---

[20] Bohannon, "'A Bold Fighter,'" 253; Giles, "The Tom Green Rifles," 21; Crozier, "A Private with General Hood," 557 ("looked sublime"); Morris, "Richmond's Fate in the Balance," 41; Hamby, "Fourth Texas in Battle of Gaines's Mill," 183–84 ("wheeled his horse").

The battle of Gaines's Mill was Lee's only tactical victory in the Seven Days. Indeed, it was Robert E. Lee's first major battlefield success—"the battle which taught the Confederate troops in Virginia how to win victory," as one Southerner later put it. But the triumph did not come without heavy casualties. In Hood's brigade, the three Texas regiments, the Georgians, and Hampton's infantry lost 89 killed, 470 wounded, and 4 men missing. Losses in the 4th Texas were particularly severe. Col. John Marshall was shot dead before he reached the creek; Lt. Col. Bradfute Warwick fell mortally wounded; four captains and five lieutenants were also killed or mortally wounded. Hood remarkably was not hit. That night he visited the field hospitals where his injured soldiers were being treated. Hood walked the battlefield, calling for stretcher-bearers when he came upon a wounded soldier, even giving his horse to one man so he could carry the bodies of his two brothers for burial. This exacting work took its emotional toll. In the darkness a staff officer came upon the general, "sitting on a cracker-box, crying." The next morning Hood received the casualty report from the 4th Texas: 531 entered the battle, 261 were gone. According to Chaplain Nicholas Davis, ministering to Hood's Texas Brigade:

> He rode up & inquired "Is this the 4th. Tex." Some one answered "this is all that remains." He turned his horse but not soon enough to prevent the regiment from witnessing how well he loved his men. They watched him as he rode using his handkerchief until out of sight. There was not a soldier in that line but what thought more of [him] now than ever before.

On the morning of the 28th General Hood and Senator Wigfall addressed the Texans. Wigfall, the practiced politician, had no trouble finding the right words. Not so Hood, who, as one Texan recorded, "is a better soldier than speaker." But that did not matter; Hood's personal leadership in the battle had won the firm respect of his soldiers. "He knew the men and they knew him," Chaplain Davis later recorded, "and each knew the confidence which the other had, in an hour of trial, and with a mutual confidence thus sustained, there was no fear of failure."[21]

Chroniclers since then have heralded the achievement of Hood and his soldiers on the battlefield of Gaines's Mill. "Of brigade achievements there

---

[21] Gruber, "'Texas need not feel ashamed,'" 56; Polley, "Brave Texans in the Virginia Army," 158–59; E. B. Williams, *Hood's Texas Brigade*, 66–69; R. E. L. Krick, "The Battle of Gaines's Mill," 62; Whiting report, July [undated], 1862, *OR*, vol. 11, pt. 2, 564–65; Lee report, March 6, 1863, *OR*, vol. 11, pt. 2, 493; Jackson report, February 20, 1863, *OR*, vol. 11, pt. 2, 555–56; Henderson, "General Hood's Brigade," 193 ("battle which taught"); Davis, *Hood's Texas Brigade*, 91, 119.

had been many," writes Dr. Freeman, noting that Law's and Pickett's units helped the Texas Brigade crack the Yankee line. "In the praise of the camps, highest honors went to Hood and his fine regiments." Lee made mention of "the brave Texans" leading the final charge. Jackson also noted that "the Fourth Texas, under the lead of General Hood, was the first to pierce these strongholds and seize the guns," which he called an "almost matchless display of daring and valor." The day after the battle Stonewall rode over the ground of Hood's advance and remarked, "The men who carried this position were soldiers indeed!" Postmaster General Reagan recalled in his memoirs that after Gaines's Mill General Lee on more than one occasion suggested that Reagan should help in recruiting a whole division of Texas infantry for his army, "remarking that with such a force he would engage to break any line of battle on earth in an open field."

Historians have echoed this praise. Clifford Dowdey observes that in the battle "Hood showed the instincts of the born fighter." Biographer Richard O'Connor (1949) states that Hood's charge was an "almost matchless display of daring and desperate valor." Dyer also observes that Hood after Gaines's Mill "now had the respect of both Lee and Jackson—a respect fairly won by soldierly conduct and bravery on the field of battle." Hood and his command were not engaged in the rest of the Seven Days' battles, but their record had been established. "Hood had performed splendidly in his only assigned task," McMurry writes. "He had been ordered to attack, he had done so, and the enemy had been driven from the field." McMurry concludes that the effect of Gaines's Mill on Hood's reputation and prospects for promotion cannot be overemphasized: "Hood, in fact, was about the only Southern general to emerge from the Seven Days with an enhanced reputation." Freeman, as usual, offers the best evaluation: "Anyone who had followed the operations of the Army after Gaines's Mill would have said that of all the officers under Longstreet, the most likely to be a great soldier was Hood."[22]

For this reason, when General Whiting took a medical furlough on July 26 (never to return to Lee's army), Hood was chosen to command his small division: his own Texas Brigade (command of which eventually devolved to Col. William T. Wofford of the 18th Georgia) and Evander Law's, which

---

[22] Freeman, *Lee's Lieutenants*, 1:535; 2:273; Reagan, *Memoirs*, 145; Dowdey, *Lee*, 245; O'Connor, *Cavalier General*, 82; Dyer, *Gallant Hood*, 95.

Twenty-five years after Hood's death, his daughter Ida wrote a tribute to her father for the *New Orleans Picayune.* In it she claims that "in the battle of Gaines' Mill he received his first wound in the civil war," but her statement is not to be seen elsewhere in the literature ("In Memory of General J. B. Hood," 154).

had charged with him at Gaines's Mill. The division was placed in Longstreet's wing of the army and thus was posted near Richmond while McClellan's Army of the Potomac inertly held to its fortified lines at Harrison's Landing for more than a month after the Seven Days' battles. Hood and his men had time to spend in the capital, whose citizens "looked upon with special favour" any Texan wearing the Lone Star emblem on his hat.

Meanwhile, a new threat had emerged north of the Confederate capital. After Jackson's departure from the Valley to join Lee for the Seven Days, Lincoln's War Department resolved to unite the three separate forces fighting Jackson in the Valley (Maj. Gen. Nathaniel P. Banks's, Brig. Gen. James Shields's and Maj. Gen. John C. Fremont's) with Irvin McDowell's corps, then near Fredericksburg. To command this new Union "Army of Virginia," Maj. Gen. John Pope was summoned from Mississippi in mid-June, primarily for his successes at New Madrid and Island No. 10. Pope arrived in Washington and learned of his new assignment from secretary of war Edwin M. Stanton on June 26—the day Lee launched his assault at Mechanicsville. Pope was to assemble this force and march against the Virginia Central Railroad, which connected Richmond with the Shenandoah. Lee would be compelled to meet the threat as McClellan moved against the Rebel capital (or so Stanton hoped). Pope assumed command of his army, which numbered 50,000 men, on June 27, but it took two weeks to get it organized and assembled. On July 10, when Pope ordered his troops to be ready to march, his units were strung out from Front Royal to Warrenton to Fredericksburg. Nevertheless, Union cavalry were able on the 12th to occupy Culpeper Courthouse, a point twenty miles from Gordonsville and the Virginia Central. Lee responded the next day by ordering Jackson with his and Ewell's divisions to Gordonsville.[23]

During this time, Lee and Longstreet, with most of the army, kept watch on McClellan, who was busy doing nothing except calling for reinforcements. Washington's impatience was signaled by the telegram sent by Maj. Gen. Henry W. Halleck, the Union army's chief of staff, to Mac on July 30 that he should begin sending his sick soldiers back from Harrison's. Then on August 4 came orders for McClellan to start withdrawing his entire army toward Fredericksburg and Washington for an eventual uniting with Pope's forces. McClellan, reluctant to help Pope, took his time but on August 7 began his withdrawal. This allowed Lee to start moving the rest of his troops northward to help Jackson confront Pope. On the 7th Hood received orders to have his

[23] McMurry, *John Bell Hood*, 52; E. B. Williams, *Hood's Texas Brigade*, 76; Sears, *To the Gates of Richmond*, 350–53; Cozzens, *General John Pope*, 74–76, 89; Eckenrode and Conrad, *James Longstreet*, 89; Hennessy, *Return to Bull Run*, 6.

men prepare three days' rations and send their excess baggage into Richmond. Lee did not move hastily; he had to make sure McClellan's army was moving downriver. Thus it was on August 13 that Lee directed Longstreet to put ten brigades on the Virginia Central to Gordonsville. Hood's division was ordered to take the Richmond, Fredericksburg & Potomac R.R. to Hanover Junction (a point about a third of the way to Gordonsville), encamp and guard the railroad until further orders. Hood's troops eventually united with Longstreet at Gordonsville and on August 20, at the head of Longstreet's column, crossed the Rapidan, following the route Jackson's command had taken.

Lee's challenge was to engage and defeat Pope before McClellan could reinforce him. Pope's army at the time was stretched with its left at Fredericksburg along the Rappahannock, and its right guarding the river crossing at Kelly's Ford, Rappahannock Bridge and Beverly's Ford. Pope knew that Lee was moving against him. Part of his army, under Banks, had attacked Jackson on August 9 at Cedar Mountain north of the Rapidan and had been driven from the field, yet Pope could not advance farther until McClellan's troops began to arrive.[24]

On August 21, Jackson was marching northwestward, trying to find a Rappahannock ford not guarded by the enemy. The next day he found Freeman's Ford, upstream from Rappahannock Station, held by Union infantry and artillery, and so kept marching upstream toward Sulphur Springs. Hood, leading Longstreet's column and close upon Jackson's rear, was called into action on the 22nd. Brig. Gen. Isaac Trimble, commanding Stonewall's rearguard, was exchanging fire with Federals who had crossed the river at Freeman's, threatening Jackson's wagon train. Trimble asked Hood to join in an attack on the Yankees. Hood's and Trimble's troops charged and quickly drove the enemy back across the river. "Pouring a dreadful fire into their crowds of confused and broken lines, as they were huddling together to cross," wrote Chaplain Davis of the 4th Texas; "many were shot in the back, and others drowned by the crushing crowd which pressed for the other shore." After that the Confederates were content to hold the river's south bank and count casualties. Hood thought the Federals had lost from two hundred to three hundred; Trimble reported that "the water was literally covered with dead and wounded," and that he had taken over a hundred prisoners. (John Hennessy, historian of Second Manassas, totals the Union loss at between fifty and seventy-five.) Trimble's own loss was forty-seven killed and

[24] Sears, *To the Gates of Richmond*, 353; Hennessy, *Return to Bull Run*, 10, 28, 30–31; McMurry, *John Bell Hood*, 53; A. P. Mason, Special Orders No. 181, August 13, 1862, *OR*, vol. 11, pt. 3, 675.

wounded; Hood reported ten wounded in the Texas Brigade, none in Law's. At Freeman's Ford, Hood and his soldiers had again driven the enemy from the battlefield. During the fight, however, the general had come under enemy fire. "A party of Yankees concealed in the woods fired upon him," writes Bushrod Frobel, "wounding a courier who was riding by his side." "They came very near killing Hood," Frobel concluded, "who escaped by the merest accident."[25]

On August 22, Jackson found no Federals at Sulphur Springs, but an afternoon rainstorm so quickly raised the river that a crossing could not be made. Longstreet's troops, including Hood's at Freeman's, held the Rappahannock crossing-points downstream, keeping Pope's forces occupied. The Federal and Confederate armies were deadlocked, neither able to take the offensive, when on the 23rd "Jeb" Stuart brought to Lee intelligence from a raid he had led into the enemy's rear. Stuart had captured valuable papers from Pope's headquarters wagons, which confirmed his army's strength (45,000) and that McClellan's troops would be arriving in five days. More, a dispatch from Pope to Halleck dated August 20 revealed Pope's concerns that until McClellan's reinforcements began arriving from Fredericksburg, he worried that his supply line to Washington, the Orange & Alexandria Railroad, was vulnerable to Rebel raids. All this information prompted Lee on the afternoon of the 24th to devise a bold but risky plan: Jackson and three divisions to march upriver, cross the Rappahannock beyond Pope's right, then march swiftly eastward behind Pope's army, reach the O. & A. and break it. Lee and Longstreet would meanwhile hold Pope's attention until Jackson cut his supply line and forced his retreat, whereupon Lee and Longstreet would advance, catching Pope in a vise. Everything hinged on secrecy and speed. Jackson, eagerly grasping the daring opportunity, told Lee he would have his troops moving within an hour.

Actually, Jackson started his men moving before dawn of August 25. While Stonewall's foot cavalry marched to cut the Orange & Alexandria in Pope's rear, perhaps at Bristoe Station, perhaps at Manassas Junction—Lee left this to Jackson's discretion—Longstreet with his five divisions held the line along the Rappahannock. During the 25th and morning of the 26th Hood's division held the crossing at Sulphur Springs while Northern and Southern batteries fired at each other across the river. By midday of August

[25] Hennessy, *Return to Bull Run*, 66–70, 487 n.33; O'Connor, *Cavalier General*, 91–92; Everett, *Chaplain Davis*, 108–109; Hood report, September 27, 1862, *OR*, vol. 12, pt. 2, 605–606; Trimble report, January 30, 1863, *OR*, vol. 12, pt. 2, 719; [Frobel], "Field and Camp," 332.

26, Lee had learned that Pope was moving most of his forces downriver toward junction with the promised reinforcements. He decided it was time to join Jackson and instructed Longstreet to set out that afternoon. The divisions of Maj. Gen. David R. Jones, brigadiers Hood, Cadmus Wilcox, and James Kemper (Maj. Gen. Richard Anderson's would stay behind a few days) followed the route Jackson had taken: crossing the river at Hinson's Mill Ford, then a dozen miles north to Salem and the Manassas Gap Railroad, then following the rails through Thoroughfare Gap and on toward Manassas. Stonewall's troops, marching fifty-four miles in just thirty-six hours, reached Bristoe Station in the late afternoon of August 26 and in a night attack captured the huge Union supply depot at Manassas Junction.[26]

Thanks to couriers of the 4th Virginia Cavalry, Lee and Jackson were kept informed of each other's progress. On the evening of August 26, Jackson sent word that his advance had broken the Orange railroad at Bristoe's. Lee replied that Longstreet's column would reach Salem on the 27th and march through Thoroughfare Gap the next day. Sure enough, in the evening of the 28th Jackson received a courier's message that Lee and Longstreet had reached Thoroughfare, twelve miles to the northwest, and would be through the gap the next morning.

By then, Jackson's men at the Yankees' supply base had gorged on lobster salad, oysters, and other good things to which their stomachs had long been strangers. Then the Southerners filled their haversacks with provisions and set fire to all the rest. Starting at 9 P.M. on August 27 Stonewall led his troops back to the old battlefield of Manassas and took defensive position, awaiting Pope's arrival if and when he had caught on to their movements and whereabouts.

Pope eventually did, and with his characteristic, unhinged bravado announced on the night of the 27th his intention to attack Jackson at Manassas. With his 50,000 troops against Jackson's smaller force (24,000), Pope was confident that "we shall bag the whole crowd." No one yet in the war had come close to bagging Stonewall Jackson, but Pope believed, as he later reported, "that we were sufficiently in advance of Longstreet...that by using our whole force vigorously we should be able to crush Jackson completely before Longstreet could have reached the scene of action."[27] Hood's division in

---

[26] Robertson, *Stonewall Jackson*, 545–47; Hennessy, *Return to Bull Run*, 80, 93, 96, 100 (map of Jackson's route), 103–104, 107–108, 115, 118, 137, 309; McMurry, *John Bell Hood*, 53; Wert, *General James Longstreet*, 163–65.

[27] Hennessy, *Return to Bull Run*, 129–30, 135–36, 138–39, 151–52; Sutherland, *The Emergence of Total War*, 67.

Longstreet's column had been on the march since the afternoon of August 26. Lee and Longstreet did not call for the quick pace with which Jackson's command had reached Manassas Junction; on the night of the 27th, they encamped near White Plains, five miles west of Thoroughfare Gap. The men were allowed a good night's rest; they did not resume march until 11 A.M. on August 28. By midafternoon the head of the Confederate column reached the gap and found it unguarded by the enemy although Union cavalry tried to slow the Rebels' march by felling trees in the road east of the gap. Soon Longstreet's lead division, Jones's, encountered a division of Yankee infantry blocking its advance. Lee called on Hood's troops to flank the enemy position, which was accomplished by Law's brigade. By darkness the Federals were retreating. That night Hood's and Jones's divisions camped east of the gap. At six o'clock the next morning, Hood's division led Longstreet's column marching to join Jackson.

With Jackson's and Longstreet's forces on the verge of junction, only John Pope could think that he held the advantage. Once he had fixed Stonewall's position along the railroad cut of the unfinished Manassas Gap Railroad, Pope determined to attack him there. The firing started around 8 A.M. on the 29th; only once in the afternoon did a Union assault break Jackson's line before being driven back. Meanwhile, Hood's Texans connected with Jackson's line on its right. Hood remembered that he encountered General Jackson at about 10:30 that morning, and that "he rode forward and extended me a hearty welcome." After noon Longstreet's troops were deployed and ready for Lee's grand stroke; while Pope continued his assaults on Jackson, Longstreet would attack and roll up the Federals' left flank. Yet Longstreet's examination of the field led him to conclude that Pope's flank troops were so positioned as to be able to thwart an assault. Lee was unconvinced and rode out around 3 P.M. to ascertain the situation himself. An hour later, Federal movements (Fitz John Porter's corps) seemed to open the door again to a flanking attack. By the time Longstreet and Lee had met and laid their plan, it was five o'clock, with only two hours of daylight left. The two generals agreed on a reconnaissance in force that evening and the full attack the next morning.

In execution, little separates a recon in force from an actual infantry assault. Longstreet chose the division commander best suited for the work: John Bell Hood. "Though personally reserved, he was one of those rare men who genuinely loved combat," writes John Hennessy, historian of the battle. "His fierce, straightforward fighting style was perfectly suited to the brigade and division he commanded, and to the task assigned him by Longstreet that evening." Hood understood his orders: he would advance "and attack the enemy." Longstreet put Cadmus Wilcox's division in support of Hood. The commitment of troops showed that Lee and Longstreet intended the movement to set the stage for the next morning's major assault.

The advance began at 6:30, sunset. The growing darkness and the woods they marched through made it hard for the Southerners to maintain their lines. "I couldn't have told my best friend from a bluecoat five steps off," wrote a Texan. Despite the darkness, the Confederates advanced, volleyed at the Yankees and drove them back in confusion. The 4th Alabama, in Law's brigade, even captured a piece of artillery. The firing ended by 8 P.M. when Hood realized that his advance, a mile or more, had placed his troops beyond the other Southern units, at some places too close to the enemy. Hood rode back, found Longstreet and Lee, and secured permission to retire. At 2 A.M. the Confederates began withdrawing to their original positions. The howitzer Law's men had taken could not be carried back, so they smashed the carriage and managed to spike the tube.[28]

Hood and Wilcox reported their findings at Lee's headquarters: the surprisingly strong enemy resistance to their advance cast doubt on the prospects for a successful morning attack. On the 30th Lee therefore decided to cancel the assault and await Pope's next move. With Jackson's wing aligned along the unfinished railroad, Longstreet's divisions positioned southward, on Jackson's right: Wilcox's, Hood's, Kemper's and Jones's, with Anderson's in reserve. Hood chose a position for Col. Stephen D. Lee's artillery battalion near the Brawner house, between Jackson and Longstreet, so that Lee's eighteen guns could support both wings of the army. If the Federals renewed their attacks on Jackson's line, Lee was confident they would be repulsed. If the enemy for some reason did not attack, Longstreet would create an afternoon diversion so that Jackson could slip away, cross Bull Run, and march north. The maneuver would force Pope to withdraw back toward the Washington defenses.

That night Pope reasoned that his army had fought well. Afternoon attacks had not broken Jackson's line, but he believed they had inflicted on the Rebels heavier casualties than his troops had suffered. In a telegram to Washington, sent at 5 A.M. on the 30th, Pope stated that "the enemy is still in our front, but badly used up." The Northern commander was choosing to believe

[28] Dyer, *Gallant Hood*, 116; Hennessy, *Return to Bull Run*, 153–54, 158–61, 200, 224, 226, 229–31, 288–90, 296–99, 302–303; Hennessy, "The Second Battle of Manassas," 26, 30; Hood report, September 27, *OR*, vol. 12, pt. 2, 605; McMurry, *John Bell Hood*, 54; Hood, *A & R*, 33; Naisawald, *Grape and Canister*, 158.

what he wanted. For instance, he knew that Longstreet's troops had arrived, but he assumed that they would strengthen Jackson's line, not extend it. Moreover, Pope seized every piece of information that indicated the Rebels might be retreating. Hood's division had withdrawn from its reconnaissance; this allowed Pope to include in his wire to Halleck, "the enemy was driven from the field, which we now occupy." An hour before dawn, Richard Anderson's division, having arrived by night march, was seen shifting position to the west. Pope thus added, "the news just reached me from the front that the enemy is retreating toward the mountains. I go forward at once to see."

At the same time, Pope could not decide what to do. Meeting with his corps commanders at 7 A.M., Pope considered continuing his attacks against Jackson. Several generals—Irvin McDowell, Franz Sigel, and Samuel Heintzelman—did not accept that the Rebels were retreating, so the conference broke up without any agreement on plan. Pope's indecision and vacillation continued. One of his aides, Col. David Hunter Strother, observed that Pope and McDowell "spent the morning under a tree waiting for the enemy to retreat." They were in for a long wait. At 9 A.M., Brig. Gen. James Ricketts, division commander in McDowell's corps, found Rebels in his front, came under fire, and called for help; Pope disregarded the news. An hour later, Brig. Gen. Isaac Stevens (whose division stood in the center of Pope's line) ordered a recon into the woods in his front and also skirmished with the enemy. Again, Pope dismissed this intel. He did the same when Fitz John Porter, whose corps held the Federal left, reported ominously that Confederates not only held the field but also threatened his position. After a soldier claimed that the Rebels had withdrawn—Porter urged that he not be trusted—Pope concluded that Lee had indeed begun to retreat and ordered Porter to attack the enemy's rearguard. Then he ordered all his forces to launch a pursuit. Ricketts's troops advanced and quickly came under artillery fire; he sent word back to McDowell, who then called the whole thing off. In the words of Bruce Catton, "it was probably the briefest pursuit in history."

On the left, Porter was already under fire from S. D. Lee's cannon. This activity by noontime had the effect for Pope of confirming the Rebel army's position. Thus fixed, it would be attacked. Pope ordered Porter to prepare his assault. It took two hours for Porter to assemble his forces. At three o'clock they moved forward against that part of Jackson's line held by Brigadier William E. Starke's division. Within thirty minutes they were bloodily repulsed, in large part by the devastating fire of Lee's artillery.[29]

[29] Freeman, *R. E. Lee*, 2:328–30; Freeman, *Lee's Lieutenants*, 2:138; Kelly, "Second Manassas: The Battle and Campaign," 35–40; Hennessy, "Second Manassas," 46, 48–51; D. G.

Lee now had the opportunity to launch the crushing attack he had been planning: Longstreet's 25,000 troops, all fresh, against the enemy left. The Confederate line overlapped the Federals' to the south, and Longstreet saw the possibility of enveloping the enemy flank. His four attacking divisions—Wilcox's, Hood's, Kemper's and Jones's from north to south (Anderson's in support)—would have a front of more than a mile and a half. Longstreet chose as their topographical objective the Henry house plateau, part of the old battleground of a year prior. The Henry Hill was more than a mile in front of Hood, whose Texas Brigade Longstreet made the "column of direction" for the entire assault. As Longstreet dispatched his orders, Lee signaled to Jackson that Longstreet was about to attack: "look out for and protect his left flank."

A little after 4 P.M., Hood and his Texas Brigade stepped forth from the woods concealing them and began their advance. From left to right, the line formed as 1st Texas, 4th Texas, 18th Georgia, Hampton's Legion, then the 5th Texas, a battlefront of maybe 700 yards. Hood rode at its center, in front of the 18th Georgia. The troops had just started forward when a courier called Hood back to confer with General Longstreet. Hood rode at full speed and a half mile back met Longstreet, who reminded Hood not to advance so fast that his command would get ahead of the other divisions—Pete knew the fighting spirit of Hood and the Texas Brigade. Hood rode quickly back to rejoin his division, soon to learn that this is exactly what his troops had done.

Capt. W. H. Sellers, Hood's adjutant, took command in the general's brief absence. Advancing into a field the Southerners encountered skirmishers from the 10th New York, who fired and fled with the Confederates close behind. The Yankees retired back to the 5th New York (Duryee's Zouaves), which was forming a line to contest the Rebels' advance. The zouaves stood in the open with their baggy red pants and blue jackets. Their uniforms amused Hood's men; one called them "both interesting and ridiculous in the extreme." The sight "served only to increase the merriment and enthusiasm" of Hood's soldiers as they closed in. The zouaves, hindered by the men of the 10th running into their line, fired high, ineffective volleys while the Confederates, forty yards away, riddled them with musketry. They stood this for perhaps five minutes, then turned and ran, the Southerners cutting them down. "The Texans rushed after the fugitives," one of them remembered, "and, clubbing their muskets, continued the work of destruction until every enemy in

---

Martin, *The Second Bull Run Campaign*, 206, 211–12; Catton, *Mr. Lincoln's Army*, 35; Thompson, "Second Manassas: Enfilading Thunder," 51.

sight was left prone upon the ground." Lieutenant Lemon of the 18th Georgia recalled the mauling of the 5th New York twenty-five years later.

> We halted almost in their faces & leveled our pieces & sent a brutal volley right into them, & I distinctly can remember hearing hundreds of their piercing screams above the clatter of our musketry. No sooner had we fired than we were upon them with bayonet & clubbed musket & without exception they broke & fled pel mel down the slope to their rear, dropping or flinging away their muskets, which now only served to impede their progress to the rear. We pursued them with almost reckless abandon down the valley shooting & killing them at every step. Our formation & order were hopelessly lost & we had become, I am somewhat now ashamed to say, a savage mob in our fury to destroy this hated bunch.

The Southerners chased the New Yorkers down the slope, across Young's Branch, and back up the south bank hill. "Hundreds of Yanks were madly running up the hill with our men but a few feet behind and shooting them down & bayoneting them in the back," Lemon recorded. By the time the 5th New York stopped to rally at Henry Hill, they numbered but sixty men. The regiment had been destroyed in just ten minutes—so fast, Colonel Wofford remembered, "that no halt was perceivable" in the Confederates' advance.[30]

"The ground was perfectly strewn with the bodies of the red breeched fellows," remembered a soldier of the 5th Texas. Hood rode through the dead and wounded zouaves as he hurried to catch up with his troops. By the time he did, the 4th Texas and 18th Georgia had overrun Capt. Mark Kerns's Pennsylvania battery and driven the Yankees farther back. Hood halted his men for realigning; ahead lay high ground (Chinn Ridge) teeming with enemy infantry and artillery. He called for Brig. Gen. Nathan G. Evans's brigade to come up. When it did Hood's and Evans's troops attacked the Federals, pushed them back in places, but were soon repulsed. Col. Montgomery Corse's brigade of Kemper's division joined the fight, and the Confederates of all three brigades then drove the enemy from their position. Two more of Kemper's brigades and two of D. R. Jones's added strength to the Southerners' seizure of Chinn Ridge, the high ground a third of a mile west of Henry Hill. By then it was 6 P.M. Hood's and Kemper's troops were tired and scattered;

---

[30] Hennessy, *Return to Bull Run*, 361–70; Davis, *Hood's Texas Brigade*, 118; Hood, *A & R*, 36–37; Ural, *Hood's Texas Brigade*, 112 (Hood's "poor decision" to place the inexperienced Sellers in command of his brigade); Patchan, *Second Manassas*, 23; Lemon, *Feed Them the Steel!*, 31; Polley, "Second Manassas," 219; Wofford report, September 8, *OR*, vol. 12, pt. 2, 609.

they took no part in the last fighting of the day, as Confederates tried unsuccessfully to take Henry Hill. Holding that important point allowed Pope to order retreat for his beaten army, which withdrew back toward Washington in a far more orderly manner than had Irvin McDowell's from the same field a year before. Lee had not destroyed the Federal army, as had been his hope, but he had certainly disciplined the miscreant Pope.

The cost of this achievement, according to Surg. Lafayette Guild, the ANVA's medical director, was 1,090 killed and 6,154 wounded for a total of 7,244 casualties. Dr. Guild listed no missing, but Livermore calculated a mere 89 Confederate soldiers absent at the end of the battle. On the other side, Pope proved in his final report as incompetent to compile his army's losses as he had been in leading it to battle: he deferred to his corps commanders. But an appended "Grand Recapitulation" for August 16–September 2 (including the fight at Chantilly) shows 929 killed, 4,389 wounded, and a whopping 2,787 captured or missing, for 8,105 soldiers lost. A sometimes unheralded aspect of the Confederates' victory at Second Manassas is the number of Northern pieces of artillery they captured, which General Lee in his report put at thirty. The authority on the subject is L. VanLoan Naisawald, author of *Grape & Canister: The Story of the Field Artillery of the Army of the Potomac, 1861–1865.* He concedes that "official Union reports do little toward verifying this figure." But after much research in the official records and other sources, Naisawald is able to document the loss in Pope's army of twenty-one to twenty-three guns. Hood was very proud that his soldiers had captured five of them (counting the gun left wrecked on the field).[31]

In helping to bring all this about, writes Dr. Freeman, Hood had simply been "magnificent." He had placed S. D. Lee's artillery just where it could mow down Porter's infantry on the afternoon of the 30th. Keith Bohannon

---

[31] John F. Schmutz, *"The Bloody Fifth,"* 175; Hood, *A & R*, 36–37; Patchan, "Second Manassas," 25; Hennessy, *Return to Bull Run*, 378–406 *passim*; Hennessy, "Second Battle of Manassas," 56; Guild report, [no date], *OR*, vol. 12, pt. 2, 562; Livermore, *Numbers and Losses*, 89; Pope report, September 3, *OR*, vol. 12, pt. 2, 43; "Grand Recapitulation," vol. 12, pt. 2, 262; Lee report, June 8, 1863, *OR*, vol. 12, pt. 2, 558; Naisawald, *Grape and Canister*, 120, 122, 129, 139, and especially 455 n.52.

As example of variance in the literature on casualties at Second Manassas, see Wert, "Return to the Killing Ground," 25, where he gives Confederate losses as 1,305 killed and 7,048 wounded for a total of 8,353. Wert states that Pope's numbered 1,716 killed, 8,215 wounded and 3,893 missing = 13,824.

A story circulated that after Hood's attack on August 30 Longstreet came up to him and asked, "Hood, where are your Texans?" "They slipped the bridle and are gone," Hood answered; "the last I heard of them they were going toward Washington" (Hunter, "Lieut. Gen. John B. Hood," 258).

deems this act alone one of Hood's "most valuable contributions to the Confederate victory at Second Manassas." In his memoirs, Hood also recalled proudly how he had warned Richard Anderson on the night of the 29th that his division had marched into an unsafe position and pointed it back. In the grand assault, he had righted his troops' alignment, steered Evans's brigade into the fight, and given D. R. Jones the advice he asked for as he brought his division up. Biographer McMurry thus aptly states that Hood had once more, as at Gaines's Mill, "performed well as a combat leader, both in directing the fighting of his own men and in skillfully deploying other troops as they came into the battle." And once more, at Second Manassas, the Texas Brigade earned as much praise as their leader. "Hood's Texans had set the pace in pursuit," Freeman writes of their attack, "with all the contempt of danger they had shown in the swamp at Gaines's Mill." Hood praised his troops in his after-action report. "With their usual daring and enthusiasm they charged gallantly on," he wrote, noting that his Texas Brigade had driven the enemy a mile and a half, captured four guns, and caused "terrible slaughter in their ranks." Hood's troops had also suffered their share. In the two days of battle, August 29–30, the Texas Brigade's casualties were 75 killed, 540 wounded, 13 missing. McLaws's were 56 killed and 264 wounded, for a total of 935 *hors de combat* within the division. It was a stiff price to pay for one of Robert E. Lee's greatest victories of the war.[32]

Hood had been a hero at Second Manassas, but he may not have been above promoting himself further. The *Richmond Enquirer* on September 13 printed an article written by "An Eye Witness," praising the Texas Brigade for its achievements during the battle. Hood—rather, "the gallant Hood," a phrase that would resonate for many years—was hailed for his "courage and genius." McMurry suspects that Hood may have had one of his officers write the column, dated September 3, as its author demonstrated "such knowledge so soon after the battle" that he could only have been in Hood's close circle. McMurry also notes that the article was given to the *Enquirer*, the Richmond paper known to be the organ of the Davis administration. Because of all this, McMurry concludes, "Hood may have been more sophisticated in army politics than has been supposed." Sophistication does not generally characterize John Bell Hood, but ambition—the drive for fame and glory—certainly does.

Nonetheless, the hero soon found himself surprisingly under arrest. Late on the 30th Hood had seen several Federal ambulances in his front and sent

---

[32] Freeman, *Lee's Lieutenants*, 2:138; Bohannon, "'A Bold Fighter,'" 254; Hood, *A & R*, 35–37; McMurry, *John Bell Hood*, 55; Hood report, September 27, *OR*, vol. 12, pt. 2, 605–606.

forward some of his Texans to bring back the brand-new wagons, fine battlefield prizes. He distributed the trophies among his regiments. The next day, Brig. Gen. Shanks Evans ordered Hood to turn over the ambulances to his own brigade. By seniority Evans outranked Hood and was therefore his superior on that part of the field, but Hood refused. "Whereas I would cheerfully have obeyed directions to deliver them to General Lee's Quarter Master for the use of the army," he later wrote, "I did not consider it just that I should be required to yield them to another brigade of the division, which was in no manner entitled to them." Evans reported this to Longstreet, who placed Hood under arrest, removed him from command, and ordered him to Culpeper Courthouse to await court-martial. General Lee learned of the fracas and stepped in to overrule Longstreet, returning Hood to his troops although he was not reinstated to command them.[33]

August 31 was a day of heavy rain, hindering Lee's hopes for an aggressive pursuit of Pope's beaten army, which retreated back into the fortifications near Centerville. That morning, Lee, covered in poncho and in rubber overalls, was standing and talking with Longstreet and Jackson when someone cried out, "Yankee cavalry!" This proved to be a false alarm, but it created such commotion that Traveller, Lee's horse, reared; the general grabbed at the reins but tripped over his raingear, fell, broke his right wrist, and sprained his left. Unable to ride for several days, Lee was confined to an ambulance. But he was by no means idle, as he directed Jackson with three divisions, Stuart's cavalry in the lead, to attempt a turning movement: march north eight miles to the Little River Turnpike, then swing southeast, intending to get around Pope's right near Centerville. On September 1, Jackson's troops slogged through the rain toward Fairfax Courthouse. But Union cavalry alerted Pope to the threat, and, when Jackson's troops neared Chantilly on the turnpike, they found Un-

---

[33] McMurry, *John Bell Hood*, 56, 60–61, 213 n.2; "An Eyewitness," and "Texas Brigade," *Richmond Enquirer*, September 13, 1862; Hood, *A & R*, 38; Freeman, *Lee's Lieutenants*, 2:147; Dyer, *Gallant Hood*, 127–28.

On Hood's suspected involvement in the *Enquirer*'s laudatory column, McMurry states that other officers in Lee's army "had publicized themselves through the press" (213 n.2), citing an article in John M. Daniel's *Richmond Examiner* that expressed adulation to General Powell Hill for his role at Frayser's Farm. Longstreet, angered by a statement in the article, thought that one of Hill's officers wrote it and had Moxley Sorrel, his AAG, so tell the press (Hassler, *A. P. Hill, Lee's Forgotten General*, 68). Robertson, however, states that Hill "had had nothing to do with the manufacture of Daniel's stories" (Robertson, *General A.P. Hill*, 96). Gallagher, in *Lee and His Generals in War and Memory*, termed the Hood-Evans disagreement "a silly quarrel" (32).

ion infantry ahead of them. Several hours of charge and countercharge resulted in the little battle of Chantilly, fought in torrents of afternoon rain. The affair brought 500 Confederate casualties and 675 Federal before both sides backed off the field around 6:30. Chantilly also resulted in the deaths of two Union major generals, Isaac I. Stevens and Philip Kearny. They were the highest ranking Northern officers yet who died as result of combat.[34]

The battle was not resumed on September 2 as Lee, with his army thirty miles from Washington, planned his next initiative. That day Lincoln and Halleck met with McClellan in Washington and placed him in command of all troops in the area, including the Army of the Potomac and those formerly in Pope's "Army of Virginia." Pope himself was relieved of command (soon to be sent out west to fight Indians), but the Federal army was as large as ever. Though it was shaken, Lee knew that McClellan would soon restore its spirit. In the meantime, as Col. Porter Alexander phrased it, "he could not afford to sit down before Washington and await the enemy's pleasure." Near Dranesville, Virginia, Lee wrote President Davis on the 3rd, "the present seems to be the most propitious time since the commencement of the war for the Confederate army to enter Maryland." He outlined his logic. The enemy in his capital fortifications could not be attacked or besieged. "Still, we cannot afford to be idle," so an advance across the Potomac was the only option, despite the fact that many of his soldiers lacked proper clothing and shoes, and that the army as a whole needed supplies, wagons, and animals. But Virginia had largely been cleared of enemy forces, and the Confederate capital was safe. For all of these reasons, on September 4 Lee wrote the president, "I shall proceed to make the movement at once, unless you should signify your disapprobation." By the time Davis responded on the 7th, suggesting the themes of proclamations to issue to the people of Maryland, Lee and his army had already crossed the Potomac.[35]

Hood was still under arrest but with his troops. At Leesburg on the evening of September 4, according to Lieutenant Lemon of the 18th Georgia, the Texas Brigade was marching past General Lee, on Traveller at roadside. The cry rose up from the Texans, "Give us Hood! General Lee, give us Hood!" Lee was visibly moved. Raising his hat, he called out, "You shall have him,

---

[34] Harsh, *Confederate Tide Rising*, 165–68, 170–73; E. M. Thomas, *Robert E. Lee*, 255; Naisawald, "A Nasty Little Battle in the Rain—Chantilly," 12–13, 16; R. James, "Stonewall's Surprise at Ox Hill," 61; Hennessy, "Thunder at Chantilly," 55 (General Stevens's death).

[35] Murfin, *The Gleam of Bayonets*, 64, 71; Lee to Davis, September 3 and 4, 1862, *OR*, vol. 19, pt. 2, 590–91; Davis, *Man and His Hour*, 468; Davis to Lee, September 7, *OR*, vol. 19, pt. 2, 598–99.

gentlemen!" The next day the army began crossing the Potomac at White's Ford. On the Maryland side the troops marched north along the Monocacy River. General Evans ordered Hood's division to wade across, but the men would obey no orders save their commander's. As Hood put it, "my division had become restive and somewhat inclined to insubordination on account of my suspension." Hood evidently came along and told the men to ford the river; they went in, clothes and all, to wash away the dirt and grime from their march.[36]

After a couple of days along the Monocacy, Hood and his division, part of Longstreet's command, marched through Frederick on September 10. By then, Lee, Jackson, and Longstreet had met there and discussed Lee's plan for the days ahead. He decided to divide the army, sending Jackson to capture the enemy garrison at Harpers Ferry while Longstreet proceeded to Boonsboro. When Harpers Ferry fell, presumably on the 12th, the army would reunite at Boonsboro or Hagerstown, and march into Pennsylvania. Lee was displaying his trademark audacity, giving little worry to McClellan, confident that he would have "the slows." Hood's division marched through the Catoctin Mountains on September 11, then South Mountain and Boonsboro, arriving in Hagerstown on the 13th.

That was the fateful day when a copy of Lee's campaign plan was found by Federal soldiers outside Frederick. Jeffry Wert rightly calls this event "one of the greatest intelligence coups in American military history." Knowing that Lee's army was so spread out, McClellan laid plans to march his forces against Lee and bring on battle. Thanks to a spy, on the night of the 13th, Stuart was able to report the activity at McClellan's headquarters to Lee, who counseled with Longstreet. The Army of Northern Virginia was clearly in a tight spot. Jackson at Harpers Ferry had still not forced the Union garrison to surrender. Lee and Longstreet were at Hagerstown, twenty miles away. And just about as far from each of the Confederate contingents was McClellan at Frederick, a dozen miles east of South Mountain. If Mac got his forces through the mountain gaps, he would have moved between Lee and Jackson. Only D. H. Hill's division and Stuart's cavalry at Boonsboro were close enough to defend South Mountain. Lee would not have known McClellan's strength (70,000–80,000), but he would definitely have known that it vastly exceeded his own (reduced by straggling to some 37,000). Fortunately for Lee and the Southerners, the Young Napoleon acted on his great intelligence coup with un-Napoleonic lethargy. Having received the lost order by noon on the 13th,

---

[36] Lemon, *Feed Them the Steel!*, 33; Wert, *Longstreet*, 180; O'Connor, *Cavalier General*, 107; Hood, *A & R*, 39; E. B. Williams, *Hood's Texas Brigade*, 96.

McClellan issued orders for his troops to start marching westward at daylight the next day. Stephen Sears astutely observes, "no doubt had the situation been reversed and a Lost Order presented to Robert E. Lee, that general would have had Jackson's foot cavalry on the march within the hour and every other man in the army who could carry a rifle moving after them within two."[37]

On the morning of the 14th Longstreet and Lee thus marched their troops from Hagerstown back to Boonsboro, thirteen miles, to help Hill hold South Mountain. Hood's men marched fast, not so much anxious as angry that they were going into battle without their general. Cpl. Joe Polley of the 4th Texas recalled the mood: "Feeling that the commander they most trusted was deeply wronged, the officers and men of the division had given expression to their indignation, and now as they marched toward what might be another battle, their wrath grew intense. The Texans, naturally, felt most aggrieved, and were most outspoken." That afternoon, as they neared the mountain slope, Hood's men again shouted, "Give us Hood!" Lee sent for him; Hood rode up and dismounted. "General," Lee said, as Hood later recounted the conversation, "here I am just upon the eve of entering into battle, and with one of my best officers under arrest. If you will merely say that you regret this occurrence, I will release you and restore you to the command of your division." Hood respectfully explained why he could not do so. Lee asked once more, and Hood again declined. Despite Hood's refusal, Lee had no choice but to give in: "Well, I will suspend your arrest till the impending battle is decided." When they heard this good news, Hood's soldiers cried out, "Uncle Robert has released General Hood!" and "Hurrah for General Lee! Hurrah for General Hood! Go to hell, Evans!" The men cheered louder as Hood rode to his rightful place at the head of the column. "His face proud and joyful," one recalled, "the cheers deepened into a roar that drowned the volleys of the hundred cannon that were even then vengefully thundering at the Gap."[38]

---

[37] Holsworth, "Uncommon Valor," 9–11; Wert, *Longstreet*, 182–84, 187–88; Murfin, *Gleam of Bayonets*, 161–62; Sears, "Fire on the Mountain," 10; W. C. Davis, *The Deep Waters of the Proud*, 270 ("Little Mac had over 80,000 men in his army"); McPherson, *Battle Cry of Freedom*, 537 ("McClellan with 70,000 men [soon reinforced to 80,000])"; Hartwig, "Lee's Shrinking Army," 16 (Lee's army reduced by straggling and sickness 41 percent to "37,351 officers and men present in the battle"); Sears, *Landscape Turned Red*, 113, 120, 125–26.

Joseph L. Harsh reviews the historiography of Lee's famous Lost Order No. 191 in *Sounding the Shallows*, 170–75, and concludes that Lee knew by the morning of September 14 that his campaign plan had fallen into McClellan's hands.

[38] Holsworth, "Uncommon Valor," 11; Hood, *A & R*, 39–40; Wert, *Longstreet*, 188–89; Polley, "Crossing Over into Maryland," 275; York, "General Hood's Release from Arrest," 423; Silverman, Thomas, and Evans, *Shanks*, 112, 198 n.11 (no mention of Evans's response to Hood's release).

Harvey Hill's division was hotly engaged when Longstreet's troops approached South Mountain around 2 or 3 P.M. September 14. There were three passes through the mountain, Turner's, Fox's, and Crampton's, and McClellan's forces were attacking them all, trying to get through. Brig. Gen. Robert E. Rodes's brigade stubbornly held onto Turner's Gap (farthest to the north, on the Confederate left). To the south, at Crampton's Gap, Maj. Gen. William B. Franklin's VI Corps pushed back Southern defenders and took the gap, routing Brig. Gen. Howell Cobb's Georgia brigade and taking several hundred prisoners. Fortunately, the Union advance petered out by nightfall.

Fox's Gap, the middle of the three passes, became the pivotal point. In the morning Federals of Brig. Gen. Jacob Cox's Ohio division (in Maj. Gen. Jesse Reno's IX Corps) had attacked and forced back Brigadier Samuel Garland's defenders; Garland was mortally wounded in the fighting. But by 11 A.M. Cox's troops were spent and did not press farther. In a couple hours, Reno's other divisions came up, attacking just as Longstreet's column arrived on the field. The division of David Jones (called "Neighbor" by friends) was in the lead, and two of his brigades went in to help hold the line. Hood's division was next. As enemy troops descended the western slope of the mountain, Hood led Law's and Wofford's brigades through woods toward the Union lines. "Finally," as he recounted, "I gave instructions to General Law and Colonel Wofford, directing the two brigades, to order their men to fix bayonets; and, when the enemy came within seventy-five or a hundred yards, I ordered the men to front and charge." Hood's troops responded with their usual *elan* and attacked "with a genuine Confederate yell." Hood wrote that "the Federals were driven back pell mell, over and beyond the mountain, at a much quicker pace than they had descended." The attack of Hood's division

---

According to Major York of the 6th North Carolina, writing in March 1875, the Lee-Hood exchange at Boonsboro on September 14 was an emotional one. Lee explained, "Here I am going into an important battle," York remembered, noting that "the big tears were rolling down Gen. Lee's cheeks, plowing their little furrows through the dust upon his noble face." After the war, in 1873, York encountered Hood and the two had a chance to exchange memories. York recalled the time when Hood had refused to apologize to Lee and asked if he too remembered it. "'Yes,' said he, 'I do, and shall as long as I remember anything.' He alluded feelingly to his great Commander being unable to restrain his tears, and as Gen. Hood related the incident to me, the big tears came unbidden to his own eyes" (York, "General Hood's Release," 423).

Angelina Winkler offers a variant of this story. Approaching Boonsboro, General Evans ordered the Texans to form for battle. When they balked, "General Evans became angry, and threatened to turn a brigade on them. This did not frighten them into submission, so he sent to General Lee the information that the Texas brigade had mutinied," whereupon Lee restored Hood to command (Winkler, *The Confederate Capital*, 102).

stabilized the sector at Fox's Gap. Confederates had held onto the South Mountain passes—barely. To buy time for Jackson's return, they lost some 2,900 casualties to the Federals' 2,300 (among them was Major General Reno, mortally wounded). As the fighting ceased with the coming of night, Lee ordered his tired South Mountain defenders to withdraw westward across Antietam Creek. In the words of Antietam's chronicler James V. Murfin, the Confederates' retreat was "a masterpiece of quiet, skillful, undetected, disengagement." As to be expected, Hood's Texas Brigade marched in the "post of honor," the rear of the rear guard. The Southerners marched west to Boonsboro by the National Pike, then south to Sharpsburg. Hood's division crossed Antietam Creek at its Middle Bridge around noon of September 15 and took position just south of it.[39]

By that time Lee had reversed himself and had determined to stand on the west side of the creek and give defensive battle. At 8:00 on the night of the 14th, with the fight at South Mountain ending, Lee wrote McLaws, "the day has gone against us and this army will go by Sharpsburg and cross the river." Sharpsburg was just five miles from the Potomac and Lee's planned fording point at Shepherdstown. But then, twelve hours later, Jackson wrote Lee, stating that the garrison at Harpers Ferry was surrendering—all 11,500 men, rifles, wagons, and artillery. Jackson explained that he would leave A. P. Hill's division to parole the prisoners and send the booty back to Richmond. He and the rest of his troops would set out to join Lee that evening.

Jackson's message arrived about noon on the 15th. Lee had with him only 15,000 troops along Antietam Creek, but Jackson would more than double that strength after his seventeen-mile march. Just as important, Lee had marked his man in the Seven Days; he knew McClellan was timid and would not execute boldly in battle. He therefore told his officers, "we will make our stand." The Confederate line took shape as Lee positioned the three divisions

[39] Wert, *Longstreet*, 188; Holsworth, "Uncommon Valor," 11–12; Jamieson, *Death in September*, 38 (fine map of the three passes); Grimsley, "On the Edge of Disaster," 21–23, 44, 46; B. Welsh, "Firing the Gap," 44; Sears, *Landscape Turned Red*, 135; Murfin, *Gleam of Bayonets*, 173–77, 186; Hood, *A & R*, 49–50; Sears, "Fire on the Mountain," 20; Stinson, "The Battles of South Mountain," 17; Priest, *Before Antietam*, 324–27.

Priest is the most detailed authority on the casualties at South Mountain, September 14, which he counts as 317 killed + 1,300 wounded + 1,265 missing/captured = 2,882, plus 27 in Stuart/Pelham for a Confederate total of 2,909; and 437 K + 1,822 W + 87 M/C = 2,346 for the Federals (*Before Antietam*, 324–27). Some writers give different numbers; e.g., James M. McPherson, *Crossroads of Freedom: Antietam*: "The Confederates lost between 1,900 and 2,700 men and the Federals 1,813" (111).

he had with him (Hill, Hood, and Jones). He and Longstreet put in every piece of artillery they had.

Meanwhile, as always, McClellan took his time. On the 15th he spent until around noon at his headquarters, then followed his advance division's march toward Sharpsburg. At 3 P.M. he was before Antietam Creek, beginning to inspect the enemy positions across it. Of course, he did nothing beyond this. "I found that it was too late to attack that day," he later explained.[40]

Hood's division had marched hard without much to eat. "We had had issued to us no meat for several days, and little or no bread," he recalled; the commissary trains had not arrived. Hungry soldiers scoured about for green corn, green apples, anything. For some time there was nothing to eat at all. As one Southerner wrote, some men were chewing straws "merely to keep their jaws from rusting and stiffening entirely."

Despite their exhaustion, Hood's troops were ordered into new position at the extreme left of Lee's line, a mile north of Sharpsburg itself along the Hagerstown Pike into some woods near the "Dunker Church," so named because of the German Baptists' belief in baptism by complete immersion. The place may have been called "St. Mumma's Church," as both Hood and Law so refer to it in their reports; nearby resident Samuel Mumma was a German Baptist and had given the land for the church site. But to locals the Brethren were simply the Dunkers, or Dunkards. Hood placed his two brigades—Wofford to the left, Law to the right—in the forest behind, or to the west of the Dunker Church. The area, filled by hickory and oak, was called the West Woods.

On the morning of September 16, both army commanders rode their lines. When the morning mist burned away McClellan studied the Rebels' positions. At 7 A.M. he had wired Halleck, "Will attack as soon as situation of enemy is developed." In the time it took for Mac to size up the situation and shape his plans (by two o'clock), Jackson's troops began arriving from Harpers Ferry. Until they could get into position, Hood's division held the Confederate left. McClellan's plan called for a main attack on the enemy's center, with diversionary assault against Lee's left. In preparation for the morning's work, Hooker's I Corps advanced around 4 P.M. and marched across the creek at Upper Bridge. Southern videttes spotted the movement and sent word; Lee ordered Hood to form his two brigades, advance, and determine the Federals' intentions. It was near sunset when Hood's division

[40] Murfin, *Gleam of Bayonets*, 191, 193 (R. H. Chilton to McLaws, 8 P.M., September 14), 198, 202, 394–95 (Jackson to Lee, 8 A.M., September 15); Sears, *Landscape Turned Red*, 159–61; Stackpole, "Showdown at Sharpsburg," 7; Sears, "The Bloodiest Day," 5–6.

marched out of the West Woods, northward several hundred yards toward a cornfield owned by one David Miller. The Texas Brigade probed into the cornfield, Law's on its right. Bordering the cornfield was the "East Woods," holding some Pennsylvanians who were also probing for the enemy. Hood's men fired and drove back the Union skirmishers until darkness and want of ammunition brought a close to the first infantry firefight of the battle of Antietam. After dark, Hood pulled his units back for rest and rations. Brig. Gen. Alexander R. Lawton's division relieved them so they could retire and be fed, yet the commissary wagons did not arrive until well in the night, leaving Hood's men tired and famished.[41]

While Hood's troops—numbering 2,300 on the eve of battle—moved back behind the Dunker Church, Federals were shifting, too. Maj. Gen. Joseph K. F. Mansfield's XII Corps troops got orders to march toward Hooker's left flank; they would support the morning assault. Hooker's advance that afternoon had already given away McClellan's plan, so Lee knew that the battle would begin on his left. He placed Jackson in charge of that wing; Longstreet would hold the right.

At dawn of the 17th, Jackson had four brigades positioned on the northern end of the field, running west-east from the West Woods, across the Hagerstown Road south of Miller's cornfield, then southeast to the farm owned by Samuel Mumma: those commanded by Cols. Bradley T. Johnson, Andrew J. Grigsby (the Stonewall Brigade), Marcellus Douglass (formerly Lawton's), and James A. Walker (commanding the brigade of Isaac Trimble, wounded at Second Manassas). Four more formed a second line behind: Brig. Gens. Jubal Early's, Starke's, William B. Taliaferro's (commanded by Col. Edward T. Warren), and Roswell S. Ripley's. Farther back were Brig. Gen. Harry Hays's Louisiana brigade, Wofford's Texas Brigade and Law's brigade, now composed of the 4th Alabama, 6th North Carolina, 2nd and 11th Mississippi. The eleven brigades came from four divisions: Ewell's/Lawton's, John R. Jones's (previously Stonewall's), D. H. Hill's, and, of course, Hood's. Together they numbered a little over 7,500 infantry present. Hooker's corps alone had 8,600.

Artillery opened fire at 5 A.M. on the 17th: S. D. Lee's near the church, Northern guns across the creek. All three divisions of Hooker's infantry—

[41] Hood, *A & R*, 41–42; Hood report, September 27, *OR*, vol. 19, pt. 1, 923; Law report, October 2, *OR*, vol. 19, pt. 1, 937; Sears, *Landscape Turned Red*, 162; Holsworth, "Uncommon Valor," 13–14; Murfin, *Gleam of Bayonets*, 206–208; Bohannon, "'Dirty, ragged, and ill-provided for': Before Sharpsburg," 25; Bohannon, "Dirty, Ragged, and Ill-Provided For: Confederate Logistical Problems," 117.

Brig. Gen. Abner Doubleday's, Brigadier George Meade's, and James Ricketts's—marched into the Miller cornfield from the North and East Woods at 6 A.M. The Federals came under fire from Lee's artillery, then volleyed with the Rebel infantry. Colonel Douglass was killed as his troops stood in front of Union Brig. Gen. Abram Duryee's brigade trading salvos. Starke's and Hays's brigades advanced in support; Starke was hit by three balls, was carried from the field, and died within an hour. In charge and countercharge the cornfield changed hands repeatedly amid fierce musketry and cannonading. "In the time I am writing," wrote General Hooker, "every stalk of corn in the northern and greater part of the field was cut as closely as could have been done with a knife, and the slain lay in rows precisely as they had stood in their ranks a few moments before. It was never my fortune to witness a more bloody, dismal battlefield."

The Southerners suffered heavy losses among both soldiers and officers: Brig. Gen. J. R. Jones was stunned by a shell fragment to the head and taken from the field; generals Lawton and Ripley were also wounded. A third of Douglass's, Hays's, and Trimble's brigades were killed or wounded. The outnumbered Confederates gave ground grudgingly as they retreated toward the West Woods. The Federals fed in more troops to sustain their assault against Confederate efforts to counterattack. But too many of Jackson's troops were streaming to the rear, calling out to Stonewall, "No ammunition, no ammunition!"

Within an hour, Jackson's line was in danger of breaking. Hood's two brigades were his only reserve; they had to go in. The night before Jackson had permitted Hood to retire his men for rations if he committed to enter battle when needed. Now, between 6:30 and 7, Hood delivered on his word. His troops, camped on the south and west fringes of the West Woods, had finally received three days' rations: a little bacon and flour issued to each man around 4 A.M. "We had begun to cook our dough on sticks and ramrods when the enemy commenced to shell the woods and spoiled most of our hoecakes," recalled a soldier of the 4th Texas. Then orders came to fall in. Some wolfed down their greasy biscuits or half-cooked pones and bacon; others got nothing to eat at all. "I have never seen a more disgusted bunch of boys, and mad as hornets," wrote Lt. James Lemon of the 18th Georgia. At 7 A.M. Wofford's and Law's brigades formed a line facing north, with the Texas Brigade on the left: Hampton, 18th Georgia, 1st, 4th, then the 5th Texas, connecting with Law's four regiments. The line stretched from the Hagerstown Pike to the East Woods.

As Hood's troops advanced toward the cornfield, Starke's and Warren's troops limped back toward them. Hood saw General Lawton, wounded, being borne back to the rear. He rode up to Harry Hays and heard that the Louisianans were out of ammo. Hood said that they should retire to resupply as his division was coming up. "They moved forward in gallant style, making the air ring with the well-known 'rebel yell,'" recalled one Confederate, "and soon met the on-coming tide of the Federals, flushed with victory, and rolled it back like a wave is shattered and beat back when it strikes a rock."

The Southerners' musket volleys and bold advance into the cornfield drove back the Yankees. Their blood-curdling yell was an event Maj. Rufus Dawes of the 6th Wisconsin long recalled "in all its terror." Although the enemy they faced—Brigadier George Hartsuff's and Col. William Christian's brigades of Ricketts's division—always maintained that they fell back in good order, some Federals moved back faster than others. "Fear gave us wings, and strength as well," as one retreating Yankee put it. Swept from the cornfield, Ricketts's division lost 40 percent of its strength. The field was so thickly strewn with dead and wounded that Hood, mounted, worried that his horse would trample injured men. Hood directed the 4th Texas toward the left to face the enemy massing along the Hagerstown Pike. The 18th Georgia was stopped by fire from Capt. Joseph B. Campbell's Battery B, 4th U.S. Artillery; the Georgians charged several times but could not capture the guns. Hood did all he could to encourage his troops; Capt. Ike Turner of the 5th Texas remembered how "General Hood rode up ordering me...to press forward."

Hood had hoped for support, but Jackson had none to send him. At one point Stonewall sent his aide Capt. Sandie Pendleton to Hood; he told Pendleton, "Tell General Jackson unless I get reinforcements I must be forced back, but I am going on while I can." A South Carolina officer heard Col. Martin Gary, commanding the Hampton Legion, plead to Hood for more troops. "I have sent for them," came the reply, "but they are not to be had." According to one source, tears streamed down his cheeks as he answered Geary.

To worsen the odds against the Southerners, Mansfield then brought in his XII Corps to aid Hooker's troops, who were badly used up. In the fighting, General Mansfield was shot from his horse, mortally wounded by a volley from the 5th Texas, 4th Alabama, and 21st Georgia. But Hood's attack was spent. "Our losses up to this time had been heavy," recorded General Law. "The troops now confronting the enemy were insufficient to cover properly one-fourth of the line of battle; our ammunition was expended; the men had been fighting long and desperately, and were exhausted from want of food and rest." Hood had no choice but to order a withdrawal of his troops from the field, back to the Dunker Church. "By 8:00 A.M.," writes John M. Priest, micro-chronicler of the battle, "both sides had fairly shot each other to shreds." As reinforcements from McLaws's division came up, the soldiers left in Hood's division fell back and regrouped near the church, resupplied themselves with ammo, and began assessing what they had achieved—and at what cost. It was 8:30 in the morning, and Hood's division's role at Sharpsburg was over.[42]

The battle of Sharpsburg or Antietam—the two sides could not agree on a name—resulted in a tactical draw. Lee's army bent but did not break under the heavy Federal attacks. The fortunate arrival of A. P. Hill's division from Harpers Ferry in the afternoon helped Confederates on Lee's right beat back the last Union assault, that by Maj. Gen. Ambrose E. Burnside's IX Corps. For his part, McClellan did not coordinate the attacks on the Rebel left, center, and right, nor did he commit all his troops to the fight. The standoff allowed Southerners to claim a tactical victory on September 17. Lee in fact held his lines the next day, effectively daring McClellan to attack him again (Mac did not). The Army of Northern Virginia retreated across the Potomac on the 19th. This allowed Northerners to claim tactical victory as well, or maybe even a strategic one, as Lee's first raid into Northern territory had been put to an end.[43]

September 17, 1862, was the bloodiest day of the Civil War and remains the deadliest one in American history. Returns for the Army of the Potomac show 2,108 killed, 10,449 wounded, and 753 missing, totaling 12,410. For the entire campaign, including South Mountain and Harpers Ferry, Surgeon Lafayette Guild, medical director for the Army of Northern Virginia, counted 1,567 dead and 8,724 wounded for 10,291 men lost—but this did not count

---

[42] Walsh, *"Damage Them All You Can,"* 155–56; Priest, *Antietam*, 32 (map), 42–80 *passim*, 323 (858 present for duty in Wofford's brigade and 1,474 in Law's); Murfin, *Gleam of Bayonets*, 210–11, 221; Sears, "Bloodiest Day," 13–14, 16–17; Chiles, "Artillery Hell! The Guns of Antietam," 25, 41; Holsworth, "Uncommon Valor," 18, 20, 50, 54; Cheeks, "Carnage in a Cornfield," 33–35; E. B. Williams, *Hood's Texas Brigade*, 103, 108; "The Texans at Sharpsburg," 555; Johnston, "A Reminiscence of Sharpsburg," 527; Carson, "Hampton's Legion and Hood's Brigade," 342; Jacobs, *The Gray Riders*, 33 ("General Hood wiped his eyes"); Law report, October 2, *OR*, vol. 19, pt. 1, 938;, *Gallant Hood*, 139–40; O'Connor, *Cavalier General*, 115; Hood, *A & R*, 43–44; William Allan, *The Army of Northern Virginia in 1862*, 388.

[43] Gallagher, Engle, Krick, and Glatthaar, *The American Civil War*, 60–62. The Confederates' repulse of the Union attacks allows William C. Davis to exclaim, "certainly McClellan did not defeat Lee at Antietam." Possessing two-to-one numerical superiority, McClellan failed to destroy the Rebel army, an outcome that "can only be regarded as an enormous tactical and moral victory for Lee" (Davis, *Deep Waters*, 277).

missing or captured. Two reputable sources estimate the Confederate missing/captured as 1,844, possibly up to 2,000. In any event, the 3,600 soldiers killed at Sharpsburg is a higher death toll than that of the September 11, 2001, terrorist attacks; the June 6, 1944, D-Day invasion; or the December 7, 1941, bombing of Pearl Harbor.[44]

For John Bell Hood, the battle of Sharpsburg added further to his and his men's already bright reputation. Southerners knew it immediately. An article in the *Richmond Enquirer* claimed that the Texas Brigade had won eternal laurels, and hailed General Hood's "courage and genius." Several weeks later, Capt. Thomas J. Goree of Longstreet's staff predicted that Hood would be promoted to major general. "He is one of the finest young officers I ever saw," Goree wrote on October 10; "no man deserves it more." Since that time chroniclers of Sharpsburg have added their echoes. "It is unlikely that many of the ragged veterans who assembled behind the Dunker Church after their role in the great battle," writes one student, "realized that they had made history in those ninety minutes in Miller's Cornfield." Another calls the attack of Hood's division perhaps "the grandest charge of the war." Given some of the valiant Confederate assaults yet to occur, this praise may be extravagant, but Hood's division had halted Hooker's advance, pushed the Federals back, and held Lee's left until help could arrive. "For the third time that summer," writes McMurry, "Hood had played a crucial role in a major battle. The fight on the Confederate left at Antietam had been extremely close, and Hood and his men had saved Lee from a calamitous defeat."

But they had paid dearly for it. "Hood's division was simply shot to pieces," writes Stephen Sears. In his report, Hood counted for Law's brigade 50 killed, 379 wounded, and 25 missing for a total of 454 casualties. For Wofford's brigade the figures were higher: 69 killed, 417 wounded, 62 missing = 548. Many of the missing were wounded, left on the field. Among the individual regiments, the 1st Texas suffered most severely. Lt. Col. Philip A. Work stated that the unit took 226 men into the battle and came out with 44—a casualty rate of 80.5 percent, one of the highest regimental losses for a

---

[44] "Return of Casualties in the Union forces at the battle of Antietam, Md.," *OR*, vol. 19, pt. 1, 200; "Report of Surg. Lafayette Guild…of killed and wounded," *OR*, vol. 19, pt. 1, 813; "The Opposing Forces in the Maryland Campaign," in Johnson and Buel, eds., *Battles and Leaders*, 2:603 (1,844 Confederate captured or missing); Livermore, *Numbers and Losses*, 92–93 ("missing, about 2,000"); Sears, *Landscape Turned Red*, 295–96 (3,654 dead at Antietam); Lipton, "Struggle to Tally All 9/11 Dead by Anniversary," (2,983); Axelrod, *America's Wars*, 517 (2,403 dead at Pearl Harbor); Ryan, *The Longest Day*, 279 (1,465 Americans killed on D-Day). Sears accepts the Federal army returns but places Confederate losses at 1,546 killed, 7,752 wounded, and 1,018 missing for a total of 10,316 (*Landscape Turned Red*, 295–96).

single engagement in the war. After the battle as he and other officers reported to General Lee, Hood was seen to express great emotion when asked about the condition of his men. "They are lying on the field where you sent them, sir," was his trembling reply.[45]

In the weeks after the battle, Hood and his command were comfortably encamped in the lower Valley near Winchester. The troops rested from their arduous campaigns, stragglers returned to their units, and convalescing soldiers arrived from the hospitals. Hood prepared his report on the campaign, completing it on September 27. That was the day that General Jackson recommended him for promotion to major general. Stonewall had seen Hood's distinguished performance at Gaines's Mill. At Sharpsburg, he wrote, "his duties were discharged with such ability and zeal, as to commend my admiration." He added, "I regard him as one of the most promising officers of the army." Others felt the same. Col. Jerome B. Robertson of the 5th Texas urged Texas Sen. Williamson S. Oldham to support Hood's promotion: "I have seen him in camp, on the march, and on the battlefield; and confidently declare that he is one of the best officers we have," Robertson wrote. Lee would eventually endorse Jackson's recommendation, as it fit well with the reorganization of the army which he began and which was announced on November 6. What had been Longstreet's and Jackson's "commands" now became corps, with

[45] Bohannon, "'A Bold Fighter,'" 256; Cutrer, *Longstreet's Aide*, 100; Holworth, "Uncommon Valor," 54; Cheeks, "Carnage in a Cornfield," 35; McMurry, *John Bell Hood*, 58–59; Sears, *Landscape Turned Red*, 202; Hood report, September 27, *OR*, vol. 19, pt. 1, 925; Work report, September 23, *OR*, vol. 19, pt. 1, 933; Dyer, *Gallant Hood*, 143.

Texans were understandably proud of their courageous charge and even of their heavy casualties. Val Giles stated that the 1st Texas lost 183 men out of 226, "the heaviest loss of any regiment, either Union or Confederate, in any battle of the war" (Giles, "The Flag of the First Texas, A. N. Virginia," 417). See also Hamby, "Hood's Texas Brigade at Sharpsburg," 20.

A battlefield marker standing today on the south side of Cornfield Avenue states that the Texas Brigade fighting here "maintained a contest rarely equalled in warfare" (Large and Swisher, *Battle of Antietam*, 102).

The 1st Texas's loss at Antietam is generally accepted as the largest percentage of men lost in a battle among Confederate regiments, 82.3 percent (Wert, "Regiment," 621). Scholars of the Texas Brigade, such as Susannah J. Ural, have concurred: "with over 86 percent casualties, the 1st Texas had suffered the highest percentage of losses of any regiment in a single battle in the entire war" ("Fighting 'Too Fast,'" 40).

The Texans' claim has not gone unchallenged. Archie Davis accepts William F. Fox's count of casualties in the 26th North Carolina at Gettysburg—86 killed, 502 wounded and approximately 120 missing (and probably killed or wounded)—for a total of 708 men lost, and Fox's conclusion that this regimental loss was the severest of the war. Davis computes the regiment's casualty total as over 88 percent (A. K. Davis, *Boy Colonel of the Confederacy*, 352).

both officers promoted to lieutenant general. Division commanders not already major general were also promoted. On October 27 Lee recommended a number of officers for advancement in rank; Hood's elevation was approved and made effective October 10. In the process, the charges against Hood brought by Evans were quietly dropped, and Shanks himself was transferred from the army. "At thirty-one," writes McMurry, "the Texan was the youngest and most distinguished of Lee's nine major generals of infantry."

In the reorganization the Hampton Legion and 18th Georgia were taken from Wofford's brigade and with them went General Wofford himself. Texas colonel Robertson was promoted to brigadier, and to the three regiments of the Texas Brigade was now added the 3rd Arkansas. Colonel Law was also promoted, but from his brigade were taken the two Mississippi regiments. Law's brigade henceforth comprised the 6th, 54th, and 57th North Carolina, and the 4th and 44th Alabama. Hood's small division was further expanded by the addition of Brig. Gen. George T. Anderson's and Howell Cobb's brigades, the latter now under Col. Henry L. Benning. The division's strengthening was as much a product of the overall reorganization as recognition by Lee of Hood's talents as officer.[46]

But those talents had limitations, as Hood was soon made to discover. Like other soldiers in Lee's army, Hood's men wore ragged clothing and some had no shoes. Chaplain Nicholas Davis placed an appeal in the *Richmond Whig* on November 5 for citizens to send shoes and socks to the Texans, whose distance from home hindered relatives' sending such items. Gifts eventually poured in, but not before Hood's superiors took notice of the shabby state of his division. On November 7, the day after Hood's promotion was

---

[46] E. B. Williams, *Hood's Texas Brigade*, 117–18; Freeman, *Lee's Lieutenants*, 2:257–58, 264; S. Hood, *The Lost Papers*, 43 (photograph of Jackson's handwritten recommendation of Hood's promotion); Hood Service Record, National Archives ("Announced—November 6 1862 as Major General"); Robertson, *Touched with Valor*, 29 ("one of the best officers we have"); Lee to George W. Randolph, October 27, 1862, *OR*, vol. 19, pt. 2, 683; McMurry, *John Bell Hood*, 61–63; Hood, *A & R*, 45–47; O'Connor, *Cavalier General*, 120–21. South Carolina Lt. Col. Martin W. Gary on November 15 wrote directly to Hood, stating, "the announcement of the orders separating the Hampton Legion, from the Texas Brigade, has fallen heavily upon the hearts of both officers & men. It is truly like 'breaking up of family ties,' and in leaving you, we feel that we are separating from the father of the family" (Gary to Hood, November 15, 1862, #552, folder 3, Landingham Papers,).

The departure of the 18th Georgia (to join T. R. R. Cobb's brigade) was a time for sentimental farewells. Lieutenant Lemon long remembered Hood's final visit to the regiment and his personal goodbyes. "I had occasion to shake his hand, and looking into my eyes he said, 'Lieutenant, you all will be missed. Fight hard and do your duty, as you have in all ways done for me'" (Lemon, *Feed Them the Steel!*, 36).

announced, the army's inspector general, Lt. Col. Edwin J. Harvie, visited Law's and Robertson's brigades. Harvie saw men poorly clothed, counted those who were barefoot, noted which regimental camps were indifferently policed, and took down the units whose arms were mixed and in bad order. To be sure, supply scarcity was widespread in Lee's army, but to an extent, haphazard or sloppy camp conditions reflected lack of attention by officers—starting with Hood. Based on Colonel Harvie's report, General Lee took the unusual step of admonishing Hood in a letter over the signature of his chief of staff, Col. Robert H. Chilton, dated November 14. Chilton reeled off the inspector's findings, regiment by regiment. The 1st Texas was in the worst shape: "arms mixed and in very bad order; two thirds badly clad and shod; 60 barefooted; camp in bad order, and the regiment showing inexcusable neglect on the part of its officers.... While the commanding general sees much in your management to commend," Chilton concluded, "he deems it but necessary to lead to corrections to advise you of deficiencies." Hood was told directly to have the men in the division exchange muskets so that at least those within a company would have the same caliber for purposes of ammunition supply. Weapons had to be procured for those men without arms.

This was as close to reprimand as the gentlemanly Lee could deliver, particularly to a talented young officer whose battlefield prowess he had come to admire. But there was a difference between battlefield and campsite, as all three of Hood's major biographers have emphasized. Chilton's "bill of particulars," observes O'Connor, "especially the section referring to the First Texas and its neglectful officers, reflected seriously on the administrative ability of General Hood." "The deficiencies in clothing and shoes could not be charged to Hood, for this deficiency existed throughout the army," reasons Dyer. "His chief fault as commanding officer appears to have been the fact that he did not exercise close supervision over his subordinate officers and insist that they keep their regiments in good condition." McMurry offers at once the most academic and critical explanation. From the Harvie inspection and Chilton/Lee letter, "one senses that Hood, once he rose above a position in which he was kept busy with the daily tasks of caring for and leading his men, began to slacken in his duties." McMurry's conclusion is telling: "Hood seems not to have made the transition from personal leadership to thinking in the broader supervisory terms appropriate for a major general whose work involved administrative and housekeeping chores rather than personal combat leadership."[47]

The Harvie/Chilton incident, of course, is not mentioned by Hood in his memoirs, in which his focus stays in the field. In late October, movement of McClellan's army led Lee to begin shifting his army from the Valley toward the Rappahannock. At Culpeper Court House, Hood's troops spent two weeks. Their commander took lodging in town; Lt. Halsey Wigfall, son of the Texas senator, wrote his family on November 15, "I was at Genl. Hood's Head Quarters this morning. Since he has got to be a Major Genl. he has moved into a house and I suppose intends to live in style!"

There the Confederates learned that McClellan had been relieved of command—Lincoln had had enough of his insolence and indolence. Promoted from corps command was Major General Burnside, who resolved to advance on Richmond by the most direct and obvious route: right across the Rappahannock at Fredericksburg. The Northern army's advance was expected. On November 17 Hood wrote his friend Louis Wigfall, "Well I think Mr. Burnsides is coming in a few days." He predicted, "I think we will whip him badly. Our army is in good trim for an old fashioned fight." To prepare for the coming battle, Lee deployed his army at Fredericksburg. Longstreet's corps held the left, enjoying high ground behind the town and a protective stone wall along a sunken road. Jackson held the right, lower ground which included a swampy woods. When Burnside's long-expected attack came, on the morning of December 13, this area proved to be the only place during the battle where the Federals achieved a temporary break of the Confederate lines. Elsewhere, all along the line, the Yankees met with bloody repulse. Hood's division was in the center of the Southerners' line—on Longstreet's right, connecting with Jackson. The Federals did not strike hard at Hood's sector. At one point in the afternoon, to blunt a Federal assault Hood sent two North Carolina regiments forward. They so zealously drove the enemy that Hood had to send couriers to call them back. Otherwise the battle of Fredericksburg ended as a lopsided defensive victory for Lee. Burnside's failed assaults brought 12,563 Union casualties; Lee's came to 5,309. Hood's losses reflected the slight role his division had played in the fighting: 252 casualties, 172 of them from the 54th and 57th North Carolina.

---

[47] Davis, *Hood's Texas Brigade*, 177–78; Chilton to Hood, November 14, 1862, *OR*, vol. 19, pt. 2, 718–19; O'Connor, *Cavalier General*, 123–24; Dyer, *Gallant Hood*, 144–45; McMurry, *John Bell Hood*, 63–64.

It should be noted that Hood seems to have come under mild criticism. In his report for the battle, General Longstreet stated that he had told Hood "that the enemy would attack General Jackson beyond his right; that he should watch carefully the movements and when an opportunity offered he should move forward and attack the enemy's flank." Longstreet gave the same instructions to General Pickett, telling him to cooperate with Hood. Longstreet claimed that when the Federal attack against Jackson presented the anticipated opportunity, Hood did not aggressively exploit it. Indeed, Pickett is said to have come to Hood advocating attack, but that Hood wanted to double-check with Longstreet first. Valuable time was lost; Hood sent forth the two North Carolina regiments anyway. In his after-action report Longstreet merely stated that "General Hood did not feel authorized to make more than a partial advance," and chose not to make anything more of it. Five years later Longstreet wrote Porter Alexander that the Confederates' victory at Fredericksburg had been "so handsome that I did not care to mar the enjoyment of it by making official trouble," but he expressed regret that he had not brought Hood to court martial. Three decades later, writing his memoirs, Longstreet was just as caustic about "General Hood's failure to meet his orders." Old Pete observed that because Hood "was high in favor with the authorities, it did not seem prudent to attempt to push the matter." In this episode there is certainly room for disagreement. Dyer concludes, "actually there appears to have been no neglect of duty on Hood's part." McMurry does not mention Longstreet's carping at all.[48]

---

[48] Hood, *A & R*, 48–49; McMurry, *John Bell Hood*, 64–66; [L. W.] Wright, *A Southern Girl*, 95 (Hood to Wigfall, November 17, 1862; also in S. Hood, *The Lost Papers*, 142), 114 (Halsey); Dyer, *Gallant Hood*, 156–59; Rable, *Fredericksburg! Fredericksburg!*, 250–51, 288; Stackpole, *Drama on the Rappahannock*, 188–90; Hood report, [no date] 1862, *OR*, vol. 21, 623; E. P. Alexander, "The Battle of Fredericksburg," 463 (54 killed, 185 wounded and 12 missing as Hood's division loss); Longstreet report, December 20, 1862, *OR*, vol. 21, 570; L. J. Gordon, *General George E. Pickett in Life & Legend*, 94; Bohannon, "'A Bold Fighter,'" 257; Longstreet, *From Manassas to Appomattox*, 317; O'Reilly, *The Fredericksburg Campaign*, 362; O'Connor, *Cavalier General*, 131–32; McMurry, *John Bell Hood*, 65. Some of Hood's Texans had acquired a good reputation for their marksmanship. The day before the battle of Fredericksburg, Jeb Stuart and Maj. Heros von Borcke were riding along Confederate lines when they drew fire from a group of Federal cavalrymen across the field. Hood, drawn to the sound of the shooting, rode up. "This will never do," he declared. "I must send up some of my Texans, who will make short work of these impudent Yankees.... Soon a select number of these dreaded marksmen," von Borcke remembered, "crawling along the ground, after their wild Indian fashion," came up and opened fire on the enemy, who "broke into confused and rapid flight" ([Frobel], "Field and Camp" [April 1867]: 271 ("our Texas sharp-shooters"); Borcke, *Memoirs of the Confederate War for Independence*, 2:105–106.

A month before the battle, in the army's return of November 10, Hood's division numbered 7,761 officers and men, third largest in the Army of Northern Virginia. Nonetheless Hood sought to increase the size of his command. By furloughing details, he encouraged the men to bring back new recruits from home. Confederate mobilization practice favored feeding new enlistees into existing regiments; placed in the ranks with veterans, the "fresh fish" gained combat effectiveness more quickly than their counterparts from the North, where recruitment practices favored the formation of whole new regiments. Hood was particularly concerned about his three Texas regiments, for which finding new recruits was difficult. In late February 1863, an officer and two privates traveled back to McLennan County, Texas, on a sixty-day furlough to recruit. Despite their efforts, they brought back just one man. On February 5, Hood wrote Senator Wigfall, asking for help in getting Texas Confederates captured in other theaters to be exchanged in Richmond so they might join Hood's units. Hood imagined leading a whole division of Texas troops, but did not want to leave Lee's army to do so. As he acknowledged to Wigfall, "I am very fond of General Lee." But there would be no Texas division in the ANVa. A large number of Texas soldiers, captured at Fort Hindman, Arkansas, in January, were eventually exchanged in Richmond. Confederate authorities took no action to add them to Lee's army.

The army went into winter quarters spread out along the south bank of the Rappahannock. Whiling time in their log and mud-caulked cabins, men of the 1st Texas built a chapel that served also as venue for entertainments by "Hood's Minstrels." Halsey Wigfall attended one of their performances, held on Christmas Eve. "One of the best performances was 'We are a band of brothers' sung by three make-believe darkies," Wigfall related to his family; the minstrels were "dressed entirely in black, with tall black hats and crepe hatbands, looking more like a deputation from a corps of undertakers than anything else." The "band of brothers" song, as mentioned by Lieutenant Wigfall, was Harry McCarthy's famous "The Bonnie Blue Flag," the patriotic air popular among Southerners throughout the war. It was a staple of performances in New Orleans concert halls; Texans passing though the city en route to Virginia heard it, as doubtless did Daniel Collins's 4th Texas band, the musicians behind "Hood's Minstrels." During winter '62–'63, General Hood often attended the minstrel shows; even General Lee came to one.

On February 18, Hood's and Pickett's divisions were ordered with Longstreet to the Richmond area; the enemy at Newport News on the Peninsula were reportedly reinforcing for a possible campaign against the Confederate capital and needed to be watched. The men of Hood's division marched through the streets of Richmond on February 22, Washington's birthday. Secretary of war James A. Seddon wrote General Lee that day, "Pickett's and Hood's divisions have passed through, and their general appearance, spirit and cheerfulness afforded great satisfaction." Festive citizens lined the streets, cheering the soldiers, waving flags and kerchiefs. Some offered tobacco and whiskey to the veterans, who quickly accepted. Regimental bands played "Dixie" and "La Marsellaise." The Texans, enjoying their reputation among the citizens for their rowdy behavior, marched down Broad Street "screaming, holloring, and yelling like so many demons," according to A. B. Hood of the 5th Texas. Hood's division made camp at Falling Creek, six miles southeast of Richmond on the south bank of the James. Lee was familiar with the liberal foraging habits of Hood's veterans, so he urged Longstreet to station his two divisions in areas where they were "not liable to prove injurious to the agricultural interests of the country." Hood was sensitive to allegations that his men robbed civilians of their provisions. Once, Lee remarked to him that when his "Texans [came] about, chickens [had] to roost mighty high." Hood took up for his soldiers but to little avail; their reputation for rowdiness was fixed.

During this time, Hood visited Richmond and as the popular hero was introduced into the city's social circles. Mrs. Davis opened the White House every couple of weeks for receptions, and General Hood made his appearance at them. One observer pointed out that "the blonde head of Hood towered over the throng of leading editors, senior wranglers from both Houses of Congress, and dancing men wasting their time in the vain effort to talk."[49]

The medical director for Hood's division was Dr. John Darby, a South Carolinian who had entered the Confederate service as chief surgeon for Hampton's Legion. Darby had become friends with other Carolinians in the capital, the Chesnuts and Prestons. (Col. James Chesnut served as aide to President Davis; Col. John S. Preston headed the Richmond conscription bureau.) Indeed, Darby was courting one of the Preston daughters, Mary. Mary Boykin Chesnut recorded that Mary Preston's sister Sally Buchanan ("Buck") Preston dubbed John Darby "P. V." because of his response when Buck asked him if he thought Mary ("Mamie") would ever marry him: *perseverantia omnia*

---

[49] "Abstract from field return of the Army of Northern Virginia for November 10, 1862," *OR*, vol. 19, pt. 2, 713; McMurry, *John Bell Hood*, 66–68; Hattaway and Jones, *How the North Won*, 728; H. B. Simpson, *Hood's Texas Brigade: Lee's Grenadier Guard*, 214–15, 219, 223; E. B. Williams, *Hood's Texas Brigade*, 130–31; Wright, *Southern Girl*, 116 (Hood's Minstrels); Abel, *Singing the New Nation*, 52–53; Seddon to Lee, February 22, 1863, *OR*, vol. 18, 889–90; Hood, *A & R*, 51; Dyer, *Gallant Hood*, 163–64 ("the blonde head of Hood").

vincit—"persistence wins all things." In mid-March Dr. Darby asked if he could introduce the ladies to General Hood. P.V., wrote Mrs. Chesnut, "bragged of him extensively, said he had won his three stars &c under Stonewall's eye and that he was promoted by Stonewall's request." The ladies invited Darby to bring the general to evening tea at the Chesnuts' apartment, at 12th and Clay streets, downtown near the Confederate White House. Before this soiree, Mrs. Chesnut had not met Hood but definitely knew of "the famous colonel of the Fourth Texas." She also knew that his friends called him "Sam" from his West Point days, although she did not know why. She later described him at some length in her diary. She was struck by "his sad Quixote face"; "tall—thin—shy. Blue eyes and light hair, tawny beard and a vast amount of it covering the lower part of his face—an appearance of awkward strength."[50]

The introduction evidently went smoothly, perhaps warmly, for Hood invited everyone to a picnic on the morrow down at Drewry's Bluff, near his division's campsite. The next morning the ladies were climbing into their carriage when Dr. Darby rode up. "Stop! It's all up. We are ordered back to the Rappahannock. The brigade is marching through Richmond now." The group decided to walk to see the troops marching out by way of the turnpike: Dr. Darby, Mamie Preston, Mary Chesnut, Johnny Chesnut (Mary's nephew, captain in the cavalry), Capt. Sam Shannon (aide to Maj. Gen. R. H. Anderson), and Buck Preston, who had not attended the previous evening's tea. The group watched Hood's troops march by. "We had seen nothing like this before," recorded Mrs. Chesnut. Hitherto they had viewed new regiments, "spic and span in their fresh, smart clothes," marching through the city on their way to the front. Now, two years into the war, they were seeing veterans. "Such rags and tags—nothing alike—most garments and arms taken from the enemy—such shoes!" Buck gave out a moan: "Oh, our brave boys!" Mary saw pans and pots tied to the men's belts; bayonets held aloft bread loaves and bacon chunks. The men were in high spirits. "They did not seem to know their shabby condition," she recorded. "They laughed and shouted and cheered as they marched by." Amongst the civilian crowd captains Chesnut and Shannon seemed to shrink back, as if they knew what was coming. "Captain, either take off your shoulder straps or come on to the Rappahannock," a soldier called out. "Maybe you did not know there was fighting going on there." Captain Chesnut was offended. "This is my first furlough," he complained to Mary. "I have not missed a battle—not one." Darby admonished

---

[50] Manning, "Surgeons of the Confederacy," 141; McMurry, *John Bell Hood*, 68; Woodward, *Mary Chesnut's Civil War*, 441; Muhlenfield, *Mary Boykin Chesnut*, 117.

him not to reply. "Those rough and ready Texians will make you rue it, if you give them a chance."

Suddenly General Hood and staff officers rode up and dismounted. Hood caught his first glimpse of Buck; Darby introduced the two. After he mounted up, Mrs. Chesnut noted that Hood took another look at Buck and said something to the doctor. Then he rode off. "What was that he said to you?" Buck asked Darby. She was eager to know: "About me?" P. V. answered. "Only a horse compliment—he is a Kentuckian, you know. He says, 'You stand on your feet like a thoroughbred.'"[51]

The date was March 18, 1863. Joe Hooker—the same I Corps commander whom Hood had faced in the Miller cornfield—had replaced Burnside as commander of the Army of Potomac, and Lee had to keep watch for an enemy advance. Lee became concerned that the enemy were about to cross the Rappahannock and called upon Longstreet to send back Hood's division. The troops had gotten started on their march right away. John Beauchamp Jones, clerk to the secretary of war, was among those in downtown Richmond to view Hood's soldiers pass by. Citizens lavished special attention on the three Texas regiments. "Some aged ladies ran out as they passed," Jones observed, "calling the bronzed Texans their 'children,' and distributed loaves of bread and other food among them." Some soldiers had three or four loaves on their bayonets. Jones recorded, "I never saw a merrier set than these brave soldiers."

Turner Vaughan, of the 4th Alabama in Law's brigade, recorded in his diary for the 18th that they marched through the city and camped at Ashland, fifteen miles north of the capital. They got no farther than that; on March 19 a heavy snowstorm struck and covered everything in nine inches of snow. Moreover, General Lee determined that Hood's division was not needed after all. Some enemy forces had indeed sallied across the Rappahannock, but Jeb Stuart's cavalry had driven them back. On the 19th Longstreet wrote Lee, "General Hood is ordered back to his old camp." So back through Richmond Hood's troops again tramped although this time through snowy streets. On the 20th war clerk Jones recorded in his diary that Hood's soldiers "were all the morning defiling through Main Street, in high spirits, and merrily snowballing each other." General Robertson (called "Aunt Pollie" by his men for the good care he took of them) rode at the head of the Texas Brigade, and did not notice his men slipping out from the march column, looking for taverns and grog shops downtown. At a certain point Aunt Pollie turned around and saw just a thin column of troops—only the teetotalers were left. "Where in

[51] Woodward, *Mary Chestnut's Civil War*, 441–42.

the hell is the Texas Brigade?" he shouted, whereupon Hood rode up. "Let 'em go, General—let 'em go," was Hood's advice. "They deserve a little indulgence, and you'll get them back in time for the next battle."[52]

Time spent in the Richmond area gave Hood more opportunity to share the company of the Chesnuts and Prestons. Dr. Darby had introduced them to the card game of casino. Hood got in on the fun. In one game he watched Darby place two fingers over his heart, and later asked what it meant. "Oh, so you caught me," the doctor replied. "I put two fingers on my heart, as a hint she should play the deuce of hearts." The general later recalled this card game to Darby when the two were under enemy fire. "I say, doctor," Hood said, "this is not as pleasant as the corner of Clay St. *and casino*."

Soon Hood's and Pickett's divisions were sent to southeastern Virginia. The worrisome presence of a Union force at Suffolk and the Confederates' prospect of gathering provisions in the relatively untapped area of northeastern North Carolina led the War Department to send Longstreet and the two divisions toward Suffolk (about seventy miles southeast of Petersburg) possibly to attack and capture the enemy garrison there. By March 27, Lee was ready to send Longstreet off. "I consider it of the first importance to draw from the invaded districts every pound of provision and forage we can," he wrote. "It will lighten the draught from other sections and give relief to our citizens."

Hood's division was camped at Falling Creek when orders to march arrived. First leg of the march was to Petersburg April 2–3; then to the Blackwater River April 8–10; then to Suffolk April 11–12. The Southerners dug in before Suffolk on the left bank of the Nansemond River. Thus began a Confederate "siege" of Suffolk that was conducted by Longstreet without much earnestness. The historian of the Suffolk Campaign summarizes Longstreet's three weeks there: "Having seen that the Federal defenses were too strong for a frontal assault and that a turning movement would be too dangerous as long as the river teemed with gunboats," writes Stephen Cormier, "Longstreet turned his attention away from capturing the garrison and concentrated, instead, on gathering subsistence while he held the Federals tightly within their circle of forts and breastworks." Meanwhile Confederate purchasing agents

---

[52] Bohannon, "'A Bold Fighter,'" 257 (Hood-Preston introduction March 18); H. B. Simpson, *Hood's Texas Brigade: Lee's Grenadier Guard*, 220, 222; J. B. Jones, *A Rebel War Clerk's Diary* [1958], 177–78; Vaughan, "Diary of Turner Vaughan," 575; McMurry, *John Bell Hood*, 69; Wert, *Longstreet*, 232; E. M. Thomas, *The Confederate State of Richmond*, 119 ("nine inches of snow fell in Richmond on March 19 and 20"); Longstreet to Lee, March 19, *OR*, vol. 18, 927; Ural, *Hood's Texas Brigade*, 140 ("Aunt Pollie").

across the North Carolina line scoured the surrounding countryside and collected thousands of pounds of corn, bacon, and fodder, plenty to feed Longstreet's men and to load onto wagon trains back to Richmond. "O, my, ain't we in? We just swim in bacon," wrote a delighted Texan. "Big rations, and all the time we have an immense wagon train hauling out bacon, corn, wheat, flour and great droves of beeves. It's a big haul we're making."[53]

For their part, Hood's men sometimes shot at crews of the enemy gunboats. Capt. Ike Turner, commanding Co. K in the 5th Texas, was placed in charge of a special battalion of sharpshooters to pop away at the Yankees. But just as often, they fired back at the Texans. On April 14 Captain Turner was at a Confederate battery dueling with the Northern vessels when, standing on the parapet, he was shot and mortally wounded by a Federal sniper across the river. The Southerners' sharpest setback occurred on the evening of April 19, when Union troops crossed the river and attacked a Confederate fort. No pickets had been posted and the garrison regiment, the 55th North Carolina (Maj. Gen. Samuel G. French's division, Department of Southern Virginia), was well to the rear of the fort, called Battery Huger. The Federals quickly overwhelmed the gun crews and the tiny garrison, soon hauling off 137 prisoners and five field pieces. The event led to a rare duel between Confederate officers. Col. John K. Connally of the 55th heard that General Law (Hood's division) had criticized his North Carolinians for failing to help defend the fort (some of Law's troops had been captured in the Yankee attack). He went to Law's headquarters and asked for confirmation; yes, the general said, so he had been told by his officers. Connally then confronted Law's two officers, captains Leigh Terrell and John Cussons, who affirmed their statements. Connally returned to his quarters and convened his subordinates, telling them that the regiment's honor had been sullied. This called for a duel among gentlemen. Maj. Alfred H. Belo of the 55th agreed to serve as Connally's second. The contest was arranged for the next day: Connally vs. Terrell with shotguns at forty yards; Belo vs. Cussons with Mississippi rifles, also at forty. A crowd gathered to watch the Belo-Cussons duel. After two volleys, both were still standing; one had a hole through his hat, the other was bleeding from a nicked neck. Before third rounds were fired, someone rode up and announced that Colonel Connally and Captain Terrell had settled their differences and that there was no need for further fighting.

General Longstreet may have known of this altercation, but he did nothing about it. Same for Hood: he may have known about it but did nothing to

[53] Woodward, *Mary Chesnut's Civil War*, 432–33, 443; Cormier, *The Siege of Suffolk*, 9–10, 14–18, 42, 51, 56, 66, 69, 101–104, 174–77.

stop it. This conduct brings forth a rather harsh judgment from biographer McMurry: "It is not known if Hood was aware of the affair, but, aware or not, he stands indicted for negligence. If he did not know, he was ignorant of what was transpiring in his division; if he knew and did nothing to stop the duel, he was guilty of irresponsibility. In either case, the Texan, unhappy and bored, did not exercise effective command." As we shall see, John Bell Hood's future record as Confederate general will offer instances for which accusations of negligent behavior may be raised, but the Belo-Cussons rifle duel at Suffolk in April 1863 is of far less significance.[54]

Hood was indeed "unhappy and bored"; he could not wait to get back to the main front. His style of warfare was certainly not *sitzkrieg*, and siege lines made him impatient. "Here we are in front of the enemy again," Hood wrote on April 29. "The Yankees have a very strong position, and of course they increase the strength of their position daily. I presume we will leave here so soon as we gather up all the bacon in the country." The fellow officer whom Hood wrote on the 29th was none other than Gen. Robert E. Lee: "When we leave here it is my desire to return to you." Then he added, "If any troops come to the Rappahannock please don't forget me." This unusual missive—unusual in that in writing a letter to the commanding general of the army, Hood was going over the head of his superior, Lieutenant General Longstreet—was made more so by Hood's puzzling purpose. What did he mean by "please don't forget me"? When an officer jumps rank, as Hood did here, it is usually because he is unhappy in his current command situation. In *Advance and Retreat*, Hood claimed that his letter demonstrated that he "was most anxious to rejoin my old chief, General Lee." He added, "I even proceeded so far as to apply for permission to move with my division to his support." Whatever Hood's intent, Longstreet would have been miffed. Instead of criticizing Hood's breach of military etiquette, Freeman excuses it as an expression of the young general's boredom at Suffolk. Freeman intriguingly calls Hood's letter "boyish," prompting one to wonder whether he meant an adolescent's fawning over his hero or an amateur's thoughtlessness in breaking chain of command. Lee thought little of Hood's letter-writing as well, using it as opportunity to write back—not scolding Hood for writing, but scolding his troops for their boisterous misconduct. He enclosed a clipping from the Richmond papers and wrote that he had heard some of Hood's men had robbed and even killed two citizens. "I trust these charges are not true," he

[54] Cormier, *The Siege of Suffolk*, 192, 249, 251–52, 265–66; Parker, *Touched by Fire*, 56 (death of Captain Turner); H. B. Simpson, *Hood's Texas Brigade: A Compendium*, 243; Freeman, *Lee's Lieutenants*, 2:486–90; McMurry, *John Bell Hood*, 70.

added. In any event, Lee strongly urged Hood to keep his soldiers under good discipline. "The true men of your division demand that they should not be classed with thieves & cutthroats."

So Lee excused Hood for breaking channels. The biographers have followed suit. O'Connor concedes that "Longstreet probably would not have been pleased," but judges "there was nothing really improper in this letter to Lee." Dyer quotes it in full and like Freeman terms it "an almost boyish letter asking to be relieved"—presumably either from siege at Suffolk or out from under Longstreet's authority. McMurry also calls it "a boyish letter." Old Pete would surely have characterized it differently though no response by him has come to light.[55]

Back on the Rappahannock, Lee watched the movements of Hooker's army, which was stirring. "This looks as if he intended to make an aggressive movement," Lee wrote Longstreet on April 27, "but by what route I cannot ascertain." "I am glad to learn," he added, "that you are getting everything out of North Carolina as rapidly as possible. Can you give me any idea when your operations will be completed and whether any of the troops you have in North Carolina can be spared from there?" Hooker's advance in the next few days brought more anxious communications from Lee and the War Department. "The Secretary...directs the return of your command...with the utmost dispatch," came a telegram of the 29th to Longstreet. Longstreet got this message on April 30; despite its urgency, it would take Pete at least a few days to withdraw his troops safely from their lines, organize the last trains, and get them moving for Richmond. Longstreet's command did not depart the Suffolk area until nightfall of May 3. In so doing they missed participating in Lee's great victory at Chancellorsville, fought May 1–3. The troops marched for several days, entering Petersburg on May 6 and Richmond two days later. On the 10th, Hood and the rest of the army heard the tragic news that Stonewall Jackson, grievously wounded at Chancellorsville, had died of pneumonia.[56]

Everyone knew that Jackson would have to be replaced. Someone would be promoted to lieutenant general, perhaps two officers—Lee had been thinking for some time that both Jackson's and Longstreet's corps were too big and

---

[55] Hood to Lee, April 29, 1863, *OR*, vol. 51, pt. 2, 697; Freeman, *Lee's Lieutenants*, 2:492; Lee to Hood, April 2, 1863, S. Hood, *Lost Papers*, 187, 189; O'Connor, *Cavalier General*, 136–37; Dyer, *Gallant Hood*, 175; McMurry, *John Bell Hood*,

[56] Lee to Longstreet, April 27, 1863, *OR*, vol. 18, 1024–25; Cooper to Longstreet, April 29, 1863, *OR*, vol. 18, 1029; Wert, *Longstreet*, 238; E. B. Williams, *Hood's Texas Brigade*, 140; Lively, *Calamity at Chancellorsville*, 155.

unwieldy and that a third corps was needed. Hood would not have been the only officer speculating over his own chances of advancement. But he was indeed speculating. One of Hood's aides told a friend the month before that Hood was "very anxious to be sent to Tennessee." Other officers knew that their general was pondering promotion. "His eyes fixed on other heights in rank to climb," is how one put it. A soldier in the 5th Texas wrote after Chancellorsville that "since Jackson is gone," Hood was seeking "a Lt. Gen.'s stars." If corps command could not be obtained in Lee's army, elsewhere would do. Decades afterward, Confederate artillerist John C. Haskell remembered that before Gettysburg Hood "expected to be sent to the Western army."

In this atmosphere, Hood's letter of May 13 to General Lee—Stonewall's body was not yet even in the ground at Lexington—can only be understood as a shameless display of raw ambition, an unsubtle angling for promotion. To be fair, we do not know exactly what Hood wrote Lee on the 13th. In *Advance and Retreat*, he states only that in writing his "brief note" he was "giving expression to my sorrow [at Jackson's death] and at the same time, to my regret at our failure to join him before the great battle he had just fought and won." Just two weeks before Hood had written Lee from Suffolk, "if any troops come to the Rappahannock please don't forget me." Here, too, it appears that Hood was appealing to Lee that he not be forgotten. Of Hood's three biographers, O'Connor and McMurry take Hood at his word and reconstruct from Lee's written reply (quoted fully by Hood in *Advance and Retreat*) that, as McMurry puts it, "Hood had expressed delight at being back with the army, his regret at having missed the recent battle, and his sorrow at the death of Jackson." Yet in his letter Hood did more than that. Lee took the unusual step of responding to Hood with a letter of his own, replying, "I agree with you as to the size of the corps of this army." This prompts Dyer to conclude that "Hood suggested a reorganization plan for the army which would have divided Jackson's corps and perhaps Longstreet's as well." Such a reorganization would have afforded opportunities for two major generals to be elevated to corps command.

If Hood was hoping for that promotion, he would be disappointed. By the time Lee wrote back on May 21, he had already recommended to the president that Powell Hill and Dick Ewell be made lieutenant generals of a new Second and Third Corps. To be sure, Lee told the president that "R. H. Anderson and J. B. Hood are also capital officers," but they were not yet ready for greater responsibility. "They are improving," Lee wrote, and at some time in the future "will make good corps commanders, if necessary." With his letter of May 21 Lee may have had the ulterior motive of letting Hood down gently, hinting that he would not be promoted (which Lee could not fully divulge until the president and Congress acted on his recommendations). "I rely much on you. You must so inspire and lead your brave division"—note the reminder here of Hood's role in the army—"as that it may accomplish the work of a corps." Hood seems to have taken in stride the eventual news of the army's reorganization, which began to be announced on May 23. There would be more battles ahead—and more opportunities to win laurels.

At the same time, if a number of soldiers in Lee's army knew of Hood's ambition, some apparently believed that he would contrive to advance his reputation. In spring 1863, Chaplain Nicholas Davis of the 4th Texas had published in Richmond his memoir, *The Campaign from Texas to Maryland, with the Battle of Fredericksburg.* Amid his lofty praise for the Texas Brigade, General Hood received a generous share of compliments. When Davis suggested, however, that at Sharpsburg McLaws's division had been slow in coming to support Hood's troops, General McLaws took offense, and in letter of May 20 objected to Hood's having fed such slanderous information to Davis. Hood denied having done so, to the point of writing McLaws on the 24th that the chaplain's book "was not written under my auspices, and today is the first time my attention has been called to the paragraph you refer to." McLaws in turn thanked him for this assurance, but he could not refrain from adding, "I tell you frankly that it has been the opinion of many who have read it, that you must have known its contents before publication."[57]

As he had after his victory the previous summer at Manassas, after Chancellorsville Lee again thought of taking the war into Northern territory. At meetings in Richmond with the president and cabinet during May 14–17, Lee made and won his case for a movement across the Potomac into Pennsylvania. The decision was made on the 16th, with only Postmaster General Reagan dissenting—he thought the defense of Vicksburg should be top priority. Years later in his memoirs, Reagan commented on the fateful meeting:

---

[57]Lee to Davis, May 20, 1863, *OR*, vol. 25, pt. 2, 810–11; Robertson, *Stonewall Jackson*, 761 (burial on May 15); Bohannon, "'A Bold Fighter,'" 258, 282 n.34; Hood, *A & R*, 52–53 (Lee to Hood, May 21); O'Connor, *Cavalier General*, 142; McMurry, *John Bell Hood*, 71–73; Dyer, *Gallant Hood*, 177–78; Robertson, *A.P. Hill*, 193 (Ewell's announcement to corps command, May 23); Special Orders no. 146, May 30, 1863, *OR*, vol. 25, pt. 2, 840; McLaws, *A Soldier's General*, 182–83 (McLaws to Hood, May 31, 1863).

The version of Lee's letter to Hood in Dowdey and Manarin, *The Wartime Papers of R. E. Lee*, deletes Lee's statement, "I agree with you as to the size of the corps of this Army.... I am obliged to you always for your opinion. I know you give it from pure motives. If I am not always convinced, you must bear with me" (490).

"I will not now repeat the expression I made when this conclusion was reached."

Lee's reorganization was completed on May 20, and the army, 75,000 strong, started moving June 3. Lee's plan called for his forces, then quartered around Fredericksburg, to march westward via Culpeper Court House, cross the Blue Ridge into the Valley, then march down and across the Potomac while Stuart's cavalry guarded the gaps and passes. Hood's division at the time was at Verdiersville south of the Rapidan. On the morning of June 4, it crossed at Raccoon Ford and made its way to Culpeper. There it stayed until June 15 while Ewell's corps marched past and took lead in the line of march. During this time, Jeb Stuart held a celebrated review of his cavalry, June 8. General Lee was invited; Fitz Lee invited Hood, as well as "any of his people"—meaning staff officers. On June 8, Hood brought *all* of his people—his entire infantry division, marching to the parade grounds. "You invited me *and my people*," Hood ribbed Fitz, who enjoyed the joke. "Well, don't let them halloo, 'Here's your mule!'" (infantrymen's contemptuous catcall to the cavalry). "If they do, we will charge you!" piped in Wade Hampton. As the troopers rode by, Hood's men behaved themselves except when the breeze blew off troopers' hats—they ran forth and scooped them up as trophies of war. The Southerners' celebrations ended the next day when Yankee cavalry attacked Stuart at Brandy Station. "Stuart prevailed, but just barely," notes Emory Thomas, in the largest cavalry battle ever fought on the continent.

On the 15th Ewell routed the Federal garrison at Winchester, capturing 4,000 prisoners, 23 cannon, and loads of supplies, wagons, and teams. That day the head of Ewell's column marched toward the Potomac ford at Williamsport. Lee's route into Maryland would have Confederate troops wading the river between ten and twenty-five miles above Harpers Ferry—this time there would be no need to worry about the enemy garrison there. A. P. Hill's corps was the last to leave Fredericksburg, marching for Culpeper on the 14th. Between Ewell and Hill marched Longstreet's corps. After hot, tiring marches Hood's troops entered the Shenandoah on the 18th, and on the 26th crossed the Potomac. To celebrate their return to Maryland, the men were given captured Northern whiskey. "I drank my portion to the dregs, and it was a bumper for me," remembered William A. Fletcher of Co. F, 5th Texas. As always, some got more than their share. "The temperance fellows gave their shares to friends," recalled J. B. Polley, such that many in the Texas Brigade had a hard time walking straight. "The breadth of the road over which they marched that evening," Polley noted as politely as he could, "was more of an obstacle to rapid progress than its length." Others were blunter. "There were more drunk men," wrote John Stevens of the 5th Texas, "than I think I ever saw in my life." But the march pace did not slacken as Hood's column passed into Pennsylvania. One Confederate recorded that day, "we had breakfasted in Virginia, dined and wined in Maryland, and taken supper in Pennsylvania." As a result, "we were in four states that day, the fourth being in a state of intoxication."[58]

The mere act of marching into Pennsylvania delighted many Confederates. "We got back into the Union, you see," cried some Southerners to rather glum civilian onlookers. Rebel yells went up from Hood's troops. One Texan joined arms with his captain as together they stepped past the stone marking the state boundary. Part of the troops' satisfaction came from the knowledge that they were in the enemy's country and could pay back the excesses of the Yankees in Virginia. To be sure, on June 21 General Lee issued an order protecting civilians' property, but the men were determined to have their way. One even interpreted General Hood as saying, "Boys, you are now on the enemy's soil, stack your arms and do pretty much as you please." J. Mark Smither of Co. D, 5th Texas, wrote home from camp near Chambersburg, June 28: "It is true that we are living on the people and living well." He admitted that the Confederates were taking all the spare horses they could, "but we are treating the citizens as well as an army can." The invading soldiers, however, were not complimentary of the local folk they saw. A Georgian in Hood's division commented that "the features of the women would have made vinegar ashamed of itself for sourness." One surly madam chose to advertise her disdain for the Rebels as Hood's troops marched into Chambersburg on June 27. She "had seen fit to adorn her ample bosom with a huge Yankee flag," recorded Arthur J. L. Fremantle, English officer then accompanying Lee's army into Pennsylvania.

> She stood at the door of her house, her countenance expressing the greatest contempt for the barefooted Rebs; several companies passed her without taking any notice; but at length a Texan gravely remarked, "Take care,

[58] Freeman, *R. E. Lee*, 3:19; Woodworth, *Davis and Lee at War*, 231, 234; Cooke, *Wearing of the Gray*, 317; B. Davis, *Jeb Stuart*, 304–305; McPherson, *Battle Cry of Freedom*, 647–48; Kegel, *North with Lee and Jackson*, 233, 241–42; Thomas, *Lee*, 291 (Brandy Station); Sears, *Gettysburg*, 59 (Culpeper, thirty miles northwest of Fredericksburg); Gottfried, *Roads to Gettysburg*, 8–9, 27, 59, 96–97, 150, 152; Grimsley and Simpson, *Gettysburg*, 4 (Ewell's victory at Winchester "opening up the valley"); Guelzo, *Gettysburg*, 46–47, 61–62, 66; Nevins, *The Organized War 1863–1864*, 80; Fletcher, *Rebel Private Front and Rear*, 54; Polley, "Hood's Texans in Pennsylvania," 377.

> madam, for Hood's boys are great at storming breastworks when the Yankee colors is on them." After this speech the patriotic lady beat a precipitate retreat.

Other women held U.S. flags in prominent view to show contempt for the Rebels marching through town. Hood's Texans, of course, took it all in fun. "The ragged Confederate," recalled Col. Robert Powell of the 5th Texas, "pretended to think that each lady intended to distinguish her knight by the presentation of her colors." "This honor they unanimously declined," Powell observed, "declaring that they had taken all the flags they wanted from the Yankee army."

Fremantle commented on "Hood's ragged Jacks." "They certainly are a queer lot to look at," he wrote. "They carry less than any other troops; many of them have only got an old piece of carpet or rug as baggage; many have discarded their shoes in the mud; all are ragged and dirty; but full of good humor and confidence in themselves and in their general, Hood." When taunted by the ladies of Chambersburg, Fremantle noted that they responded "with cheers and laughter." Such high spirits prevailed among the officers and men throughout the army. One predicted that with the army able to subsist off the country, it could stay in Pennsylvania and "probably remain there two months."

The Confederates' hopes were aided by Joe Hooker's sluggish response to their incursion. He did not order his seven infantry corps away from the Rappahannock until June 14, after he learned that Hill's corps had left its positions. Lincoln, meanwhile, grew impatient, remarking that with Lee's army strung out from Fredericksburg to the Potomac, "the animal must be very slim somewhere." "Could you not break him?" he asked. During June 23–24 Hooker dithered in indecision before concluding on the 25th that Lee's whole army was heading across the Potomac. By then, with Rebel columns scattering in Pennsylvania, Hooker's troops had to make forced marches to catch up. But it was too late; Lincoln had concluded that Hooker was another McClellan, afraid to fight Lee. "Fighting Joe" realized he had lost the administration's confidence and on June 27 asked Halleck to be relieved of command. Request was quickly granted, and the next morning V Corps commander Maj. Gen. George Meade was named his successor.

Meade's imperative was to keep the Army of the Potomac—84,000–88,000 strong, then around Frederick—between the Rebels and Washington. Of Lee's army, on June 28 Longstreet's and Hill's corps had marched to Chambersburg. There Lieutenant Colonel Fremantle was introduced to General Hood. "He is a tall, thin, wiry-looking man, with a grave face and a light-colored beard," he wrote, "and is accounted one of the best and most promising officers in the army." "His troops are accused of being a wild set," Fremantle added, even as he commended some Texans for going into Chambersburg and dutifully destroying some barrels of whiskey that they would have rather consumed. Meanwhile, Ewell's divisions were at Carlisle and York, heading for Harrisburg. But where was Stuart? With three brigades he had ridden off in a raid around the Federal army, depriving Lee of his "eyes." If Lee was worried, he didn't show it when he told Hood, "Ah! General, the enemy is a long time finding us. If he does not succeed soon, we must go in search of him." That night, a Southern spy, Henry T. Harrison, reported to Longstreet and Lee that the Army of the Potomac had a new commander and that at least three of its corps were at Frederick marching north. The news caused Lee to order his corps commanders to march toward a concentration of the army at Cashtown or Gettysburg.[59]

Hood's division, camped near Chambersburg, started its march on the 29th. Its route, southeastward through Greenwood and Cashtown, covered some twenty-five miles to Gettysburg. In front of Hood marched McLaws's division; behind him was Pickett. Ahead of Longstreet's three divisions were A. P. Hill's, all marching through Cashtown, which was ten miles to the west of Gettysburg. On June 30 Hill's lead division, Harry Heth's, approached Gettysburg; without cavalry, Heth had to send Johnston Pettigrew's brigade of infantry ahead to screen for any enemy. They found some: Pettigrew's troops brushed into Brig. Gen. John Buford's two brigades of cavalry, which had passed through Gettysburg. Buford sent word back to the nearest Union infantry corps, Maj. Gen. John Reynolds's. Pettigrew reported back to generals Heth and Hill. Hill directed Heth to march the next day, "to ascertain what force was at Gettysburg," as Col. Walter Taylor remembered, "and, if he found infantry opposed to him, to report the fact immediately, without forcing an engagement." Heth looked forward to his assignment on July 1; there was reported a good quantity of shoes in the town, and he wanted to get them. Neither Lee nor Meade knew exactly where the other's army was, and neither was looking to bring on a battle. But the opposing forces were close enough to allow even civilian onlookers to expect a big fight soon. "There is a

---

[59] Guelzo, *Gettysburg*, 64–65, 69–72, 78, 81, 114; Nevins, *Organized War*, 81, 98; Gottfried, *Roads to Gettysburg*, 61; Coddington, *The Gettysburg Campaign*, 121–22; Pfanz, *Gettysburg*, 9, 21, 32; Stackpole, *They Met at Gettysburg*, 95 (map showing all seven Union corps across the Potomac); R. M. Powell, "With Hood at Gettysburg," 240; Fremantle, *The Fremantle Diary*, 191, 193, 199; McPherson, *Battle Cry of Freedom*, 652; Walsh, *"Damage Them All You Can,"* 249–50.

probability," predicted the *Baltimore American* on June 30, "that a great battle will be fought in the course of the present week in the neighborhood probably of Hanover or Gettysburg."[60]

"Any full rehearsal herein of the details of the Gettysburg campaign would rank high among exercises in futility, for the mountains of books piled upon the battlefield equals in weight the stone and bronze of its serried monuments." So wrote Allan Nevins more than forty years ago, and his words are even truer today. Suffice it to say that on July 1, Heth's infantry moved against Buford's dismounted horsemen; the Union I Corps came up and its commander, Reynolds, was shot dead by 10:30 in the morning; Maj. Gen. Robert Rodes's division began to approach from the north and joined the fight; and the XI Corps arrived and deployed on the I Corps's right. Lee rode up around 2 P.M., irritated that a battle had begun against his wishes. He initially considered breaking it off, but based on reports of some success, decided to commit more troops. Major generals Jubal Early's and Dorsey Pender's divisions attacked; the XI Corps broke north of town, and with it the I Corps retreated too. Confederates took 5,000 prisoners as the Federals withdrew hastily through Gettysburg to the high ground south of it: Culp's Hill, Cemetery Hill, and Cemetery Ridge. The opportunity to drive the Yankees from these places was present, but Ewell failed to act (Jackson surely would have). Came nightfall, and both armies waited for the rest of their units to arrive. The battle would continue on the morrow.[61]

When the fighting started on the morning of July 1, Hood and his division were at least ten miles away, west of Cashtown. Lee was riding with Longstreet; the commanding general was "in his usual cheerful spirits," Longstreet later recalled. Maj. Gen. Edward "Allegheny" Johnson, accompanying Ewell's wagon trains, was behind Longstreet's three divisions. But Lee wanted Ewell's corps together and told Longstreet to halt his column for Johnson and the wagons to pass ahead of him. Ewell's train proved to be fourteen miles long, and Longstreet's troops had to wait. Lee and Longstreet rode on until at Cashtown they heard distant cannonfire. General Lee hurried ahead; Longstreet and his staff followed more slowly. By the time Ewell's wagons got out of the way, it was 4 P.M. McLaws's and Hood's divisions resumed their march and kept at it well into the darkness. Around midnight

---

[60] McMurry, *John Bell Hood*, 74; Guelzo, *Gettysburg*, 121–22, 127, 129–31; Nevins, *Organized War*, 89; W. H. Taylor, "The Campaign in Pennsylvania," 307.

[61] Nevins, *Organized War*, 82; Coddington, *Gettysburg Campaign*, 260–322; Andre Trudeau, *Gettysburg*, 152–272 *passim*; Sears, *Gettysburg*, 154–225.

McLaws bivouacked at Marsh Creek more than a mile northwest of where the battle had started that morning; Hood was farther behind.[62]

Shortly after McLaws's and Hood's divisions resumed their march, after having been "completely blocked up" by Ewell's trains, Longstreet reached Lee's headquarters, pitched on the western slope of Seminary Ridge south of the Chambersburg Pike. It was around 5 P.M. Lee and Powell Hill reviewed the day's events for Longstreet, who from Seminary Ridge had seen for himself the retreating soldiers of the I Corps being pursued by eager Confederates. Longstreet and Lee discussed possibilities for the next day's continuation of the battle. Longstreet, who had advocated defensive battle throughout the campaign, suggested a turning movement around the enemy left, interposing between Meade's army and Washington and forcing Meade to attack. According to Longstreet's recollection, Lee countered, "No, the enemy is there and I am going to attack him there." In his report of the campaign, General Lee summarized the situation in his characteristic style: "A battle had, therefore, become in a measure unavoidable, and the success already gained gave hope of a favorable issue." Longstreet's corps would be Lee's main attacking force against the enemy left flank. Old Pete sent a courier back to hurry Pickett's division forward. He left Lee's headquarters and camped with McLaws's division at Marsh Creek. All were to awake very early in the morning of July 2 for a hard day's work.[63]

Both Lee and Longstreet were up and had breakfasted before sunrise. Lee sent his topographical engineer, Capt. Samuel Johnston, to reconnoiter the enemy's left and bring back a report as soon as he could. Longstreet arrived at Lee's headquarters shortly after daybreak, and discussions resumed about the proposed assault. Present, besides staff officers, were generals Hill, McLaws, and Hood. The latter recalled a decade and a half later how Lee paced back and forth, stopping at times to look across toward the enemy at Cemetery Hill, obviously deep in thought. He had his sword buckled to his belt and binoculars hanging to his side. It was early and the morning still cool (or maybe he wasn't feeling well); Lee had his coat buttoned up to his neck. Lee, Longstreet, and Hill began earnest conversation; it was apparent that Lee had not completed his planning for the day, especially on what role Ewell's corps on the left would have. He sent Maj. Charles Venable over to confer with Ewell about his possibly launching an assault. (Lee had already

---

[62] Longstreet, *Manassas to Appomattox*, 351; Guelzo, *Gettysburg*, 167, 174; Pfanz, *Gettysburg*, 23.

[63] Guelzo, *Gettysburg*, 229–31; Pfanz, *Second Day*, 25–27, 104; Lee report, January [undated], 1864, *OR*, vol. 27, pt. 2, 318.

done so earlier in the morning.) At the time, the Confederate line was still taking shape. Ewell's three divisions formed the left, facing Culp's and Cemetery Hills. Heth's and Pender's divisions followed down Seminary Ridge; Anderson's was marching to take its place and extend the line.

Lee, Longstreet, and Hill were still conferring—the commanding general was seated on a log, pointing on a map—when Captain Johnston brought in his report after some three hours of reconnaissance. It was between 7 and 8 A.M. Johnston said that he had ridden south along Seminary Ridge and ridden out to Little Round Top, a big hill south of the enemy line. Aside from a few Yankee cavalrymen on the Emmitsburg Road, Johnston claimed to have seen no enemy troops. The news that there were no Federals south of Cemetery Hill confirmed Lee in his plan to have Longstreet attack up the Emmitsburg Road. Captain Johnston thought that for Longstreet to get into position would require a march of some five miles, taking up to two hours. Longstreet at some point expressed his reluctance to launch an assault when Pickett's division was still miles away. Hood recalled Longstreet as having said to him, "The General is a little nervous this morning; he wishes me to attack; I do not wish to do so without Pickett. I never like to go into battle with one boot off." Lee did not want to wait that long and told Longstreet he could use Anderson's division, which had not been bloodied the day before. Longstreet then asked if he could at least delay his advance until Law's brigade, bringing up the rear of Hood's division, had arrived. Lee consented. Then, when Lee said, "I think you had better move," Longstreet brought Hood and McLaws over for final instructions. At some point Lee apparently conversed with McLaws himself while Longstreet paced back and forth. "General, I wish you to place your division across this road," said Lee, pointing to the Emmitsburg, "and I wish you to get there if possible without being seen by the enemy." This latter part was crucial: secrecy. Captain Johnston's prescribed route for Longstreet to get his two divisions into position had presumably taken into account the topography needed to keep the marching column from being observed by the enemy. To make sure, however, Longstreet sent off his artillery chief, Col. Porter Alexander, to bring up his guns and to "get an idea of the ground." Longstreet particularly cautioned Alexander "to keep all movements carefully out of view of a signal station whose flags we could see wig-wagging on Little Round Top."

McLaws and Hood rode off to their divisions, probably by nine o'clock, when General Lee left to confer again with Ewell. By noon Anderson's division was on Seminary Ridge, in position to take up Longstreet's attack. But the attack encountered delay after delay. First Law's brigade took longer than expected to come up; Law thought it was around noon. Then, probably around 12:30, Longstreet's flank march began as McLaws's, then Hood's divisions stepped off southward from Herr Ridge, which was 2 ½ miles northwest of Seminary Ridge. Guiding the column were Captain Johnston and McLaws's engineer, Lt. Thomas Moncure. General Longstreet may have suspected he was behind schedule, although Lee is not known to have given a peremptory time for the start of the attack. The march pace may have been brisk, or not. One soldier in McLaws's division called it "a slow and halting gait"; another referred to it as a half-trot. Either way, McLaws had barely passed Black Horse Tavern, a mile and a half into the march, when he ran into trouble. Longstreet was riding along his column when it suddenly stopped. Pete rode on and met McLaws. "What's the matter?" he asked. "Ride with me," McLaws answered, "and I will show you that we can't go on this route, according to instructions, without being seen by the enemy." Sure enough, the two generals rode to the top of a rise and saw that if their troops crested it, they would be noticed by the enemy observation posts Longstreet had already warned Alexander about. McLaws was angry, spluttering words, as one Confederate recorded, that he "would not like my grandson to repeat." But there was no alternative to a backwards countermarch. The feat, as Guelzo points out in his recent history of the battle, entailed more than a mere about-face; companies and regiments had to turn, pivot and rearrange to maintain the order in which they had been drilled to deploy into lines of battle. Hood, eager for the coming fight, was leading his troops so close upon McLaws that the rearrangement was made more difficult still. The maneuver was nevertheless accomplished, and Longstreet's column was back on the march in what Jeffry Wert calls "a comedy of errors": down and across Willoughby Run and finally up the west slope of Seminary Ridge, to the right of Anderson's division. There, as if he needed another surprise, McLaws got yet another one. As soon as his column showed itself atop the ridge, "the enemy at once opened on it with numerous artillery." It was clear that since Captain Johnston's recon earlier that morning, Federal troops had moved into the area east of Seminary Ridge and below Cemetery Hill. Longstreet had just told McLaws that no enemy was in his front; clearly he was wrong as well. As Lt. Col. Moxley Sorrel of Longstreet's staff later wrote, the Federal position "had been heavily reinforced while we were pottering around in sullen inactivity." In "one rapid glance" McLaws observed on the east side of the

Emmitsburg Road—General Lee's route of attack, prescribed eight hours earlier that day—Federal infantry and artillery, "in force much greater than I had, and extending considerably beyond my right."[64]

The Yankee troops were those of Maj. Gen. Daniel E. Sickles's III Corps. Sickles had been ordered by Meade to place his two divisions on Cemetery Ridge, extending the Union line southward from Maj. Gen. Winfield S. Hancock's II Corps and anchoring its left at Little Round Top. Having taken this position ahead of his line, however, Sickles saw that Cemetery Ridge was actually lower than terrain to the west, toward the Emmitsburg Road. So without seeking Meade's approval, Sickles ordered his troops out to the road (Brigadier Andrew Humphreys's division on the right) into Joseph Sherfy's peach orchard and then south into a large, rocky area called Devil's Den (held by Maj. Gen. David Birney's division), five hundred yards west of Little Round Top.

Federal artillery in the peach orchard opened fire on the Confederates as soon as they topped Seminary Ridge. McLaws called forth his own artillery and saw to bringing his troops up. Hood, behind him, did the same, extending from McLaws's right. Longstreet had asked McLaws, "How are you going in?" McLaws answered, "That will be determined when I can see what is in my front." McLaws took more time positioning his division than Old Pete liked. Three times Longstreet sent his staff officer, Maj. Osmund Latrobe, to prompt McLaws to launch his assault. The last was a peremptory order. But within minutes another courier rode to McLaws, telling him not to go forward until Hood had launched his own attack.

By then, 4 P.M., Hood's division had become the extreme right of Lee's army. Hood deployed his four brigades in two lines: Robertson and Law in front, Benning and Anderson behind. "Hood was aggressive, but also prudent," writes Harry Pfanz, historian of Gettysburg's second day. Therefore, while his division moved into position, Hood sent maybe a half dozen of his Texans to scout the enemy's position. They brought back word that there were no Federals on Little Round Top, and that the division could move around it to attack the enemy flank and rear. About that time General Law reported news from III Corps prisoners and some of his own scouts that the Union left was not strongly held and that U. S. wagon trains behind the Round Tops

---

[64] Pfanz, *Second Day*, 104–107, 110–23; Guelzo, *Gettysburg*, 235–42; "Leading Confederates on the Battle of Gettysburg: Letter from General John B. Hood," 147 (description of Lee), 148 ("one boot off"); McLaws, "Gettysburg," 69–70; Wert, "Little Round Top and Devil's Den," 45; Sorrel, *Recollections of a Confederate Staff Officer*, 168. Hood's letter to Longstreet, June 28, 1875, is quoted fully by Hood in *A & R*, 55–59.

were only lightly guarded. Law went to Hood and proposed a maneuver around by the right. Hood agreed with Law but cited his orders. Law protested so strongly that Hood called a staff officer, Captain James Hamilton, to ride to Longstreet and deliver their request for a change of plans: that instead of attacking up the Emmitsburg Road, he lead his division in a flanking maneuver by way of Little Round Top and attack from that direction. Hamilton returned in ten minutes with Lieutenant Colonel Sorrel of Longstreet's staff, who told Hood, "General Longstreet orders that you begin the attack at once." Hood tried three times to persuade Longstreet to change the plan. After Maj. John Fairfax of Longstreet's staff brought an order that the attack must begin, Hood again explained his idea. Fairfax brought back what Longstreet had said: "It is General Lee's order, the time is up—attack at once!" Hood tried once more, sending Maj. Harry Sellers to Longstreet. Sellers returned with the corps commander's answer: "we must obey the orders of General Lee." Hood acceded, but according to Lt. Col. Philip A. Work, commander of the 1st Texas, he told Sellers, "Very well. When we get under fire I will have a digression." Hood's "digression" would have led his division toward Little Round Top and the flank assault he had advocated. After the war Lafayette McLaws wrote of Hood's "moving towards the enemy's left." "He never did go far enough to envelop the left," McLaws explained, "not even partially. It was said at the time, on the field, that he would have done so," but scouts and guides leading the column had lost direction "and missed carrying the head of the column far enough to the right" before the division became engaged with the enemy.

In the voluminous literature of Gettysburg "what ifs," Hood's petition to redirect Longstreet's assault has received much attention. General Hood himself has been lauded for battlefield initiative and tenacity—Hood called his appeal to Longstreet for a change of orders "the first and only one I ever made during my entire military career." Hood always believed he was right. A decade after the war he wrote Longstreet, "I shall ever believe that had I been permitted to turn Round Top Mountain, we would not only have gained that position, but have been able finally to rout the enemy." Hood was not alone. Three decades after the battle, McLaws too indulged in the what-if. "Why General Longstreet did not assault early on the morning of 2 July when he could have occupied Round Top without opposition, is still an open question!" he exclaimed. Years after the war General Law contributed to the literature, writing in *Battles and Leaders* about the second day's fighting. Recounting his and Hood's failed efforts to secure Longstreet's approval of their planned maneuver around Little Round Top, Law concluded, "Just here the battle of Gettysburg was lost to the Confederate arms."

Pfanz acutely points out that, if implemented, Hood's attack by way of Little Round Top would have taken even more time to set up: more marching for Hood's troops (already hot and tired) and further shifting of McLaws's division. It was already four o'clock, daylight was limited, and Longstreet and Lee were impatient to begin. At the same time, if Hood did indeed plan a "digression"—Trudeau even had a map prepared in his book to show Hood's digressing route—then instead of commendation for battlefield initiative, Hood would be censurable for willful disobedience of orders—the kind of disobedient willfulness that has beclouded Dan Sickles's reputation at Gettysburg.

As he prepared to give the order to advance, Hood rode along the front of his division and stopped before Robertson's brigade, his old one. Pvt. John West of the 4th Texas remembered how Hood "took his hat, held it above him in his right hand, [and] rose to his full height in his stirrups." "Fix bayonets, my brave Texans!" he shouted; "forward and take those heights!"

As his division advanced, Hood stayed with Robertson's troops for a few minutes, then rode to an orchard from where he could watch the progress of Law's as well. Just then a shrapnel shell burst overhead, and a ball struck his left arm. "I saw him sway to and fro in the saddle," wrote an officer in the 1st Texas, "and then start to fall from his horse, when he was caught by one of his aides." As Pfanz puts it, "the day's assault had hardly begun and the Army of Northern Virginia's key division commander was already out of the fight."[65]

---

[65] Pfanz, *Second Day*, 122–25, 152–53, 158–67, 172–73; Law, "The Struggle for 'Round Top,'" 3:321–22; Korda, *Clouds of Glory*, 577 ("Three times he tried to persuade Longstreet"); Trudeau, *Gettysburg*, 332, 337; Chilton, *Unveiling and Dedication of Monument to Hood's Texas Brigade*, 339 ("digression"); McLaws, "Gettysburg," 73; Hood, *A & R*, 58–60; "Lafayette McLaws on Gettysburg," 40; Purifoy, "The Lost Opportunity at Gettysburg," 215–17; Ural, *Hood's Texas Brigade*, 153 (John West). Hood never wrote a report on Gettysburg; his wound and lengthy convalescence prevented it. He apparently deemed his letter to Longstreet of June 1875, printed in his memoirs, a sufficiently thorough narrative of his role in the battle such that he embellished it little.

Longstreet has long been criticized for his unhelpful conduct at Gettysburg. Characteristic of the condemning judgments is the observation of Robert K. Krick: "Longstreet had the genetic equipment to be naturally, as well as intentionally, obtuse" ("If Longstreet...Says So, It Is Most Likely Not True," 160–61). Some have speculated on what might have happened had Hood not been wounded on July 2 and had been allowed by Longstreet to march to the enemy flank and attack Round Top. William R. Forstchen offers a fascinating scenario in which Hood's assault coincides with Ewell pushing around Culp's Hill: "the two flanking forces link up and close the trap in a classic double envelopment which then triggers a panic in the

As Hood, dazed and bleeding, was carried to the rear, Longstreet's attack rolled forward. The Confederates on July 2 drove back the enemy but did not roll up his line. Ewell's assault at the other end did not wrest the Federals' hold of Culp's Hill. A mighty assault by the divisions of Pickett, Pettigrew (Heth) and Trimble (Pender) on July 3 was bloodily repulsed. Such were the outlines of the rest of battle after Hood received his injuries.

Gettysburg ranks as the bloodiest engagement of the entire war, as casualties attest. After exhaustive study, Busey and Martin count for the Federals 3,149 killed, 14,503 wounded, and 5,161 missing/captured for 22,813 lost. In Lee's army, 4,637 men were killed, 12,391 fell wounded, and 5,846 were lost as missing or captured, for a total of 22,874. Hood was to be counted among thirty general officer casualties on both sides. In Meade's army five generals were dead, beginning with Major General Reynolds; seven more were wounded, including Dan Sickles, and one wounded and captured. At the end of the campaign, six Confederate generals were dead, among them Maj. Gen. Dorsey Pender. Hood was the most grievously wounded major general; Harry Heth less so; Isaac Trimble, injured and taken prisoner. Eight more brigadiers were wounded and/or captured.[66]

Surgeons treating the wounded gave general officers their closest attention, but initially all wounded men got the same treatment. Comrades or litter bearers carried them to an aid station just out of enemy firing range. There, an assistant surgeon examined the wound and administered painkillers (whisky, opium pill, or morphine, either dusted or rubbed into the wound). Bandages were applied to stop the bleeding before the patient was sent farther back to a brigade ambulance depot. Patients unable to walk were put aboard "ambulances"—meaning any carriage, dray, or wagon that the medical officers had pressed into service. Thus was Hood borne to the rear. F. M. Colston, a Southern soldier, later wrote that he saw the general in an ambulance, "sitting in front with the driver, his arm in a bloody bandage." Hood had not yet left the battlefield; Colston recalled a shell struck the roof of a nearby building, "but the General merely looked up."

Union army." The Confederates bag 40,000 prisoners; Meade offers his sword to Lee; fragments of the Army of the Potomac flee toward Taneytown ("Lee's Victory at Gettysburg...and Then What?" 326–27). For an early survey of counterfactual writing about the war, see S. Davis, "The Literature of Confederate Victory," 44–50.

[66] Busey and Martin, *Regimental Strengths and Losses at Gettysburg*, 239, 280; "The Opposing Forces at Gettysburg, Pa.," in Johnson and Buel, eds., *Battles and Leaders*, 3:434–39; S. Davis, "The Death and Burials of General Richard Brooke Garnett," 107; Hoffsommer, "The Aftermath of Gettysburg," 35; J. Welsh, *Medical Histories of Confederate Generals*, 105; King, *Louis T. Wigfall*, 171; Wright, *Southern Girl*, 148.

Field hospitals could be any kind of structure—civilian homes, barns, churches, and other shelters in which surgeons could work. They were designated by divisions, and whenever possible, officers and men were transported to their respective divisional hospital. The one assigned to Hood's division was the big, two-story brick house on the farm owned by John E. Plank, three miles southwest of Gettysburg, west of Willoughby Run and south of the Fairfield-Hagerstown Road. The Plank family occupied the house on July 2 when, as Elizabeth Plank remembered, "an ambulance arrived at the house and without any ceremony forced open the front door and carried in a wounded officer and placed him in the guest's room and the best bed in the house." Dr. Darby and Maj. Moses B. George, the division quartermaster, organized the hospital. Some dozen regimental surgeons and assistant surgeons swarmed about, helped by over fifty nurses and stewards.

Dr. Darby attended to General Hood, and after administering chloroform, examined his arm. A lead ball from a shrapnel/spherical case shell had struck the radius in Hood's left arm 2½ inches below the elbow but had not broken the bone entirely. Darby consulted with Dr. Lafayette Guild (medical director for Lee's army), Dr. R. J. Breckinridge (medical inspector for the Department of Northern Virginia), the chief surgeon of McLaws's division, as well as the brigade surgeon of Law's brigade. The doctors agreed that an operation would be needed to extract bits of bone from the arm, but that amputation was not called for. Darby, assisted by two other surgeons, performed the operation. They made a four-inch incision, removed some bone, closed the wound and attached wooden splints on both sides of the arm. Meanwhile word spread through Hood's division of the severity of the general's wound. "I heard that his Texans are in despair," observed Lieutenant Colonel Fremantle in his diary.

That night and the next day, as the fate of Pickett's charge decided the battle, doctors continued to tend to Hood. Likely the general was removed from the Plank house, if only because of its crowding. Over 1,500 wounded were brought to the Plank farm, so many that the barn and all outbuildings were overwhelmed; even seriously wounded men were left outside in the apple orchard. Miss Plank recorded, "it was not long before all the beds were filled with wounded, and the floor covered with straw carried from the barn, all over the floors in the halls on the porches in the outbuildings, on the barn floor and every place were wounded men."

After conferring with his senior officers, General Lee on the evening of the 3rd ordered preparations for the army's retreat. Medical officers scrambled for every wheeled conveyance to carry the thousands of wounded men back to

Virginia. "Throughout the early hours of 4 July there was activity at all of the hospital sites and wagon parks of Lee's army," writes Kent Masterson Brown, the historian of Lee's retreat. One of the hardest chores was determining who among the severely wounded would have to be left behind; ultimately over 6,700 Confederates remained with surgeons; all would become prisoners (Dr. Thomas A. Means of the 11th Georgia stayed with the soldiers left at the Plank farm). Lee was able to carry away more than 8,500 other sick and injured. Darby saw that Hood was placed in "a small single horse carriage" with Brig. Gen. Wade Hampton, who had been struck in the forehead by a saber in the cavalry fight of July 3. (Darby learned that Hampton, his old commander, was wounded and saw to the general's transport so he could take care of both Hampton and Hood.) Hampton lay jostled in the wagon; he could not sit up, whereas Hood could not lie down.

The forty-two–mile route back to Williamsport, the crossing point of the Potomac, took the long line of wagons back through Cashtown and Greencastle. The trains, stretching seventeen miles, headed out around 4 P.M. on the 4th. Those of Hill's corps were in the lead, then Hood's, Pickett's, and McLaws's division wagons, all escorted by Brigadier John Imboden's cavalry. The pace of the retreat was harsh and demanding; Imboden was under orders that there would be no halt. A torrential rainstorm that afternoon added further misery for every wounded Confederate soldier making his way on the two hundred-mile southward trek. On July 6 Hood, Hampton, and Darby crossed at Williamsport. Then, farther south at Martinsburg, the generals had their wounds dressed at a receiving hospital set up in a hotel. At Winchester Hood and Hampton were taken by residents into their homes. In ambulance, the three officers traveled a hundred miles more to Staunton. Shortly thereafter, still with General Hampton, Hood was taken by rail to Charlottesville, a large Confederate hospital complex. From there, Hampton wrote Louis Wigfall on July 15 about Hood: "He is doing well and his arm will be saved. All he needs now is good nursing, together with cheerful company and generous living." Hood himself predicted a speedy recovery. A member of his old Texas brigade wrote home on August 7 that the general "sent us word that he would be out about the 15th" of that month. Senator Wigfall had a summer home in Charlottesville and Hood spent two weeks there. Louise Wigfall, or "Louly," the senator's daughter, provided him company and care. She was an attractive young lady, and Hood later confessed to having developed strong feelings for her. Louly in turn later described the general most favorably: "he was a man of singular simplicity of character and charm of manner—boyish in his enthusiasm—superbly handsome, with beautiful blue eyes, golden hair and flowing beard—broad shouldered, tall and erect—a noble man of undaunted courage and blameless life."[67]

At some point Longstreet visited Hood in Charlottesville to discuss his ideas for the coming campaigns. In May 1863 Longstreet had proposed to War Secretary Seddon that a corps from Lee's army be sent to reinforce Bragg's Army of Tennessee. With reinforcements also from Joe Johnston, Bragg could defeat Rosecrans's army in middle Tennessee and advance again into Kentucky, as he had done in fall 1862. Gettysburg and Vicksburg, of course, had changed things radically, but Longstreet did not give up on his idea. On August 15 he wrote to the secretary of war. "Strict military formality—never a fetish in the Southern army," writes Glenn Tucker, "would have necessitated forwarding the letter through Lee," so Longstreet reminded Seddon that the secretary had earlier solicited his ideas on strategy. Again he proposed sending part of the Army of Northern Virginia to strengthen that of Tennessee. President Davis had been interested in this idea of western concentration since earlier in the year. Other high-ranking Confederates also favored it, notably Gen. G. T. Beauregard. To be sure, part of Longstreet's motivation in promoting western concentration was personal. "I am not essential here," he complained in a letter of August 18 to his friend, Senator Wigfall (meaning, in Lee's army—especially after his ideas for the recent battle had been spurned). "I hope that I may get west in time to save what there is left of us." Longstreet told Lee of his interest in going west before the latter was called to Richmond in late August for a strategic conference with the president. Lee objected to the weakening of his army and not until September

---

[67] S. Davis, "An Avalanche of Wounded," 7–8; Colston, "Gettysburg as I Saw It," 552; S. Hood, "Dr. John T. Darby report of J. B. Hood's Gettysburg Wound," *Lost Papers*, 22, 24; Coco, *A Vast Sea of Misery*, 141–42; Coco, *A Strange and Blighted Land*, 86–87, 217; Fremantle, *Diary*, 209 (entry of July 2); K. M. Brown, *Retreat from Gettysburg*, 52, 95–96, 112, 116, 146–47, 176, 283, 358, 367, 381–82; S. French, "'Hurry Was the Order of the Day," 36–37 (42-mile retreat route); Wellman, *Giant in Gray*, 120–21; S. C. Tucker, *Brigadier General John D. Imboden*, 153–54; Welch, "Gettysburg Finale," 53; Dyer, *Gallant Hood*, 198–200; O'Connor, *Cavalier General*, 152, 155–56; McMurry, *John Bell Hood*, 75; Ural, *Hood's Texas Brigade*, 167; Woodward, *Mary Chesnut's Civil War*, 509 (Hood says he was engaged to Louly Wigfall).

Dr. Darby's medical report on Hood's Gettysburg wound, written in December 1863, is detailed by Stephen M. Hood in "A Shot in the Arm," 42–43. Darby's finding of a shrapnel ball through the forearm contradicts Dyer's statement that Hood "had been shot through the left hand, forearm, elbow and biceps" (Dyer, *Gallant Hood*, 198). Dyer states that he obtained this information from "a member of the Hood family who prefers to remain anonymous" (348 n.60). McMurry repeats Dyer's statement regarding the nature of the wound, but without citing Dyer as his source (*John Bell Hood*, 75).

5 was the decision made: Longstreet and the divisions of Hood and McLaws would be put on the train and transferred to the west.

By September 9, when Hood's and McLaws's troops started boarding the cars at Orange Court House, the situation out west had changed. In mid-August Maj. Gen. William S. Rosecrans and his Army of the Cumberland set out from Tullahoma in middle Tennessee, aiming to maneuver Bragg's army out of Chattanooga. He succeeded. Feinting left, signaling intent to cross upriver, the Federals crossed the Tennessee downstream at Bridgeport. Outnumbered and outflanked, Bragg ordered a retreat from Chattanooga, which Rosecrans's troops entered on the 9th. For the next week, Bragg's and Rosecrans's armies maneuvered about as the Federals advanced into northwest Georgia and the Confederates tried to catch and pounce on separate enemy columns. Bragg was unable to deliver a telling blow even as his army received reinforcements. From Johnston came two infantry divisions with artillery, leaving Mississippi by train August 24 (Maj. Gen. William H. T. Walker) and the 25th (Maj. Gen. John C. Breckinridge). These troops' transit took them traveling by way of Montgomery and Atlanta. Walker's division began joining Bragg's army at Chattanooga on August 27. Breckinridge's troops arrived September 2 after an eight-day trip. Some troops from Mississippi came later. On September 12, Col. Moses H. Wright, commander of the post of Atlanta, wired an officer in Resaca that Brigadiers Evander McNair's and John Gregg's brigades, 3,400 men, left the city at 9:30 that night bound northward. Maj. Gen. Simon B. Buckner, bringing two divisions from Knoxville, also arrived in time for the coming battle. The recent Union seizure of Knoxville forced Buckner's troops to move by train through Charlotte, Wilmington, and Atlanta.[68]

The train transfer of Longstreet's force—eight infantry brigades and six batteries—would be a huge operation. One such Confederate troop movement had already taken place in summer 1862, the hauling of Braxton Bragg's army from Tupelo, Mississippi, to Chattanooga before his raid into Kentucky that fall. Bragg's trip had covered 776 miles and involved six different railroads through Mobile, Montgomery, and Atlanta. This one would be the war's "longest and most famous Confederate troop movement by rail," according to the chronicler of Confederate railroads. The shorter, more direct route via

[68] G. Tucker, *Chickamauga*, 86–89; Connelly and Jones, *The Politics of Command*, 123–26; Cozzens, *This Terrible Sound*, 28, 31–33, 37–38, 40, 56–57, 59, 62; Woodworth, *Chickamauga*, 4–6; R. K. Brown, *To the Manner Born*, 62; W. C. Davis, *Breckinridge*, 368; Wright to Gen. H. C. Wayne, September 12, 1863, *OR*, vol. 30, pt. 4, 643; Stickles, *Simon Bolivar Buckner*, 226–27; Sehlinger, *Kentucky's Last Cavalier*, 149.

Knoxville and Chattanooga was blocked by the Federals' capture of those two points. Longstreet's route would now be through North Carolina—actually two routes, for some of his troops went through Raleigh and Charlotte, others through Wilmington. Each route ran over 700 miles of track to Atlanta, employing six rail companies. Added to this would be the more than one hundred miles via the Western & Atlantic from Atlanta to Bragg's army south of Chattanooga. As always, especially at this time of the war, Confederate railroads faced shortages of rolling stock. Not surprisingly, according to Longstreet's aide Moxley Sorrel, "never before were such crazy cars—passenger, baggage, mail, coal, box, platform, all and every sort wabbling on the jumping strap-iron—used for hauling good soldiers."[69]

Hood was back in Richmond by August 6, renting a residential suite for which the government treasury paid ten dollars a day. Dr. Darby took rooms with him, noting that "use of the arm rapidly commenced." It is likely that in the capital the two young Confederate officers also visited their belles: Doctor Darby, Mamie Preston; General Hood, her sister Buck. That phenomenal chronicler of Richmond's social doings, Mary Boykin Chesnut, would have doubtlessly provided us rich portraiture of these summer dalliances, had her journals for this period survived. But she destroyed her diary entries from May–October 1863, when Yankee raiders got too close.

In every capital city there is much talk and gossip, and Hood figured in some of it. There was speculation on why Gettysburg was lost; Hood admitted that he "suffered himself to criticize very freely our operations in Pennsylvania." General Stuart came in for his share of the blame. Robert G. H. Kean, an official in the War Department, recorded in his diary on August 9, "it is also rumored that Stuart is to be relieved of the command of the cavalry of the Army of Northern Virginia and Hood put in that position." Despite the criticism he took regarding the Gettysburg campaign, Stuart in summer 1863 thought he would be promoted to lieutenant general if General Lee approved designating the army's cavalry division as a corps. Jeb seemed unconcerned by the gossip in the capital about Hood possibly being put in charge of Lee's cavalry. Indeed, the *Petersburg Express* in the first week of August published a column stating the Hood had been appointed chief of cavalry for the Army of

[69] Black, *The Railroads of the Confederacy*, 181–83, 185, 191; G. Tucker, *Chickamauga*, 91–94; Cozzens, *This Terrible Sound*, 59; Sorrel, *Recollections*, 189. The railroad troop movement of Longstreet's two divisions is sometimes said to have been of nine infantry brigades (G. Tucker, *Chickamauga*, 92). To be sure, Longstreet's command at the time of Chickamauga did consist of nine brigades (*OR*, vol. 30, pt. 2, 17–18), but Tige Anderson's brigade was sent to Charleston, not north Georgia (Freeman, *Lee's Lieutenants*, 3:224).

Northern Virginia, and that Stuart and Wade Hampton were reporting to him; the *Richmond Dispatch* mentioned the article in its issue of August 8. "I suppose you have seen mention of Hood's assignment to the command of the Cavalry as Lt Genl," Stuart wrote his wife, Flora, on August 11. "It is *all bosh*, gotten up by some malicious scamp striving to undermine & vilify me." The next day in another letter he commented further on the Hood rumors: "Gen Lee knows nothing of it & I believe entre nous *he* has recommended me already for the place." Lee had not, but Stuart remained hopeful. "I feel very confident that Gen'l Lee has recommended me for Lt Gen'l of Cavalry," he wrote Flora on the 18th. "The newspaper talk on the subject is an effort (which I believe will fail) to manufacture public opinion in favor of Hood." "The L.G. is still in suspense," he wrote her on August 28. A week later he beamed, "Rumor is quite rife that I have been actually appointed Lt. Gen'l." Jeb was further heartened by word from Hood that he would not accept promotion over Stuart's head. Indeed, Stuart credited Hood with no ambitious maneuvering, and that "the Hood imbroglio" was the work of "some of his eager friends." In the end, Stuart would be disappointed. On September 9, Lee's headquarters issued its reorganization of the army's cavalry, and while some officers received promotion, Stuart did not. Dr. Freeman observes that this had nothing to do with Stuart, for whom General Lee maintained a genuine affection. Rather, "the most probable explanation is that Lee did not believe the command of the cavalry carried with it a responsibility equal to that borne by the men who led the infantry Corps."

Hood may have indeed visited Buck Preston during his period of medical convalescence. Months later he confessed to her how he remembered the very hat she had worn that March day when he first saw her. "You wore it at the turnpike—where I surrendered at first sight." But Hood's sights were also on the front. Thus, when the men of his division marched into Richmond, September 9–10, Hood was there. His officers called upon him, pleading that he accompany them to the new front, even though his arm was in a sling. Hood quickly agreed. He wanted action, and the theater in Virginia, with Lee's army recuperating from the Pennsylvania campaign, offered little prospect. At the same time, as we have seen, Hood sensed that rosier prospects for promotion lay out west. Besides his troops' entreaties, Longstreet may have also asked Hood to return to his division. Upon Hood's wounding at Gettysburg, General Law had taken command of it, but only temporarily. Longstreet had another issue. For the move to Georgia, Anderson's brigade was detached. Replacing it was Brig. Gen. Micah Jenkins's brigade. Jenkins outranked all other brigadiers in the division and would have taken charge of it—an outsider, sure to provoke anger among Hood's stalwarts. When Sam agreed to go with them, Longstreet's command problem was resolved.

Hood's men were exhilarated to have their commander back. Adding to their satisfaction was the cordial reception they received from the citizens of Richmond as they marched through. To boot, they got new uniforms from the Quartermaster Department to replace the bewildering array of worn-out colors and clothes they had carried into Pennsylvania.[70]

Hood's division boarded the trains in Richmond before McLaws's. Hood was not immediately with them. Mrs. Chesnut and Buck Preston were traveling home to South Carolina, and Hood rode with them as far as Petersburg. Before he parted the general asked Buck if she would consent to engagement with him. As Hood later related to Mary Chesnut, "she half-promised me to think of it. She would not say yes, but she would not say no, that is, not exactly." The soldier would have none of this indecision. "At any rate," he explained, "I went off saying, I am engaged to you, and she said, I am not engaged to you."

At Petersburg Hood joined his troops onboard the train. As they made their way, at every stop citizens greeted them with cheers. "The journey through the States from Virginia was a continuous ovation to the troops,"

[70] Dyer, *Gallant Hood*, 200–202, 348 n.67; McMurry, *John Bell Hood*, 75–6; Woodward, *Mary Chesnut's Civil War*, 425, 565; Kean, *Inside the Confederate Government*, 90; J. E. B. Stuart to "My Darling Wife," August 11, 12, and 28, 1863, Stuart Papers; Stuart to Flora, August 18, The Gilder Lehrman Institute of American History, GLCO2334.01 (my thanks to Alinda Borell, archivist, for her assistance); E. M. Thomas, *Bold Dragoon*, 259 (quoting Stuart to Flora of September 4, 1863); "Appointment," *Richmond Dispatch*, August 8, 1863 (citing the *Petersburg Express* story about Hood and Stuart); Wert, *Cavalryman of the Lost Cause*, 304; "Reorganization of the Cavalry Corps of the Army of Northern Virginia, Maj. Gen. J. E. B. Stuart, C. S. Army, commanding, September 9, 1863," *OR*, vol. 29, pt. 2, 707–708; Freeman, *Lee's Lieutenants*, 3:211–12; O'Connor, *Cavalier General*, 158; E. B. Williams, *Hood's Texas Brigade*, 169.

I here wish to thank John McClure, reference department manager of the Virginia Historical Society, for his invaluable help concerning the Stuart letters relating to the "Hood imbroglio," which have been cited in Stuart's biographies but not in Hood's. It is not farfetched to infer that greater communication between "eastern theater" and "western theater" historians of the war would improve the scholarship of both camps.

Stuart's letter of August 28 seems to refer to Hood as "the friendliest man to me imaginable." Mitchell quotes this letter, noting that a "torn" passage prevents us from reading fully what Jeb wrote to his wife about Hood (A. H. Mitchell, *The Letters of Major General James E. B. Stuart*, 336). Mr. McClure at the VHS has kindly examined Stuart's manuscript letter and found that the missive is not torn. The back face of the page has been cut by someone seeking to clip General Stuart's autograph. The front of it would fill in Stuart's assessment to Flora of Hood. Again, I thank Mr. McClure for this valuable research assistance.

remembered Moxley Sorrel. "Old men slapped their hands in praise, boys threw up their hats in joy, while the ladies fanned the breeze with their handkerchiefs," recalled one soldier. Best of all was the food and good things offered up. "The ladies are out to-night to welcome us," wrote one Confederate in his diary; "they have many nice viands." At Sumter, South Carolina, long tables were laid out spread with "the bounties of the land" for the troops of Robertson's Texas Brigade. After only a fifteen-minute stop, they took with them what they had not eaten. "Everybody appeared to be overflowing with patriotism," recalled Pvt. Robert Campbell, Co. A, 5th Texas. The veterans especially appreciated the attention lavished by the pretty girls. As Sorrel recorded, "kisses and tokens of love and admiration for these war-worn heroes were ungrudgingly passed around." Best of all, favors were bestowed regardless of rank. As John West of the 4th Texas put it, "rags and dirt seemed to be a recommendation where gilt and brass failed to excite attention."

Stopover in Raleigh on the night of September 9–10 allowed the men of Rock Benning's brigade, all Georgians, to engage in some political vociferation. A newspaper, William W. Holden's *Standard*, had developed a reputation for lukewarmness toward the Confederate war effort. Around 10 P.M. some fifty to a hundred soldiers, officers among them, broke into the newspaper offices downtown and dumped the type into the street, but they did not touch the printing presses. Governor Zebulon Vance arrived, addressed the soldiers, and drew cheers from them before they dispersed.

As always, the Texans made the most of opportunities for fun on their long trip. In Wilmington some members of Robertson's brigade found their way to a waterfront section called "Paddy's Hollow" and availed themselves of the bars and taverns there. There was "considerable noise and disturbance," whereupon the local constabulary arrived and tried to impose order—unsuccessfully. Three cops, all elderly men exempted from the service, were beaten, knocked on the head or stabbed in the side. No arrests were made. "The pummeled nocturnal guardians," writes Colonel Simpson, "limped a hasty retreat carrying their wounded with them and left the waterfront to the Lone Star victors."

Sleeping on boxcar floors and open platform cars, Longstreet's troops traveled across the Carolinas and thence to Atlanta. Mrs. Chesnut saw the trains passing through Kingsville, South Carolina. "It was a strange sight—miles, *apparently*, of platform cars—soldiers rolled in their blankets, lying in rows, heads all covered, fast asleep. In their gray blankets, packed in regular order, they looked like swathed mummies," she wrote. "God bless the gallant fellows."[71]

Benning's brigade was first to arrive in Atlanta on September 12. Previous passage through the city of some of Brig. Gen. Bushrod Johnson's troops meant a day's stopover for the Georgians, as they awaited the return of locomotives and cars. The respite allowed the men to get new shoes before heading north on the Western & Atlantic. Other units arrived and also faced delays. Law's brigade arrived in the city at dawn of the 15th and had to wait more than a day for cars to arrive. During their stay the men were issued hats, shoes, and other clothing. As always, officers were entitled to the city's finer quarters. General McLaws took a room at the Trout House, the best hotel in Atlanta. He got a good night's sleep on the night of the 18th–19th, disturbed only by a rustling rat, which he disconcertingly found under his pillow in the morning.

Bragg's army at the time was encamped some twenty miles northwest of Dalton, east of Chickamauga Creek and ten miles from the railroad. Closest depot was Ringgold, and it was there that Longstreet's troops began stepping off the cars on the afternoon of the 17th, heading to join the main body. Hood's division was already in the field when Hood arrived at Ringgold in the early afternoon of the 18th. General Bragg had an aide at the station, ready to lead Hood to his command. (One student writes that Bragg had prepared a signed order giving Hood command of troops on the Confederate right.) In his memoirs, Hood recalled how his roan horse, named "Jeff Davis," jumped from the train; he mounted "with one arm in a sling" and rode toward the sound of musketry, which ominously had already begun.

Bragg's and Rosecrans's armies had been in such proximity to one another for several days that everyone knew a big battle was brewing. By September 17 Rosecrans's three infantry corps were within supporting distance

---

[71] Dyer, *Gallant Hood*, 202; Chesnut, *A Diary from Dixie*, 307 ("from Richmond I came back to Camden"); Woodward, *Mary Chesnut's Civil War*, 469–70, 516; Sorrel, *Recollections*, 191; Moore, *A Life For the Confederacy*, 166 ("many nice viands"); H. B. Simpson, *Hood's Texas Brigade: Lee's Grenadier Guard*, 302–304; Skoch, "A Test of Rebel Rails," 18 (Campbell); Cozzens, *This Terrible Sound*, 95; G. Tucker, *Chickamauga*, 95–96; Harris, *William Woods Holden*, 138–39; "An Emeute in Raleigh—two Newspaper Offices mobbed" and "The two mobs in Raleigh—additional particulars," *Richmond Dispatch*, September 11, 14, 1863.

The mobbing of the *Raleigh Standard* office by Hood's soldiers was still making news two months later. Correspondence among North Carolina governor Zeb Vance, President Davis, Brig. Gen. Benning, and Lt. Col. W. S. Sheppard (2nd Georgia) was reprinted in the *Atlanta Intelligencer*, November 4, 1863, under the headline, "The Correspondence Relative to the Recent Mob at Raleigh."

of each other in an area ten to twenty miles south of Chattanooga between Missionary Ridge and Pigeon Mountain, in a valley coursed through by Chickamauga Creek. The Army of the Cumberland numbered some 58,000 men. Bragg's forces, thanks to reinforcements, would total 68,000 troops—and this was without all of Longstreet's command. (Three brigades and all of Alexander's artillery were still en route.) Bragg may have been aware of his numerical superiority—a distinct rarity for Confederate forces entering a major battle—but he also knew he would have to assume the offensive. Advancing north from La Fayette along the east side of Chickamauga Creek on September 17, his plan was to pass in front of Rosecrans's front, whose left flank was posted at Lee and Gordon's Mill, cross the Chickamauga downstream (i.e., to the north), then attack the enemy left. His assault, called for on the morning of the 18th, would employ several crossing points to the creek, and develop into an echelon attack:

Bushrod Johnson's division, with Brig. Gen. Nathan Bedford Forrest's cavalry, to cross at Reed's Bridge, turn to the left "and sweep up the Chickamauga, toward Lee's and Gordon's Mills";

W. H. T. Walker's two divisions to cross at Alexander's Bridge, join in Johnson's advance "and push vigorously on the enemy's flank and rear in the same direction";

Simon Buckner's two divisions to cross at Thedford's Ford and "join in the movement to the left";

Lt. Gen. Leonidas Polk, with Maj. Gen. Frank Cheatham's division, would try to cross at Lee and Gordon's Mill; while

Lt. Gen. D. H. Hill's two divisions would guard the left, maintain pressure, and attack if the enemy shifted troops to the north.

The Confederates' advance, begun at 7:30 A.M. on the 18th, got thrown off at the start by stubborn enemy resistance at Reed's and Alexander's bridges. Johnson was not able to cross at Reed's until 4:30 that afternoon. Walker also had trouble, faced by Col. John T. Wilder's mounted infantry, fighting with seven-shot Spencer repeating rifles. By the time Wilder was pushed back, it was 5 P.M. The Southerners' slow start allowed Rosecrans to shift his lines northward; Maj. Gen. George Thomas and his XIV Corps were given the responsibility of halting the Rebels on the Union left.[72]

---

[72] Moses Wright to Col. G. W. Brent, September 12, 1863, *OR*, vol. 30, pt. 4, 643; G. Tucker, *Chickamauga*, 92–93, 96–97, 105, 110–15, 118; Cozzens, *This Terrible Sound*, 95, 98–99 (map, "September 17, Nightfall Situation"), 534; McLaws, *Soldier's General*, 202 (McLaws

Bragg's eagerness to call Hood into the battle stemmed from his impatience at Bushrod Johnson's inability to get across the creek at Reed's Bridge. Around 3 P.M., Hood arrived on the field, received "a touching welcome" from his Texans, conferred with Johnson, and took command. The Federals had partly burned the bridge and were deployed on the west side of the creek. "I ordered forward some pieces of artillery, opened fire, and, at the same time, threw out flankers to effect a crossing above and below and join in the attack," Hood recalled. "Our opponents quickly retreated." The Southerners repaired the bridge and, as ordered, pressed Wilder's troops southward until darkness forced a halt east of the Viniard farm, 2½ miles southwest of the Reed Bridge. During the night Law's and Benning's brigades arrived. General Law took command of the division when Hood, until Longstreet's arrival, assumed overall charge of the corps. General McLaws, back in Atlanta, would miss the battle entirely.

The battle would continue on the 19th. Bragg planned to pursue his plan of attacking Rosecrans's left flank. During the night, however, Rosy had marched three of Thomas's divisions farther north to the Kelly farm, three miles from the army's previous left flank at Lee and Gordon's Mill. Thus when Bragg sent forth his opening movement after dawn, Confederates ran stoutly into enemy positions. As a result, "over the years Bragg has been criticized for the use of such a straightforward and unimaginative plan," writes Steven E. Woodworth. We note his contention because later in the war, the same would happen to John Bell Hood: what had been planned as a flanking attack turned into a frontal one.

Both Bragg and Rosecrans shifted troops around as the battle slowly spread southward. To the extent that the opposing generals held control of what was becoming a "soldiers' battle," Rosecrans's objective was to hold his

---

at Trout House); Laine and Penny, *Law's Alabama Brigade*, 142; LaFantasie, *Gettysburg Requiem*, 138 (Law's brigade arrival); Hood, *A & R*, 61; Brew, "Death Deep in the Woods," 17 ("signed order"); O'Connor, *Cavalier General*, 158–59; Dyer, *Gallant Hood*, 201–203; McMurry, *John Bell Hood*, 76; Spruill, *Guide to the Battle of Chickamauga*, 3, 6.

Brig. Gens. William T. Wofford's, Goode Bryan's, and Micah Jenkins's brigades arrived too late to join in the battle of Chickamauga (*OR*, vol. 30, pt. 2, 17–18). Bryan's and Wofford's arrived in Atlanta on the morning of September 19 but were delayed in their departure (Oeffinger, *Soldier's General*, 202). Jenkins and his brigade did not leave Petersburg until September 17 and arrived in Ringgold in the afternoon of September 20 (Baldwin, *The Struck Eagle*, 216). Alexander's artillery battalion of 26 guns was last to arrive, boarding at Petersburg at 4 P.M. September 17. Its route took in Wilmington, Kingsville, and Augusta. Colonel Alexander kept close eye on his schedule: arrived in Atlanta 2 P.M. on the 23rd, and Ringgold 12 miles from the battlefield at 2 A.M. on September 25. All told, Alexander covered 843 miles in seven days and ten hours (E. P. Alexander, *Military Memoirs of a Confederate*, 449).

north-south lines intact along the La Fayette Road. For his part, with his initial plan for a flanking attack now ruined, Bragg determined to attack in any way that would damage the enemy.

Hood's role in all this came on the afternoon of September 19, when both Bragg and Rosecrans decided to attack the other's southern flank. By this time Hood's position east of Viniard's was on the south part of the battlefield. At 1:30 Federals of Maj. Gen. Jefferson C. Davis's division advanced, ran into John Gregg's Tennessee brigade and beat it back, before Southern artillery forced the Northerners' withdrawal. Then Hood ordered Johnson's division to advance. The Confederates did not get far when a Union counterattack stopped them. At 2:30 Hood ordered three brigades of his divisions into the fight. As Robertson's Texans marched forth in their new uniforms (their dark shell-jackets and light blue trousers made them look like Yankees to some), they heard from comrades who had already been in the fight, "You fellers'll catch hell in thar. Them fellers out thar you ar going' up again, ain't none of the blue-bellied, white-livered Yanks an' sassidge-eatin' forrin hirelin's you have in Virginny that'll run at the snap of a cap—they are Western fellers, an' they'll mighty quick give you a bellyful o' fightin'."

In the thick woods, visibility was limited and formations fell apart, yet the Confederates managed to drive U. S. Brig. Gen. William Carlin's and Col. Hans Heg's brigades into the east Viniard field up against the La Fayette Road. Counterattack was followed by another attack, and Hood's troops pushed Davis's across the road. Then, around 4:30 or 5, Hood's reserve, Benning's brigade, charged and drove back Davis's tired, depleted forces. Colonel Heg was mortally wounded, and the Georgians captured four guns of the 8th Indiana Artillery. But Benning was stopped by Wilder's Spencers and Capt. Eli Lilly's 18th Indiana Battery. Then, as the afternoon faded away, a final Union charge retook the four guns and pushed the Rebels back across the road into the east Viniard field. After so much blood, the situation in this sector at 5:30 was essentially the same as when the fighting had started four hours earlier. On the battlefield as a whole the same prevailed: the opposing lines, despite some temporary Confederate breakthroughs, were pretty much the same as at dawn of the 19th. The difference, of course, could be counted in the casualties for the day's fighting: 6,000–9,000 Confederates, maybe 7,000 Federals.[73]

---

[73] Cozzens, *This Terrible Sound,* 111–14, 119–20, 198–229 (including "Them fellers," 203), 294, 407; Hood, *A & R*, 61; D. A. Powell, *The Chickamauga Campaign—A Mad Irregular Battle*, 246 ("Robertson's Texans, who raised their hats in lieu of cheering as the wounded general trotted past"); D. A. Powell, *The Chickamauga Campaign—Glory or the Grave*, xii ("until

Later in the evening Hood made his way to headquarters to report on his day's work. There he met several of the army's top officers, "and, to my surprise, not one spoke in a sanguine tone regarding the results of the battle in which we were then engaged." The gallant Hood, fresh from Lee's army, was experiencing the high-level malaise which for the past year had infected the Army of Tennessee. At length Hood encountered his old friend John Breckinridge, major general commanding a division in Harvey Hill's corps. He was seated under a tree, "a heavy slouch hat upon his head," Hood remembered. The two began talking about the next day's battle, with Hood articulating prospects for routing the enemy and winning the day. "My dear Hood," Breckinridge cried, happy to hear such unfamiliar optimism, "I am delighted to hear you say so. You give me renewed hope; God grant it may be so." Hood had nothing with him save what Jeff Davis carried, so he was without any camp equipment. He chanced upon Col. J. Stoddard Johnston of General Buckner's staff, who invited him to their bivouac. After a late supper, Hood piled some leaves and, sharing Johnston's blanket, fell asleep.

Late on the 19th two more of Longstreet's brigades joined the army: those of Brigadiers Joseph B. Kershaw and Benjamin G. Humphreys, who had taken over from William Barksdale (dead at Gettysburg). Peter himself arrived and, despite no guide from Bragg's staff at the rail depot, got to headquarters by 11 P.M. By then Bragg had met with his generals and laid out the plan for the 20th, which, rather unimaginatively, was to do more of the same: dawn attack by his right wing (under Polk) against the Federal left, with Longstreet, commanding the army's left wing, taking up the assault. As Bragg later phrased it, "the whole line was then to be pushed vigorously and persistently against the enemy throughout its extent." Longstreet spent an hour talking with Bragg and confirming his role in the day's battle. He would have use not only of his five brigades, but three more of Bushrod Johnson's for his assault. It was midnight before Pete ended his meeting with Bragg. Then he and his two staff officers, Moxley Sorrel and P. T. Manning, found a place near Thedford's Ford to finally get some rest.

Across the La Fayette Road at the Widow Glenn's house, General Rosecrans held his own council of war and announced his plan for the morrow, which was simply to stand on the defensive. Before the battle, the Federals were aware that part of Lee's army would reinforce Bragg—the Confederates had not tried hard to keep the troop transfer secret. Prisoners taken

---

Longstreet arrived, Maj. Gen. John B. Hood led the corps"); Woodworth, *Chickamauga*, 11; G. Tucker, *Chickamauga*, 164, 175, 211–12; Woodworth, "Soldier with a Blunted Sword," 131; Noblitt, "A Flank Unturned at Chickamauga," 36.

during the day confirmed the arrival of Longstreet's troops. "We were greatly outnumbered," Rosecrans later reported, so his main goal was to hold his army intact, covering the roads back to Chattanooga. Thomas believed that the main enemy attack would fall upon his corps. "I would strengthen the left," he suggested to Rosecrans. Rosy's anxieties were expressed in his orders to Thomas: "you will defend your position with the utmost stubbornness. In case our army should be overwhelmed it will retire on Rossville and Chattanooga."

Thomas was right about the Rebels' intent, yet the Confederate assault on the Union left did not begin at dawn. Polk had been indifferent in issuing orders during the night. Not until 9:30 A.M. did Breckinridge's division advance, successfully overlapping the Union left, driving the enemy back and turning south until it was stopped by counterattacks. When Confederate brigades to the south took up the echelon assault, they ran into Thomas's strong lines, strengthened by log and rail barricades constructed during the night. Polk's attacks were poorly coordinated, piecemeal, and repulsed.[74]

Longstreet, up at dawn, supervised the movement of his command into the assault formation he wanted. During the morning he sought out Hood, asking about his troops' position and spirits. "I informed him that the feeling of officers and men was never better," Hood remembered, "that we had driven the enemy fully one mile the day before, and that we would rout him before sunset." Longstreet responded enthusiastically: "we would of course whip and drive him from the field." Old Peter had good reason for confidence. As Glenn Tucker has pointed out, in his last big assault, on the third day at Gettysburg, he had nine brigades, four of them severely weakened by earlier combat; here on the second day at Chickamauga, he had eight, most of them relatively fresh; a few had not even seen action yet. The field in front of him on July 3 was long and open; here he could mass his columns in protective forests three hundred yards from the Brotherton house, a cabin on the La Fayette road which would serve as a guide to the attacking column, much as the "clump of trees" had been for the troops of Pickett, Pettigrew and Trimble. For good measure, Longstreet had with him the best scout possible, Confederate Private Tom Brotherton, whose father owned the house in Longstreet's front, and who explained the lay of the land behind it, where the Yankee line was.

Longstreet would soon possess another advantage, though he did not yet know it: a colossal, breathtaking enemy blunder. At his headquarters

---

[74] Hood, *A & R*, 62; G. Tucker, *Chickamauga*, 211–13; Johnston, "General J. B. Hood and Chickamauga," 552; Cozzens, *This Terrible Sound*, 292–304; Woodworth, *Chickamauga*, 41–53.

Rosecrans had been receiving messages throughout the morning about the fighting on his left. Thomas had been calling for reinforcements and Rosecrans had been sending them. One of the courier officers riding behind the left-center of the Union line thought he saw a gap between the division of Maj. Gen. Joseph J. Reynolds and Brig. Gen. Thomas J. Wood, and so reported to HQ. Rosecrans responded at 10:45 by having a staff officer write an order to Wood that he should "close up on Reynolds as fast as possible, and support him." There was no gap in the Union line; the division of Brig. Gen John M. Brannan was properly in line, to Reynolds's right and Wood's left. But Brannan's troops, well into the forest, were hidden from view of the unsuspecting staff officer. When he was handed Rosecrans's "close up" order at 10:55, General Wood knew that there was no gap between him and Reynolds. But just an hour and a half previous, a tired and irritable William S. Rosecrans had dressed down Wood before his entire staff: "You have disobeyed my specific orders…and, by God, I will not tolerate it!" This time Wood made certain there would be no such reprimand. He carefully placed the paper in his pocket notebook and ordered his troops to begin marching toward the left at double quick. Wood's officers knew the Rebels were before them. "We well knew that the enemy was immediately in our front," recorded Col. Charles Harker. Harker and Col. George Buell got their brigades moving, but only after Buell protested that his skirmishers were even then actively engaged. Wood made clear the order was imperative, and Buell fell into line. The withdrawal of Wood's division left a 600-yard gap in Rosecrans's line.[75]

Longstreet had Johnson's and Hood's troops in place by 10 A.M. when Humphreys's and Kershaw's brigades arrived. The formation offered "an attacking column of unprecedented depth," writes Peter Cozzens: two of Johnson's brigades up, one behind them; then Law's brigade (now under Col. William F. Perry, 44th Alabama) of Hood's division, with Robertson's and Benning's behind; Humphreys and Kershaw were at the back. This concentration of almost 11,000 Confederates, five brigades deep, offered a front of 500 yards—about the width of the Brotherton field, through which they would be marching. Everything was ready when, at 11:10, Longstreet gave the order to advance. Out from the woods the Confederates marched, Col. John Fulton's and Brig. Gen. Evander McNair's brigades in the front. They advanced across the La Fayette Road and into the field, shooting and routing Yankee skirmishers. Sweeping around the Brotherton house, they struck the

---

[75] Hood, *A & R*, 63; G. Tucker, *Chickamauga*, 253–59, 260–62; Cozzens, *This Terrible Sound*, 315–17, 358–67; Woodworth, *Chickamauga*, 57–58; Welch, "'Old Rosy' and the Fatal Order," 46.

Federal breastworks in the timber west of the field and drove away what few Northern troops were left in them. They kept going into the woods, approaching the Dyer house, near which were Rosecrans's headquarters. "Only twenty minutes after stepping off from their line of departure," writes Cozzens, Confederates "were on the edge of the Dyer field, with nothing but stubble and sedge standing between them and the Yankee rear."

The right of Rosecrans's line just melted away under Longstreet's attack. "The sudden withdrawal of troops from my left," Brig. Gen. Jefferson Davis reported (meaning Wood's division), "and the absence of any support on my right, just as the attack was being made, made my position little better than an outpost and perfectly untenable against the overwhelming force coming against it." The Confederates charged through the gap, but Federal resistance stiffened on McNair's right, where Buell's column on the march now turned to stand and fight. Hood, riding with his division, led it to extend Johnson's two-brigade front so that about 11:45 five Confederate brigades, including Robertson's and Law's/Perry's, charged into the Dyer field. "The scene now presented was unspeakably grand," recorded General Johnson.

> The resolute and impetuous charge, the rush of our heavy columns sweeping out from the shadow and gloom of the forest into the open fields flooded with sunlight, the glitter of arms, the onward clash of artillery and mounted men, the retreat of the foe, the shouts of the hosts of our army, the dust, the smoke, the noise of fire-arms—of whistling balls and grapeshot and of bursting shell—made up a battle scene of unsurpassed grandeur.

It was here that Hood caught up with Johnson. Exchanging salutes, Hood said simply, "Go ahead, and keep ahead of everything."

North and west of the Dyer field, Union artillery opened on the Southerners. As Robertson ordered his brigade to attack those guns, the Texans were flank-attacked on their right and pushed back. Benning, too, was roughly handled by a Federal charge. Before long, Hood's troops were retreating in disordered groups, back across the Dyer field into the woods.

"Hood was there," writes Cozzens, "trying through sheer force of will to restore a semblance of discipline." "Move up, men," he shouted, "those fellows are shooting in the tops of the trees."

But they were not. As General Kershaw brought his brigade into the field shortly after noon, Hood rode with them. General Robertson rode to join him, seeking orders. "Aunt Pollie" saluted, then watched as Hood suddenly dropped his reins and reeled in his saddle. A minie ball had struck his right leg just below the hip. Courier Abner Wilkins helped the general from his horse; some Texans hurried forth to help. Wilkins called for stretcher bearers, and Hood was carried from the field.[76]

Borne to a dressing station, Hood would have had bandages applied to his wound. He was then carried farther to the rear, to the divisional field hospital, located at the Reed house and farm near Reed's bridge over the Chickamauga. Earlier historians—O'Connor in 1949 and Dyer in 1950—have only had Hood's memoir as their source, in which the general relates, "when wounded I was borne to the hospital of my old division, where a most difficult operation was performed by Dr. T. G. Richardson, of New Orleans. He was at the time Chief Medical Officer of the Army of Tennessee." Actually, Hood's medical care that afternoon was supervised by a team of physicians, and a very impressive one. Dr. Darby, divisional surgeon and Hood's good friend, eventually took over the general's medical care, but he was not at Chickamauga. Darby arrived in north Georgia on the 24th, having left Virginia several days later than Hood and the division (Mary Chesnut recorded in her diary that "P. V.," enamored of Mamie Preston, "stayed sweethearting a day too long in Richmond, basking in the smiles of love and beauty"). Upon his arrival, the conscientious doctor began writing a daily, extensive medical journal of Hood's convalescence. From it we learn that at the "field infirmary" fully four medical doctors examined the general's wound and discussed their course of treatment: acting chief surgeon W. O. Wilson of Hood's division (a Dr. Hudson, of the 4th Alabama in Law's brigade, was "acting chief surgeon"

---

[76] Cozzens, *This Terrible Sound*, 316, 368–70, 407–408, 411–12; Powell, *Glory or Grave*, 112; Spruill, *Guide to Chickamauga*, 184 ("unsurpassed grandeur"); Furqueron, "The Bull of the Woods," 120 ("little better than an outpost"). In *A & R*, Hood places his wounding at around 2:30 P.M., but Cozzens's reckoning of sometime after noon is convincing (*A & R*, 64; *This Terrible Sound*, 411–12). Hood mentions his Texans' superstition that in battle when he was not riding his favorite roan, "Jeff," he was wounded, as at Gettysburg and here at Chickamauga; "Jeff Davis" had been wounded on September 19 (*A & R*, 64–65).

On the basis of Dr. Darby's medical report of Hood's surgery, Stephen M. Hood conjectures, "it seems highly likely that Hood was a victim of friendly fire" ("John Bell Hood: Pain & Prescriptions," 32). I am not persuaded. Rather, I like the statement of William Lee White that, after his Texans began to give way, Hood grabbed the colors of his old regiment, the 4th Texas, as a means of rallying his troops. "Among his milling men, holding the flag and the reins of his horse with his disabled arm, Hood made an excellent target"—for enemy fire (W. L. White, *Bushwhacking on a Grand Scale*, 115).

Woodworth makes the interesting observation that Hood was the only Confederate division or corps commander wounded at Chickamauga, and he suggests that at least part of the reason lay in Hood's practice of riding with his troops into battle. Though he was exercising "a sort of de facto corps command," he "was still trying to exercise command much like the commander of a very, very large brigade. He was well to the front, visible to his men, trying to inspire them, but also visible to the enemy" (*No Band of Brothers*, 120, 122).

of the division until Darby's arrival); acting brigade surgeon John C. Jones of Robertson's brigade; Tobias G. Richardson, medical inspector, Army of Tennessee; and Edward A. Flewellen, assistant medical director for the Army of Tennessee. The doctors noted that Hood was in shock, but "not so great as is usually seen in such a serious fracture." The ball had exited with two wounds; Dr. Darby thought it may have been a minie ball with detachable base. The team agreed upon amputation because five inches of bone had been shattered. "Chloroform was given and the operation performed by Surgeon Richardson assisted by Surgeons Hudson and Jones at 4 P.M," Darby entered in his journal. Civil War surgeons generally used two amputative techniques, the circular and flap methods. The circular was easier and faster to perform; the general's surgical team employed it in their operation.

Borrowing information from the other physicians upon his arrival, Darby recorded the details of Hood's postoperative care, beginning on September 21: "Remained at the infirmary from period of operation until 3 PM of this day. Very little excitement from the operation. Some nausea from chloroform. At 3 P.M. was borne on a litter five miles en route for residence of Col Little of 11th Ga. Regt. Appetite & condition good."

Col. Francis H. (Frank) Little of the 11th Georgia lived south of La Fayette, some twenty-five miles from the battlefield. His regiment was in Tige Anderson's brigade, and Little volunteered his father William's home as a safe place where the general could receive close attention. Hood's journey there took two days. On the 21st, the party carrying him, under supervision of Dr. Jones, covered only five miles before they stopped to rest for the night. Darby recorded that the general "passed a quiet night. Appetite for breakfast good, spirits cheerful." On the 22nd Hood was borne another twenty miles farther south to the Little residence; clouds of dust kicked up by wagons on the road caused the patient some coughing. Arriving at 4 P.M. on the 24th, Darby took charge of the case from Surgeon Jones and recorded that he was "much surprised to see so little constitutional disturbance." The patient was evidently recovering from his surgery well. Darby noted Hood's "spirits fine—never has thoughts otherwise than he would recover."[77]

---

[77] W. C. Davis, *The Orphan Brigade*, 191 ("Some of the most severely wounded were carried to a house near Reed's Bridge"); Bohannon, "'A Bold Fighter,'" 261 ("When Hood arrived at his division's field hospital at the Reed Farm"); O'Connor, *Cavalier General*, 165; Dyer, *Gallant Hood*, 211–12, 350 n.5; Hood, *A & R*, 65 (Hood noted that Dr. Richardson "is now the President of the Medical Association of the United States" (Richardson presided over the American Medical Association, 1878–1879); S. Hood, *Lost Papers*, 27–29 (Darby, Hood medical journal, entries of September 20–24); Woodward, *Mary Chesnut's Civil War*, 443; Jones, "Roster of the Medical Officers of the Army of Tennessee," 192 (Dr. Darby as "Chief

Hood's initial recovery from surgery, September 20–24, bespoke the excellent medical care he was receiving. Medical statistics for amputation in the upper third of the femur show death resulted in slightly more than 50 percent of all cases. (Indeed, just a decade before in the Crimea, 86.8 percent of British soldiers amputated in the upper third of the thigh died.) According to *The Medical and Surgical History of the War of the Rebellion*, of the 268 Union and Confederate soldiers recorded as having died from Hood's type of amputation, more than a third perished within five days of the operation, doubtless from hemorrhage or shock. No cause is given for more than 180 of the reported fatalities. Not surprisingly, given the grossly unsanitary conditions of all Civil War surgery, the second most prominent cause of death was pyaemia, a "surgical fever" similar to gangrene. Post-operative infection was the chief cause of surgical mortality in the Civil War and, as the name implies, "hospital gangrene" commonly occurred in the crowded wards where soldiers' unclean wounds festered. The main reason General Hood escaped infection of his leg stump was that he was not taken to a hospital. As was the custom for general officers of both sides, at Colonel Little's home Hood received personal care and commanded the attention of the army's best medical officers.

While he recovered, Hood would have received final news of the battle which had laid him low. Longstreet's attack had sent Rosecrans's army packing back to Chattanooga, although in a reasonably organized retreat, thanks

---

Surgeon Hood's Division"); 199 (Dr. Flewellen); 216 (Dr. W. O. Hudson); 222 (Dr. Jones, supervising Hood to the Little farmhouse); 252 (Dr. Richardson and "his melancholy duty to amputate the thigh of the gallant Major General Hood"); Bollet, *Civil War Medicine*, 149.

Decades ago local tradition developed that the Clisby Austin house near Tunnel Hill, Georgia, was the site of Hood's amputation and that the severed leg was buried in the Austin family cemetery. See notes by Wilbur Kurtz, August 21, 1937, in his notebook no. 11, 259–60, Wilbur G. Kurtz Collection, Atlanta History Center, and Brown and Elwell, *Crossroads of Conflict*, 11. The legend is manifestly untrue, as Wilbur Kurtz affirmed in a notebook entry after a visit to the Resaca area in June 1946: "This is the little Austin family cemetery where tis said that Gen. John B. Hood's leg is buried. The story is that Hood was carried to the Austin house after being severely wounded at Chickamauga, and at this house, his leg was amputated." Then Kurtz adds boldly, "Not a word of truth in this story about amputation of Hood's leg at the Austin house—K," (notebook no. 20, 212). Besides, Civil War surgeons customarily performed amputations within hours of the wound at field hospitals not far from the battlefield. The Austin house is more than ten miles from Chickamauga (see Davis, "Avalanche of Wounded," 8). Finally, Dr. Darby's medical log, which has recently come to light, proves that Hood was taken September 21–22 from the field hospital to Col. Little's house in the Armuchee Valley, not to the Austin house.

Dr. David Bosshardt of Fort Oglethorpe, Georgia, confirms the local knowledge that Hood's amputation occurred at the field hospital set up in the Reed house, near Reed's Bridge at Chickamauga Creek (conversation with Dr. Bosshardt, October 26, 2013).

to Thomas's stubborn hold of the Union left, anchored on Snodgrass Hill. Nevertheless, the Southern victory came at a heavy cost. In his memoirs Colonel (later General) Alexander made a point of tabulating the Confederate army's losses: 2,074 killed, 12,797 wounded, and 1,328 missing for a total of 16,199, to which he added another 250 cavalrymen. Later writers put Southern casualties at higher figures: Livermore (1901) 18,500; Horn (1941) 17,800; Glenn Tucker (1961) almost 21,000; Connelly (1971) 18,000. On the Union side, General Rosecrans reported a little over 16,000 casualties: 1,657 killed, 9,756 wounded, 4,757 captured or missing.

During the battle nine Confederate general officers had been taken from the field. Brigadiers Preston Smith and Ben Hardin Helm were killed. Among those wounded, Hood had been the most seriously injured. Moreover, his removal from the battle had occurred at the most critical time. "Hood's wounding was an incalculable loss to the Confederate effort at Chickamauga," writes Cozzens. "His presence on the field at the height of the breakthrough, when total success depended on a rapid exploitation of the breach, was critical; no other field commander in Bragg's army had the sort of reckless, self-assured aggressiveness so appropriate to the moment."

Two days after the battle a rumor circulated that Hood would not survive his surgery. From its headquarters in Atlanta, the Press Association of the Confederate States of America (which telegraphed news articles to member newspapers) on the 22nd sent out a telegraphic report, "Further from the battlefield," which stated, "Our loss in general officers is very great. Brig. Gen. Helm, of Ky., was killed while leading a charge. Gen. Hood's wound is reported to be a mortal one." The *Atlanta Intelligencer* ran this news: "Gen. Hood's leg was amputated some distance above the knee, and it is our painful duty to state that he died after the operation." Papers as far away as Richmond carried this incorrect news on the 23rd. "We understand that Major General Hood is dead," lamented the *Richmond Enquirer*. "If true, the victory, however great and decisive, has been most dearly purchased." Col. Josiah Gorgas of the Ordnance Bureau noted in his diary, "Our job is damped by the loss of General Hood, who died under the operation of amputating his leg.... His loss is a severe one, second only to that of Jackson." General Lee took the news hard. "I grieve for the gallant dead," he wrote Longstreet, "and mourn for our brave Hood."[78]

---

[78] S. Davis, "John Bell Hood's 'Addictions,'" 30; Trammell, "'Life Is Better Than Limb,'" 46–51; Freemon, *Microbes and Minie Balls*, 91; Woodworth, *A Deep Steady Thunder*, 89–97; Alexander, *Memoirs*, 463–64; Livermore, *Numbers and Losses*, 105–106; Horn, *The Army of Tennessee*, 273; G. Tucker, *Chickamauga*, 388; T. L. Connelly, *Autumn of Glory*, 227; "Return

Then came word that Hood was not dead after all. The Press Association sent out a correction: "*Gen. Hood is not dead.* His right leg was amputated. He says he will live to fight the Yankees at least another battle." "We learn to-day," wrote War Department clerk John B. Jones on the 24th, "that Gen. Hood is not dead, and will recover." John S. Thrasher, superintendent of the Press Association in Atlanta, later felt compelled to explain how he had gotten the story wrong. Several days before the big battle in north Georgia, General Bragg had decided to send back the Press Association reporter, Will Woodson, who would have been reporting from the Army of Tennessee. As a result, Thrasher later complained to his Board of Directors, "I was compelled to make up in Atlanta the Press reports of the interesting events on the Chickamauga from intelligence obtained from wounded brought from the field, and information received from persons coming from that vicinity." In the *Atlanta Intelligencer* Thrasher publicly criticized "the unwise course of Gen. Bragg in not permitting the Press Association to remain in the front." Bragg later relented, and Thrasher's news service could circulate more reliable telegraphic dispatches from the Army of Tennessee.

Despite his wound, Hood had shown more battlefield leadership at Chickamauga, September 18, 19, and 20, than he had at Gettysburg, July 2. With good reason, General Longstreet on the 24th sent a letter through Bragg to the Confederate Adjutant General in Richmond, Gen. Samuel Cooper, recommending Hood for promotion to lieutenant general. In the recent battle he explained, "General Hood handled his troops with the coolness and ability that I have rarely known by any officer, on any field." Bragg added a warm endorsement and sent Longstreet's letter on to the War Department. There Secretary Seddon on October 3 wrote, "I cannot too warmly express my appreciation of the character and services of this distinguished officer, and cordially concur in recommending his promotion, if only as an appropriate testimonial of the Gratitude of the Confederacy." That same day President Davis endorsed the recommendation for promotion: "The services of Major

---

of Casualties in the Army of the Cumberland…Sept. 19 and 20, 1863," *OR*, vol. 30, pt. 1, 171–79; Cozzens, *This Terrible Sound*, 276, 321, 534; "The Opposing Forces at Chickamauga, Ga.," 3:673–76; "TELEGRAPHIC. Further from the battlefield" (Atlanta, September 22), *Atlanta Intelligencer*, September 23; "TELEGRAPHIC. The Victory of Chickamauga," *Atlanta Intelligencer*, September 24; S. A. Meredith to E. M. Stanton, September 23, *OR*, vol. 30, pt. 3, 790 (quoting the *Richmond Enquirer*, September 23, and the Press Association dispatch of September 22); Dyer, *Gallant Hood*, 211 (quoting the *Enquirer* of September 23); McMurry, *John Bell Hood*, 79; Lee to Longstreet, September 25, *OR*, vol. 29, pt. 2, 749.

General Hood, and his character as a soldier and patriot, are equal to any reward, and justify the highest trust."[79]

On September 25 Dr. Darby dressed Hood's wound for the first time. At nine that night the general was restless—"considerable arterial excitement"; Darby gave a quarter-grain of morphine. In the several days following the doctor observed morning nausea, afternoon fever, fatigue, and loss of appetite. On the 27th he wrote, "These symptoms increased to an extent that gave little encouragement for hopes of a recovery." At nighttime Darby gave Hood a quarter-grain of morphine; the drug was not for postsurgical pain but to help the patient sleep. Eventually Hood's appetite improved, and the wound showed signs of healing. Darby informed others of the patient's improvement, including apparently Wade Hampton. On October 2, Hampton was able to write Senator Wigfall, "My latest news from Hood is favorable & I have written to beg him to come to my home as soon as he can move." On the 4th the patient "awoke feeling well—breakfasted heartily." Darby gradually removed ligatures from the stump, which was oozing. But in the medical practice of the day the appearance of "laudable pus" was no cause for alarm; "discharge healthy," the doctor recorded. Some bleeding occurred on the night of the 6th; Hood awoke anxious as to his state, and experienced "intense pain in stump and groin." Two abscesses formed on the wound; Darby dressed it with water, administered iron and quinine, and saw to a good diet: "chicken, beef steak, squab, mutton chops, rice eggs bread butter, white potatoes." Nightly morphine infusions sometimes increased to half-grains, not enough to concern the doctor. The patient was definitely improving. Texan Henry P. Brewster had attached himself to Hood's staff and stayed with him during his convalescence and reported on it to Senator Wigfall. On the 8th Wigfall knew that Hood was already asking when he could return to service in the field. On the 11th Darby noted, "Slept well and awoke refreshed—Appetite good—Little leakage of blood, pus from abscess decreasing and more laudable." The general that day even enjoyed some sherry and punch. On the 13th, Darby noted "wound commencing to heal," but in the following days pus continued to be discharged from an abscess.

---

[79] "TELEGRAPHIC NEWS," *Richmond Sentinel*, September 25 ("Gen. Hood is not dead"); J. B. Jones, *Rebel War Clerk's Diary* (1958 ed), 281; *Minutes of the Board of Directors of the Press Association*, 10, 23–24; Risley, "Wartime News over Southern Wires," 157; Hood, *A & R*, 65–66 (Longstreet's letter and endorsements, Hood Service Record; also published in S. Hood, *Lost Papers*, 46–50).

Meanwhile, many Southerners continued to wish Hood a good convalescence. On the 14th Brig. Gen. G. W. C. Lee, serving on the staff of President Davis, wrote Hood, "The President desires me to express his gratification at your well doing and his prayers for your speedy recovery." On October 16 the patient sat up for the first time—twenty minutes while his bed was being made. This so encouraged Darby that the next day he wrote General Bragg, stating he hoped to see Hood soon able to walk on crutches. General Bragg passed on the good news to President Davis. "He is reported much improved," he wrote on October 17. "Still improving" was the word on the 23rd. But the recovery was still rocky. "Oct 22nd—Passed a bad night, suffering from much pain in the stump due to abscess and also from sweating." Then, four nights later, the patient actually enjoyed "a good nights rest." Darby recorded every improvement in Hood's condition. On the 27th he sat up for almost an hour and did so every day after that. It was good that the patient was able to move, because on October 29 Darby received a letter from General Bragg recommending that Hood be moved farther to the rear. The doctor reluctantly agreed; he wanted to wait another ten days, but Hood had been expressing anxiety that he might fall captive to enemy raiders. A day of rain delayed the departure, but on the morning of November 1, after dressing the wound ("found it looking well"), Dr. Darby and several accompanying staff officers had the general placed on a cot and carried on the shoulders of slaves owned by the Little family (overseen by seventeen-year-old Randolph Little) to Tunnel Hill. (Dalton was the nearest railroad depot, but the trek there over rough Rocky Face Ridge would have been more dangerous to the patient; remember Jackson's fall from the litter on the night of May 2.) The dozen slaves, splitting the work in teams of six, at Hood's request sang spirituals as they walked along. The party reached the Tunnel Hill station at 6:30 P.M. They spent the night and on the 2nd left for Atlanta at 7 A.M. on a train car specifically ordered up by Bragg for its array of beds ("make him comfortable," the general ordered). They covered the hundred-plus miles in nine hours and took lodging on Whitehall Street in Atlanta at the home of the superintendent of the Press Association, John Thrasher. Thrasher was a hospitable man and had taken in as guest Lt. Gen. Leonidas Polk when the latter was relieved by Bragg after Chickamauga. Polk moved out so that Hood could move in. The patient continued to convalesce. "Nov 4th—Rested well. Awoke with strong appetite for breakfast," Darby recorded. "Wound much improved. Slept without morphine for first time." On the 7th the doctor noticed that the wound was almost entirely closed; "very healthy pus in small quantity" was still being discharged by the abscesses.

By this time, almost three weeks after his battlefield injury and amputation, Hood had been talking about having his stump fitted for an artificial leg. The prosthesis could be obtained in Richmond, which meant Hood and Darby traveling there. Indeed, as early as October 25, Charlotte Wigfall, the senator's wife, was writing her daughter about the possibility that Hood would be in the capital sometime that winter ("he is going first to pay Gen. Hampton a visit in Columbia"). On November 9, with the patient "sufficiently improved to commence the journey to Richmond," the general and his physician set out on the 800-mile train ride. The master of transportation for the Western & Atlantic let them use the railroad superintendent's car. The route took them through Augusta (reached morning, November 10) and Wilmington, where they stayed from the 11th to the 15th. The *Wilmington Daily Journal* offered a glimpse of the wounded warrior. "Like many other men of signal bravery," the paper observed, "General Hood, in personal appearance, is the very opposite of the conventional idea of a fire eater. Somewhat bleached by illness rendering his delicate features more delicate, with soft hair of light brown and silky beard nearly golden, he looks like a mild, sensitive and amiable man."

Hood and Darby arrived in Richmond at 2 on the afternoon of the 16th; Darby noticed that Hood was "in good spirits and in far better condition than when he left Atlanta." The general was strong enough to begin walking on crutches. "Walked across the room three times today," Darby recorded on November 21. From then on Hood's progress was so steady and unremarkable that Darby ceased his medical journal on the 24th. The reason could also have been that Darby was preparing to leave for Europe. Prosthetic limbs were manufactured in the states, but the best were European-made. The doctor knew well the condition of the general's stump and could have artificial legs made to order if he went himself. Money was no object—even the soldiers chipped in. Hood's three Texas regiments started a fund. Placing a limit of $20 per soldier, the Texans in one day raised $3,100; their total sum approached $5,000. Darby left Richmond sometime between November 26 and December 5, as Senator Wigfall's wife Charlotte wrote her daughter Louise on the 26th, "We hear that Dr. Darby is going to Europe to buy a leg for him." On the 5th Mrs. Wigfall wrote Louly, "Dr. Darby has gone to Europe to procure a leg for the General with the money contributed by the Texas Brigade." While in London, where he spent at least several weeks, Darby supervised the manufacture of "Anglesey" legs, which he had shipped with him back through the blockade.[80]

[80] S. Hood, *Lost Papers*, 25–30 (Darby medical journal, entries of September 25–November 21); Bragg to Hood, October 31, 1863 and G. W. C. Lee to Hood, October 14, 1863, S.

In Richmond, Hood stayed awhile at the home of Gen. and Mrs. G. W. Smith, but he remembered in his memoirs that he eventually took a room at the Spotswood Hotel, Richmond's trendiest and, with five brick stories, largest. President Davis, Secretary of War L. P. Walker and other cabinet officers had lived there, so many dignitaries that the place, at Main and 8th streets, could be called the unofficial Confederate capitol. He eventually expanded his quarters. In the first half of February he billed the army's Quartermaster Department for five rooms, each costing $30 a month. During his several months in Richmond, the famous general received all sorts of visitors. A reporter for the *Richmond Examiner* described him as looking "hale and hearty, despite the suffering and exhaustion occasioned by the loss of a limb." Noting that Hood "expresses himself in the best of health and spirits," the writer implied Hood's eagerness to return to the field. After Longstreet's recommendation for his promotion to lieutenant general, the administration proposed to do just that,

---

Hood, *Lost Papers*, 192, 194; Wright, *Southern Girl*, 149, 152; Bragg to Davis, October 17, 23, 1863, in Crist, Williams, and Dillard, eds., *The Papers of Jefferson Davis*, 10:26, 32; Wilbur G. Kurtz, "Some Unofficial Glimpses of General John B. Hood" (handwritten ms.), 5–6; Dixie Little to Wilbur G. Kurtz, September 3, 1946, box 79, folder 13, Kurtz Collection; Lively, *Calamity at Chancellorsville*, 148–50 ("Number of Falls from the Litter"); McMurry, *John Bell Hood*, 79–80 (*Wilmington Daily Journal*); "General Hood," "Major-General Hood," and "Serenade to General Hood," *Memphis Appeal*, November 2, 4, and 5, 1863; S. Davis, *What the Yankees Did to Us*, 174–75 (on Thrasher); Parks, *General Leonidas Polk C.S.A.*, 351–53; "We learn that Major-General Hood...," *Atlanta Intelligencer*, November 11, 1863 (re: superintendent's special car, stating that Hood was leaving Atlanta on the afternoon of the 11th); "Arrival of General Hood," *Richmond Examiner*, November 17, 1863; "Hood and His Men," *Augusta Constitutionalist*, November 3, 1863 (the Texans' fund); Freeman, *Lee's Lieutenants*, 3:231–32; [L. W.] Wright, *A Southern Girl*, 161; S. Hood, *John Bell Hood*, 278.

Miss Little's letter to Kurtz, cited above, is another example of the latter's fastidious research on the war. Kurtz used it as basis for his paper read in 1950, "Some Unofficial Glimpses of General John B. Hood." Dixie was granddaughter of Colonel Little; her father had been a teen at the time of Hood's arrival at the Little home. "I have heard my Father, John Randolph Little tell many times (my Father was only 17 yrs of age then) of taking twelve Negro slaves & carrying Gen. Hood on a litter to Tunnel Hill, Ga., to catch a train away. Father was the only white person along, besides some of Gen. Hoods staff. Gen Hood had these Negroes to sing Corn songs all the way." With typical modesty, Kurtz never sought to publicize the information he got from Miss Little, only using it in a talk he gave in Atlanta in 1950.

"Did anyone tell you that a night or two after Gen. Hood was moved from there, that a bunch of Yankees came to Grandfather Littles looking for Gen Hood?" she asked in another letter (Dixie Little to Kurtz, September 21, 1946, box 79, folder 13, Kurtz Collection). If the Little family story is true, one may imagine the consequences upon Civil War history if Maj. Gen. Hood had possibly fallen into enemy hands. The Confederates' capture of Maj. Gen. George Stoneman, even the Federals' capture of Brig. Gen. Rooney Lee, convalescing from his wound at Brandy Station, would have paled before such an event.

but the only question was when. President Davis was very mindful that by law lieutenant generals could only be appointed to fill an opening for infantry corps command. Promotion solely for gallantry and laudable service was not proper. "The recommendation to confer additional rank, as a testimonial, must have been hastily made," Davis wrote October 3 on Longstreet's letter. "The law prescribes the conditions on which Lieutenant Generals may be appointed." Indeed, on the 6th John B. Jones, the War Department clerk, recorded, "Gen. Bragg and others recommend Gen. Hood for promotion to a lieutenant-generalcy; but the President says it is impossible, as the number authorized by Congress is full." On the 14th Secretary Seddon confided to Senator Wigfall, "(*entre nous* strictly)" that "the President could not be brought to think he could confer the appointment unless he could appoint to a corps."

General Bragg continued to do his part. After he had relieved Harvey Hill of corps command on October 15, Bragg appointed Major General Breckinridge as successor. But Breckinridge, too, was one of the "anti-Bragg men" in the Army of Tennessee, and Bragg wished for him to go, too. Thus he wrote President Davis on October 23, again recommending Hood's promotion to lieutenant general, pointedly stating that Hood could take over Breckinridge's corps. More, he pointed out that Hood, besides being a true soldier, "will cordially sustain me" in the ongoing arguments among his army's high-ranking officers. Indeed, Bragg's most recent biographer, Earl J. Hess, observes that in recommending Hood's promotion, the unpopular army commander "was eager to see fresh blood infused into the poisonous atmosphere of the Army of Tennessee—fresh blood he could count on to support his tenure as its commander." Davis did not object. "I will promote Major-General Hood to be a lieutenant-general," he wrote Bragg on October 29. Bragg passed this good news on to Hood in a letter two days later: "The President has just informed me he shall promote you at my request to Cmd a Corps in this army." Bragg added that the corps in question was being temporarily led by John Breckinridge.

The real question was when Hood would be able to take to the field. Dick Ewell, whose leg was amputated after Groveton, took nine months to recover, returning to Lee's army only after Chancellorsville. If Hood needed that length of time after his amputation, he would not return to duty until summer 1864. But whereas Ewell was reluctant to return—"he should have waited a few more months," writes his recent biographer—Hood was impatient to do so. Everyone seemed to know that he would return to service in the Army of Tennessee. After the disaster at Missionary Ridge, Bragg was relieved, and Lt. Gen. William J. Hardee assumed temporary command of the army. A few weeks later Breckinridge left for Virginia and Maj. Gen. Thomas C. Hindman took charge of his corps. Hardee made sure Hindman knew that this would be only temporary, until John Bell Hood returned to the army.[81]

During winter 1863–1864, the time of his convalescence in Richmond, the best source we have for Hood's activities is not his memoir, or even his three biographies, but the diary kept by the wife of one of President Davis's aides. Mary Boykin Chesnut faithfully recorded the comings-and-goings of Richmond's high-class set, and in them Hood flowed, albeit awkwardly. Setting for much of this activity was the house on Clay and 12th streets in which the Chesnuts rented upstairs rooms. On December 9, Mary Chesnut and Mary Preston were preparing luncheon when someone called out, "Come here, Mrs. C. They are lifting Hood out of his carriage, here, at your door." Hood was coming to see Sally Buchanan Preston, "Buck," with whom he was ardently infatuated. Luncheon was served, which the general enjoyed with some German wine. "Heavens, how he was waited upon!" Mrs. Chesnut observed. Everyone, it seemed, dropped by, including the Wigfalls. Some of the ladies caught Sam's eye. To Mary Chesnut he whispered that Mary Elizabeth Randolph, wife of the former secretary of war, was "handsomer than ever."

Hood, lying on the downstairs sofa with a blanket over him, was the center of much attention. Angela Mallory, wife of the Navy secretary, called him "a martyr to his country." There were pathetic whispers, too, and some tears, as visitors could not help fixating on Hood's missing leg. "My God, it is hard!" the general confessed. "So many strangers scare me always. I can't run as I did before."

Hood indeed found it hard to fit in with Richmond's upper crust, try as he might to impress Buck Preston. At one dinner party General Simon Buckner was present, and he and Hood recalled the night they shared a campfire at Chickamauga. That afternoon they had enjoyed a carriage ride with the ladies. Hood paid an odd compliment to his fellow Kentuckian: "I prefer a

---

[81] "Arrival of General Hood," *Richmond Examiner*, November 17, 1863; Woodward, *Mary Chesnut's Civil War*, 492; Hood, *A & R*, 66–68; Thomas, *Confederate State of Richmond*, 44–45; Richard M. Lee, *General Lee's City*, 148–49; "Commutation of quarters and fuel," February 1–15, 1864, Hood Service Record; J. B. Jones, *Rebel War Clerk's Diary* (1958 ed.), 288; Davis to Bragg, October 29, 1863, *OR*, vol. 52, pt. 2, 555; S. Hood, *Lost Papers*, 191 (Seddon to Wigfall, October 14, 1863, excerpted in *A & R*, 67), 194 (Bragg to Hood, October 31, 1863); W. C. Davis, *Breckinridge*, 382 ("anti-Bragg men"), 383, 385; Crist, Williams, and Dillard, eds., *The Papers of Jefferson Davis*, 10:32; Hess, *Braxton Bragg*, 189; Pfanz, *Richard S. Ewell*, 271–74; Neal and Kremm, *The Lion of the South*, 181.

ride with you to the company of any woman in the world." Buckner replied, "I prefer your company to that of any man, certainly." "This was the standing joke of the dinner," Mrs. Chesnut wrote. "It flashed up in every form. Poor Sam got out of it badly, if he got out of it at all." Buckner explained, "Hood never gets out of any scrape; that is, unless he can fight out." Not surprisingly, Hood felt ill at ease in the merry society of Richmond's elite. At one soiree a week before Christmas, he complained, "No more parties. The country is in danger. There is too much levity here." Yet the Yuletide season brought even more socializing. On Christmas Day 1863, the Chesnuts were again entertaining. At one point James read aloud Tennyson's "The Charge of the Light Brigade." "Hood was excited beyond anything I ever imagined," Mary recorded. "He sat straight up. His eyes grew flaming, scintillating. And he made a gesture, which J. C. said was like the motion of a soldier receiving his orders in battle, at the end of every line."

That day, however, was a hard one for Hood; Buck Preston told him that she would not marry him. It seems that the refined young lady found fault with Hood's coarse ways. "She told me to say South Carolina," Hood told Mary, "to rattle the *r* and not to pronounce it as if it had two *l*'s and no *r*." Buck even objected to the way he asked for more Burgundy at dinner. "He is only a rude soldier," Buck complained to Mary. "He can't talk of literature and high art." Mary tried to pass off the cutting comment by saying that neither could she.

Fortunately, Hood the hero had won the heart of the public. Riding with Mary Chesnut to the Prestons' home on Franklin Street, a crowd of boys caught sight of the wounded general. "Boys," one cried, "here's a fellow lost his leg at Chickamauga—cheer with a will." At some point during the winter Dr. Darby returned from England with the prostheses—maybe a half-dozen wooden legs. Standing on one, with crutches, Hood allowed himself to be photographed, probably at the studio of Julian Vannerson at Main and Fourteenth streets in Richmond. Vannerson took pictures of a number of prominent Confederates, notably General Lee. "I am improving daily," he wrote Jerome Robertson on January 2, "and hope soon to be able to ride and walk again." Indeed, in her diary entry of January 18, Mrs. Chesnut noted that the *Examiner* reported that Hood had been seen in the city riding horseback.

The "gallant Hood" received adulation at all sorts of parties. "Half the youth and beauty in Richmond brooded over his convalescence," writes a historian of the wartime capital. Hood "was a lion at such affairs." At one, hosted by Mary Randolph, the entertainment featured "a series of charades in pantomime," recalled Constance Harrison, wife of the president's personal secretary. In one, Mary Preston read a poem on knighthood by John Reuben Thompson, Richmond's man of letters (and former editor of the *Southern Literary Messenger*). General Hood, seated in the front, was unprepared for Mamie's recitation:

> HOOD, whose keen sword has never known a stain,
> Whose valor brightened Chickamauga's plain;
> Well might he stand in glory's blazing roll
> To represent to future times my whole....

"Looking like the hero of a Wagner opera," Mrs. Harrison wrote, Hood "was compelled by a tumult of handclappings to arise and bow, blushing to the roots of his hair, and it was several minutes before the performance could go on."

It certainly did not hurt that the president himself took Hood under his wing. The two were often seen riding together in the city. One Sunday after services at St. Paul's, the president gave the hobbling general his arm as they walked down the church steps. The two men developed a genuine friendship during the winter. In cultivating it, Hood could be a shameless flatterer. Once when Davis complained about some of his generals, Hood asked, "Mr. President, why don't you come and lead us yourself? I would follow you to the death." To this Mary Chesnut could only comment, "if you stay here in Richmond much longer, you will grow to be a courtier. You came a rough Texian!" For dinner at the Chesnuts' in early January, General Hood arrived in the Davises' presidential carriage. A month later Hood was the honored guest at a reception given at the Confederate White House where he received "a perfect ovation," as he carefully walked his way through the crowd, assisted by Colonel Preston, Buck's father.[82]

President Davis was able to observe Hood's recovery. To mount his horse, Hood was helped by three aides. One held his crutches. Another supported him from behind as he lifted his left foot into the stirrup. A third helped fit the general's wooden leg into the other stirrup. (The prosthesis helped him keep balance in the saddle.) Then he was strapped to the saddle and was ready to ride off. Most important, the president noticed that Hood

---

[82] Woodward, *Mary Chesnut's Civil War*, 425, 502, 514–16, 527, 544, 551–52, 565, 570, 576 (Darby and legs); O'Connor, *Gallant Hood*, 215 (Prestons's home on Franklin Street); McMurry, *John Bell Hood*, 92; Hood, *Rise, Fall, and Resurrection* (cover photograph of Hood); Robertson, *Touched with Valor*, 56; Bill, *The Beleaguered City*, 188; Harrison, *Recollections Grave and Gay*, 172.

was not taking morphine or any other opiate for postsurgical pain. "Davis would almost certainly have known of any narcotic addiction through the medical officers," Dr. Paul Steiner writes. With his health so attested, the government prepared the official process of promotion to lieutenant general. On December 28, the Confederate House of Representatives unanimously passed a resolution honoring Hood, asking the speaker to invite him to attend any of its sessions.

Everyone expected Hood to return to duty. For some time, there was speculation that he might be sent farther west, to Texas. A friend, D. S. Terry, wrote him in mid-November, "I am now on my way to Texas where I hope to see you before the end of the winter." He had heard that Hood "would be raised to the grade of Lieut. Genl (which you have so nobly earned) & assigned to that department." Others suggested elsewhere in the Trans-Mississippi. Senator Augustus Garland of Arkansas wrote President Davis in late December suggesting that Hood be sent to his state, to replace either Theophilus Holmes or Sterling Price. In early February 1864, it was rumored in Richmond that Hood would be put in command of the capital's local defense troops; later came the newspaper's correction: "He will resume command in the army of Tennessee."[83]

---

[83] Sword, *Embrace an Angry Wind*, 114 (citing Confederate officer Joseph B. Cumming on Hood mounting); Dyer, *Gallant Hood*, 218, 238, 354 n.36 (Hood talking postwar of his difficulties with his artificial limb); Steiner, *Medical-Military Portraits of Union and Confederate Generals*, 229; O'Connor, *Cavalier General*, 218; S. Hood, *Lost Papers*, 195 (D. S. Terry to Hood, November 16, 1863); Garland to Davis, December 28, 1863, *OR*, vol. 22, pt. 2, 1114–15; "From Richmond," *Augusta Constitutionalist*, February 10, 1864.

In "John Bell Hood's 'Addictions' in Civil War Literature," I did my best to counter the decades of speculations that Hood continually used laudanum. I term it a "stupid, baseless derogation," in *Atlanta Will Fall*, 203. Inexplicably, the speculation has been continued by recent writers who seem to be repeating the speculation raised by older generations. One example is Alan Axelrod: "his wounded arm and the stump of his leg gave him continued agony, which he sought to dull with laudanum, an opiate that almost certainly impeded his judgment—to what degree, we cannot know" (*Generals South, Generals North*, 121). Then there is a recent biography of Forrest, in which the authors declare, "Hood's wounds should have disqualified him for command, not to mention his addiction to alcohol and laudanum" (Davison and Foxx, *Nathan Bedford Forrest*, 348). Worse is Allen C. Guelzo's statement that Hood "sustained himself on alcohol and opiates" which had the effect of "sapping his strength and deadening his judgment" (*Fateful Lightning*, 267).

If there remains any serious scholarly adherence to this gross fabrication, Stephen M. (Sam) Hood should have destroyed it in his chapter, "John Bell Hood: Laudanum, Legends, and Lore," in *Rise, Fall, and Resurrection*, 267–90. But he has not, as sadly evidenced by Robert O'Connell's recent statement that Hood was "frequently dosed with the opiate laudanum for his almost constant pain" ("Sherman Closes In," 46; *Fierce Patriot*, 145).

In January the president prepared to submit Hood's name to the Senate for confirmation as lieutenant general. By then, following Braxton Bragg's resignation as commander of the Army of Tennessee and William J. Hardee's brief tenure in that position, Davis had placed Gen. Joseph E. Johnston in command of the army. In it there was need for an infantry corps commander. "I respectfully suggest the division of this army into three corps," he wrote the president on January 2, "&...the appointment of Lieut: Generals from some other army." Two weeks later he asked Adjutant General Cooper for Maj. Gen. Mansfield Lovell, "whose military qualities I know." But Secretary of War Seddon disapproved. "It would be injudicious," recorded the War Department clerk J. B. Jones, "and it would not give confidence to the army." The president concurred; General Lovell's reputation still suffered from his command of New Orleans when the enemy took that city. Johnston gradually came around in favor of Hood for the new corps position.

Among some there was still concern, however, about Hood's ability to return to duty. On February 6, Sen. Herschel Johnson of Georgia wrote the president, "I have thought much about your sending Genl. Hood to Genl. J. I know his gallantry and his merit." Johnson added his reservations: "But is it not well to consider seriously whether he is physically equal to the arduous duties of the position?" The administration had asked Hood that very question and on January 28 Hood assured Adjutant General Cooper that he was ready, but he would rely much on his staff members. "Owing to my condition, I will need their services much more than I should have done before I received my wounds." A carriage sent to his new field of duty would also help.

The government was satisfied. The president endorsed Hood's letter to Cooper, "Propose nomination for Genl. Hood to be Lieut. Genl." The nomination was submitted; the Senate confirmed it on February 4, to rank from September 20, 1863. (The War Department often conferred promotions "to rank" from the date of an officer's signal service; such dating was important

---

Stephen (Sam) Hood has further sought to dispel the legend of Hood's laudanum use in his article "John Bell Hood: Pain & Prescriptions," which draws from his *Rise and Fall* as well as from my essay in *Blue & Gray*, October 1998. Emphasizing previously unpublished details from Dr. Darby's medical report on Hood's Chickamauga wounding and recovery, Hood concludes that Darby's report irrefutably proves that John Bell Hood did not fall victim to laudanum abuse. Darby's journal, September 20–November 24, is now published in S. Hood, *Lost Papers*, 27–38, as cited above.

Sam's reserved conclusion in *CWTI* contrasts with the magazine editors' fiery front-cover headline: "Myth Busted! New revelations of his drug use will change history." Actually, for some of us, the myth of John Bell Hood as opium-eater has long been "busted." I guess we should just be glad that the mainstream Civil War press has caught on.

in determining his seniority.) Five days later, on the 9th, came the order from the Secretary of War: "Lieut. Gen. J. B. Hood will proceed without delay to Dalton Ga., and report to General Joseph E. Johnston, commanding, for assignment to the command of an army corps." Seddon added, "He expects to leave in a few days."

Sam Hood had to say his goodbyes in Richmond. When he announced his promotion to Mary Chesnut, she noticed how his eyes "blazed." She could not help thinking, "All that ambition still—in spite of those terrible wounds." But she kept her thoughts to herself as Hood said that this was the happiest year of his life. "Again his eyes blazed up."

Hood's promotion nevertheless caused others to grumble. Jeb Stuart was one; in a letter of February 6 to his friend Congressman Alexander Boteler, Stuart complained that he had been denied promotion, "the acknowledged evidence of appreciation." At least Hood had not been given command of Lee's cavalry, as had been rumored during the past summer; Jeb consoled himself, "I know that General Lee would not have another in my place." Though he had recommended Hood after Chickamauga for a lieutenant generalship, James Longstreet is said to have remarked, "Hood is one of the best division commanders I ever knew. He would fight anybody, anywhere, at any time. But he has no more discretion than any pugnacious schoolboy might be expected to manifest." Mary Chesnut recorded conversation with her nephew who said that he had heard complaints from C. S. officers over whom Hood had been elevated. "There are all the generals whose heads he has stepped over with his one left," Johnny Chesnut said. "Already every man jack who was a general while he was a colonel grumbles and says he is going up too fast." One of these oversloughed officers may have been Daniel Harvey Hill. The president had intended to promote Hill to lieutenant general in summer 1863 but changed his mind after Chickamauga and Chattanooga. "No other Confederate officer had been so humiliated," writes his biographer. Still, Hill was very gracious about Hood's advancement. "I cordially concur in the preference for this mutilated hero," he wrote on February 23 to General Cooper, "and unhesitatingly admit his claim for promotion to be stronger than my own."[84]

---

[84] McMurry, *John Bell Hood*, 86 ("J. E. B. Stuart was disappointed that Hood was promoted to lieutenant general"); Johnston to Davis, January 2, and February 6, 1864, Crist, Williams, and Dillard, eds., *Papers of Jefferson Davis*, 10:146, 225; Johnston to Cooper, January 16, 1864, *OR*, vol. 32, pt. 2, 563–64; Jones, *Rebel War Clerk's Diary* (1958 ed.), 330–31 (entry of February 1); Bergeron, "Mansfield Lovell," 4:102; Hood to Cooper, January 28, 1863 [*sic*: 1864], Hood Service Record, National Archives (accepted February 9; confirmed by the Senate February 11); M. J. Wright, *General Officers*, 17; Special Orders No. 33, February 9, 1864, *OR*,

Most understood that Hood was ambitious and would do almost anything for fame and glory. "Maggie said people who knew General Hood before the war said that there was nothing in him," Mrs. Chesnut wrote, referring to her friend Margaret Howell, President Davis's sister-in-law. "It was only military glory." Whereupon Burton Harrison, the president's secretary, exclaimed, "Now, *only* the military glory! I like that! The glory and the fame that he has gained during the war—that is Hood."

Hood's biographers have realized that military ambition was the salient characteristic of their subject. McMurry quotes Burton Harrison, and uses his words, "The Glory & the Fame," to head his chapter on Hood in Richmond.

---

vol. 32, pt. 2 699; Woodward, *Mary Chesnut's Civil War*, 551, 557; Wert, *Cavalryman of the Lost Cause*, 329; G. C. Eggleston, *The History of the Confederate War*, 2:274 (Longstreet, "pugnacious schoolboy"); Bridges, *Lee's Maverick General*, 193–94, 254–55; Hill to Cooper, February 23, 1864, *OR*, vol. 53, 312.

It is worth noting that President Davis convened his cabinet to discuss appointing Johnston as commander of the Army of Tennessee, probably on December 2. Several secretaries voiced opposition. Years later War Secretary Seddon recalled the meeting. "A majority of the Cabinet, with Mr. Benjamin specially decided…felt at first repugnance to his selection," he recounted. President Davis voiced his own reluctance to the appointment. Seddon, despite his recollection of Johnston's "absence of enterprise" in Mississippi earlier that year, nevertheless held a high regard for the general's "military sagacity." He therefore advocated for Johnston, believing that given the opportunity he would "direct all his thoughts and skill to an offensive campaign." As other officers (such as Gen. G. T. Beauregard) were discussed and found wanting, the choice of Johnston eventually won over a majority of the secretaries (Secrist, "Prelude to the Atlanta Campaign," 18 [noting the *Richmond Whig* of December 4 reporting the cabinet meeting]; Crist, Williams, and Dillard, eds., *Papers of Jefferson Davis*, 10:97 [Davis to Seddon, December 2, meeting "to day at 12. M"]; Rowland, *Jefferson Davis Constitutionalist*, 8:351 [Seddon to W. T. Walthall, February 10, 1879]).

Johnston was not immediately notified of his appointment to army command; President Davis evidently had trouble swallowing his misgivings. Beauregard's name continued to swirl about. "The appointment of Beauregard to succeed Bragg is not officially announced," John B. Jones, the Rebel war clerk, entered into his diary on December 2, "and the programme may be changed." The very next day General Lee wrote the president, offering his own opinion on "the condition of affairs in Georgia and Tennessee." Tellingly, he did not mention Johnston as replacement for Bragg, but recommended that Beauregard, then at Charleston, be given the responsibility. As we shall see, this was not the last time Lee recommended the Creole general for command of the Army of Tennessee (*OR*, vol. 29, pt. 2, 858–59 [Lee to Davis, December 3, 1863]; J. B. Jones, *Rebel War Clerk's Diary* [1958 ed.], 314).

As for Davis's interest in promoting the maimed John Bell Hood to lieutenant general, in a paper delivered at an academic conference in 1989, Grady McWhiney wondered if "the Confederate President had a psychological affinity for sick people." Dr. McWhiney placed the president's recommendation of Hood's promotion in context of his sustained support of generals Bragg, Johnston, and even Robert E. Lee when their health hampered their performance of duty ("Jefferson Davis and Confederate Military Leadership," 21–22).

"It appears that fame and popularity had wrought a change in the man," writes Dyer. "No longer was he the gay, happy-go-lucky combat officer who asked nothing more than orders to lead his men into battle. The virus of ambition was now strong in his veins." Ambition as something slightly poisonous is also how O'Connor phrased it: "the glory-bug had bitten Hood and he was determined to gain higher command."[85]

What we do not see in the literature is how General Lee felt about Hood's promotion. The young officer had earned his fame in Lee's army and had only gone west as part of Longstreet's transfer. Now, back in Richmond and ready to return to duty, even though there was no vacancy at the corps leadership level then in the Army of Northern Virginia, Hood could probably have been reassigned back to the army if Lee had requested it. There is no evidence that he did. Let us recall that Lee had passed over Hood, naming instead Ewell and A P. Hill for corps command the previous spring. Now that the government had promoted him, it is at least plausible to infer that Lee viewed Hood as having risen to a level which Lee himself had not viewed as prudent. We shall later see more evidence of Lee's halting confidence in Hood.

Thus at the apogee of his career (McMurry's phrasing), having risen from lieutenant to lieutenant general in the thirty-three-plus months from April 1861 to February 1864, John Bell Hood had demonstrated personal strengths of courage and leadership. He had demonstrated weaknesses, too: the sloppy camp administration chided by Chilton and the lax discipline criticized by Lee himself. He had also jumped the chain of command, writing General Lee over Lieutenant General Longstreet, his corps commander. His most evident characteristic, seen by many, was his overriding ambition. It had brought him to this point, a most enviable one: command of an infantry corps in one of the Confederacy's two most important armies. Whether Hood's ambition would prove to be a heroic virtue or a tragic flaw remained to be seen. All could agree, however, that now Hood's ambition would be put to the test.

---

[85] Woodward, *Mary Chesnut's Civil War*, 568; McMurry, *John Bell Hood*, 92; Dyer, *Gallant Hood*, 225; O'Connor, *Cavalier General*, 172.

Part 1

Hood as Corps Commander, February–July 1864

Chapter 1

February–March

After Secretary Seddon wrote on February 9 that Hood would leave for Georgia "in a few days," Hood stayed in Richmond for almost another week. Mary Chesnut saw him at church on Sunday the 14th. He might have left on the 15th, as that was the day he apparently gave up the carriage which he had secured for his riding around the capital (and was reimbursed $250). He traveled by train via Columbia, where the *Daily South Carolinian* on the 19th printed an admiring column. By that time Johnston at Dalton expressed impatience for his arrival: "Lieutenant-General Hood is much wanted here," he wired General Cooper in Richmond. That day Johnston telegraphed Hood in Columbia, evidently to speed him along: "We want you much." But Hood tarried a little more, staying at the home of General and Mrs. Wade Hampton. There Mary Preston hosted a big dinner for the visiting general. Hood enjoyed the change of scenery from Richmond. "If you went to ride," he complained, "all the little boys followed as if you were the circus come to town—cheering, running alongside, saying such things." Then if he gave Buck a goodnight hand-kiss at her doorway, "as you looked up you could see eyes from every window of every house in Franklin Street peering down on you." He eventually departed Columbia and on the evening of the 23rd was in Atlanta. Two days later he reported to Johnston's headquarters at Dalton and was received warmly by his fellow officers. (General Joe Wheeler had written to Bragg on February 14 that Hood "will be appreciated by this army.") On the 28th orders went out announcing Hood's assignment to the command of Hindman's corps, consisting of the infantry divisions of Thomas Hindman, Carter Stevenson and Peter Stewart. Officers and men present for duty numbered 18,517.[1]

---

[1] Woodward, *Mary Chesnut's Civil War*, 566–67, 574–75; McMurry, *John Bell Hood*, 93; Johnston to Cooper, February 18, 1864, *OR*, vol. 32, pt. 2, 763; Johnston to Hood, February 18, *OR*, vol. 52, pt. 2, 624; Wright, *Southern Girl*, 152 (on Hood's intent to visit Hampton); *Memphis Appeal*, February 24, 1864 ("Major-General Hood arrived in this city last evening");

Up to this point in Hood's career, as he asserted later in his memoir, "my relations with my superiors, as well as with officers of lesser rank, had been of a most friendly character." Indeed, at Dalton General Johnston and Hood formed a close relationship. "General Hood writes that he dines with General Joe Johnston very often and finds him cordial and kind," recorded Mary Chesnut in her diary. Hood took to his new role quickly. On March 3 he, Johnston and Brig. Gen. William W. Mackall, the army's chief of staff, rode out to recon the area around Dalton. Three days later Johnston wrote Senator Wigfall, "Hood's arrival gave me much comfort." Three weeks later he wrote Wigfall again: "My greatest comfort since getting here—indeed the only one in a military way—was Hood's arrival."[2]

Unfortunately, a rift eventually developed between the two officers, beginning with what was probably a quiet, possibly unspoken disagreement on strategy and tactics for the army in the upcoming spring campaign. In brief, the administration—President Davis and General Bragg, his military adviser—wanted Johnston to launch an offensive into Tennessee, but Johnston saw too many difficulties standing in his way to do so. He was seriously outnumbered by the enemy forces in his front; they were so strongly fortified at Chattanooga that he would have to march around them; and he had neither enough provisions nor wagons to supply the army in a move forward. For these reasons he recommended to Davis as early as January 2, in his very first week of army command, that he first "beat the enemy when he advances, and then move forward." This was not what the president wanted to read. Some days before, Davis had written Johnston that "to regain possession of the Territory from which we have been driven" required "prompt and vigorous action." Hood was not involved in this correspondence. Davis's biographers have not documented conversations between the president and Hood during winter 1863–1864 concerning the situation in Georgia, but Hood later referred to them, and knew the president's position before he had left Richmond.

Even more, the administration's desire for "prompt and vigorous action" coincided perfectly with Hood's own aggressive temperament as a soldier.

Gen. Orders No. 26, February 25, 1864, *OR*, vol. 32, pt. 2, 804 ("Lieutenant-General Hood having arrived"); Wheeler to Bragg, February 14, 1864, T. L. Connelly, *Autumn of Glory*, 317; Special Orders No. 57, February 28, 1864, *OR*, vol. 32, pt. 2, 812; Hindman to J. P. Wilson, April 3, 1864, ("I will resume command of my division this evening"), Wilson Papers, MSS 176, box 1, folder 2; Neal and Kremm, *Lion of the South*, 191; abstract from return, March 10, 1864, *OR*, vol. 32, pt. 3, 603.

[2] Hood, *A & R*, 68; Woodward, *Mary Chesnut's Civil War*, 583; McMurry, *John Bell Hood*, 95.

The idea of sitting on the defensive awaiting the enemy's advance later in the spring was unacceptable to Hood. The year before, holding static lines at Suffolk, he had written Wigfall, "Here we are near Suffolk with the enemy confined within his stronghold, and we amusing ourselves gathering bacon & corn."

The prospect of "amusing ourselves" on the defensive again was distasteful enough for Hood on March 7, 1864, to once more write a friend in Richmond. But this time that friend was not Senator Wigfall; it was President Davis. "I am exceedingly anxious, as I expressed to you before leaving Richmond," he opened, "to have this army strengthened so as to enable us to move to the rear of the enemy and with a certainty of success." If but 10,000 or 15,000 reinforcements arrived—say, from Polk in Mississippi—the Army of Tennessee would number "60,000 or 70,000 men, which, I think should be sufficient to defeat and destroy all the Federals on this side of the Ohio River." "I am eager for us to take the initiative," he affirmed.

Such confidence, contrasted with Johnston's pessimism and hesitation, must surely have pleased Davis. Hood, "the awkward flatterer," as Mrs. Chesnut called him, may have deliberately told the president what he wanted to hear in order to rise further still in his estimation. In any case, without doubt such expressions were at Johnston's expense. There is no record of a written reply by Davis, which can be taken to mean that he did not regard Hood's writing as censurable. Indeed, as William C. Davis has reckoned, the president may have welcomed a high-level source of information from Johnston's army, even though "Davis probably did not ask it." And as far as Hood's swipe at his superior officer, Davis judges that "the president would see in Hood's carping commentary on Johnston nothing more than a brave fighting general's frustration with a hesitating commander. Davis never recognized Hood's underlying ambition."

More recent presidential biographer William J. Cooper states that during their winter in Richmond, Hood "talked to Davis about the need for an offensive course in Georgia." As for Hood's going over Johnston's head in writing the president, Cooper asserts, "surely Hood's conduct was unprofessional, but the administration shared responsibility," because neither the president, nor General Bragg nor Secretary Seddon—whom Hood would also write in the coming weeks—"ever alerted Johnston about Hood's back-door dispatches."[3]

---

[3] Hallock, *Braxton Bragg and Confederate Defeat*, 2:163 (Bragg as military adviser, February 24, 1864); S. Davis, *Atlanta Will Fall*, 27–28; Johnston to Davis, January 2, 1864, *OR*, vol. 32, pt. 2, 511; Davis to Johnston, December 23, 1863, Crist, Williams, and Dillard, eds.,

In his March 7 letter to the president, Hood included a rather uncomplimentary remark regarding his command: "The divisions of Stewart, Stevenson, and Hindman make up my corps. You perceive I have all the untried troops of this army; I hope however, to do good work." Hood expounded upon this in a letter he wrote Col. James Chesnut that same day: "I enclose to you a copy of a letter I have written to the President. I have written this from my deep interest in our affairs throughout our country." Then he added, "My Corps is composed of all the untried troops of this army. I have all the Vicksburg warriors, the Divisions I command, are Stuarts Stevensons & Hindmans. So if I dont do as well as may be expected, I wish my friends to know, what troops I command...."

The new corps commander was incorrect in his characterization of his troops as "all the untried troops of this army." All four brigades of Hindman's division had fought more than a year before at Murfreesboro. Those four brigades, then part of Maj. Gen. Jones Withers's division, suffered 2,500 casualties. Hood was also wrong that he had "all the Vicksburg warriors." To be sure, some four of his brigades had surrendered with the river garrison and had been paroled (along with Maj. Gen. Carter Stevenson, now one of Hood's division commanders). But these were less than a third of the 29,000 officers and men who capitulated on July 4, 1863. Finally, in his letters to Davis and Chesnut, Hood failed to take note of the fighting ardor and patriotism of his soldiers. In winter 1864, as Hood was assuming command of his corps in Georgia, many of his troops volunteered to re-enlist for the war, resolving to do so with martial resolutions adopted by their regiments. In Manigault's brigade, three-fourths of the 10th and 19th South Carolina regiments re-enlisted for the duration, as did all but thirty men of the 28th Alabama. In Baker's brigade Alabamans of the 42nd Infantry resolved to continue resisting "the wicked and Abolitionist majority" of the old Union. Soldiers in Deas's brigade took vows against "base submission" to the enemy who would enslave them.[4]

Papers of Jefferson Davis, 10:120–21; Hood to Wigfall, April 22, 1863, S. Hood, *Lost Papers*, 143; Hood to Davis, March 7, 1864, *OR*, vol. 32, pt. 3, 606–607 ("I am eager for us to take the initiative"); Woodward, *Mary Chesnut's Civil War*, 519; Davis, *Man and His Hour*, 548; W. J. Cooper, *Jefferson Davis, American*, 474.

[4] Hood to Chestnut, March 7, 1864, GLC #02703, Gilder Lehrman Institute of American History; Cozzens, *No Better Place to Die*, 109–11, 125–25, 152–53 (Hindman's brigades at Murfreesboro); "Return of Casualties in the Confederate Forces" (Murfreesboro), *OR*, vol. 20, pt. 1, 678; Fullenkamp, Bowman, and Luvaas, *Guide to the Vicksburg Campaign*, 468–49; "Consolidated statement of prisoners of war captured and paroled" (Vicksburg), *OR*, vol. 24, pt. 2, 325; Bohannon, "'Witness to the Redemption of the Army,'" 114–15.

Hood's letter to the president had probably not arrived in Richmond before he wrote both General Bragg and Secretary Seddon. Under date of March 10, Hood repeated the argument he had made to Davis. "I fear more of our territory will be given up to the enemy.... If we can concentrate and fall upon the enemy before he is ready we shall beat him badly" (to Bragg). "We have a sufficient number of troops, if thrown together, to defeat his entire army" (to Seddon). Hood enclosed to the secretary copy of his letters to Davis and Bragg, so as to reinforce his points. "I am an earnest friend to the President," he added to Seddon (as if to justify his extraordinary correspondence), "and am ever willing to express to him my ideas in regard to the approaching campaign."[5]

There is no record of a reply from either Bragg or Seddon back to Hood. In truth, officers in the Army of Tennessee sent letters to Bragg in such number as to lead one to assume that everyone thought the practice acceptable. Scrolling through the Bragg Papers at the Western Reserve Historical Society, one sees that from the time of his relief from command, November 30, 1863, to the time of his assumption of his role as President Davis's military adviser, February 24, 1864, Bragg received at least five letters from general officers of the Army of Tennessee. Encouraging this epistolary stream were not only the genuine bonds of friendship between these officers and Bragg—very effusively expressed, given Bragg's embarrassing departure from the army—but also Bragg's essentially officeless position during these three months, which he spent in convalescence and leisure at Warm Springs, Georgia, and other places. Thus Brig. Gen. John K. Jackson wrote him on January 10, and Brig. Gen. Edward Walthall on the 21st, trying to set the record straight about their roles in the disaster at Missionary Ridge. On January 25, Brigadier William B. Bate wrote him as a friend, consoling him about his recent humiliation. "I thought at the time and think now you erred in asking to be relieved from command," Bate asserted, acknowledging that within the Army of Tennessee, "you had enemies as every positive & independent character must have." Still, a number of officers continued to support him. Moreover, Bate added, "the same officers who sustained you are the ones on whom

---

Hood's letter to Colonel Chesnut has generally gone unnoticed in the literature. Its first page is pictured in Denkler, *Voices of the Civil War*, 17. McMurry does not mention it in his biography. In her diary, under date of March 24, 1864, Mary Chesnut wrote, "Read Hood's letter to J.C.," but editor Woodward adds no explanatory note. I express my appreciation to Alinda Borell, archivist at the Gilder Lehrman, for her help in obtaining the text of Hood's letter.

[5] Hood to Bragg, March 10, 1864, and to Seddon, March 10, *OR*, vol. 32, pt. 3, 606–608.

Genl Johnston will *ultimately* rely." One of those sustainers was cavalry commander Maj. Gen. Joseph Wheeler, who on February 14 wrote Bragg that many in the army were pleased at his appointment as Davis's adviser: "you will see that this army is not so much neglected."

After Bragg assumed his new office in Richmond, the letters kept coming. A notorious example of a general going around Johnston and sending letters to Richmond involved Maj. Gen. William H. T. Walker. The latter had alerted the administration in January of Pat Cleburne's proposal, announced at a meeting of Johnston's generals, that slaves should be encouraged to fight as C. S. troops and then freed after the war. Walker had asked Johnston's permission to send Cleburne's paper to Richmond. Johnston refused, but Walker sent it anyway—directly to the president. Walker explained the whole incident in a letter to Bragg of March 8. The next day Brig. Gen. States Rights Gist also wrote Bragg about "the 'Abolition Document.'" "It is really a rich affair," he observed, "and I am delighted beyond expression to know that the traitors will meet with their just deserts." Back-biting among officers in the Army of Tennessee was a long tradition, and Gist could not refrain from snitching: "You will see that Hindman is one of the Chief Offenders, and is I think the most dangerous man of all."

Lt. Col. Felix Robertson, commanding Wheeler's horse artillery, felt obliged in late January 1864 to give General Bragg a report on the condition of the army. "The Army here is in a fair state. Its *morale* seems good," he wrote. "The troops are undoubtedly in good condition and I am glad to say general confidence exists in Gen Johnston." Other officers wrote Bragg asking for favors. Patton Anderson, promoted to major general in February and sent to Florida, stated under date of March 8 that he looked to the reinforcement of Johnston's army before the spring campaign opened. "I would like above all things," he added, "to take a Division to his Army and share with them the toils of the coming campaign." (Anderson would return to the army later that summer.) Col. Alpheus Baker wrote Bragg on March 2, asking for a brigade command; Bragg evidently helped his friend with a brigadier's commission effective March 5.

Apparently, some of Johnston's officers, and not just Hood, knew that Johnston was making a case against the Richmond authorities' hopes that he and his army take offensive action in the spring. Several correspondents emphasized that the Army of Tennessee was in good condition and ready to fight. "The Army is in line, ready for the enemy," wrote General Gist on February 27; "they will make a good fight." Robertson, commanding artillery attached to Wheeler's cavalry, claimed his troops were ready "to do good service." Bate wrote that his division was "in good condition." Generals expressed their desire for the army to take the offensive in various ways. On March 17, General Bate wrote Bragg, "Gen'l Wheeler…informed me that there is a Division and a half of Infantry & a small force of cavalry of the enemy at Ringgold, & about the same number of each at Cleveland. It seems to me, one or the other party might be captured such is his opinion too—I suggested the propriety of his having an interview with Gen'l Johnston in relation to it—he left me with the intention of doing so." On March 19 Stewart stated that the army as a whole was in "fine condition." "We all flatter ourselves that through your instrumentality," he added, "our army will be increased until it can take the initiative in the campaign and become aggressive—and that all our future operations will be on the offensive." Then, in a more exasperated tone, he asked, "Are we to hold still, remaining on the defensive in this position until Grant comes down with his combined armies to drive us out?"

Cavalry chief Joe Wheeler sent four letters to Bragg before the campaign had begun, during February–April 1864. In his of March 3, he made a point of informing Bragg of the positions of the enemy's various infantry corps, adding, "I will write to you from time to time but beg you will not take the time and trouble to acknowledge them as I know you must be very much occupied in getting matters arranged in your departments." He continued the stream of information, as when he wrote on April 16, "all of Howard's Corps are now at Cleveland." Wheeler seemed to be making the same point as Hood: if reinforced, the Army of Tennessee could take the offensive. Yet Hood's letters would have been more damaging to Johnston than those of other officers, given that he knew he had the ear of the president. "During the spring of 1864 Hood did more than all other correspondents from the army," writes Connelly, "to discredit Johnston's version of conditions at Dalton." Hood's motive was self-serving, he continues: "His extreme ambition was recognized even by his friends."[6]

---

[6] Hood, *Rise, Fall, and Resurrection*, 23 ("other Johnston subordinates also corresponded with Johnston's superiors"); Jackson to Bragg, January 10, 1864; Walthall to Bragg, January 21, 1864; Bate to Bragg, January 25 and March 17, 1864; Wheeler to Bragg, February 14, March 3 and April 16, 1864; Gist to Bragg, February 27 and March 9, 1864; Robertson to Bragg, January 28, 1864; Anderson to Bragg, March 8, 1864; Baker to Bragg, March 2, 1864; Stewart to Bragg, March 19, 1864, Braxton Bragg Papers, box 2, folder 12; I thank Anne Bridges of the Hodges Library and Helen Matthews of the Atlanta History Center for making the invaluable Western Reserve Bragg Papers available to me on microfilm through interlibrary loan); McMurry, "Patton Anderson: Major General, C. S. A.," 15–16; S. Davis, "'That Extraordinary Document'," 18; M. M. Hull, "Concerning the Emancipation of the Slaves," 151–52; T. L.

Thus it became evident that the Army of Tennessee would stay on the defensive until the enemy advanced against it. Moreover, there were unsettling signs that General Johnston was looking to his rear even before the campaign had begun. On March 12, Capt. Lemuel P. Grant, supervising construction of Atlanta's fortifications, received orders that the next morning General Johnston was traveling from Dalton to Atlanta to "make an inspection of the works in this vicinity"; Grant was to guide the tour. The advent of the campaign was almost two months away, and already Joe Johnston was inspecting the fortifications almost a hundred miles to his rear. Captain Grant got other hints that Johnston was looking to his rear when in April 1864—again, before the campaign had even begun—Johnston began calling for the engineers to begin constructing duplicate bridges, able to be hauled to the rivers between Dalton and Atlanta over which the (retreating) army would have to cross.

Before all that happened, the men needed drilling, training and disciplining. Within several days of taking command of his corps, Hood announced that there would be days of hard work ahead: "two thirds of the elements of success in battle consist in preparation for it." On March 7 a circular prescribed tactics to be used in battle. Orders a few days later called for regulation Confederate battle flags to be displayed by all units. A few days after that, instructions were issued on what to do with captured guns and colors—always highly coveted trophies. On March 17, more instructions appeared on how infantry in the open field were to break the enemy's lines. Hood was clearly showing his troops that he gave little thought to defensive tactics.[7]

Connelly, *Autumn of Glory*, 317–21; Hallock, *Braxton Bragg*, 167–68; McMurry, *John Bell Hood*, 97–98; Elliott, *Soldier of Tennessee*, 171.

After the war, some Confederates criticized Johnston for not having assumed the spring offensive that the administration wanted. Victor Rose, who had ridden with Sul Ross's cavalry in the Atlanta Campaign, asserted in his history of the Texas Brigade during the war, "the end came; and the commencement of the end dates from the day that General Johnston inaugurated his ignoble retreat by retiring from Dalton, Georgia. Had he assumed the offensive there, the Confederacy would have been spared the sad catastrophe that befell it" (Rose, *Ross' Texas Brigade*, 170).

[7] S. Davis, *Atlanta Will Fall*, 34; L. P. Grant to Lt. Col. A. L. Rives, Actg. Chf. Engr. Bureau, Richmond, April 4, 1864 ("duplicate bridges are now being framed under my direction from the W. & A. R.R."), MSS 100, box 7, folder 5 (Letters Received, 1864), Lemuel Grant Papers; Lash, "A Yankee Bridge Builder for the Confederacy, 1862–1865," 123–24; Hood to Davis, March 7, 1864, *OR*, vol. 32, pt. 3, 607; McMurry, *John Bell Hood*, 98–99; Gen. Orders No. 31, March 1, 1864; No. 36, March 7; No. 38, March 11; No. 40, March 16; No. 41, March 17, *OR*, vol. 32, pt. 3, 575, 592–93, 612–13, 644, 647; Buell, *The Warrior Generals*, 352.

The soldiers liked this kind of attention and liked even more the general who bestowed it upon them. They did not even mind all the drilling that Hood ordered. A Southern soldier calling himself "Lone Jack" wrote from Dalton on March 22: "Brigade drills are of daily occurrence. Division ditto semi-weekly, and corps drills bi-frequently, not to mention battalion and company drills, weekly and occasional reviews, and four roll calls a day, wherein all men turn out under arms," the correspondent explained. "Not withstanding—or rather because of—all this, Generals Johnston and Hood are exceedingly popular." For the new corps commander recently arrived from Virginia, to be matched in popularity with the army's commanding general was certainly a compliment to Hood. Some even complimented the way he looked. Within a week of Hood's assumption of command, Neil Gillis wrote home that the new lieutenant general "reviewed us to-day, he is a fine looking officer." On March 6 Dr. George W. Peddy, surgeon in Stevenson's division, wrote his wife, Kate, that Hood "is a good officer & the finest looking man I think I ever saw."

Pvt. Philip D. Stephenson of Slocomb's Louisiana Battery, attached to Bate's division, recalled Hood vividly in his memoir, written thirty years after the war:

> Hood's personal appearance was commanding and striking. He was over six feet and of splendid proportions. His hair and beard were a light yellow and worn long. The eyes were large and bold, but of a singularly light grey. The general effect of his presence was impressive.... With his yellow waving hair and great tawny beard and big, bold blue eyes and herculean frame he looked like a gigantic old Saxon chieftain come to life again.

Stephenson noted in the two months or so from Hood's arrival in Dalton to the start of the campaign, that the "junior corps commander" became very popular with the men. "Even Hardee, our old stand by, the Old Reliable of the army," Stephenson commented, "was for the moment overshadowed." "Indeed," he added, "Hood, for a time, was looked upon by the men as Johnston's ablest officer."

Hood made a point of riding and being seen on horseback among the troops. One officer wrote that Hood "has a wooden or cork leg, but he wears a boot and spur on it as he does on the other." General Wheeler wrote to Bragg on March 3 that Hood "is doing well. He rides fourteen or fifteen miles a day without much fatigue." Halsey Wigfall, the senator's son, was now on Hood's staff and accompanied the general on his rides. "He brought a carriage up from Atlanta when he came," Wigfall wrote on March 21, "but has sent it back and rides everywhere on horseback." On the 24th Hood wired Col. Moses H. Wright in Atlanta, asking that he have saddles sent to him in Dalton. He even released to the press the text of a letter he had written a friend, conveying his ease on horseback. "I am told some of the old women in trowsers in our country fear I am not in good health and that I have to be tied or fastened on my horse, etc.," he began. But Hood explained that he had been "riding all over this country with Gen. Johnston...without feeling any inconvenience whatever from it....I tell you I am in good health and as young as ever, and am as ready and in as good condition to fight in battle as I ever was; so do not trouble yourself about me."

The men loved this kind of spirit, and it showed. On March 16 Hood held a grand review of his corps. Samuel C. Reid, writing as "290" for the *Atlanta Register*, described the event. Despite the cold weather, "a number of ladies were in attendance," and so were a number of general officers. "The evolutions and discipline of the various regiments evinced a high state of *esprit du corps*," Reid noticed. The reporter waxed further, "Every man evinced the ring of the true metal in him, and seemed panting but for a chance to 'Lay the summer's dust with showers of blood/Rained from the wounds of slaughter'd Abolitionists.'" After the review, a sham battle was held, with loud musket volleys and artillery blasts "waking up the enemy's scouts," Reid reckoned. Then, as another recorded, "Hood rode through the corps, amid deafening cheers." Hood also rode in the grand review of the army held at Dalton on April 19. Each corps commander—Hardee and Hood—rode with his staff at the front of his troops; looking on, according to one Confederate, was "General Johnston, dressed in an elegant new uniform and mounted on a superb, thoroughbred, dapple gray mare."[8]

---

[8] McMurry, *John Bell Hood*, 99–100; "A Letter from Lone Jack," *Atlanta Intelligencer*, April 11, 1864; Neil Gillis to Mary, March 4, 1864, #300, microfilm reel 1, MARBL, Gillis Family Papers; Peddy, *Saddle Bag and Spinning Wheel*, 224; Stephenson, *Memoir*, 209–210; Wheeler to Bragg, March 3, 1864, Bragg Papers, Western Reserve; Hood to Wigfall, April 5, 1864, S. Hood, *Lost Papers*, 146 ("Halsey is near me at present & well"); O'Connor, *Cavalier General*, 189; Wright, *Southern Girl*, 173; "General Hood in the Saddle," *Memphis Appeal*, March 14, 1864; "From Our Special Army Correspondent '290.' Grand Review of Hood's Corps" (Tunnel Hill, March 17), *Atlanta Register*, March 19; "TELEGRAPHIC. FROM THE ARMY OF TENNESSEE" (Dalton, March 16), *Atlanta Register*, March 18; Roberts, "Review of the Army of Tennessee at Dalton, GA.," 150.

According to Mary Chesnut, Hood was angry that his "women in trowsers" letter was published—a statement that is puzzling given Hood's obvious intent to spread word about his fitness for field service (Woodward, *Mary Chesnut's Civil War*, 592).

Hood's popularity extended beyond the Army of Tennessee. In spring 1864, someone sent him a cane carved from a Fort Sumter flagstaff, which had been shot down during the bombardment by Union gunboats (Connelly and Jones, *Politics of Command*, 85).

Chapter 2

April

The spring campaign would not begin, however, without one more round of sneaky letter-writing by Hood and Bragg. This time, in early April, Hood was even more critical of General Johnston than he had been the month before. On April 3 Hood thanked Bragg and the president for their willingness to reinforce Johnston if he launched offensive operations. "How unfortunate for our Country," he wrote, "for the Generals in the field to fail to Cooperate thoroughly with the Authorities of the Government and not act together." "You know how anxious I have been to advance," he added, with this for good measure: "You know I am fond of large engagements and hope you will not forget me." Ten days later, after he had apparently received a letter not only from Bragg but also General Lee, Hood was even more critical. "I have done all in my power to induce General Johnston to accept the proposition you made to move forward," he wrote Bragg on April 13; "He will not consent." Hood made clear where he stood on the issue. "I regret this exceedingly, as my heart was fixed upon our going to the front and regaining Tennessee and Kentucky." Hood stated that on the 12th he had received a letter from General Lee in which Lee also called for "driving the enemy in your front to the Ohio River." If this were not enough, Hood added that General Hardee had also written a long letter to President Davis. A decade later, writing his memoirs, Hood did not apologize for his correspondence with the higher-ups, and in *Advance and Retreat* even included the texts of his letters to Davis, March 7, and to Bragg, April 13.

Meanwhile Hood and Senator Wigfall were conducting their own correspondence. After Wigfall wrote him on March 29, Hood replied on April 5, beginning with "I hasten to answer your questions which must be purely between us." Wigfall in Richmond also wanted to know why Johnston was refusing to lead his army in a forward movement. "Genl J__ talks quite freely with me," Hood wrote, so he knew the situation. The Richmond authorities promised to reinforce Johnston with 10,000 troops from Beauregard's coastal garrisons and another 5,000 from Polk in Alabama and Mississippi. Then, if he could unite with Longstreet in east Tennessee, Johnston would have "an army of about 75,000, which I think enough to thrash all the Yankees from here to the Ohio River." Johnston, however, was of no mind to do this.

> While J__ did not entirely refuse this he failed to meet it with his arms open. He proposed to have all the troops assemble here, and allow him to move to the left in lieu of the right. But he wished (as he believed would be the case) the enemy to advance on him. I think he should have said to Bragg, "I am delighted to have the chance to regain Tenn & Ky and am ready to move as you propose, but would prefer the other route." So Bragg replied that as he had failed to accept the plan to move forward, nothing could be done, and no troops furnished.

"You know I urged all this before I left Richmond," Hood added, further confirming the probability that Hood and President Davis had discussed the western situation before Hood had left for Georgia. Further, Hood offered his critique of Johnston: "He is too petulant to administer an army well." Hood preferred General Lee much more. "There is more fight in Lee than J__." As things stood, Hood closed, "I am & have been very much worried about our affairs."

So Hood, Wheeler, Hardee, and other generals were writing Davis, Bragg, Seddon, and Wigfall, all complaining about Joe Johnston, but Hood was hurting him the most. For good reason John P. Dyer observes, "Had Johnston known about this correspondence he might justifiably have asked who was commanding the army, he or Hood."[1]

Indeed, the general in control of the situation in north Georgia was neither Joseph E. Johnston nor John B. Hood, but William T. Sherman. In March, President Lincoln had brought Ulysses S. Grant east, promoted him to lieutenant general, and made him general-in-chief of all Union armies.

---

[1] Hood to Bragg, April 3, Bragg papers, Western Reserve (quoted in McMurry, *John Bell Hood*, 96–97); Hood, *A & R*, 92–94 (Hood to Davis, March 7, and to Bragg, April 13, 1864, which are also in *OR*, vol. 32, pt. 3, 606–7 and 781); Hood to Wigfall, April 5, 1864, S. Hood, *Lost Papers*, 145; Dyer, *Gallant Hood*, 231.

I am indebted to Sam Hood for transcribed copy of the Hood to Wigfall letter, which at the time of this writing had not been published; it now appears in *Lost Papers* as cited above. Louise ("Louly") Wigfall Wright had it among her father's papers when, decades after the war, probably between 1897 and 1915, she sent it to Lillian Hood Post, one of Hood's daughters. Because the letter was "of an intimate personal nature," Mrs. Wright did not use it in her memoir, *A Southern Girl in '61* (1905), "but I feel that I am safe in giving it to you," she explained in her accompanying letter to Mrs. Post (Louise Wigfall Wright to Lillian Hood Post, March 6 [c. 1905], S. Hood, *Lost Papers*, 141).

Grant in turn left Maj. Gen. Cump Sherman in command of the Military Division of the Mississippi, all Federal forces between the Appalachians and the Mississippi River (actually beyond it). Sherman's objective in the upcoming campaign, as Grant wrote it out on April 4, was "to move against Johnston's army, to break it up and to get into the interior of the enemy's country as far as you can, inflicting all the damage you can against their war resources."

In preparation, Sherman was able to move troops about and concentrate them in ways that Confederate leaders in the western theater could only dream about. As a result, around Chattanooga in spring 1864, Sherman had organized 110,000 men and 254 cannon in three armies: George Thomas's Army of the Cumberland, Maj. Gen. James B. McPherson's Army of the Tennessee, and Maj. Gen. John M. Schofield's Army of the Ohio. Moreover, Sherman was a logistical master, particularly in employing railroads to supply his forces. The story may be apocryphal, but it is realistic: how Sherman addressed a quartermaster in Nashville before the campaign opened. "I'm going to move on Joe Johnston the day Grant telegraphs me he is going to hit Bobby Lee," Sherman snapped, "and if you don't have my army supplied, and keep it supplied, we'll eat your mules up, sir—eat your mules up!"[2]

Against Sherman's army group, Johnston had half as many troops: in its returns of April 30, the Army of Tennessee numbered 54,500 men present for duty: 42,400 infantry split between Hood's and Hardee's corps; 7,800 cavalry, about 3,200 cannoneers for 144 guns (and another 1,100 staff, escort, and engineers). He had them deployed in a strong defensive position along Rocky Face Ridge, west and north of Dalton. Johnston and his soldiers knew they had hard work ahead. On April 28, Confederate Andrew Edge wrote his wife that General Johnston had prohibited any further furloughs. "He sed he kneaded all men he had know in front tha think that we will have a big fight

---

[2] Johnston report, October 20, 1864, *OR*, vol. 38, pt. 3, 614 (Mercer's brigade); Jenkins, "Dalton: The Opening of the Georgia Campaign," 42–43, 49; Mackall, *A Son's Recollections of His Father*, 208–209 (letters of May 2, May 12); Castel, *Decision in the West*, 54, 132–50; Sherman report, September 15, 1864, *OR*, vol. 38, pt. 1, 64; Sherman, "The Grand Strategy of the Last Year of the War," 4:252; Bogle, "The Western & Atlantic Railroad and Its Boom Town," 14; Cohen and Bogle, *The General & the Texas*, 148 (photograph of present-day tunnel); Bonds, "The W & A: Sherman's Lifeline," 35 (the Chetoogeta as the "first railroad tunnel in the South"); S. Davis, *Atlanta Will Fall*, 21; Castel, *Decision in the West*, 129; McDonough and Jones, *War So Terrible*, 96; Longacre, *Cavalry of the Heartland*, 270; McMurry, *Atlanta 1864*, 211 n.1; Butkovich, *The Battle of Pickett's Mill*, 33–34; Lash, *Destroyer of the Iron Horse*, 104–30; The "W. and A. Railroad," *Memphis Appeal*, May 19, 1864; McMurry, "Atlanta Campaign: Rocky Face to the Dallas Line," 17; D. A. Brown, *The Bold Cavaliers*, 107; Mackall to Wheeler, May 7, *OR*, vol. 38, pt. 3, 672–73.

here rit off." On May 1 the commanding general suggested that the ladies who had been sharing quarters with their officer-husbands should leave.

One of the departees was Mary Hardee, the lieutenant general's bride (they had been married a few months before). She left on May 5 for LaGrange, Georgia.[3]

[3] S. Davis, *Atlanta Will Fall*, 19–20, 26, 39; Grant to Sherman, April 4, 1864, *OR*, vol. 32, pt. 3, 246; "Abstract from returns showing the effective strength of the army in the field under Maj. Gen. William T. Sherman, during the campaign against Atlanta," April 30, *OR*, vol. 38, pt. 1, 115 (110,123 men); Sherman report, September 15, *OR*, vol. 38, pt. 1, 62–63 (98,797 men at the start of the campaign, which some sources accept); Sherman to Grant, May 4, 1864, *OR*, vol. 38, pt. 4, 25; Burt, "Sherman, Railroad General," 49; "Abstract from returns of the Army of Tennessee," April 30, *OR*, vol. 38, pt. 3, 676; McMurry, *John Bell Hood*, 101–102; Andrew P. Edge to wife, April 28, 1864, Edge Correspondence; Hughes, *Hardee*, 198.

Grant's omission of *Atlanta* as Sherman's objective in his orders of April 4 has led some students to call the Federal offensive the *Georgia* campaign. During it, however, at least some Federals began calling their operation the "Atlanta campaign." Capt. Henry Dwight of the 20th Ohio, writing a column that would be published in *Harper's* in August 1864, described the manner in which his comrades constructed field fortifications: "thus at least we fight in Georgia, in the Atlanta campaign" (Dwight, "How We Fight at Atlanta," 940).

Chapter 3

May

The enemy was visibly very busy on the front as Sherman completed preparations for the onset of his campaign. On the night of May 4 he wrote Grant, "Everything very quiet with the enemy. Johnston evidently awaits my initiative." Johnston was also waiting for reinforcements. On May 4 he learned that the administration was going to send troops from Polk's department. They would travel by train to Blue Mountain in northeast Alabama, then march via Rome to join the army. Other troops came from coastal garrisons. Brig. Gen. Hugh W. Mercer's brigade of 1,400 infantry arrived at Dalton on May 2.

The Confederate line ran to seven miles, north-south along Rocky Face, then perpendicular to it, eastward into Crow Valley. Johnston believed Sherman would demonstrate against the ridge but throw his main force against the Confederate right. Lt. Thomas B. Mackall, staff officer, recorded in his diary on May 5, "Scouts think attack will be on our right." It was therefore quite telling that Johnston placed Hood's three divisions in the more threatened sector of his line, the right. Not only that, but he temporarily assigned two of Hardee's divisions to Hood for strengthening that sector. From this demonstration of trust and confidence it may be inferred that Johnston was unaware that Hood had been criticizing him in his private letters to Richmond. It may even be inferred that Johnston esteemed Hood more highly than his other corps commander, Lieutenant General Hardee, who had been with the Army of Tennessee from its earliest days.

Johnston was wrong as to Sherman's intent: he planned to dislodge Johnston not on the Confederates' right, but on their left. During the winter General Thomas had learned that twelve miles south of Dalton was a cut through the ridge, Snake Creek Gap, which offered a good route for flanking the Rebels' left. Sherman intended to send McPherson's army—22,300 infantry of the XV and XVI Corps—through Snake Creek with instructions to march east and cut the Western & Atlantic near Resaca, a railroad town about a dozen miles south of Dalton. When it did so, Johnston would have to retreat. While McPherson made his flank march, Thomas and Schofield were to keep the Rebels' attention on them with feints and demonstrations.

Sherman's plan worked very well—up to a point. While Thomas's three corps, the IV, XIV and XX, approached Rocky Face from the northwest and Schofield's XXIII Corps from the north during May 5–7, McPherson advanced toward Snake Creek Gap. On the 8th, while Schofield's Federals skirmished with Hood's troops and Thomas on the other end of the line launched a sharp demonstration at Dug Gap, McPherson's advance troops occupied Snake Creek Gap. The next morning, they pushed back some Confederate cavalry and marched toward the railroad. But McPherson lost his nerve, fearing a Rebel attack, and called his columns back to the gap.

Johnston spent May 9 with Hood well to the north, observing his men easily turning back several demonstrations against the Confederate right. ("On the 9th five assaults were made on Lieutenant-General Hood's troops," Johnston later wrote; "all were repulsed.") Reports of McPherson's movements, however, forced the commanding general to look to his other flank. Just as he had called on Hood for trouble on the right, now Johnston called on him for trouble on the left, ordering him to take the train to Resaca, and sending three divisions to march and join him there. At 8 A.M. on the 10th, Hood wired from Resaca that the enemy had withdrawn: "R all right. Hold on to Dalton." Johnston accordingly ordered Hood to return to the main force. As for any remaining threat against Resaca, Polk and his infantry were expected any day. On the 10th Lieutenant General Polk reached Rome. Johnston ordered him to Resaca and to take command there. The next day Johnston finally concluded that Resaca was Sherman's objective and began shifting his army there. The troop movement continued well into May 12, the day on which General Mackall wrote to his wife, "don't be made unhappy if reports are put in circulation that we are retreating." That evening Johnston took one of the last trains south from Dalton. By that time Sherman had met with McPherson and rebuked him for his timidity. "Well, Mac, you have missed the opportunity of a lifetime." "I was somewhat disappointed at the result," he later wrote, with considerable understatement. But Sherman got over it, ordered McPherson's force back through the gap toward Resaca, and directed Thomas's and Schofield's to follow.

Given the importance of the Western & Atlantic Railroad for both armies, students of the Atlanta Campaign have long wondered why General Johnston did not order the destruction of the 1,477-foot tunnel through Chetoogeta Mountain north of Dalton. Sherman had calculated that a soldier needed two pounds of food a day, and a horse ten times that amount. To bear that quantity of provisions for 100,000 men and 50,000 animals (not to mention ammunition and other supplies), Sherman reckoned on the need for at least 65 to 80 cars daily bringing freight and food from the army's supply base at Nashville. Demolition or blockage of the tunnel would have stopped this logistical flow. In the opening days of Sherman's advance, Confederate forces withdrew from Tunnel Hill. Troops of the Federal XIV Corps occupied the area on May 7. "The tunnel itself, much to Sherman's relief, has not been either blocked or destroyed; only the track has been removed," writes Albert Castel.

Wheeler's cavalry, the last Southerners to pass through the tunnel, have sometimes been blamed for not having blown it up. "Joseph Wheeler's failure to blow up the tunnel when the Federal advance began, which he assured Johnston would be done, constituted a major shortcoming of 'Fighting Joe's' Civil War career," write McDonough and Jones. Similarly, Edward G. Longacre has recently observed that Wheeler's abandonment of Tunnel Hill on May 7 was done "so precipitately that he lacked the time to close the railroad tunnel by detonating munitions at its mouth, as he had intended."

Richard McMurry shrewdly discerns that all the writing about Tunnel Hill ignores the fact that if the Confederates had indeed blocked the Chetoogeta underpass, Sherman had an alternative logistical resource: the East Tennessee & Georgia Railroad, which ran from Cleveland, Tennessee, to Dalton, bypassing Tunnel Hill.

Yet the literature focuses on the unblown tunnel. If there is any blame to be laid for leaving it operable, it should not go to Wheeler, but to General Johnston himself. Johnston's strategy, articulated to the Richmond authorities throughout the winter, had always been predicated on retreating from his Rocky Face Ridge position, so abandoning Tunnel Hill would not have been a sudden decision. Brad Butkovich is one recent writer who faults Johnston for failing to destroy the tunnel, which "would have significantly hindered Sherman's ability to supply the Federal forces south of Chattanooga."

Johnston may have had several reasons to leave the tunnel intact. The Western & Atlantic Railroad was owned by the State of Georgia, and, throughout winter 1863–1864, Johnston and Georgia governor Joseph E. Brown had been in frequent communication about the railroad's logistical service to the Army of Tennessee. Johnston would have known that the Chetoogeta tunnel was a prized possession of Joe Brown. It had taken the state almost two years to build, and its completion was the reason that the W. & A. did not begin operation until May 1850. In not ordering its destruction, Johnston may not have wanted to anger Governor Brown. Besides, there was the remote possibility that the Confederates could defeat Sherman, push him back, and reoccupy the territory in north Georgia they had abandoned (and regain use of the rail tunnel).

This is exactly what the editors of the *Memphis Appeal* stated in a column of May 19. "In making the late retrograde movement from Dalton, we are informed our forces were ordered to leave the track of the Western & Atlantic Railroad uninjured," the paper declared, offering the rationale:

> We have been asked why a general destruction of the road was not affected by our troops, as they moved along the line south. Our opinion is that it was not done for the good reason that our authorities expect soon to use it. We do not profess to be posted as to General Johnston's purposes, but can easily conceive that after the Federal army is routed, as it will be whenever a general battle occurs, he will not remain idle, but at once advance and reoccupy the country recently abandoned, as well as much more. In this event the road will be needed. The uneasy may rest satisfied with what has been done.

Some scholars have pointed to technical difficulties facing those wishing to blow a tunnel. Richard McMurry states that destroying a railroad tunnel before the invention of dynamite was no easy task; drilling and placing tons of black powder would have taken days or weeks, and the Confederates did not have the luxury of time when Sherman's forces began their advance May 5–7. Nevertheless, at least once during the war Confederate cavalry had blocked a long railway mountain tunnel. In August 1862, Colonel John Hunt Morgan led his 2nd Kentucky Cavalry in a raid to Gallatin, Tennessee, where they blocked an 800-foot underpass on the Louisville & Nashville Railroad by building a cross-tie fire inside the tunnel, then crashing a locomotive into the flaming barrier. The boiler explosion ignited the roof supports, and the slate rock overhead crashed down. It took Federal work crews more than three months to reopen the tunnel.

Thus at least some Southerners knew how to block a tunnel and deny railroad use to the Yankees. But there is no indication that Johnston thought that blowing Tunnel Hill was important. On May 7, when Confederate cavalry was fighting the enemy at the underpass before giving it up, General Mackall at headquarters sent Wheeler at least three dispatches. Not one mentioned the need to destroy the tunnel.

Johnston had narrowly averted disaster, but Sherman had succeeded in fooling him as to his objective and tactics. The key had been the Federals' knowledge of Snake Creek Gap, about which Johnston seemed to have no knowledge at all. Fortunately for the Confederates, McPherson had withdrawn his infantry after passing through the gap and had faltered when cutting the Rebels' railroad lay in his grasp—a prospect which had led Sherman to cry, "I've got Joe Johnston dead!"

That Joe Johnston, or rather his army, came so close to being "dead" merits some scrutiny. Officers in the Southern army started asking why Snake Creek Gap was left unguarded until enemy infantry started marching through it. "How this gap, which opened upon our rear and line of communication, from which it was distant at Resaca only five miles, was neglected," asked General Cleburne in his report, "I cannot imagine." Cleburne stated that General Mackall, Johnston's chief of staff, had told him, "it was the result of a flagrant disobedience of orders, by whom he did not say." "Certainly the commanding general never could have failed to appreciate its importance," Cleburne concluded, trying to exonerate Johnston from any charge of negligence or dereliction. Years later, writing for *Battles and Leaders*, W. C. P. Breckinridge, who also had fought in the Army of Tennessee, asked, "why these gaps were left unguarded...I neither pretend to know nor venture to guess....Nor do I offer any criticism," he added.

Indeed, friends of the commanding general sought to place blame elsewhere. In telling Cleburne that an unnamed officer had been guilty of "flagrant disobedience of orders," General Mackall most likely was referring to Wheeler. Midday on May 6 Mackall had sent message to the cavalry chief: "General Johnston wishes you to take measures to acquire the most accurate information of the present position of the enemy this side of and to the east of Taylor's Ridge. Get precise information as to where his flanks rest." Taylor's Ridge is several miles west of Rocky Face, running parallel to it; Sherman's forces were using it to screen their southward marches. The next day, however, Lieutenant Mackall recorded, "Not a thing has been ascertained by Wheeler's cav—inactive....But few scout reports rcd. from Genl Wheeler—generally unsatisfactory." On the other hand, in their after-action reports neither Johnston nor Wheeler made any statement that there had been neglect concerning Snake Creek Gap, much less that a blunder had been committed in leaving it unguarded.

Ten years later Johnston went to great length to demonstrate his mastery of the campaign and belittle the whole Snake Creek Gap incident, to the point of claiming that "we had examined the country very minutely; and learned of its character thoroughly." As if to exculpate himself for the loss of the gap, Johnston goes on even to say,

> We could calculate with sufficient accuracy, therefore, the time that would be required for the march of so great an army from Tunnel Hill to Resaca

> through the long defile of Snake Creek Gap, and by the single road beyond that pass. We knew also in how many hours our comparatively small force, moving without baggage trains and in three columns on roads made good by us, would reach the same point from Dalton.

Of these remarks Atlanta historian Wilbur Kurtz writes, "If this sort of rationalizing sounds naïve, it should be understood as deliberately so.... This is like unto the good man of the house who, knowing that thieves are abroad, bars the front door and all the windows but leaves the back door open, reassured by the knowledge that he is only a dozen paces from the open portal."

Nevertheless, if leaving Snake Creek Gap unguarded was indeed a blunder, Johnston is not blamed for it, at least by some. The general's biographers Govan and Livingood conclude, "the evidence points to the cavalry as the source of the failure." More recently, Craig L. Symonds agrees: "the eyes of the army" had let Johnston down. Others, though, are not convinced. Thomas Lawrence Connelly in 1971 charged that "Johnston completely ignored Snake Creek Gap, and made no inquiries as to its status." In 1990 the renowned Edwin C. Bearss asked why Johnston, a trained topographical engineer, could have ignored the vulnerable terrain south of his left flank. Roy Morris and Phil Noblitt make the same point: "although he had been in Dalton for nearly five months, the normally scrupulous master of defense had somehow neglected to have the countryside below Dalton adequately mapped or scouted." To some, the blame lay not with Johnston but with his mapmakers. One student states recently that Wheeler "had the best maps in the Army. Yet key terrain was missing on several of his maps. For example, vital mountain passes such as Snake Creek Gap were missing." This assertion runs counter to observations one can make from the maps carried by Brigadier General Henry D. Clayton (Stewart's division, Hood's corps). They show that the engineers carefully drew ridges, streams, gaps, and roads. One rivulet in particular conforms to Snake Creek and Gap, though the features are unnamed on the Clayton map. If an infantry brigadier knew of the ridge gaps, the commanding general knew as well.[1]

---

[1] Cleburne report, August 16, 1864, *OR*, vol. 38, pt. 3, 721; Breckinridge, "The Opening of the Atlanta Campaign," 4:277; W. W. Mackall to Wheeler, May 6 (12 noon), *OR*, vol. 38, pt. 4, 664; McMurry, "Mackall Journal," entry of May 6, 37 n.68 (Taylor's Ridge); S. Davis, *Atlanta Will Fall*, 42–43; Johnston report, October 20, 1864, *OR*, vol. 38, pt. 3, 614–15; Wheeler report, June 1, 1864, *OR*, vol. 328, pt. 3, 943; J. E. Johnston, *Narrative*, 315–16; Kurtz, "Why Was Snake Creek Gap Left Unguarded?" unpublished typescript, c. 1935, box 49, folder 14, 20–21, Kurtz Collection; Key, *The Battle of Atlanta and the Georgia Campaign*, 16 (unquestioningly accepting Johnston's statement about knowing the Federals' rate of march through the gap); Govan and Livingood, *Different Valor*, 266; Symonds, *Johnston*, 272–73; T.

Regardless of who was to blame, by his flanking maneuver at Snake Creek Gap Sherman had forced Johnston to retreat from his strong position at Rocky Face. It would be the first of six more such retrogrades by Johnston in the next two and a half months, but the point is that for the Confederates, had McPherson been bolder on May 9, it could have been so much worse.

In this first round of thc Atlanta Campaign, the week of May 5–12, Johnston had clearly relied on Hood as his most trusted corps commander. Staff officers saw it. Lieutenant Mackall, aide and cousin to Brig. Gen. William W. Mackall, Johnston's chief of staff, kept a detailed diary and pointedly included Hood's encounters with the commanding general. As Dr. McMurry states, the Mackall diary "is almost the only detailed source for studying the army's high command." That Hood's name appears in its entries for the opening week of the campaign more often than William J. Hardee's is telling.[2]

On their last night in Dalton, that of May 11–12, Hood, Johnston and the just-arrived Polk met at Johnston's headquarters for a couple of hours. Then Polk left with Hood to the latter's quarters and there performed a solemn rite the two had been discussing that evening: Hood's Christian baptism. There had been in the Southern army during winter 1863–1864 a dramatic increase in religious activity among the officers and soldiers. At revival meetings men converted to Christianity in great numbers. "There will be more Christians under the leadership of General Johnston in the next great battle than have ever faced the foe in this army," wrote one observer. Chaplains and missionaries lived and worked throughout the camps. From Johnston's army came the call for more Bibles and Testaments as the men applied themselves to faithful reading. Flavel Barber, a soldier of the 3rd Tennessee in Hood's corps, wrote in March and April 1864, "the religious revival still goes on…producing some good fruits." In Tucker's Mississippi brigade, also in

---

L. Connelly, *Autumn of Glory*, 336; Bearss, "Sherman and Hood at Atlanta"; Morris and Noblitt, "The History of a Failure," 40; Elam, "The Road to Atlanta," 86; Clayton, maps of Alabama, Georgia, and Tennessee. See also R. S. Davis, "'Every Crossroads and Farm'," 151–67.

Amidst all the fault-finding among Confederates for having left Snake Creek Gap unguarded was the finger-pointing in Sherman's army over the lost opportunity to assail the Rebel army in its rear. The fingers pointed to General McPherson, who lost his nerve when he could have advanced from the gap and exploited, as Sherman said, "the opportunity of a lifetime." In McPherson's defense, John R. Scales points out that this was "McPherson's first experience at the army level"—before the opening of the Georgia campaign he had commanded the XVII Corps (*Sherman Invades Georgia*, 163).

[2] Castel, *Decision in the West*, 141; McMurry, *John Bell Hood*, 103; McMurry, "Mackall Journal," entries of May 2, 9, 12. I am indebted to Richard for giving me copy of his typescript of Mackall's writings, located at the Earl Gregg Swem Library of the College of William and Mary. They are too rarely cited in studies of the Georgia Campaign.

Hood's corps, nearly two hundred men converted to the faith in just a few weeks.

The army's generals helped with this work. When Brig. Gen. William N. Pendleton, an Episcopal priest, visited the army on inspection tour in April, he delivered a sermon to the soldiers. Maj. Gen. Peter Stewart, division commander, was also a Presbyterian elder; he helped serve Communion. Brigadier Mark Lowrey, a Baptist preacher, closed services with prayer.

In this atmosphere Hood committed himself to be baptized. He does not mention religion in his reminiscences and is not quoted by others as conversing about his religious beliefs to the extent, say, of a Stonewall Jackson. But he occasionally went to St. Paul's in Richmond with the Davises and Chesnuts and was therefore comfortable with the Episcopal prayer book and rites of service. He had apparently been thinking about baptism for days. On May 8 the Rev. Dr. Charles Quintard, army chaplain in Atlanta, received a telegram from Maj. Henry Hampton: "Can't you come up? General Hood wishes to be baptized." Quintard was then ministering at the newly founded St. Luke's Church and in the city's hospitals and could not. So on the evening of May 11, when Hood met Polk, Hood made his request to the bishop. Polk was a West Pointer, lieutenant general and a friend of President Davis. He had been first commander of what would become the Army of Tennessee, but he was also Episcopal bishop of Louisiana. In June 1861, as he stepped down the stairs of the Virginia capitol following his meeting with Davis, bearing his new major general's commission, Polk was asked if he had thrown off the bishop's gown for the soldier's sword. "No, sir," came the reply, "I buckle the sword over the gown."

Thus Polk had served both roles in the army, and with Hood's baptism at Dalton he was doing so here, too. It was after midnight; Henry Brewster, Hood's aide, stood in Hood's quarters, a simply furnished room, holding a candle. Polk sanctified the holy water—really rain-spill scooped up outside by the bishop and borne by Brewster in an unceremonious horse-bucket. Hood rose to receive the sacrament though Polk told him he need not. But Hood insisted on standing with his crutches. William D. Gale, the bishop's son-in-law and aide, was among the staff officers present and recorded that "Lt Gen H seemed much impressed," as he bowed his head, leaning on his crutches as the bishop crossed the general's forehead. Brewster later told Mary Chesnut that the scene was very homespun—she noted that Brewster was a good Catholic, accustomed to rich formality. "There stood the battered old hero (barely thirty years old). There the Warrior Bishop Polk. And there stood your humble servant, with a flaring tallow candle in one hand and a horse bucket of water in the other." Yet the bishop was moved. "His heart was fully in it," Polk wrote his wife days later. "He with one leg leaning on his crutches—a veteran in the midst of his & my officers & I the officiating minister." To Col. Walter H. Rogers, also in attendance, Hood "looked happy and as though a burden had been lifted" from him.[3]

The next day, however, it was back to war. Hood accompanied Johnston on the evening train out of Dalton; Hood got off at Tilton, a few miles north of Resaca, to join his troops taking position nearby. Throughout May 13, the Confederates dug in to the north and west of Resaca. Johnston's line had its flanks anchored by two rivers. Polk's troops, now considered the third infantry corps of the army, held the left, extending northward from the Oostanaula River. Hardee was in the center, and Hood on the right, his flank stretching toward the Conasauga. When finished, the Southerners' line was four miles long as they waited for the enemy to come up.

One has to pause and realize that Johnston had drawn his army into a defensive position with a broad and deep river to its back. Such a tactic is theoretically taboo, so one wonders why he, an experienced general, would do it. Steven Newton, well known advocate for General Johnston, offers an answer that contrasts with "the modern mythical interpretation of Joseph Johnston as timorous and noncombative." Johnston dug in with a river behind him, Newton concludes, "because he thought he could win." Another reason might be that Johnston knew that the Oostanaula formed the first of only three river barriers impeding the enemy's approach to Atlanta. One may thus judge that he did not want to give it up readily.

During the 13th, Sherman's line also formed. McPherson's two corps marched again through Snake Creek Gap, Maj. Gen. Grenville Dodge's XVI anchoring the Union right on the Oostanaula; then northward was Maj. Gen. John A. Logan's XV. Of Thomas's army, Maj. Gen. Joseph Hooker's XX Corps formed on Logan's left, then Maj. Gen. John M. Palmer's XIV. Schofield's XXIII Corps held the Union left. All were in position by evening. On the 14th Maj. Gen. Oliver O. Howard's IV Corps, marching southward following the Confederates' retreat, would fall in on Schofield's left, extending the Federal line eastward, like the Rebels', toward the Conasauga.

---

[3] W. Bennett, *A Narrative of the Great Revival*, 344, 347, 351, 383; Woodworth, *While God is Marching On*, 236; Daniel, *Soldiering the Army of Tennessee*, 120; J. W. Jones, "The Morale of the Confederate Armies," 12:182–83; Norton, "Revivalism in the Confederate Armies," 417, 421; Quintard, *Doctor Quintard*, 84; Parks, *General Leonidas Polk*, 170, 374; Polk, *Leonidas Polk Bishop and General*, 2:329; O'Connor, *Cavalier General*, 192–93; McMurry, *John Bell Hood*, 103; Woodward, *Mary Chesnut's Civil War*, 616; Dyer, *Gallant Hood*, 234; Robins, *The Bishop of the Old South*, 189.

Sherman ordered an assault on the enemy center-right for the 14th, but the charge was not intended to be of such force as to break Johnston's line in the manner that, say, Col. Emory Upton's had been the week before, five hundred miles to the north, at Spotsylvania Courthouse, Virginia. There Federal troops had massed in twenty lines deep, 275 to 300 men across. They crouched into position in the predawn darkness, then stormed at first light over and through General Lee's lines. That was a decisive infantry attack. On the other hand, Sherman planned his assaults more as demonstrations in force, intended to draw the Rebels' attention while the real action took place beyond their left, which he intended to flank, just as he had done at Rocky Face Ridge. On the Federals' far right, four miles downriver from Resaca, Sherman ordered two pontoon bridges laid across the Oostanaula at Lay's Ferry. When his troops—Brig. Gen. Tom Sweeny's division of Dodge's corps—got across, he expected Johnston to retreat.

Skirmishing between the opposing lines opened at daylight on May 14, but the planned morning assault was delayed until 11 A.M. Then two of Schofield's divisions moved forward, across rough terrain, down into the valley of Camp Creek, and up against fortified positions, on which the Southerners had been working on the previous day and night. Predictably the Federals were repulsed by Maj. Gen. William B. Bate's division. Hindman's division also came under fire. Then Brig. Gen. Richard W. Johnson (1st Division, XIV Corps) sent in his troops. They at least got halfway across the field on the east side of the creek before they, too, were driven back by heavy fire from Pat Cleburne's infantry. Sherman's assault had failed by 3 P.M.[4]

Then Johnston perceived an opportunity on his right. Earlier in the day, after Howard's corps extended the Federal left, he had sent cavalry to probe the enemy's position. They reported that the Yankees had not stretched their left to the Conasauga. Better, Hood's line extended farther to the east, almost to the river, thus overlapping the enemy line. The situation invited a flank attack, and around four o'clock Johnston ordered Hood to launch it. Hood committed Carter Stevenson's and Peter Stewart's divisions to the effort. An hour later, around five o'clock, the Confederates advanced. According to one soldier of Maj. Gen. David Stanley's division, holding the extreme Union left,

---

[4] McMurry, *John Bell Hood*, 103; Castel, *Decision in the West*, 150–53, 156, 160–61; McMurry, "Resaca: 'A Heap of Hard Fiten,'" 11; Newton, "Joseph Johnston and Snake Creek Gap," 66; Matter, *If It Takes All Summer*, 189; Scaife, *The Campaign for Atlanta*, 30–31, map f.p. 38; Walthall report, June 2, 1864, *OR*, vol. 38, pt. 3, 796; Luvaas and Nelson, *Guide to the Atlanta Campaign*, 1094; Waters, "Lines of Battle," 207; Cleburne report, August 16, 1864, *OR*, vol. 38, pt. 3, 722; S. Davis, "Sherman in North Georgia: The Battle of Resaca," 22–23.

the Southerners "formed in admirable order, their flags floating gaily, many of their officers mounted." Brig. Gen. John C. Brown, leading a Tennessee brigade in Carter Stevenson's division, rode down the line of his troops, calling out, "Boys, when I give the command to move forward, I want every one of you to step as though your legs were hung on hinges."

Hood's attack enjoyed initial success. His line so outstretched the Federals' that Stewart's division on the far right did not even come into contact with the enemy. Stevenson's troops, however, did more than their share, driving back the Yankees so fast that some threw away their rifles and knapsacks. But then the attack slowed. The Southerners entered an open field and came under fierce artillery fire from Capt. Peter Simonson's 5th Indiana Battery, posted on a hill on the other side. Northern canister rounds and the timely arrival of reinforcements (Brig. Gen. Alpheus Williams's division of Hooker's corps) helped Stanley's troops drive the Southerners back, eventually to their own lines.

Casualties for the Confederate attack on the afternoon of May 14 are hard to come by. General Stevenson reported 575 killed, wounded, and missing for the entire battle. Stanley and Williams also provide theirs for the whole battle: 200 and 417 respectively. One of these casualties was Lt. John H. Waddey of the 32d Tennessee (Brig. Gen. John C. Brown's brigade), who was shot in the throat during the afternoon's charge but was able to speak before he died. "He said tell the boys to do nothing in life they will regret in death," according to a comrade. "He was noble in death."

Hood's attack had failed to roll up the enemy line—both he and Johnston had watched it—but Johnston was sufficiently impressed to order a renewal of the assault in the morning. Moreover, this would be a bigger effort. Lieutenant Mackall wrote in his journal, "Stewart to take the enemy in flank, Stevenson and Hindman to take up the movement." Johnston even ordered two brigades from Polk's and Cheatham's sectors to come over and support the attack.[5]

Events on the other flank were unfolding. Sherman planned for Sweeny's infantry to cross at Lay's Ferry, but the enemy was found to be on the south bank there in force, so a downstream point was chosen for the pontoon bridges. Near where Snake Creek empties into the Oostanaula, Federal infantry crossed and drove off the small number of Rebels. Yet Sweeny called

[5] McMurry, "The Atlanta Campaign: December 23, 1863 to July 18, 1864," 100–101; Castel, *Decision in the West*, 163–66, 168; Davis, "Resaca," 24, 26; Elliott, *John C. Brown of Tennessee*, 94; Richard A. Irvin to William D. Irvin, May 22, 1864, typescript courtesy of Gary Waddey, Nashville, TN; McMurry, "Mackall Journal," entries of May 14–15.

his bridgehead back upon news that the Rebels might try to flank it. Sherman was not too displeased as McPherson made gains elsewhere. His troops had charged across Camp Creek and taken high ground within artillery range of the railroad and wagon bridges over the river. Southern troops seeking to cross would be vulnerable to fire from McPherson's position.

News that Federals had crossed the Oostanaula and were near the river bridges reached Johnston after dark. He ordered Walker's division on a night march toward Lay's Ferry to ascertain the situation and sent word to Hood canceling his morning attack. At the start of the 15th, Sherman therefore held the initiative. His orders were for Schofield to withdraw from the middle of the line, march east and reposition on the extreme left so that the Federal line rested securely on the Conasauga. Thomas would have Palmer's corps extend to the left, taking Schofield's place. Hooker and Howard would then make a "steady and strong attack" against the center-right of the enemy lines. But Sherman added the caveat that they "will not assault fortified positions unless sure of success"—a rather limp attempt to absolve himself of blame for issuing the order. While Hooker and Howard attacked, McPherson's artillery was to shell the bridges across the Oostanaula, making them "too hot for the passage of troops." Otherwise McPherson's main responsibility was to put down the overdue pontoon bridge at Lay's Ferry and make "a good lodgment on the other bank." Sherman wanted everything to start happening around 8 A.M.

At that hour Sweeny's troops obediently got moving, crossed the river on a big flatboat and chased off the few Rebels on the south bank. Then the pontoons were laid. By noon a brigade of infantry had crossed. A brisk Confederate attack against the Union bridgehead failed; Maj. Gen. W. H. T. Walker reported this disconcerting news to headquarters. Then Sherman's "steady and strong attack" fell around noon on Hood's sector of the line, held by Stevenson's and Hindman's divisions. During the morning Hood had ordered Stevenson to place a battery "in such a position as would enable it to drive off a battery that was annoying General Hindman's line." Counterbattery fire is different from counterinfantry fire, and to see the annoying enemy guns the Confederates actually had to set theirs—Capt. Maximilian Van Den Corput's "Cherokee Battery" of four Napoleons—some twenty yards in front of the Confederate entrenchments. Soldiers dug an earthen embrasure. Stevenson may have planned to connect the exposed battery to his main line with infantry trenches, but "before the necessary measures for the protection of the artillery could be taken," Stevenson later wrote, "I received repeated and peremptory orders to open."

The Cherokee Battery had barely gotten into position before the Yankees attacked. Elsewhere Stevenson's and Hindman's troops repulsed the assault without having to call for reinforcements. But here the Confederates ran into trouble when the Federals charged up and overran the battery. Col.—later U. S. President—Benjamin Harrison's 70th Indiana "entered the embrasures," Harrison later reported, "striking down and bayoneting the rebel gunners, many of whom defiantly stood by their guns till struck down." After overwhelming the battery, the Northerners quickly came under fire from Stevenson's infantry, and were forced to withdraw and find shelter down the hill. This left Van Den Corput's cannon around three o'clock sitting in no-man's land, as neither side chose to send troops forth to try to secure them. "Neither Federals nor Confederates could gain possession of them," recalled a chronicler of the 32d Tennessee. "Come on—take those guns!" yelled the Southerners. "Come on and take 'em yourselves," the Northerners yelled back. The stalemate at Van Den Corput's battery was the only positive outcome of the Union attack. General Hood summarized the fighting: "about the middle of the day on the 15th the enemy made assaults on Stevenson's front and the right in several lines of battle, each successive line being repulsed with loss." Hood's characterization was more accurate than Sherman's; in his campaign report he later stated that on the afternoon of the 15th "General Hooker drove the enemy from several strong hills, captured a 4-gun battery and many prisoners."[6]

General Johnston changed his mind about the attack that he had canceled the night before. Around 3 P.M. he rode to Hood's headquarters and learned that Stevenson and Hindman had easily repulsed the enemy charge. That news, plus General Walker's morning report that he saw no Yankees on the south bank of the Oostanaula, led Johnston to reissue orders for Hood to launch the flanking attack they had envisioned the previous evening: Stewart

---

[6] Castel, *Decision in the West*, 162–63, 166–69, 173–5, 178; SFO 6, May 15, 1864, *OR*, vol. 38, pt. 4, 199; Sherman to Thomas, May 15, *OR*, vol. 38, pt. 4, 190; R. K. Brown, *To the Manner Born*, 227–32; Walker to Mackall, May 15, *OR*, vol. 38, pt. 4, 713; Stevenson report, May 30, *OR*, vol. 38, pt. 3, 812; Harrison report, May 20, *OR*, vol. 38, pt. 2, 371; McGuire, "Thirty-second Tennessee Infantry," 1:479; Hood report, undated, *OR*, vol. 38, pt. 3, 761; Sherman report, September 15, 1864, *OR*, vol. 38, pt. 1, 64; Davis, "Resaca," 42–46.

Thirty years after the war, Frank Anderson of Nashville, who had served as one of General Hood's couriers at Resaca, recalled how Hood had told him that Van Den Corput's battery had to be defended to the last. "The kind and courteous treatment made the men of the escort company all love Gen. Hood," Anderson wrote. "He was a born gentleman. He never failed to salute a courier, and usually had a kind word for him, no matter where he was or what his surroundings" ("A Courier at the Battle of Resaca," 297).

to advance against the enemy flank, Stevenson and Hindman supporting. But in the afternoon Johnston learned that the enemy had indeed crossed the river and were just a few miles from the W. & A. He quickly cancelled the assault, but the order did not reach Stewart in time. Around 5 P.M. Stewart formed his brigades in a double battle-line and the men stepped forth. Instead of advancing around and against the left of the Federal line as they had the afternoon before, the Southerners found Yankees square ahead of them. Hooker had extended his front by placing Alpheus Williams's division there, on high ground overlooking open fields across which the Rebels would have to advance. Williams's line was strengthened with two batteries.

From their side of the field, Confederates could see the Yankees. Capt. George W. Welch, commanding the 38th Alabama, observed three heavy enemy lines 500 yards to his front when, at five o'clock, his regiment was ordered to form for the advance. "Notwithstanding the almost utter impossibility of success, apparent to every one, the order was promptly obeyed," Welch recorded. The Southerners moved forward and drove back enemy skirmishers, enduring a punishing fire of artillery and musketry. They covered several hundred yards and at maybe 75 yards opened their own volleys. "We held our position for several minutes," recalled Col. Bushrod Jones of the 58th Alabama, and "fired four or five rounds, when the men began to give way" under the Northerners' intense fire. The Confederates withdrew back into their lines "broken and confused" after about twenty minutes, according to one Union officer. General Stewart rode forth and tried to reform his men but could not. After a brigade from Stevenson's division had also advanced and been beaten back, the fighting was over by 6:30 P.M. In his report Hood put the best face he could on the repulse, which probably cost between 600 and 800 killed, wounded, and missing: "At 4 o'clock in the afternoon General Stewart moved forward from the right with his division, driving the enemy before him, but was subsequently forced to resume his original position before largely superior numbers."

The battle of Resaca, May 14–15, cost Johnston close to 3,000 killed, wounded, and missing, as best as we can count them from the Confederate officers' sketchy reports. Sherman's loss was a third higher, as his troops had done more of the attacking. To Halleck he reported the wounded from all three armies at Resaca totaled 3,375, adding on the 16th, "the number of dead not yet reported, but will not exceed the usual proportion." The "usual proportion" averaged one man dead to every five wounded, as General Johnston himself later affirmed. This leads Dr. Castel to do the math and conclude that

Sherman's loss was "about 4,000, of whom 600 are dead or mortally wounded."

On the evening of May 15, Johnston had to make a decision. Report from General Walker had come to his headquarters during the day that Federal infantry had finally gotten across the Oostanaula and threatened to advance on the Confederates' railroad in their rear. In council with Hood, Hardee, and some division commanders, Johnston announced that the army had no choice but to retreat across the river to a new position. Around 8:30 P.M. the corps commanders discussed their night march across the wagon and train bridges, plus two pontoon spans the engineers had laid upstream.

About midnight Hardee's troops crossed the river on the wagon bridge; Polk's used the railroad bridge. Hood's troops formed the rearguard; they pulled in their pickets between 1 and 3 A.M. Those in Stevenson's sector may not have known that during the night the Yankees were quietly removing the four cannon of the Cherokee Battery that had been left out in no-man's land. Brig. Gen. John W. Geary, division commander in the XX Corps, had ordered a detachment to sneak forward in the darkness, dig through the earthwork, and rope-drag the four guns down the hill. Their success was celebrated in the Northern press; *Harper's* carried a woodcut drawing of the feat. The Federals who did the work were probably unaware that the Rebels were retreating that night. What few pickets remained in the Confederate works were not about to start a firefight, even if they heard the Yankees' toiling beyond their line.

By 3:30 A.M. all infantry, artillery and wagons were across. The engineers took up their pontoons and set fire to the bridges. The blazes alerted skirmishers in Logan's front, who advanced to catch stragglers not yet across the river. They fired at Rebels burning the wagon bridge and drove them off. As morning dawned, the Federals used their haversacks as water pails and managed to save the bridge before it was ruined.

The Confederates marched that night to Calhoun, five miles south of Resaca via the W. & A. There General Johnston sent a telegram to President Davis on what had transpired: "The enemy laid a pontoon-bridge below Calhoun, under protection of two divisions. I was compelled to fall back on this place. His attacks on our troops yesterday were repulsed."

Sherman, on the other hand, had brighter news to wire to General Halleck in Washington: "We are in possession of Resaca.... We saved the common road bridge, but the railroad bridge is burned.... Our columns are now crossing the Oostanaula.... We will pursue smartly to the Etowah."[7]

At Resaca as he had at Dalton, Johnston had relied significantly on Hood. Biographer McMurry makes much of this: "The battle of Resaca furnishes further evidence of the close relationship between Johnston and Hood." Johnston used Hood and his corps for the only offensive strikes he ordered in the two days of fighting, May 14–15. To be sure, these attacks had failed, but for the usual tactical reasons, and Johnston was on hand to watch both of them. He would have regretted their failure, but there is no evidence that he criticized Hood for it.

Ten years later, however, such dislike—*hatred* would not be too strong a word—had developed between the two men that in their memoirs they leveled harsh accusations and counter-accusations against each other. Johnston's *Narrative of Military Operations* was published in 1874. Hood, who had already begun to assemble material for his own memoirs, read Johnston's text closely, noting the aspersions made against him. One of them involved Resaca: "no material was lost by us in the campaign, but the four field-pieces exposed and abandoned at Resaca by General Hood," Johnston claimed in his *Narrative.* Hood was very sensitive to artillery lost and won; he had told his troops often that cannon were among the greatest prizes soldiers could win on the battlefield. He therefore took it upon himself in *Advance and Retreat* (which he wrote in the years after 1874 until his death), to downplay the gravity of the battery's loss by making no fewer than five statements.

---

[7] Castel, *Decision in the West*, 175–81, 188; McMurry, *Hood*, 106; Ridley, "The Battle of Resaca," 36–37 (staff officers trying to cancel the attack); Robinson report, July 24, 1864, *OR*, vol. 38, pt. 2, 86–87; Hood report, undated, *OR*, vol. 38, pt. 3, 761; Secrist, *The Battle of Resaca*, 52–57 (Van Den Corput's battery); Elliott, *Soldier of Tennessee*, 178–79; Mackall "Journal of Operations," May 16, *OR*, vol. 38, pt. 3, 981; Davis, "Resaca," 47–48; Dolzall, "Opportunity Lost," 26 (Resaca casualties); J. E. Johnston, *Narrative*, 358 (5:1); Johnston to Davis, May 16, 1864, *OR*, vol. 38, pt. 4, 717; Sherman to Halleck, May 16, 1864, *OR*, vol. 38, pt. 4, 201; Secrist, "Resaca: For Sherman a Moment of Truth," 38 ("because the Confederates abandoned the field, the Union commander could claim victory at Resaca").

On May 16, Sherman wired General Halleck that at Resaca his forces had captured "about 1,000 prisoners and 8 guns" (*OR*, vol. 38, pt. 4, 201). Both of these claims appear to be exaggerated (Davis, "Resaca," 50). In his campaign report he claimed that after the Confederates retreated, "we found in Resaca another 4-gun battery and a good lot of stores" (Sherman report, September 15, 1864, *OR*, vol. 38, pt. 1, 65). In the Confederate sources, however, I see no mention of any artillery loss beyond that of the Cherokee Battery.

1) Johnston approved leaving the guns behind ("The whole matter was laid before General Johnston, and the guns were abandoned with his concurrence; at least such is my recollection").

2) An advance by Stevenson's infantry to retake the guns would have been costly.

3) His division had more pieces of artillery than it needed.

4) The cannon were "four old iron pieces, not worth the sacrifice of even one man." Some of these statements may have been wide of the mark, as in "four old iron pieces." Everyone agrees that Van Den Corput's guns were bronze Napoleons. (Indeed, decades later one of the Federal officers involved in their capture recalled that "the guns appeared to be quite new—twelve pounders [bronze])." There is a story that they were even cast in Atlanta, and that one of them was inscribed "Minnie, the Belle of Alabama." But the point is that after the war Hood was determined to defend his record.

5) For good measure, and to throw it back at Johnston, Hood quoted Col. Hypolite Oladowski, Johnston's chief of ordnance, in a letter he had written to Braxton Bragg in May 1874. Oladowski claimed that from Dalton to Atlanta the army had lost 19,000 muskets—a material loss, Hood implied, that was as significant as the loss of the Cherokee Battery, but one to which Johnston made no reference in his memoirs.[8]

All of this—Hood's statements in his memoir—would be in the future. As the Army of Tennessee retreated southward after giving up the Oostanaula, Hood continued to be held in high esteem by his superior. Retreating from Resaca during the night of May 15–16, the Confederates passed through Calhoun the next morning. Sherman's forces followed cautiously, harried by Confederate cavalry. Johnston hoped to find a position to make a defensive stand but could not. About this time the press reported the alleged exchange

---

[8] J. E. Johnston, *Narrative*, 351; Hood, *A & R*, 96–97; Kirkpatrick, "The Fifth Ohio Infantry at Resaca," 360; Kurtz, "Data on the capture and removal of the four brass twelve-pounder Napoleon guns…" unpublished, handwritten manuscript, box 46, folder 29, Kurtz Collection; Castel, "The Battle That Should Not Have Been," 83 ("Minnie, the Belle of Alabama").

As if to underscore Hood's contention that the loss of Van Den Corput's four guns was immaterial, army returns dated June 10 (after Resaca) show the same number of artillery pieces attached to Hood's corps (36) as the returns of April 30, before the campaign had begun ("Abstract from returns of the Army of Tennessee," April 30 and June 10, *OR*, vol. 38, pt. 3, 676–77).

between a soldier in the ranks and the commanding general of the army. "General, don't fall back any farther, we are getting mighty tired," the man called out. "I am not retreating," Johnston retorted. "The enemy is on our flank and rear, and we must face the foe." "Bully for you," the soldier cried; he and his comrades raised three cheers for Old Joe. In the same vein, the Confederate press made out that Johnston's retreat from Dalton had not been a retreat at all. The *Columbia South Carolinian* opined on May 17 that in shifting his position, Johnston "merely changed his line of battle." But retreats were retreats.

South of the Oostanaula, the Georgia countryside changes. Gone are the mountains and ridges of the Dalton area that gave the Army of Tennessee the defensive positions Johnston always searched for. The rolling hills and wide valleys of the piedmont afforded no suitable terrain for a line. So in its second consecutive night march, May 16–17, the army continued toward Adairsville. The generals at least got some sleep while the soldiers trudged. Lieutenant Mackall recorded in his journal that Hood had ridden with General Johnston on May 16 and spent the night at Johnston's headquarters, rising early and leaving around 4 A.M. That day, the 17th, Johnston again found the topography at Adairsville unsuited for a new defensive position. Hood rode with Johnston during the afternoon; that evening the commanding general called in his senior officers to discuss whether a stand could be made at Adairsville. General Hardee's acting assistant adjutant general, Maj. Henry Hampton, recorded in his journal that Hardee "advocated giving battle to the enemy in the position we then held in front of Adairsville." In his memoir written a decade later, General Johnston only stated that "the breadth of the valley here exceeded so much the front of our army properly formed for battle, that we could obtain no advantage of ground." He consequently ordered his troops to continue their retreat.

Johnston does not mention how Hood got involved in the story. According to Lieutenant Mackall's journal as printed after the war in the *Official Records*, when the question of taking a stand came up, Hood said, "can't be done" while Hardee countered, "we will have to fight." The lieutenant added, "[Hood] has been anxious to get from this place south of Etowah."

A number of historians have accepted Mackall's statement. "Hardee advocated making the fight at Adairsville," Connelly wrote in 1971; "Hood argued that the army should retreat across the Etowah River." Since that time, however, Richard McMurry has examined the original wartime journal kept by Thomas B. Mackall and compared it to its much different version appearing in the *Official Records*. In a seminal article published in *Civil War History* in 1974, he observed in the *OR* both additions to and deletions from the original manuscript and discerned that their upshot was to strengthen the record of Joseph E. Johnston's conduct during the Atlanta campaign and to weaken that of John B. Hood. He assembled convincing evidence that Mackall, with another former Johnston staff officer, Edwin J. Harvie—who just happened to be working on the *Official Records* project in the 1880s—colluded with Johnston himself to prepare the Mackall "journal" so as to favor the general. As a result, McMurry closed his investigation with some strong advice: "historians must cease citing the published Mackall Journal as a contemporary document."[9]

Some writers, however, have not gotten the message. As recently as 2016, Dr. James Lee McDonough asserts in his biography of General Sherman that when Hardee urged Johnston at Adairsville to stand and fight, "Hood advised Johnston to retreat." His source, we see in endnotes, is Connelly, whose *Autumn of Glory* came out in 1971, before McMurry's Mackall research. In truth, Hood may or may not have urged a retreat from Adairsville (like Johnston, he is mum on the subject in his memoir). The lesson here is that in discussing the Atlanta Campaign, historians should be cognizant of an evolving literature.

On the evening of May 17, the generals took time out from their strategizing for a spiritual respite: the baptism of Joseph E. Johnston as administered by Leonidas Polk. Mrs. Johnston, writing from Atlanta, had asked the bishop to perform the rite. She was sure that Polk, "even whilst leading your soldiers to victory," would not be too busy to "lead my soldier nearer to God." Polk obliged and baptized Johnston that evening before the army left Adairsville. General Mackall was present, describing it as "a very solemn scene," though noting to his wife rather oddly, "as the ceremony went on I felt as if I was parting from an old friend."

Johnston reached Cassville by 10 A.M. on May 18. That evening Johnston again met with his three corps commanders and discussed a new plan as

---

[9] McMurry, *John Bell Hood*, 107, 218 n.38 (Johnston and Hood together); Andrews, *The South Reports the Civil War*, 439 ("Bully for you," as reported in the *Montgomery Advertiser*, May 20); untitled editorial, *Columbia South Carolinian*, May 17 ("changed his line"); Secrist, "The Role of Cavalry in the Atlanta Campaign, 1864," 514 ("the country between the Oostenaula and the Etowah is open"); McMurry, "Cassville," 4, 6; McMurry, "Mackall Journal," entry of May 17; Mackall, "Journal of Operations, May 14-June 4," *OR*, vol. 38, pt. 3, 982–83; Maj. Henry Hampton, "Itinerary of Hardee's Corps May 15-June 14," *OR*, vol. 38, pt. 3, 704–705 (entry of May 17, 1864; "with enthusiasm"); J. E. Johnston, *Narrative*, 319–20; T. L. Connelly, *Autumn of Glory*, 345; McMurry, "The Mackall Journal and Its Antecedents" (*Civil War History*), 322–28.

he looked over a map. The two roads leading south from Adairsville *finally* gave the Confederates an opportunity to attack part of Sherman's forces. One road led to Kingston and the other to Cassville, a town eleven miles southeast of Adairsville. Johnston figured that Sherman would use both routes and decided that he would fall upon the column marching for Cassville. He ordered Hardee's corps, the army's cavalry and trains toward Kingston, giving the impression that this was the army's main route.

On May 18, Sherman took the bait, sending most of his troops toward Kingston on a "plain, well-marked trail" left by the Rebels. The Federals followed Hardee closely, engaged by Wheeler's cavalry and Cheatham's infantry. "Some hot fighting on our front," as one soldier put it.

Only Hooker's XX Corps, with Schofield's XXIII behind it, marched for Cassville. North of the town Johnston positioned Polk's corps to face the enemy with Hood's to its right. As at Resaca, Hood's corps would be the pivot wing, the one Johnston intended to send in an attacking battle on the 19th. Polk would demonstrate against the Federals' front while Hood assaulted their flank.

Johnston wanted a fight. Word had begun to come in of Confederate armies' victories elsewhere. In Louisiana, Lt. Gen. Richard Taylor had defeated Banks's Red River expedition. Lee was repelling Grant's bloody charges at Spotsylvania. Indeed, by the time this news got to Col. Taylor Beatty, one of Hood's staff officers, it was exaggerated. "News of Bank's surrender said to be confirmed," he recorded in his diary on May 18. "News from V-a," he added, "Lee has destroyed 45000 of Grant's army." Indeed, the Press Association had issued a telegraphic column titled, "LATEST FROM LEE'S ARMY…YANKEE LOSS 45,000!" The *Atlanta Intelligencer* just a few days before had trumpeted the headline, "Banks and his Army Reported to have Surrendered at Alexandria." Then, on the 17th the paper reported, "Lee has driven Grant's forces across the Rapidan with great slaughter." Lieutenant Mackall recorded that these reports made some of Johnston's officers eager to fight and win a victory, lest their army in its retreats ruin the good news coming from other theaters.

Johnston was confident that he had caught Sherman with his forces divided. Moreover, the Army of Tennessee was at its strongest. Maj. Gen. Samuel French's division, the last of Polk's Army of Mississippi, arrived on May 18, and by some accounts, Johnston's army now numbered over 73,000—larger than Lee's in Virginia (and the largest army ever raised in the western theater by the Confederacy). Further, Johnston had entrusted the attacking battle to the corps commander who had most clearly shown a talent for just this assignment: John B. Hood. "In planning his ambush at Cassville, Johnston reserved the crucial role for Hood's corps," writes Johnston's biographer Craig Symonds. "Hood's enthusiasm for the offensive was his salient characteristic.... It was only fitting that he now be assigned the responsibility of delivering the blow that would send Sherman reeling from Georgia."

Prospects for victory were so bright that on the evening of the 18th Johnston had a bellicose general order drawn up for his troops.

> General Orders Headquarters Army of Tennessee
> No.__Cassville, Ga., May 19, 1864
> Soldiers of the Army of Tennessee, you have displayed the highest quality of the soldier—firmness in combat, patience under toil. By your courage and skill you have repulsed every assault of the enemy. By marches by day and by marches by night you have defeated every attempt upon your communications. Your communications are secured. You will now turn and march to meet his advancing columns. Fully confiding in the conduct of the officers, the courage of the soldiers, I lead you to battle. We may confidently trust that the Almighty Father will still reward the patriots' toils and bless the patriots' banners. Cheered by the success of our brothers in Virginia and beyond the Mississippi, our efforts will equal theirs. Strengthened by His support, those efforts will be crowned with like glories.
> J. E. Johnston, General.

The order was read to the troops early the next morning. They responded with understandable enthusiasm. "The cheers that went up from each command as 'Old Joe's' ringing battle order was read to the troops," one veteran recalled, "reverberated again in their ears."[10]

---

[10] McDonough, *William Tecumseh Sherman*, 483, 484 ("largest army that the western Confederacy ever marshaled"), 522; Hood, *A & R*, 99 (stating that he urged an attacking battle only after the army had retreated to Cassville); Symonds, *Johnston*, 288–91; Park, *General Leonidas Polk*, 377–78; Mackall, *Son's Recollections*, 210 (letter of May 18, 1864); Beatty diary, entry of May 18, 1864 (I thank Matthew Turi, manuscripts research and instruction librarian, for his assistance in making this important source available to me); "LATEST FROM LEE'S ARMY...YANKEE LOSS 45,000!," *Columbia South Carolinian*, May 19, 1864; "Latest from Virginia," *Atlanta Intelligencer*, May 14; "TELEGRAPHIC," *Atlanta Intelligencer*, May 17, 1864; S. G. French, *Two Wars*, 194; S. Davis, *Atlanta Will Fall*, 58; Stone, "From the Oostenaula to the Chattahoochee," 71 (French's arrival "increasing his force to fully 75,000 present for duty"); Castel, *Decision in the West*, 200; General Orders, May 19, 1864, *OR*, vol. 38, pt. 4, 728; "General Joseph E. Johnston's Campaign in Georgia," 314.

At 8 A.M. on May 19 Hood's infantry started marching to the area east of the Cassville-Adairsville road to be in a position for their attack. Polk's troops began skirmishing with the Union cavalry in their front, thus setting the stage for Hood's assault. About 10:30, however, Colonel Beatty, riding with Hood's column, saw Yankee cavalry on the right flank coming in from a road to the northeast and reported it to the general. Sherman had ordered a raid upon the Western & Atlantic; by chance these horsemen had come upon Hood's troops. Just at that time Brigadier General Mackall rode up with clarifying orders from headquarters. Hood told him of the enemy threatening his flank and that he would have to withdraw. Mackall sent word of this back to headquarters by courier; Lieutenant Mackall later recorded "enemy in heavy force close to Hood on Canton road." Hood halted his forces and drew back. At headquarters Johnston responded to these developments with disbelief, crying out (according to Lieutenant Mackall), "It can't be!" Confederate cavalry that morning had reported no enemy in the area. But Johnston was forced to accept the situation. He called for a map, studied it, and concluded that if this were the case, Hood would have to fall back. He cancelled the attack and as Hardee's corps joined the army from Kingston, ordered it to retire to a ridge line southeast of Cassville.[11]

For the rest of his life, Joseph E. Johnston refused to believe that Hood had encountered an enemy force on the morning of May 19. Lieutenant Mackall dutifully recorded in his journal for the 19th, "Report of column road not afterwards confirmed." The next day Johnston wired Richmond that "while the officer charged with the lead was advancing he was deceived by a false report that a heavy column of the enemy had turned our right and was close upon him, and took a defensive position. When the mistake was discovered it was too late to resume the movement." Johnston used the terms "false

---

In his memoirs General Johnston acknowledged receiving 21,600 reinforcements when he published the letter of Kinloch Falconer, former A. A. G., who on May 1, 1865, wrote that on May 12,

> Loring's division, Army of the Mississippi, and Canty's division, joined at Resaca, with about eight thousand effectives. French's division, same army, joined near Kingston several days later (about four thousand effectives). Quarles' brigade from Mobile (about twenty-two hundred effectives) joined at New Hope Church on the 26th. The cavalry of the Mississippi Army, which joined near Adairsville, was estimated at three thousand nine hundred effectives; and Martin's cavalry division, which joined near Resaca, at three thousand five hundred (*Narrative*, 574).

[11] McMurry, *John Bell Hood*, 108; Castel, *Decision in the West*, 201; Beatty diary, entry of May 19; Mackall, "Journal of Operations," *OR*, vol. 38, pt. 3, 983.

report" and "mistake," but he did not name Hood as "the officer in charge": he knew General Hood had friends in high places. Thus when an unnamed correspondent for the *Columbus Sun* (possibly a Confederate officer) wrote a column titled, "Why Johnston did not Fight at Cass Station," the incident was ascribed to "a failure upon the part of his corps commanders to comply with the plan of battle." The writer described how the march-to-attack was proceeding "when a staff officer, charging up to one of Johnston's distinguished Lieutenant Generals, made the startling announcement that the enemy were on his flank in large and threatening numbers." The "Lieutenant General" is unnamed but is noted as "placing infinite confidence in the announcement of a staff officer." The writer concludes that "the Lieutenant General, fearing to move, failed to comply with the orders of his commanding General."

Johnston was not quite so critical. In his report for the campaign, written in October 1864 near Macon, Johnston did not accuse Hood of poor performance; instead he blamed the staff officer (without naming Colonel Beatty) who had allegedly seen the enemy column. "When it was discovered that the officer was mistaken, the opportunity had passed."

About that time, under date of September 22, writing from Vineville, Georgia (he was apparently staying with Johnston), General Mackall wrote "Memoranda of the Operations at Cassville on May 19, 1864." Mackall may have written the report at Johnston's request; in it he gives a detailed account of events, even if he composed it four months after they had occurred. While Hood's column was marching in preparation for attack, Mackall wrote, he was sent by Johnston to tell Hood that he should not veer too far to the right of Polk's line; Hardee had reported that he was being pressed by heavy enemy forces.

> He instantly said, "And they are on me, too. The cavalry gave me no warning. I only learned the fact through officers of my own staff, and I am now falling back to form a line farther to the rear." I asked him, "What road are the enemy moving on." He replied, "On both the Canton and Spring Place road; and did you not see them?" I answered that I had seen no enemy. I then rode back.

Mackall added that when he started back for headquarters, Hood was pulling his troops back, looking for a new position to redeploy.

Hood did not write his report on the Georgia Campaign until mid-February 1865. By that time, residing in Richmond, he had Johnston's report before him, and challenged its assertion about the canceled attack.

> While in motion a body of the enemy, which I supposed to be cavalry, made its appearance on the Canton road, in rear of the right of my original position. Major-General Hindman was then in that direction with his division.... It was not a mistake (as General Johnston states) that the force appeared, as is shown from the fact that Major-General Hindman had men wounded from the small-arms and artillery fired from this body.

Hood was only slightly wrong: on the morning of May 19 it was Stewart's, not Hindman's, division that suffered casualties. But he was certainly more correct than Johnston—or for that matter, Brigadier General Mackall or Lieutenant Mackall, who all sided with the commanding general that there had been no enemy force on Hood's flank.

Hood's officers subsequently supported his claim. In his report, written on June 5, three weeks after Cassville, Stewart dutifully admitted that on the morning of May 19 he had suffered losses; indeed, "I regret to say that a number of men belonging to the Eighteenth Alabama Regiment unnecessarily and disgracefully surrendered." General Stewart referred to the report of Maj. John E. Austin, who commanded the 14th Louisiana Battalion of Sharpshooters. Austin commanded the sharpshooters on the left (south) of the 18th Alabama after Hood apparently sent Stewart's division from the center of his column toward the east to face the Yankee cavalry. Austin reported "a slight skirmish with the enemy," writing, "His cavalry and infantry skirmishers turned my left flank, which was supported by a portion of General Clayton's brigade deployed as skirmishers." Austin confirmed the "disgraceful surrender" of a company of the 18th Alabama. Brig. Gen. Henry Clayton also mentioned that some of his skirmishers had been captured "by a sudden dash of the enemy's cavalry."

Hood would have seen at least some of these reports; Stewart's and Clayton's were addressed to Maj. James W. Ratchford, his A. A. G. If he had wanted to demolish Johnston's claim that the enemy force on his flank never existed, in his report Hood could have mentioned these officers; at least he should have gotten the name of their division commander right (Stevenson, not Hindman). Instead, writing only one sentence on the issue, Hood essentially allowed Johnston to continue to believe that the enemy on his flank had been, in the words of Atlanta historian Wilbur G. Kurtz, "phantom troops."[12]

[12] McMurry, "Rocky Face to Dallas," 54; McMurry, "Mackall Journal," entry of May 19; Johnston to Davis, May 20, *OR*, vol. 38, pt. 3, 728; Johnston report, October 20, 1864, *OR*, vol. 38, pt. 3, 616; McMurry, "The Atlanta Campaign: A New Look," 5 ("In his report...Johnston sought...to demonstrate that in every dispute he had been correct and that his critics—

Years after the war, when feelings against Hood had hardened, Johnston again brought up the aborted attack at Cassville and went to some length in his *Narrative* (1874) to indict Hood for faulty generalship:

> When General Hood's column had moved two or three miles, that officer received a report from a member of his staff, to the effect that the enemy was approaching on the Canton road, in rear of the right of the position from which he had just marched. Instead of transmitting this report to me, and moving on in obedience to his orders, he fell back to that road and formed his corps across it, facing to our right and rear, toward Canton, without informing me of his strange departure from the instructions he had received. I heard of this erratic movement after it had caused such loss of time as to make the attack intended impracticable; for its success depended on accuracy in timing it. The intention was therefore abandoned.

Johnston's *Narrative* came out in spring 1874; an advance review appeared on January 8 in the *New Orleans Picayune* (critical of Johnston as one of those who "whet their insatiate appetites for revenge upon the slain heroes of the late war, and spit their venom on those who survive"). Living in New Orleans at the time, Hood thus learned of the self-serving thrust of Johnston's book. Before the end of January, as he prepared to leave for Europe (for reasons of health), Jefferson Davis wrote his wife, Varina, "Hood is quite excited and anxious for *others* to fight Johnston. *Says* he will make a reply to the part especially affecting himself."

Hood had already begun collecting information from other Confederate officers for a memoir of his war service. For instance, in March 1874 he asked Col. Taylor Beatty what he remembered about Cassville. Beatty had kept a diary (currently in the Southern Historical Collection at Chapel Hill) and he consulted it before answering Hood in his letter of March 29:

> As we marched across an open valley…one of the soldiers called my attention to a dark line off to our right, saying they were Yankees. I called your attention to the fact, but you said it could not be so, but must be our cavalry, if it was a body of men. Falling back, the same soldier (to whom I had said I thought the dark line was a fence or hedge), said that they were throwing out skirmishers in our direction, and I at once called your attention to that fact. You halted the division, and ordered General Hindman

---

especially Hood—had always been in error"); Mackall, "Memoranda of the operations at Cassville on May 19, 1864," *OR*, vol. 38, pt. 3, 621–22; Hood report, February 15, 1865, *OR*, vol. 38, pt. 3, 635; Stewart report, June 5, 1864, *OR*, vol. 38, pt. 3, 818; Austin report, May 29, 1864, *OR*, vol. 38, pt. 3, 862; Clayton report, May 29, *OR*, vol. 38, pt. 3, 833; Kurtz, "General Joseph E. Johnston's Review of Sherman's *Memoirs*."

> to send out a body of skirmishers to find out who they were. In a few minutes a sharp skirmish was in progress, and several of our men were wounded and killed in your immediate proximity. I recollect very distinctly that five men were hit at one time by the fragments of a shell, which exploded not more than twenty-five yards from where you were sitting on horseback.

Publication of Johnston's *Narrative* in April 1874 prompted Hood to write a series of long articles as a "Reply to Certain Statements Contained in Gen. Joseph E. Johnston's Book," some of which he submitted to the *New Orleans Times* and which were quickly printed. In his second "Reply" (*Times*, May 31, 1874), Hood published not only Colonel Beatty's letter of March 29, but also that of Major Austin of the 14th Louisiana Sharpshooters, who lived in New Orleans at the time. The date of Austin's letter, May 26, 1874, shows that Hood was scurrying to find other officers' corroboration of his version of events at Cassville. "From my observations," Austin wrote,

> I am forced to believe that General Johnston makes an error in his book in discrediting the presence of the enemy on your right, while you were moving to the north, across the large open field, to get in position. If my memory serves me, your extreme right flank was not covered by cavalry at all, as is not only usual but most essential in a movement such as you were making, and you must have had to rely on information of the enemy in that quarter from your staff and escort.

Hood's "Reply" impressed readers. Mrs. Leonidas Polk wrote to Hood on June 10, stating that "I presume I have to thank you for a copy of the *Times* containing the second of your well discussed strictures on Gen. Johnston's book." She added that "Col. Gale who from his position near my husband is familiar with the whole campaign, endorses every word you say."[13]

---

[13] J. E. Johnston, *Narrative*, 321–22; Strode, *Jefferson Davis*, 389; "A.," "Gen. Johnston's Book," *New Orleans Picayune*, January 8, 1874 (my thanks go to Anthony DelRosario, coordinator of research help desk, Howard-Tilton Memorial Library of Tulane University for providing me with copy of this article); McMurry, *John Bell Hood*, 196–98; Govan and Livingood, *Different Valor*, 385 (publication of Johnston's memoir in April 1874); "Gen. J. B. Hood's Reply to Certain Statements Contained in Gen. Joseph E. Johnston's Book. No. 1," *New Orleans Times*, May 10, 1874; "Gen. J. B. Hood's Reply to Statements of Gen. J. E. Johnston," *New Orleans Times*, May 31, 1874 (including Austin and Beatty letters); Hood, *A & R*, 101–103 (also quoting Austin and Beatty letters); Mrs. Leonidas Polk to "My dear General," June 10 [no year, but presumably 1874], S. Hood, *Lost Papers*, 88; Miller, *John Bell Hood*, 204 ("Hood began publishing attacks at once," but without citing the *New Orleans Times* or any other "attack").

But Johnston would continue to have his say. Three years after publication of his *Narrative*, Johnston was asked by editor Alexander K. McClure of the *Philadelphia Weekly Times* to contribute an article in the newspaper's series, "The Annals of the War." Johnston agreed, and wrote his article as a rejoinder to some of Sherman's statements made in his *Memoirs*, which were published in 1875. Johnston's "The Dalton-Atlanta Operations" appeared in the *Times* issue of November 3, 1877. In the article Johnston repeated and sharpened his contention that Hood had erred at Cassville, referring to "Hood's errant movement" back to his starting position on Polk's right based on "a wild report brought to him." Johnston even claimed that "neither this information nor his action upon it was reported," which was of course not true; Brigadier General Mackall, Johnston's chief of staff, had himself sent courier back with word of the surprise enemy appearance and Hood's halting of the planned assault.

The willfulness—indeed stubbornness—with which Johnston held to his version of the events at Cassville leads biographer Symonds to conclude that Johnston's unwillingness to accept Hood's version "was primarily a measure of his own disappointment" for having to cancel the attacking battle for May 19, especially as he had so loudly proclaimed its sanguine prospects to his troops.

To be sure, in military historiography, the solidest corroboration of any version of events on the battlefield comes when the enemy says the same thing that you are. Volume 38, part 2 of the *Official Records* was published in 1891, well after Hood's death and the very year of Johnston's. Albert Castel credits

---

The *Times* printed at least two of Hood's "Reply" articles in its issues of May 10, May 31, and June 7, 1874. Beyond these columns (which form the basis of chs. 4–7 in *A & R*, 69–128), I see none further in the *Times*. My guess is that Hood wrote two more "Reply" articles but did not send them to the *Times*, saving them instead for his memoir (see *A & R*, chs. 8 and 9).

Hood's biographers have hitherto been unaware of Hood's articles for the *New Orleans Times*. I wish to express my gratitude to Sean Benjamin, public services librarian, Louisiana Research Collection of Tulane's Howard-Tilton Library, for making available to me the library's pristine, original issues of the *Times*, which at the time of this writing have not been microfilmed or digitized.

The article by "A." on "Gen. Johnston's Book" prompted the editors of the *Picayune* to observe, "We shall regret the publication of Gen. Joseph E. Johnston's book if it should revive the controversies among the gallant chiefs of the Confederate army, which arose during the war, and especially if so true, efficient and heroic an officer as J. B. Hood should be brought under criticism or censure" (untitled editorial, *New Orleans Picayune*, January 10, 1874). I thank Dr. Gordon Jones, senior military historian at the Atlanta History Center, for his kind assistance in helping me find this and other articles in the postwar *Picayune*.

Atlanta historian Wilbur Kurtz for having pored through the *OR* and finding the reports of Col. Edward M. McCook and Lt. William B. Rippetoe in volumc 38, part 2. Castel cites Kurtz's unpublished paper, "General Joseph E. Johnston's Review of Sherman's Memoirs, published in Drake's Annals of the Army of Tennessee, Nashville, April 1878" in connection with his statement that after having identified the Federal troops harassing Hood's flank at Cassville, Kurtz concluded "that Johnston lied about what happened on May 19 in order to preserve his own reputation." In his characteristic prose, Kurtz indicted Johnston ever more colorfully than that, regarding the engagement of Stewart's troops on the morning of May 19:

> Didn't Johnston wonder what all the shooting was about? ...Far from that, in Johnston's book, it didn't rise to the dignity of a simple fact. It is doubtful if any battle of its size, in world history, has been so effectually erased from its pages by the simple process of declaring a commanding officer a deluded prevaricator, and an embattled enemy mere phantoms. Hood may have had a woeful countenance but he never tilted at windmills.

Kurtz concluded that Johnston's stubborn position was "obviously a face-saving measure to rationalize his failure to fight at Cassville after his flaming proclamation," and that "it would only pass for history only as long as the complete story awaited access in certain records."

Of these "certain records," the report of Colonel McCook, commanding the 1st Cavalry Division in Thomas's army, is the most important. Writing on May 20, the day after the fighting, McCook stated that he and his troopers had been ordered to ride toward Cassville. "I got within four miles of the place, skirmishing nearly all the way, driving them back. The advance developed the fact that the enemy were in heavy force in front of Cassville. One division of them (Stevenson's) advanced on me in line of battle. I was compelled to withdraw." Of interest, McCook added that the fighting did not end on the morning of the 19th after his column withdrew and Hood pulled his infantry back to their original lines. That afternoon Major General George Stoneman ordered McCook to attack the enemy in his front. "There was nothing but infantry in my front," he reported later.

> I drove them from two lines of rifle-pits. The Second Indiana in a saber charge captured one entire company of Eighteenth Alabama. Part of the Eighth Iowa in a charge also killed and captured a number of the enemy. My artillery knocked one of the enemy's guns and one of their caissons all to pieces, which they left on the field. The prisoners inform me that I was

> fighting Stevenson's division in the morning and Stewart's in the afternoon. All the prisoners taken are infantry. We captured two of their ammunition wagons and a quantity of flour, corn, and bacon.

Such was the fighting that Johnston never referred to in his reminiscences. The artillery McCook mentions was the 18th Indiana Battery, which was attached to McCook's division for the first three months of the campaign. Lieutenant Rippetoe reported shelling the enemy at Cassville "about 10 A.M.," and continued firing well into the afternoon. So much, as Wilbur Kurtz would say, for "phantom troops."

More recently, Timothy F. Weiss has sifted through the evidence and concluded, regarding Yankee troops' presence on Hood's flank, "they did appear there." Weiss echoes several historians as to which Federal units they may have been (including Castel on Edward McCook's cavalry division). Two decades ago local historian William R. Scaife strengthened the evidence of McCook's Union cavalry at Cassville by mentioning in his *Campaign for Atlanta* (1993) "extensive ground evidence." Scaife elaborated on this in another monograph on Cassville in which he explained the ground evidence, exhumed by relic hunters, as "dropped and fired Spencer Carbine cartridges"—sure signs of McCook's cavalry.[14]

---

[14] Cozzens, introduction to Cozzens and Girardi, *The New Annals of the Civil War*, xi; J. E. Johnston, "The Dalton-Atlanta Operations," 333 (Johnston's article was reprinted in Drake, *The Annals of the Army of Tennessee and Early Tennessee History*, 1–13); Symonds, *Johnston*, 293; Castel, *Decision in the West*, 590 n.102; Kurtz, "Johnston's Review of Sherman's *Memoirs*," 9; McCook report, May 20, 1864, *OR*, vol. 38, pt. 2, 751–52; Rippetoe report, September 10, *OR*, vol. 38, pt. 2, 800–801; Weiss, "'I lead you to battle'," 443–44; Scaife, *Campaign for Atlanta*, 45 n.16; Scaife, "Affair at Cassville," 7 (I thank Greg Biggs of Clarksville, TN, for copy of Scaife's pamphlet).

Johnston's version of Cassville was accepted by some Southerners long after the war. Example is the reminiscence of Maj. Joseph B. Cumming, staff officer for Maj. Gen. William H. T. Walker (infantry division commander in Hardee's corps). Writing of Johnston's intended attack, Cumming stated, "Unfortunately this promising movement was thwarted by General Hood's failure to attack owing to wrong information communicated to him by one of his staff officers" (Cumming War Recollections, 55).

Atlanta historian William R. Scaife was no fan of John Bell Hood, but he did not argue with the *Official Records*. Based on McCook's and Rippetoe's reports—plus another, by Maj. David A. Briggs, 2nd Indiana Cavalry ("and captured 35 rebels of Company C, Eighteenth Alabama [rebel] Infantry")—Scaife composed a map of the action at Cassville, May 19, which clearly depicts Stewart's division moving to swing eastward to form line of battle against McCook's and Rippetoe's troops, riding from Wofford's Crossing on the Canton Road (Briggs report [undated], *OR*, vol. 38, pt. 2, 785; Scaife, *Campaign for Atlanta*, 45, f.p.48 [map]).

Most students of the Georgia Campaign agree with Hood. A writer on the affair at Cassville concludes, "The person to blame (or credit, depending on one's point of view) for

But that's what we know today. On the afternoon of May 19 Generals Johnston and Hood may have exchanged words about the aborted attack, but there is no record of them having done so. As word spread through the Southern ranks that the much-heralded attack had been canceled, officers and men expressed their bitter disappointment. General Mackall confessed to his wife, "I could not restrain my tears when I found we could not strike." Still, after Johnston ordered the army to fall back to high ground southeast of Cassville, the main task was laying out a new defensive line. The close skirmishing of Federal cavalry—not just in Hood's front, but Hardee's too, as his corps marched east to rejoin the rest of the army—indicated that the enemy would close up pretty fast. That afternoon Generals Johnston, Polk and Mackall supervised the laying out of a new defensive line southeast of Cassville, on a ridge behind the town cemetery. When the infantry filed in, Hood's corps took the right, Polk the center, and Hardee moved to hold the left. Sherman's troops came up and both sides engaged in light skirmishing and cannonading. As the Federals approached the town, Confederate sharpshooters fired from its buildings before falling back. In his new position Johnston hoped Sherman would attack; for his part Sherman was inclined to give battle. Both sides expected heavy fighting on the 20th.

Hood's failure to attack at Cassville was neither Hood nor Johnston, but Edward M. McCook, whose impetuous advance down the Canton Road into Hood's rear had caused Hood to pause and reflect" (Bonin, "Lost Victories," 37).

The authoritative Castel seems to have settled the matter once and for all, stating that fully two Federal cavalry divisions, McCook's, followed by Stoneman's, were bearing down on Hood's flank via the Canton road (*Decision in the West*, 201). Recent writers such as James Donnell have accepted this without so much as mentioning that there was once an argument between Johnston and Hood over "phantom troops" (Donnell, *Atlanta 1864*, 41).

Confirmation of the scholarly consensus is found today in the pavilion built by the Works Progress Administration in the mid-1930s, one of five describing the Atlanta Campaign for motorists along U. S. Highway 41 from Ringgold to Dallas. At the fourth pavilion, at Cassville, the original U. S. government marker has been replaced by one from the Georgia Historical Commission, whose text clearly states that McCook's cavalry division bearing down on his right flank caused Hood to cancel the attack. A colored map on the plaque shows the menacing column coming in from the northeast.

Nevertheless, some today are still siding with Johnston about those "phantom troops." Steven H. Newton writes that Hood on the morning of the 19th "mistook wandering Federal cavalry for a strong enemy force on his own flank." As for the Confederates of the 18th Alabama overrun and captured in McCook's cavalry charge that day, Newton claims that the Alabamans allowed themselves to be taken prisoner on the skirmish line on May 20 rather than desert openly (Newton, "What Talent Might Have Achieved," 34. I thank Mark Dunkelman of Providence, RI, for calling my attention to Newton's article.).

Sgt. Rice C. Bull of the 123rd New York got a chance to view the Rebel works on the high ground south of town. That night he and his comrades went to sleep "thinking about what the morrow might bring for those frowning breastworks looked ominous." Johnston in his memoir termed his Cassville line "the best we had occupied." But the Confederate position had a flaw. In the afternoon Brig. Gen. Francis Shoup, chief of the army's artillery, pointed out to Johnston that a sector of Polk's line held by French's division was exposed to enemy artillery enfilade fire. Johnston thought the Union guns were too distant to cause much trouble; in any case, French's troops could take cover in the dug traverses. But Hood's sector too came under fire. "I saw one battery of ours knocked to pieces," recalled Brig. Gen. Arthur M. Manigault, brigade commander in Hindman's division, "and the gunners driven from their guns in less than fifteen minutes." That evening Hood and his staff themselves came under fire when they rode on inspection of their lines. "I shall long remember," wrote Colonel Beatty, "that ride as one of the most disagreeable it has ever been my fortune to take, being, as we were, continually under a heavy cross-fire of artillery."

Polk asked Hood to meet him at his quarters to discuss the problems with their position; they agreed that it was untenable. Polk then asked General Johnston to meet with them and about 9 P.M. the two corps commanders explained that if the Yankees attacked in the morning, they could not hold out more than an hour once the enemy batteries opened fire. To be sure, Hood and Johnston differ in their reminiscences as to who said what. (Hood printed in *Advance & Retreat* the long letter written in 1874 by Polk's chief engineer, Capt. Walter J. Morris, complete with a map to prove his point.) An observer, Maj. Gen. Samuel French, later recorded his remembrance:

> Soon after supper Generals Johnston, Polk and Hood went to General Polk's office, and General Johnston asked me to go with them. The matter presented to the meeting was "Can we win the battle on the morrow?" Hood said he thought not, for if attacked in the morning, he would not be able to hold his line, because it was enfiladed by the guns of the enemy, now in position, and that General Polk's line was also enfiladed, and could not be held against a vigorous attack; or words to that effect. General Polk confirmed Hood's statement in regard to his line. General Johnston argued for the maintenance of his plans very firmly.

Johnston remembered that the conference lasted more than an hour, and that he finally gave in to Polk and Hood. Lieutenant General Hardee arrived at the meeting late; he disagreed with the need to retreat. But as Johnston put it, "I yielded at last, in the belief that the confidence of the commanders of two of the three corps of the army, of their inability to resist the enemy, would inevitably be communicated to their troops." He ordered the army to retreat that night across the Etowah. In his *Narrative* he termed the whole episode an "error."[15]

Johnston got more bad news on May 20. The telegram he had sent Richmond on the 16th announcing the army's retreat from Resaca provoked an irritated, even angry reply from President Davis: "Your dispatch of the 16th received; read with disappointment. I hope the re-enforcements sent will enable you to achieve important results." "I cannot judge of the circumstances which caused General Johnston to retire from Dalton to Calhoun," Davis wrote Lee. The president would have been even more disappointed if he had known that Johnston had retreated again from Cassville. On the 20th the Confederate army crossed the Etowah and took position along the high

---

[15] McMurry, "Rocky Face to Dallas," 56; Mackall, "Journal," *OR*, vol. 38, pt. 3, 984; McMurry, "Cassville," 9, 45; Castel, *Decision in the West*, 203; Bull, *Soldiering*, 110 ("in the houses were sharpshooters"), 111 ("frowning breastworks"); J. E. Johnston, *Narrative*, 322–24; S. G. French to "Editor [*New Orleans*] *Picayune*," December 12, 1893, reprinted in S. G. French, "Lt. General Leonidas Polk at Cassville," 4–5; French, *Two Wars*, 196 ("enfiladed Hoskin's gun"); Manigault, *A Carolinian Goes to War*, 187; Hood, *A & R*, 103 (Beatty to Hood, March 29, 1874), 106–16.

Hood's printing of Captain Morris's letter to W. M. Polk about the weakness of the Cassville line and the necessity of retreat on the night of May 19–20—more than five pages of text plus a map (printed in the *OR Atlas*, 2:plate LXII, 7)—indicates Hood's sensitivity to Johnston's allegation that he and Polk had acted irresponsibly. Hood dates the letter as June 25, 1874. He apparently got Morris's report from General Polk's son, who sent it to him in January 1878. "My intention was that it should first be published in Genl Polk's memoirs, but they have been so long delayed as to make it doubtful whether they will appear in a year," the younger Polk wrote, "feeling that it is simple justice to place the papers in your hands" (W. M. Polk to Hood, January 31, 1878, S. Hood, *Lost Papers*, 99). On the other hand, Morris's letter has been printed from the manuscript in the Leonidas Polk Papers at Chapel Hill, carrying the date of June 25, 1878 (Hewett, Trudeau, and Suderow, *Supplement to the Official Records*, part I, vol. 7, 133–39).

On May 22 Sergeant Bull of the 123rd New York got a chance to inspect Johnston's Cassville fortifications and was much impressed by their strength. "They were the finest we had seen up to that time and it must have taken much labor to build them," he wrote in his memoir. "Johnston's Army could not have reached there more than a few hours before we were in front of them so they must have been constructed prior to the start of the campaign. They were no ordinary breastworks that could be built overnight but strong elaborate fortifications with redoubts and abatis in front" (*Soldiering*, 113). Bull's reminiscences were based on his wartime diary and letters, and scholars of the Georgia Campaign have found them reliable. With this statement he contradicts the established notion that Johnston laid out his Cassville line only after Hood cancelled the assault on May 19. At this point Bull's conjecture that it had been laid out well before cannot be confirmed; it is enough at least to mention it.

ground four miles south of the river around Allatoona. Maj. Gen. G. W. Smith, in military retirement and then serving as president of the Etowah Manufacturing and Mining Company (known as Mark Cooper's iron works), oversaw the loading up of its valuable machinery. Rearguard troops burned the Etowah bridges. More, for the first time in the campaign, Johnston ordered his troops to rip up track of the Western & Atlantic. Jeffrey Lash, chronicler of Johnston's handling of railroads during the war, writes that after Cassville, "Johnston then determined to hinder Sherman's forward movement by devastating the Western & Atlantic Railroad below the Etowah." Confederates took up rails south of the river not only to Allatoona, but south toward Big Shanty and Marietta. Actually, according to New York Sgt. Rice Bull, Rebels had damaged the W. & A. north of the river as well. While Sherman allowed his troops to rest for several days at the Etowah, May 20–22, Bull recorded in his diary, "during that time the railroad could be repaired.... It was wonderful how rapidly our men could make repairs as we advanced."

Meanwhile Johnston had to let Richmond know that he had retreated across the second major river in north Georgia. On May 21 he wrote the president:

> In the past six days the enemy has pressed us back to this point, thirty-two miles.... I know that my dispatch must of necessity create the feeling you express. I have earnestly sought an opportunity to strike the enemy. The direction of the railroad to this point has enabled him to press me back by steadily moving to the left and by fortifying the moment he halted. He has made an assault upon his superior forces too hazardous, and in making this retrograde march we have lost much by straggling or desertion.

Johnston's retreats were starting to impair the men's morale. "Georgia troops dropped off; all in pretty good spirits up to falling back from Cassville," recorded Lieutenant Mackall. "Change of line not understood but thought all right," he added, "but night retreat after issuing general order impaired confidence." Tellingly, Mackall noted, "troops think no stand to be made north of the Chattahoochee." The troops were not just thinking it; officers were, too. Colonel Beatty of Hood's staff recorded in his diary for May 20, "left last night at 11 P.M. marched all night to Etowah river.... Reported now that we are to fall back to the Chattahoochee 8 miles this side of Atlanta." After Cassville, "for the first time the men begin to grumble and grow dispirited," observed a surgeon from Tennessee. "Gen'l J. *is not the man* we thought him," wrote Lt. William D. Gale, General Polk's son-in-law and aide, in his diary on May 21, adding that Johnston's failure as army commander was "beyond doubt and I fear beyond repair."

"Indications are that Chattahoochee will be the line," wired Brigadier Marcus Wright in Atlanta to Bragg on May 23. Even Johnston was feeling demoralized. "I have seen so much beautiful country given up as to be made unhappy by it," he wrote his wife, Lydia, on May 23. "You can not imagine how disheartening it is & at the same time humiliating to see apprehension of the people of a country abandoned to the enemy." "I have never been so little satisfied with myself," he confessed to Lydia; "have never been so weak."

The men in the ranks were split. "Our army is in fine spirits," wrote Lt. John Davidson to his wife on May 21, "no one but straglers think we are whipped." Capt. Irving A. Buck of General Cleburne's staff agreed. "I am happy to say our army is not at all demoralized and confidence is unshaken in General Johnston," he wrote on the 21st. Sgt.-Maj. Robert Banks agreed: "the troops are in fine spirits and eager for decisive fight," he wrote home on the 23d. "The army is in excellent spirits and has unbounded confidence in our General," wrote Capt. James I. Hall of the 9th Tennessee to his parents on May 30; "he deserves it." Artillery officer Andrew Jackson Neal wrote his mother that some of the men believed that Johnston would fall back on Atlanta; "I think he will turn upon the enemy half way between here and Marietta and we will race for the Tennessee." On the other hand, some were grumbling about Johnston's retrogrades. In Capt. Samuel Foster's company of the 24th Texas Cavalry, "some say we are going to Florida and put in a pontoon bridge over to Cuba and go over there," he recorded in his diary on May 24 before adding, "while others contend that some Yank would put a torpedo under it and blow it up." Many soldiers admired General Johnston but recognized the difficulties he was facing. Pvt. J. W. Cotton wrote that Old Joe was "determined to fite them some where he has been trying to get them to fight him now for 2 weakes but they wont do it they keep flanking him."

"Waverly"—really Henry Watterson, the veteran journalist writing for the *Memphis Appeal*—sent in a column dated, "In the Field, Monday, May 25," which told this story:

> A merchant of Atlanta said to Gen. Johnston, "There is a great fear of your falling back behind Atlanta, general." "Indeed!" replied the general, "do you think that Whitehall street will be a barrier against the enemy?' "No sir, but they are alarmed." "It is quite groundless. We shall hold Atlanta," was the laconic response of the chief as he turned away.

Watterson—this time writing as "Shadow"—got in a lot of hot water for expressing his own opinion in another column for the *Mobile Advertiser and Register*:

> ...you want to know, I suspect, my candid opinion of the real situation of Atlanta. I will tell you frankly.... Taking the fall of Nashville and Chattanooga as a reminder, the events which have led us from Dalton hither look very much the way they looked from Bowling Green to Tullahoma. I observe my own feeling on the subject, and I think Atlanta will fall.

There was also much ambivalence in Johnston's army about their commander's course of conduct. J. N. Wyatt of the 12th Tennessee expressed it under date of May 25. "There is some dissatisfaction with Gen. Johnston for retreating so much," he wrote, "but still we all repose the greatest confidence in him as a general." Sgt. John W. Hagan effectively demonstrated the same ambivalence by contradicting himself several times in a single letter, written to his wife on May 18.

> ...we have give up a large Scope of country & I fear we will never get it back. the country we are pasing through now is the finest I ever saw & the finest wheat crops growing & it will all be destroyed. I beleave when we make a Stand we will give the yankees a thrashing but I fear we will never get the country back. I beleave unless we get heavey reinforcements soon that the yanks will take Atlanta easey. the truith is we have run until I am geting out of heart & we must make a Stand Soon or the armey will be demoralized, but all is in good Spirits now & beleave Gen Johnston will make a stand &whip the yankees badley....

In truth, Johnston was making a stand. By some accounts, the Confederates' position at the Allatoona heights was as strong as their line at Rocky Face Ridge. Allatoona was noted for a deep, 175-foot cut through the mountain in which ran the Western & Atlantic. Johnston hoped Sherman would attack him there, but Sherman knew the place and its formidability; in the 1840s as a young officer he had ridden through the area. Instead of attacking Johnston at Allatoona, he planned to give his troops a few days' rest, cross the Etowah downriver, and march toward Dallas, fifteen miles south of the river, thus flanking Johnston. From Dallas he would approach Marietta, the railroad town twenty miles south of Allatoona Station, as measured by the railroad. Sherman remained supremely confident. From the Thomas V. B. Hargis house in Kingston, which was his headquarters May 19–23, Sherman wrote his wife, Ellen, "I have no doubt we must have a terrific Battle at some point near the Chattahoochee." But of the final issue he had little doubt. "If I cant take Atlanta and Stir up Georgia considerably I am mistaken."[16]

---

[16] Castel, *Decision in the West*, 209; Davis to Johnston, May 18, 1864, *OR*, vol. 38, pt. 4, 725; Davis to Lee, May 20, *OR*, vol. 51, pt. 2, 952; Hudson, *The Odyssey of a Southerner*, 158; Lash, *Destroyer of the Iron Horse*, 144; Bull, *Soldiering*, 112; Johnston to Davis, May 21, 1864,

In setting Dallas as his objective, Sherman knew he was leaving the railroad. On May 20 he ordered his three army commanders to load sufficient

---

OR, vol. 38, pt. 4, 736; Mackall, "Journal," *OR*, vol. 38, pt. 3, 984–85; Beatty diary, May 20; Blount, *The Battles of New Hope Church*, 38; Butkovich, *Pickett's Mill*, 29–30; Kennett, *Marching through Georgia*, 79 (Tennessee surgeon); T. L. Connelly, *Autumn of Glory*, 366 (Colonel Gale); Wright to Bragg, May 23, 1864, *OR*, vol. 52, pt. 2, 672; McMurry, *Atlanta 1864*, 84 (Johnston to wife, May 23); W. J. Cooper, *Jefferson Davis, American*, 483 (quoting more of same letter); Peacock, "A Wartime Story," 87; Buck, "Headquarters Staff," 108 (Buck letter, May 21); Banks, "Civil War Letters of Robert W. Banks," 211; Fleming, *The Confederate Ninth Tennessee Infantry*, 145; Neal to mother, May 20, 1864, Neal Papers, MSS 218, box 1, folder 2; Foster, *One of Cleburne's Command*, 80; McNeill, "A Survey of Confederate Soldier Morale," 7; "Waverly," "From the Front. General Johnston's Army," *Memphis Appeal*, May 24, 1864; "Shadow," "Our Army Correspondence. Letter from Atlanta," *Mobile Advertiser & Register*, May 29; S. Davis, *Atlanta Will Fall*, 3, 68–69; Wyatt, "Dalton-Atlanta Campaign," 519; Hagan, "The Confederate Letters of John W. Hagan, Part II," 272; Bonds, "Sherman's 1st March through Georgia," 36 ("[B]y a mere accident I was enabled to traverse on horseback the very ground where in after years I had to conduct vast armies and fight great battles"); McMurry, "Rocky Face to Dallas," 56–57; McMurry, *John Bell Hood*, 110; Barnwell, *Barnwell's Atlanta City Directory*, 101 (on railroad mileage from Allatoona to Marietta); "House Site—Thomas V. B. Hargis," in Scruggs, *Georgia Historical Markers*, 21–22 (also in Kurtz, historical marker texts, box 33, folder 1); Simpson and Berlin, *Sherman's Civil War*, 639 (Sherman to Ellen, May 22).

On Confederate soldiers' morale during Johnston's command in the Georgia Campaign, Richard McMurry has concluded, "What seems to have happened was that many of Johnston's men who began the campaign with high hope gradually lost confidence as the army fell back and then doubt began to creep into their minds about the safety of Atlanta, the Confederacy's prospects for victory, and the ability of their commander to defeat Sherman" ("Confederate Morale in the Atlanta Campaign of 1864," 233). Larry J. Daniel agrees: "Clearly, those who claim that Johnston's retreats did not adversely affect morale do so in the face of significant evidence to the contrary" (*Soldiering*, 141–42). More recently, Bradley R. Clampitt concludes that "the decision not to fight at Cassville momentarily dispirited much of the army," but that within days morale began to improve. "We are going to demolish them and that soon," wrote a soldier in the 17th Alabama (Clampitt, *The Confederate Heartland*, 60). Steven H. Newton adds to the discussion the astute observation that "those elements entering the campaign with the highest spirits and confidence were least affected by his retreats, while those whose will was shaky at the outset lost heart fairly quickly" ("'Formidable Only in Flight?'," 54).

Johnston was sensitive about his army's morale. In his telegram to Davis of May 21 he had stated that in retreating from the start of the campaign "we have lost much by straggling or desertion." When this was printed in the *Official Records*, someone inserted "we have *[not]* lost much" (italics added). The editors of the *Jefferson Davis Papers* confirm that Johnston's statement reads "we have lost much" (Crist, Williams, and Dillard, eds., *Papers of Jefferson Davis*, 10:434). It is plausible that the *OR* editors handling this and other Southerners' documents were former Confederates; see McMurry, "The *Official Records* of the Rebellion," 369 ("former Rebels were hired to do part of the work of compiling and editing the documents"). If so, it is also plausible that a Southerner sympathetic to Johnston may have intentionally sought to change the otherwise negative import of the general's telegram.

provisions in wagons to feed their troops for twenty days. With replacements and newly arrived regiments, Sherman figured he would have at least 80,000 men when he started moving. Moreover, he ordered the XVII Corps, then traveling through Alabama, to join him via Rome and connect with McPherson's army as it advanced toward the Chattahoochee. The Federal commander had every reason to regard his prospects as bright.

Not so the Confederate commander. In his letter to the president, Johnston had to acknowledge that Sherman had kept flanking his army by the left, forcing it back along the line of the railroad. Johnston expressed little confidence that this trend of events would change, which got Richmond talking. "Johnston is falling back as hard as he can," wrote Col. Josiah Gorgas of the Ordnance Department. "I trust the country will sooner or later find out what sort of General he is. I don't think he will suit the emergency." Robert G. H. Kean, head of the Bureau of War, recorded Secretary Seddon's displeasure. "The Secretary is dissatisfied," Kean wrote on the 22nd. "He told me this morning that General Johnston's theory of war seemed to be never to fight unless strong enough certainly to overwhelm your enemy, and under all circumstances merely to continue to elude him. This is a very just criticism." The next day, John Beauchamp Jones, clerk in the War Department, recorded in his diary "Gen. Johnston, without a defeat, has fallen back to Calhoun, Ga." On the 28th, Jones entered alarmingly, "I saw a dispatch, to-day, from Gen. Johnston to his Chief Commissary, at Atlanta, ordering him, after reserving ten days' rations, to send the rest of the stores to Augusta!" Jones's exclamation seems to suggest his belief that Johnston would make no determined stand to hold Atlanta.

On May 21, the day Johnston wrote his letter to the president, Hood was writing one of his own. To Davis Hood explained that his aide, Col. Henry Brewster, would bear this note in his visit to Richmond and personally give it to the president. "I think it would be well for you to have a conversation with him in relation to our affairs &c. &c." What Brewster eventually told Davis about the army's "affairs" can be guessed from Hood's early letters to Davis, Bragg, and Seddon. It can also be affirmed by Mary Chesnut's diary entry after conversation with Brewster. "He said Joe Johnston was kept from fighting at Dalton by no plan—no strategy," she wrote. She recalled Brewster's words: "'What is the matter with him? Overcautious.'" She added what Brewster told her: "all this delay is breaking Hood's heart." And then this: "So much retreating would demoralize even General Lee's army." Once again Hood was breaking chain of command in writing the president and sending a personal representative to meet with him. In doing so, it must be concluded that the ambitious Hood was positioning himself for promotion to still higher level, should that opportunity present itself. That time might indeed come soon. Mrs. Chesnut recorded her belief that the "Joe Johnston disaffection is eating into the very vitals of our distracted country."[17]

Johnston soon realized that Sherman would not attack him in the Allatoona hills. Even though Lieutenant Mackall wrote privately on May 22, "scouts not enterprising," videttes brought in news that in the previous night the enemy were fortifying at Gillem's Bridge across the Etowah. Gillem's is more than ten miles west of where the W. & A. crosses the river; clearly Sherman was not going to follow the railroad, but cross well to its west. In orders issued May 22, Sherman prescribed different routes for his three armies to march toward Dallas. Schofield's XXIII Corps, on the army group's left, would cross at Milam's Bridge, upstream from Gillem's, then head south toward Burnt Hickory. Thomas's would cross at Gillem's and Island Ford, then head south via Euharlee and Stilesborough. McPherson, on the right, would cross at Wooley's Bridge west of Kingston. Confederate cavalry had burned Milam's Bridge and tried to do the same at Gillem's but were only partly successful. Federal forces had little difficulty crossing the Etowah; their main cause of delay was Thomas's re-routing of Hooker's corps to Milam's, which held up Schofield.

Confederate cavalry had reported enemy activity near Milam's and Gillem's on the evening of May 22—that day, a Sunday, Bishop Polk had preached to a throng that included generals Johnston and Hood. The next morning cavalry brigadier Lawrence S. ("Sul") Ross sent in word that the Yankees were crossing at Milam's, Gillem's, and Wooley's. Johnston was ready. Lieutenant Mackall recorded that as early as 3:30 P.M. on the 22nd, "lieutenant-generals and chief of artillery notified to have everything in read-

---

[17] Special Field Orders No. 9 and No. 11, May 20, May 22, *OR*, vol. 38, pt. 4, 271–72, 288–89; Castel, *Decision in the West*, 214; Johnston to Davis, May 21, *OR*, vol. 38, pt. 4, 736; Govan and Livingood, *Different Valor*, 277; McMurry, *John Bell Hood*, 110; Eliot, *West Point in the Confederacy*, 100–101 (quoting Hood's "hitherto unpublished" letter); McMurry, *Atlanta 1864*, 130–31 (on Gorgas and Kean); J. B. Jones, *A Rebel War Clerk's Diary* [2015], 2:195 ("without a defeat"), 198; Woodward, *Mary Chesnut's Civil War*, 616.

In his 2015 edition of Jones's *War Clerk's Diary*, Robertson observes that Jones saw correspondence that was not included in the *Official Records* (vol. 1, viii). Johnston's message to his chief commissary at Atlanta (Maj. J. F. Cummings, according to the "City Military Directory," printed in the *Intelligencer*, May 11, 1864), appears to be one of those, as I do not see it in the *OR*, vol. 38, pt. 4. Jones's exclamatory comment thus has not appeared previously in the literature of Johnston's retrograde movement back to Atlanta.

iness to move at moment's notice." Maj. W. E. Moore, army chief of subsistence, was notified to have three days' rations for each soldier's haversack. The army would march southwest from Allatoona toward Dallas by at least two roads. One would entail crossing Pumpkinvine Creek to the west, and engineers were ordered to have two pontoon bridges for the troops to cross on, plus clear a road to shorten their distance. Local guides were found who knew the country south of the Etowah.

Thus on the morning of May 23, when Confederate cavalry reported the enemy movements, Johnston ordered Hardee's and Polk's corps to start marching toward Dallas, the country town where several of the roads from the Etowah bridges converged. But Johnston remained wary of Sherman's intentions. He kept Hood's corps at Allatoona ("watching crossings near railroad," noted Lieutenant Mackall) and even sent Wheeler and some cavalry riding north across the river to reconnoiter the enemy's positions. Wheeler spotted no infantry but captured a Union wagon train, burning some wagons and bringing back about eighty of them with 180 prisoners. Soon headquarters had enough information for Johnston to conclude that Sherman's whole army was bypassing Allatoona and to send Hood's troops moving to join the rest of the army on May 24. After a march of some eight miles, Hood took position near New Hope Church on the right of the Confederate line (apparently Johnston's preferred position for his corps); Polk held the center, Hardee the left. The six-mile line was in place by noon of May 25. In Hood's sector, facing northwest, Stevenson's division held the right, Stewart was at the center near the church, and Hindman held the left. All knew that action lay ahead. On the 24th Hood issued a general order to his troops: "the lieutenant-general commanding desires to say to the officers and soldiers of his command that in the coming battle their country expects of them victory." A. A. G. J. W. Ratchford closed with what Hood would himself have dictated, "Death is far preferable to defeat."[18]

---

[18] McMurry, "Mackall Journal," entry of May 22; Butkovich, *Pickett's Mill*, 40–44 (including map by author, 43); Castel, *Decision in the West*, 210 (map), 216–20; Special Field Orders No. 11, May 22, *OR*, vol. 38, pt. 4, 288–89; "Milam's Bridge," "Gillem's Bridge" and "Wooley's Bridge," Scruggs, *Georgia Historical Markers*, 16 (Kurtz, historical marker texts, box 32, folder 1); Mackall, "Journal," *OR*, vol. 38, pt. 3, 985–87; McMurry, *John Bell Hood*, 110; Roth, "Battles on the Dallas-New Hope-Pickett's Mill Line," 59; McMurry, "Rocky Face to Dallas," 60; McMurry, "Atlanta Campaign" dissertation, 148–52; McMurry, "The 'Hell Hole,'" 34; "Battle of Cassville, Ga. May 24th, 1864," 3 ("the army and medical stores which the Confederates captured with this wagon train were of immense benefit to Johnston's army"); J. W. Ratchford general order, May 24, 1864, *OR*, vol. 38, pt. 4, 741.

The battle of New Hope Church, May 25, was about to begin. Hood knew that Yankees were marching toward his front. On nearby Elsberry Mountain, observers saw columns of dust. Late in the morning Hood sent Col. Bushrod Jones's 35th/58th Alabama and Maj. John Austin's battalion of Louisiana sharpshooters forward from his main line to develop the enemy position. A firefight broke out when the Southerners encountered advancing enemy forces; after maybe twenty minutes, Hood's scouts fell back to the main line. They captured a Yankee who said he belonged to Geary's division and that Hooker's corps was headed in their direction. Upon receipt of the news, many of Hood's soldiers started building log parapets. Men of Stovall's brigade, at New Hope Church cemetery, did not fortify at all, relying on the vertical tombstones for their protection. Three batteries of artillery were brought up and placed along the line. Nevertheless, Johnston was a bit worried that the enemy would attack and break through, cutting off Stevenson's division, which was the army's far right flank. As Johnston rode among Stewart's men building their breastworks, perhaps to speed them in their work, "he told us that the enemy were 'out there' just three or four hundred yards," remembered Lt. Bromfield Ridley of Stewart's staff.

The enemy were indeed out there. Union brigadier John Geary's advance had been contested by Hood's recon; captured Confederates told him that Hood's entire corps was in front. Geary, aware that he was several miles away from the rest of the army, halted. Williams's division of Hooker's XX Corps marched toward Geary; General Sherman was informed. "Let Williams go in anywhere as soon as he gets up," he snapped, somewhat irritated at his forces' lack of progress. "I don't see what they are waiting for in front now. There haven't been twenty rebels there today." Sherman told Thomas to attack, with Maj. Gen. Daniel Butterfield's and Geary's divisions joining Williams in the assault.

The Union advance began around 5 P.M. Williams's division was on the right, Butterfield's the left, with Geary behind in reserve. The Northern infantry were formed in columns of brigades (one behind another) so that the front of Williams's and Butterfield's divisions was reduced. This had the effect of contracting the Federals' front to about the length of Stewart's line, which had three brigades (Marcellus Stovall's, Henry Clayton's, and Alpheus Baker's) in the main line and Randall Gibson's in reserve. The consequence was that at the initial point of attack the numerical superiority of Hooker's corps (16,000 men) was negated. It also meant that the Confederates' concentrated musketry and cannonfire would inflict heavier casualties among the packed assailants. The Federals had to cover a mile through woods and underbrush just to get into position for attack. Then Williams, riding behind the first line with Hooker, ordered the "double quick." First the Northerners had to clamber down into a wide ravine that they soon called "the hell hole." Climbing back up, they came under a hail of rifle and artillery fire from the Confederate line on the high ground ahead. Trying to return fire, many took cover behind rocks and trees.

The Confederates under attack appreciated their earthworks and barricades. "We had an excellent position," recalled H. J. Lea, "and our artillery and muskets were used freely and maintained our position, though the force of the enemy was much greater than ours." Lt. Andrew J. Neal wrote home later, "It is fun for our troops to stand in their trenches and mow down their lines as they advance." General Stewart, called "Old Straight" by his men, rode behind the line as his troops kept up their firing. Stewart's son, Lt. R. C. Stewart, was in line as Straight rode by. "Now, father," he called out, "you know you promised mother that you would not expose yourself to-day." The men in the ranks, even amidst the battle, took up the son's remark, laughing and repeating it down the line: "Now, father, you know you promised mother that you would not expose yourself to-day!" Johnston, at Hood's headquarters, sent a courier to ask if he needed help; Stewart declined: "My own troops will hold the position." Confederate cannon were particularly effective. Hood noted that the sixteen guns engaged, supervised by his chief of artillery, Col. Robert Beckham, "did great execution in the enemy's ranks" and, as he put it, "added much to their discomfiture." Union brigadier Williams also attested to the fire of the Rebels' artillery; "they poured into us canister and shrapnel from all directions except the rear."

After two and a half hours, darkness and rain ended the battle. Those Federals able to do so withdrew back to their start position. Hooker's corps lost 1,664 killed, wounded, and missing, including 302 missing, presumed dead and lying on the battlefield. The Confederates cheered over their smart little defensive victory at New Hope Church, which cost them around 475 men of 4,000 in Stewart's division. (Stevenson's division on the right was not engaged.) "All, both officers and men, never performed their duty better," reported Capt. R. R. Asbury, commanding the 52nd Georgia. "They exhibited by their coolness and bravery that they were determined to teach the invader that they were fighting freemen, who knew their rights and would dare maintain them." "General Stewart was with his men," noted a Confederate writing for the *Memphis/Atlanta Appeal*, "and by his constant presence set them an example which we take pleasure in saying was nobly emulated by all."

General Hood's presence on the field is undocumented; if he were at his headquarters it was because he evidently believed Stewart had the situation in hand. "General Hood repulsed the attack made upon him with great ease," wrote Thomas B. Roy, Hardee's chief of staff, who noted, "his second line was not brought into action at all." At the end of the day Hood had personally not contributed materially to the Southern victory beyond laying out his lines and positioning his three divisions. But the day had been won. "Stewart's division repulsed Hooker's corps," was Lieutenant Mackall's terse entry in his journal for May 25. That night Johnston slept at Hood's headquarters.[19]

During the night soldiers of both armies strengthened their earthworks. Pvt. Hiram Williams, one of Hood's pioneers, recorded, "the firing between the pickets was kept up all night long, so as to keep the men on either side from working on their breastworks....My regiment, however, managed to get very fair works erected." "We are getting to be expert with pick and spade," wrote another Southerner, who acknowledged that the hard work was worth it, as ditches of any kind helped "protect us from the bullets and fragments of shells which are continually flying around and about us."

On the 26th, both sides further deployed and fortified. Skirmishing and artillery fire occurred throughout the day and even though Sherman during the night of the 25th had thought of renewing his attack, he reconsidered at daylight upon seeing "a strong line of intrenchments facing us, with a heavy force of infantry and guns"—the woods between the opposing armies had been so shot up that the enemies could now discern each other's works. Thus Federals spent the day lengthening their line. McPherson marched into Dallas and took position as Sherman's right. Thomas held the center and

---

[19] Castel, *Decision in the West*, 221–26; Blount, *New Hope Church*, 45–71 (including the author's unfounded accusation that Hood "often turns to laudanum and whiskey for relief" of pain from his wounds, 47); McMurry, "Rocky Face Ridge to Dallas," 60; Lea, "In the Battle of New Hope Church," 62; W. S. Jones, "'Glory Enough for All,'" 216 (Stewart story); Ridley, *Battles and Sketches of the Army of Tennessee*, 304 ("my own troops"); Stewart report, June 5, 1864, *OR*, vol. 38, pt. 3, 818; Hood report, undated, *OR*, vol. 38, pt. 3, 761; A. S. Williams, *From the Cannon's Mouth*, 312; S. Davis, "No Hope of Success," 39; Asbury report, May 29, 1864, *OR*, vol. 38, pt. 3, 831; J. B. D., "Account of the Dallas Line," *Memphis Appeal*, May 27, reprinted in *Supplement to the Official Records*, pt. 1, vol. 7, 139; T. B. Roy endorsement, May 27, 1864, *OR*, vol. 38, pt. 4, 744; Mackall, "Journal," *OR*, vol. 38, pt. 3, 987.

Jeff Dean argues that "Hell-Hole" was Union soldiers' name for the battle of May 25 and not the name for the deep ravine through which they had to pass before attacking Stewart's troops holding the high ground at New Hope churchyard and cemetery (J. S. Dean, *The Hell-Hole in Georgia*, 13). Philip L. Secrist takes the opposite view: "A few hours later, Cummings and Bull were lying side by side in the rain and mud in the 'Hell Hole' near the Confederate battle line at the cemetery" (*Sherman's 1864 Trail of Battle to Atlanta*, 74).

Schofield the left. Aware that enemy troops were shifting strength toward their left (eastward), Hood moved Hindman's division to the right of Stevenson. Johnston ordered the line extended even farther by moving Cleburne's division of Hardee's corps to take up the far right of the Confederates' position.[20]

For the 27th Sherman ordered the artillery of Schofield's XXIII, Howard's IV and Hooker's XX corps—those holding the eastern half of the Federal line—to open fire from early morning until 9 A.M. Afterward Howard would attack the enemy. The day began according to plan when the Union guns opened up; Confederate artillerymen returned fire. On the Union right, McPherson began a demonstration but, in a few hours, reported the enemy position too strong to budge. Skirmishing continued in his sector throughout the day, especially between Cheatham's division and the XV Corps. On the other end of the Federal line, Major General Howard informed Brig. Gen. Thomas J. Wood that his division had been chosen for the "arduous and dangerous task...to find the extreme right of the enemy's position, turn it, and attack him in flank." Wood complained that his men had worked during the night building earthworks, but he accepted the assignment. Howard added that Richard Johnson's division of the XIV Corps would support him.

It took several hours for Wood's division to get moving, especially after generals Howard and Thomas realized that it would be safer for Wood to swing around Schofield's flank to the east and from there make his attack. Johnson's division followed on Wood's left. During the march behind Schofield and into their attack positions, buglers sounded their horns frequently, and the noise drew Rebel shells. "If we are expected to surprise the enemy," asked an exasperated Lt. Alexis Cope of the 15th Ohio, "why don't they stop those damned bugles?"

Marching east with his troops, General Howard discovered that the Rebels' flank would be hard to turn, as they were extending their line and digging in. Finally, at 3:15 P.M., after several miles of marching Howard jotted a note to Thomas, "I am on the east side of the creek on which Pettit's [Pickett's] Mill is, facing south, and am now turning the enemy's right flank, I think."

Notice the *think* part. Howard, Thomas, and, for that matter, Sherman had all seen what happened when a maneuver failed to turn the enemy flank and became instead a frontal assault. But Sherman was impatient for the battle to begin, and Thomas's staff relayed his impatience to Howard: "General

[20] Blount, *New Hope Church*, 67–75; Sherman, *Memoirs*, 2:44; McMurry, "Rocky Face to Dallas," 60; McMurry, *John Bell Hood*, 110–11; Butkovich, *Pickett's Mill*, 67 (map, "May 26, 1864"); J. Dean, "The Battle of Pickett's Mill," 29.

Thomas says that Major General Sherman wishes us to get on the enemy's right flank and rear as soon as possible." In other words, Howard was told to launch his attack. Wood arrayed his three brigades in column, as Hooker had done two days before at New Hope Church. At the front was Brig. Gen. William B. Hazen's brigade, followed by Col. William H. Gibson's, then Col. Frederick Knefler's. Hazen sent a staff officer, Lt. Ambrose Bierce, to recon the ground ahead. Bierce came back with tough news: a quarter-mile uphill, through thick brush and ravines before getting near the Rebel works, which were well-built and manned by maybe two divisions. The intervening woods would allow no artillery to be brought up; this would be an unsupported infantry attack. Nevertheless, at 4:30 Wood ordered Hazen to advance. General Howard was nearby, and Wood told him, "We will put in Hazen and see what success he has." By this remark General Hazen learned that his brigade would make at least the initial attack alone, that Wood's two other brigades and further support (Col. Benjamin Scribner's brigade) would be held back until Hazen determined the strength of the enemy position. Howard had decided that instead of a full-scale attack by 14,000 men in seven brigades (counting Brig. Gen. Nathaniel McLean's of the XXIII Corps), Hazen's 2,000 soldiers would go it alone. Hazen dutifully rode back to the head of his brigade. With him was Lieutenant Bierce, who later wrote, "Only by a look which I knew how to read did he betray his sense of the criminal blunder."[21]

Cleburne knew that the enemy was marching to the east across his front. A reconnaissance late in the morning by part of Govan's brigade disclosed the enemy's flanking movement; the Southerners had seen Howard's columns (and probably heard the Yankee buglers, too). Cleburne sent this news to

---

[21] Castel, *Decision in the West*, 229–41; Blount, *New Hope Church*, 81–85; McInvale, *The Battle of Pickett's Mill*, f.p. 68 (Kurtz's map, "Battles at Dallas, May 27–31"); Wood report, September 10, 1864, *OR*, vol. 38, pt. 1, 377; Butkovich, *Pickett's Mill*, 75–76, 89, 93–94; Noblitt, "The 'Crime' at Pickett's Mill," 41–42; Lee, *Thomas J. Wood*, 188–89; Howard to Thomas, May 27, *OR*, vol. 38, pt. 4, 324; E. S. Cooper, *William Babcock Hazen*, 113; Bierce, "The Crime at Pickett's Mill," 686.

After the war, Ambrose Bierce became a renowned author of short fiction, including articles based on his war service (9th Indiana, promotion to first lieutenant). "The Crime at Pickett's Mill" is widely quoted for its cynicism about the Union attack of May 27. In his story, Bierce quotes Gen. Hazen's statement on the charge of his troops: "they went in at a run, and as organizations were broken in less than a minute" ("Pickett's Mill," 692; Hazen, *A Narrative of Military Service*, 261).

On the reasons for the failure of the Union "flank" attack of May 27, and the difficulties faced by generals (of both sides) throughout the war trying to plan successful flank attacks, see S. Davis, "A Reappraisal of the Generalship of John Bell Hood," 76–77. My observations are based on Albert Castel's astute conclusions in "Mars and the Reverend Longstreet," 112–14.

Johnston, who jumped to the conclusion that Sherman was preparing to give up his Dallas-New Hope line and was marching back toward the railroad. He therefore ordered Hardee and Hood to send out several brigades, push back the enemy skirmishers, and learn what force remained in their front. Sharp fighting soon disclosed that the Yankees were still firmly in position.

Hood took no part in the battle of Pickett's Mill, Cleburne's bloody repulse of Hazen's charge, which began around 5 P.M. For a time, the Federals threatened to overlap the Confederate right, but Cleburne shuttled troops to extend his line, held at the far right by Brigadier John Kelly's cavalry. Brig. Gen. Hiram Granbury's troops of Cleburne's division did not have time to entrench when the Yankees charged. "Ah, damn you, we have caught you without your logs now," some yelled. Brig. Gen. Mark Lowrey's brigade of Cleburne's division also helped beat back the Yankees, with Lowrey riding his favorite horse, "Rebel." After Hazen was thrown back, Gibson's and Knefler's brigades went in, and they too suffered loss without breaking Cleburne's line. Scribner's brigade also advanced, driving Kelly's troopers back onto the Confederate infantry line, but was stopped there. By seven o'clock Wood and Howard concluded that further assaults were pointless.

The Confederates' defensive victory on May 27 at Pickett's Mill—so named for the farm and gristmill owned by Martha "Fanny" Pickett, a widow after her husband, Lt. Benjamin Pickett, had been killed at Chickamauga—was a boost to their morale, trumpeted in the Southern press. "Glorious news from the front," blared the *Augusta Chronicle & Sentinel*; "General Cleburne ambuscades the enemy." "Everything worked like a watch," commented Maj. Calhoun Benham of Cleburne's staff, as he paid compliment to both Hood and Cleburne for the troop dispositions which ensured the victory. To Ambrose Bierce the Union attack of May 27 was a "crime"; the Northern attack had gained nothing. Its casualties can be reckoned at 230 Federals killed, 1,016 wounded, and 319 missing for a total loss of 1,580 (Hazen lost almost a third of them). General Cleburne counted 85 dead and 363 wounded for a total loss of 448. Most of these were in Granbury's brigade, fighting without "logs." Possibly 200 Confederate cavalrymen also were KWM. Brad Butkovich, the battle's recent chronicler, concludes that "a figure of 648 Confederate casualties is a good guess, and not unrealistic."[22]

---

[22] Castel, *Decision in the West*, 230–31, 241; McMurry, "Rocky Face to Dallas," 60; Blount, *New Hope Church*, 84, 94; Cleburne report, May 30, 1864, *OR*, vol. 38, pt. 3, 724–25; Dean, "Pickett's Mill," 31–34; Butkovich, *Pickett's Mill*, 106–109, 155–56 (casualties); Secrist, "'Scenes of Awful Carnage,'" 9 ("without your logs"); Lowrey, "An Autobiography," 371–72;

Having beaten back the enemy attack on the 27th, that night Johnston planned one of his own on the next day. With the position of Sherman's left having been established by the afternoon's fighting, Johnston planned a flank march and assault in the same sector. As at Resaca and as at Cassville, he called on Hood to direct the attack; Polk and Hardee were put in readiness to advance from their lines in support. Before dawn on the morning of the 28th, Hood led his troops out on a route that placed them in position for the assault. But once there, around 6 A.M., Wheeler's videttes brought the bad news that during the night the Federals had pulled back to a safer position and entrenched too formidably to be successfully assailed. Hood sent word of this by Lt. B. H. Blanton back to Johnston, who was at Cleburne's headquarters. Johnston, understandably disappointed, ordered Hood back to his place in the line.

Cassville had been repeated. Here, as there, Johnston had entrusted to Hood responsibility for leading a flank attack on Sherman (recall his recent wire to Davis that he had "earnestly sought an opportunity to strike the enemy"), and Hood had encountered tactical surprises which compelled cancellation of the attack. Hood was a bold fighter, as General Lee would attest, and sometimes he ordered frontal attacks which later resulted in futile, bloody, and grievous repulses (more on that later). But there were times when, in the fluid nature of battle, he recognized that chances for victory were naught and called off the assault. A commander's judgment is always fallible, but on the field of battle where he was, and we are not, it should most times be respected. As Castel writes, "Not even Lieutenant Mackall's postwar 'journal,' fabricated to support Johnston in his efforts to discredit Hood, describes Johnston as being dissatisfied with Hood's conduct on May 28." Mackall's unpublished writings merely state for May 28, "At sunrise Hood has not crossed little Pumpkin Vine Ck—Attack to be made on eny's left and rear fails, on acct of their line being drawn back too far."[23]

Bohmfalk, "Cleburne's Victory at the Pickett Settlement," 230; Calhoun Benham, "Maj.-Gen. P. R. Cleburne: A Biography," *Kennesaw Gazette*, August 1, 1889, 2 ("worked like a watch").

Dr. Castel notices that Sherman's memoirs "contain not a single word" about his troops' bloody repulse at Pickett's Mill. "It is as if it never occurred," he writes, "despite its being the second bloodiest defeat experienced by the Union forces during the Atlanta campaign" (after Kennesaw Mountain). "But then perhaps that is the reason Sherman does not mention it," he concludes, adding this to his list of "offenses" committed by Sherman in his self-serving recollections ("Prevaricating through Georgia," 97).

[23] McMurry, *John Bell Hood*, 111; Castel, *Decision in the West*, 209, 211, 242–43; Lee to Davis, July 12, 1864, in Freeman, *Lee's Dispatches*, 282 ("Hood is a bold fighter"); McMurry, "Mackall Journal," entry of May 28.

Having been forced to cancel Hood's attack against the Federals' left, Johnston saw an opportunity on their other flank. Sherman was clearly moving troops northeastward and Johnston thought they might be withdrawing from their right, held by McPherson's Army of the Tennessee. He ordered General Hardee to have Maj. Gen. William B. Bate's division, on the Confederate left, "develop the enemy" near Dallas and "ascertain his strength and position, as it is believed he is not in force." Some of Bate's men already knew that the enemy lines ahead were fortified and still occupied by infantry and artillery. Lt. Col. Franklin Montgomery of the 1st Mississippi Cavalry wrote that he had taken "a good look at a part of their works, which seemed very strong"; Montgomery "had no doubt [that they] were well manned." But General Bate believed that he faced only an enemy skirmish line. Kentuckian John S. Jackman (one of the so-called Confederate "Orphan Brigade," which had lost their state to the Federals) commented in his diary, "The boys generally know what is in front, and could have told Gen. Bate better."

By 3 P.M. Bate had formed his plan: Brig. Gen. Frank C. Armstrong's brigade of dismounted Mississippi cavalry, on the left, would charge the Yankee line; if they encountered strong opposition they were to retire. If not, they would press their attack toward the right. Two artillery shots would then signal three brigades to advance: Col. Thomas B. Smith's Georgians and Tennesseans, Col. Robert Bullock's Florida Brigade (formerly that of Brigadier Jesse Finley, injured at Resaca), and Brig. Gen. Joseph Lewis's "Orphan Brigade" from Kentucky.

Around 3:45 or 4:00, Armstrong's troops sallied forth so suddenly that they drove the Federals back—men of Maj. Gen. John A. "Black Jack" Logan's XV Corps. The Mississippians even captured three cannon before Black Jack personally rode up and rallied his men. "Damn your regiments! Damn your officers!" he yelled. "Forward and yell like hell!" Logan's performance inspired one Federal officer to term it "without question the most inspiring and magnificent example of personal magnetism and personal leadership I witnessed during the War." When Logan rallied Brig. Gen. William Harrow's division, "you should have heard them cheer him," wrote one soldier. The Federals retook their guns and pushed the Rebels back across the field.

Armstrong's repulse should have ended the battle, and Bate ordered no signal guns fired. He further sent couriers to his brigade commanders cancelling the assault. The message reached Colonel Smith, but not Bullock and Lewis. Amidst the roar of battle from the left, Lewis worried that he might have missed the sound of the signal guns. After conferring, Lewis and Bullock decided to charge anyway.

They should not have. A hundred yards in front of their works, the Northerners had felled the blackjack scrub for a clear field of fire. Brig. Gen. Joseph Lightburn's brigade of Brigadier Morgan L. Smith's division, XV Corps, stood behind a log and dirt parapet, supported by artillery. The result was predictable, as recorded by Confederate Lt. Henry Reddick.

> I suppose we had gone about three hundred yards when we could see a lot of newly cut brush, and from behind it here rose the Yankees in three or four ranks. I know the air seemed blue with their uniforms. As they rose they fired volley after volley into our line of battle, and we returned fire, but soon had to fall back, leaving half our men killed, wounded, or prisoners.

"The Confederate assault never seriously challenged the entrenched Federals," writes Zack C. Waters, chronicler of General Bate's role in the campaign. "All the Yankees had to do was load and fire." General Logan reported 379 casualties in his three divisions engaged. Confederate losses totaled 600 to 800, reflecting the fact that, as one Southerner reflected, "death was feasting in our midst." "The company and entire brigade suffered immensely & accomplished nothing," lamented Florida sergeant Archie Livingston in a letter sent a few days later. Bate's soldiers were understandably angry about their repulse at Dallas, and their commander, called "Old Grits," was the subject of some harsh remarks. John Jackman recorded after the battle that Grits was catching it "from all sides and quarters." Afterward, some of the Orphans could be more charitable. Decades after the war, Kentuckian Lot D. Young wrote this assessment of the fight at Dallas:

> The reader may wonder why this attack was ordered against a force so strongly and irresistibly posted. The answer is easy to the old veteran who knows the difficulty of ascertaining an enemy's position in a heavily timbered country like this, with trees and bushes in full leaf, and how great the danger from the ever alert sharpshooter to the man attempting a reconnaissance. The object was to develop his strength at this point, the commander believing Sherman to be only feigning while he was carrying out other and ulterior plans. But so it was, we paid dearly for the desired information.

It should be mentioned that in his campaign report Johnston says not one word about the Confederate defeat at Dallas on May 28, stating only that "the usual skirmishing" occurred that day.[24]

Both sides spent May 29 watchfully waiting in lines anywhere from two hundred yards to a half-mile apart. Skirmishers were active, peppering the other side, and artillerymen kept up their fire. "Because they are so evenly matched and their defenses are so strong," writes Castel, "little ground is gained or lost by either side." "During day all rather quiet Sharpshooting," was the report in Lieutenant Mackall's diary. But Sherman held the initiative; on the 30th he wanted his forces to start withdrawing from their lines for a move back to the railroad. He had flanked Johnston out of Allatoona, and now it was time to move back to his supply line. His immediate objective was Acworth, five miles down the rails from Allatoona Station. McPherson, however, had difficulty disengaging from the Rebels (Bate's division) in his front. On June 1, the Army of the Tennessee carefully withdrew, and Sherman's forces started marching northeastward.[25]

---

[24] Castel, *Decision in the West*, 243–46; Waters, "Death Was Feasting in Our Midst," 6–14; Waters, "Lines of Battle," 210–15; Davis, *Orphan Brigade*, 221–23; Logan report, September 13, *OR*, vol. 38, pt. 3, 95–96; Harrow report, September 9, *OR*, vol. 38, pt. 3, 279; Lofland report, September 10, *OR*, vol. 38, pt. 3, 187; Osterhaus report, September 9, *OR*, vol. 38, pt. 3, 131; Ecelbarger, "A Human Hurricane on Horseback," 75 ("personal leadership"); Townsend, *Yankee Warhorse*, 154 ("you should have heard"); Sheppard, *By the Noble Daring of Her Sons*, 185 (Livingston letter); Jackman, *Diary of a Confederate Soldier*, 133; Dean, *Hell-Hole*, 77–85; Johnston report, October 20, 1864, *OR*, vol. 38, pt. 3, 616.

Brigadier General Mackall, Johnston's chief of staff, put a different cast on Grits's repulse. To his wife, on June 6, he explained that "General Bates had a fight with McPherson and lost some two hundred eighty men and I fancy McPherson about the same. Sherman reports our loss about twenty-five hundred and this was more than all the men that Bates had." Apparently aware of Northern newspapers' reports of the Dallas fight, which Southern newspapers routinely reprinted, Mackall added that Sherman "is telling pretty big whoppers" (Mackall, *Son's Recollections*, 214 [letter of June 6]).

[25] Castel, *Decision in the West*, 248–251; McMurry, "Mackall Journal," entry of May 29.

Chapter 4

June

Johnston received reports of the enemy movement from observers on Elsberry Mountain, to which the general himself rode around noon on June 2. Sherman's withdrawal was something of a victory in Johnston's eyes, as he telegraphed Bragg that in the recent days' "partial engagements," his army "has had great advantage, and the sum of all the combats amounts to a battle." (Bragg was not impressed.) The men in the ranks shared Johnston's sunny view of the recent fighting, which (except for Bate's charge at Dallas) had given them victory over the Yankees. "Our army was never in better spirits," wrote Richard Walpole, an artilleryman in Cowan's Mississippi Battery on May 29. "All seem confident of success."

In this context, a statement made about General Johnston by Hood in *Advance & Retreat* stands out dramatically:

> Just before leaving New Hope Church, his three corps commanders were assembled alone, at night, in his quarters—then a little cabin near the church—when General Johnston suggested Macon as being the place to fall back upon. If I remember rightly, this suggestion was received in silence, for I cannot recall the reply of one of us at the moment. I well remember, however, after we had left the presence of General Johnston, and were riding through the darkness of the night to our respective headquarters, that the unanimous sentiment expressed on this occasion was to this effect: "In the name of Heaven, what is to become of us? Here we are with the depots for recruits drained, from Mobile to Richmond, all the troops having been sent either to us or to General Lee, in Virginia; our Army fifty or sixty miles from Dalton, no general battle fought, and our Commander talking of Macon, one hundred miles beyond Atlanta, as being the place to fall back upon!"

In making this blatant criticism of Johnston's lack of aggression, Hood may have been incorrect in his memory, for the allegation of the commanding general's intention to retreat to Macon stands alone in the literature. Of his three biographers, only O'Connor quotes it. Neither Dyer nor McMurry even notices the statement from Hood's memoir. At its face, therefore, Hood's claim that Johnston planned to fall back on Macon seems quite a stretch indeed. Johnston clearly planned to fall back on Atlanta, with its extensive fortifications. (He knew of their strength and extent, having the previous March requested the engineer supervising their construction, Capt. Lemuel P. Grant, to give him a tour of the works, then still in construction.) Johnston had Maj. Gen. Mansfield Lovell, who was serving as a volunteer aide, inspect the ferries and crossing-points on the Chattahoochee; Lovell submitted a report on May 31. Johnston clearly was planning a retrograde back to Atlanta, but to say he was planning to retreat all the way to Macon was a stretch. Writing more than a decade after the war, claiming that in late May Johnston planned to retreat a hundred miles below Atlanta, Hood must be faulted for at least a lapsed memory, if not worse.

Meanwhile Johnston had to follow Sherman, or rather beat him to the next defensive position. "Genl. J— has been thinking of changing his position for two or three days," General Hardee wrote his wife on June 1, "but has delayed with the hope that the enemy would shift his." By maps the next front looked like Lost Mountain, six miles southeast of the Confederates' Dallas-New Hope Church-Pickett's Mill line. On June 3, when Sherman's advance reached the railroad at Acworth, Johnston issued orders for the march to Lost Mountain. The men in the ranks apparently did not view this as another retreat, for their spirits had been buoyed by their defensive victories at New Hope Church and Pickett's Mill. "The morale of the army was never better than it is now," wrote General Mackall to his wife on June 3; "the men are sanguine of success and their confidence in Johnston is undiminished." Lt. Col. Columbus Sykes of the 43rd Mississippi expressed the same cheery sentiment to his wife on June 6: "The army is...ready and anxious to fight the enemy whenever 'Old Joe' gives the word." General Mackall distinguished between the army's confidence and the people's concerns that too much territory was being given up. "This retreat will damage the General in public estimation till we beat the foe," he wrote on May 21, when the army had given up Cassville. Mackall worried that Johnston would suffer "a temporary odium" even if he ultimately won the campaign, if only because "a suffering people are always impatient," and Johnston's tactical plans were so slow in unfolding. "There are always nervous people who want quick victory," Mackall wrote his wife June 3, yet he added the consolation, "armies seldom win or lose immediately upon the opening of a campaign."[1]

Following Sherman's eastward shift, the Confederates withdrew during the night of June 4–5. The next day Capt. Samuel McKittrick, 16th South Carolina, wrote his wife, "Hearing the yanks were getting close we came some 5 or 6 miles." "My dear," he added, "it is a hard struggle for independence. The troops are much fatigued." Federals took note of the Rebels' shift of position. On June 5, Jesse L. Dozer of the 26th Illinois entered into his diary, using a peculiar misspelling of *enemy*, "during the night the enema evacuated the place."

Johnston pitched his new line from Lost Mountain northeastward to Gilgal Church, an important road junction. Then he extended it farther still, across the railroad, all the way to its right anchor at Brush Mountain. This new "mountain line," stretching for ten miles, included at its center Pine Mountain, which was situated actually a mile in front of the Southerners' main line. It thus formed a salient, though not connected by trenches to the main works; it was too commanding (three hundred feet high) not to be occupied. The corps were aligned as they had been previously: Hood on the right, Polk center, Hardee on the left. Bate's division with artillery was placed on Pine Mountain. Perhaps Sherman, Johnston thought, might recklessly attack the position, giving Johnston prospects for the kind of defensive victory he always looked for.

Bate's precarious position led to a yarn that made its way into the Atlanta newspapers. 'Bill, can you tell why we are placed out here in front of the

---

[1] Castel, *Decision in the West*, 255–59; Blount, *New Hope Church*, 154–56; Mackall, "Journal," *OR*, vol. 38, pt. 3, 990; Johnston to Bragg, *OR*, vol. 38, pt. 4, 753; A. P. Mason, "Confidential Circular," June 1, *OR*, vol. 52, pt. 2, 674 (retreat from New Hope); Richard Walpole to "Dear Anna," May 29, 1864, Bomar Family Papers, MSS 86, box 1; Hood, *A & R*, 124; O'Connor, *Cavalier General*, 195–96; S. Davis, *Atlanta Will Fall*, 34 (Johnston and Captain Grant); Lovell to Johnston, May 31, *OR*, vol. 38, pt. 4, 749; Hardee to "My own dear Mary," June 1, Hardee Letters; McMurry, "Rocky Face to Dallas," 62; Hallock, *Braxton Bragg*, 191 (Mackall to wife).

Lovell tried to boost Johnston's spirits in his message of May 31 by offering estimates of Sherman's campaign casualties to date. Based on calculations of "the head of the special service corps, an old friend of mine," Lovell toted up 13,000 Federals lost at Dalton, 5,800 at Resaca, 8,000 to 10,000 casualties at Dallas, some 10,000 men sick, "and his men report at least that number of stragglers." General Lovell did not add up these figures. Had he done so he would have been more than justified in concluding that Sherman's army "must be greatly diminished" (*OR*, vol. 38, pt. 4, 749; Symonds, *Johnston*, 300).

army?" one Johnny Reb asked another. "Well, no," came the reply, "but I understand that General Johnston has set a trap for Yankees." "Yes, that's so," commented Johnny. "Don't you know what a bait (Bate) he has on it."

After reaching the Western & Atlantic at Acworth, Sherman spent a week resting his troops and planning his next move. On the 8th he composed for the authorities in Washington a report of his operations to date. In it he was, at times, quite selective in his facts. "I gradually enveloped the enemy in Resaca," he wrote, "and pressed him so hard that he evacuated in the night of May 15"—note that there is no mention of the really decisive tactical move, the flanking maneuver of Tom Sweeny's division. He turned Hooker's sharp repulse at New Hope Church into something of a victory: "General Hooker attacked well and drove the enemy back to the very road," only to be stopped by a "pitchy dark night." Needless to say, he completely ignored the Federals' bloody repulse at Pickett's Mill.

During the week at Allatoona, the railroad bridge across the Etowah was rebuilt and Allatoona Station became a major supply depot for the army, fortified against attack from Rebel cavalry. Two divisions of the XVII Corps arrived on June 8, making good those troops recently lost to battle and sickness; the XVII became the third corps in the Army of the Tennessee. Finally, on the 10th the army group started to march, Schofield on the right, Thomas in the center, and McPherson on the left (reversing the arrangement the week before at Dallas). Sherman continued to radiate confidence: "I think thus far I have played my game well," he wrote his wife, Ellen, on June 12. Sherman's "game"—his grand tactical plan—was summarized by a Northern officer after the war: "General Grant's motto in the wilderness, and indeed everywhere, was: 'Fight'; while the motto of General Sherman was 'Maneuver when you can and fight when you must.'"[2]

Johnston was far less radiant. Despite the addition of Polk's corps, he knew his army was still outnumbered, especially with Sherman's accession of

---

[2] Castel, *Decision in the West*, 259; McKittrick, "A Confederate Officer's Letters on Sherman's March to Atlanta," 493–94; Dozer, "Sherman through Georgia and the Carolinas," 315; Hess, *Kennesaw Mountain*, 6, 8–9; McMurry, "Kennesaw Mountain," 22; "Harvey," "FROM THE FRONT," *Memphis Appeal*, June 24 ("bait"); Sherman report, June 8, 1864, *OR*, vol. 38, pt. 1, 60; Special Field Orders No. 21, June 9, 1864, *OR*, vol. 38, pt. 4, 454–46; Simpson and Berlin, *Sherman's Civil War*, 646 (Sherman to "Dearest Ellen," June 12); Adams, "Campaign for Atlanta," 204.

The Confederate leadership was aware of Sherman's reinforcement. On June 2, General Hardee wrote his wife: "they are waiting for re-inforcements, which to the number of 7,000, we have information are on their way to join Sherman" (Hardee to "My own dearly loved wife," June 2, Hardee Letters).

the XVII Corps. Indeed, Confederate and Federal returns for June 1–10 showed 69,000 C.S. to 120,000 U.S. as "present"; 60,000 Southerners to 113,000 Northerners "effective." Accustomed to being outnumbered, General Hardee affirmed to his wife that "our Army is still large and very effective." Yet Johnston was so pessimistic about his ability to hold the Lost-Brush Mountain line that he sent most of the army's wagon trains south of the Chattahoochee. He also knew that the administration was watching his every movement warily, if not critically. His staff knew this, too, as evidenced in a letter written on June 8 by the general's aide, Lt. Richard Manning: "I understand that in consequence of the General not having provoked a *General Engagement* with Sherman that the *enemy* at Richmond (whom I regard as the most dangerous that this army & its General have) are busy criticising—blaming—abusing & undermining." Lieutenant Manning was right. If there were just one individual who was Johnston's enemy in Richmond, it was Braxton Bragg. On June 4, Bragg felt compelled to bring President Davis up to date with a memorandum. "Mr. President," he began, "the condition of affairs in Georgia is daily becoming more serious." After reviewing Sherman's strength, he ventured, "Should all these forces concentrate on the Army of Tennessee we may well apprehend disaster." With this kind of pessimism, Bragg could be seen as culpable as Hood of working to undo Joe Johnston, but at least he was not as sneaky. On the 7th, Bragg sent to Johnston a copy of his "daily becoming more serious" memo to Davis. He also sent a copy to General Lee, who on June 9 demonstrated concern about Johnston and the situation in Georgia. "I do not know whence reinforcements can be drawn to our armies," he wrote President Davis, "unless Gen Kirby Smith can cross a part of his force to join Gen. Johnston and enable him to assume the offensive." Davis replied that same day, opining that any troops brought from the Trans-Mississippi could not reach Georgia "in time for the battle which must be fought for Georgia." The president added his worry that "unless Genl Johnston strikes before the enemy has brought up all the reinforcements reported to be moving his chances will be greatly diminished for the success which seemed attainable before he retreated." Others in Richmond were aware of the situation. John B. Jones, the War Department clerk, recorded in his diary on June 11, "there is discontent, in the Army of the West, with Gen. Johnston."

In Georgia, Sherman's reinforced army group was on the move. On June 10, Howard's IV Corps first encountered the Rebels on Pine Mountain; Howard was soon joined by the two other corps of Thomas's army (Maj. Gen. John Palmer's XIV and Hooker's XX). Skirmishing and artillery fire filled the next several days as the Federals tested the Southerners' position and began working their way around to the east of the mountain. By the evening of the 10th, General Hardee had begun to worry that the enemy might envelop Pine Mountain and cut off Bate's division. "I may be compelled to abandon it," he confided to his wife in a letter that day. On June 13, Hardee asked General Johnston to join him the next morning in an inspection of the threatened sector.

Lieutenant General Polk, whose corps held the Confederate center, was conferring with Johnston at army headquarters about his own part of the line when Johnston asked him to join in the next day's reconnaissance. After breakfast on June 14, Johnston and Polk met, proceeded with their staffs to Hardee's headquarters around 10 A.M., and from there rode to Bate's position on Pine Mountain.

What happened next must be reconstructed from remembrances of the Confederate officers present. In a significant letter published in the *Memphis Appeal* two weeks afterward, Col. W. S. Dilworth stated that the three generals arrived around eleven o'clock at the headquarters of Finley's Florida brigade, of which Dilworth was in command and which occupied "the most prominent position on Pine ridge." The generals dismounted and began to survey the enemy's lines. Dilworth explained what then occurred ("Captain Beauregard" was Rene Beauregard, commanding Ferguson's South Carolina Battery and the son of the famed general):

> We went to the position on the highest point on the ridge occupied by Capt. Beauregard's battery. Before reaching this point, I asked the generals not to allow more than three or four persons to go with us, as a large crowd would be sure to attract the fire of the enemy. All dropped back except Gens. Johnston, Hardee, and Polk. When we arrived at the battery I mounted the parapet and commenced pointing out the positions of the enemy's lines, batteries, &c. In about ten minutes there came a shower of minie balls from the enemy's sharpshooters. I turned and saw that a large crowd had collected around the battery. I turned and told the generals that unless the crowd scattered, their artillery would open.

Colonel Dilworth was right. Down below, from their lines some 600 yards away, Federal officers had spotted the group of Confederates on the mountain. One of them was General Sherman himself, who had noticed the Rebel battery on the crest. "Near it, in plain view," as Sherman recalled in his *Memoirs*, "stood a group of the enemy, evidently observing us with glasses. General Howard...was near by, and I called his attention to this group, and ordered him to compel it to keep behind its cover." "How saucy they are!" Howard remembered Sherman as saying, and even though his artillerists were under orders to conserve ammunition, a nearby battery was directed to fire three salvos. Sherman rode farther down his line and ordered Brig. Gen. John Geary to have his guns fire another round of salvos. Soon a story circulated that General Sherman himself directed the fire of one of the guns. After a shell arced its way toward the crest of Pine Mountain, Sherman is said to have remarked, "It has fallen about twenty-five feet too short; a little more fuse and you have it." The next shell, fuse lengthened, was closer to the mark. "That will do," Sherman remarked before riding off with his orderly.

The two Union batteries which fired at the group of Rebel generals were the 5th Indiana and Battery E, Pennsylvania Light Artillery. In the *Official Records* is the statement by Maj. Gen. David Stanley, division commander in Howard's corps, that "the second shot fired from a rifled section of the Fifth Indiana Battery" was the one that killed Polk. Brig. Gen. Walter Whitaker, one of Stanley's brigade leaders, was more specific, claiming that "a shell from the Fifth Indiana Battery, commanded by Lieutenant Morrison, fired from a 3-inch Rodman gun, from the section commanded by Lieutenant Ellison, killed Lieutenant-General Polk." On the other hand, the other division commander involved, Geary, asserted that *his* guns killed Polk. Geary, without mentioning Sherman, says that he ordered Capt. James D. McGill's battery of 3-inch rifles (the Pennsylvania battery) to fire on the group of Rebel officers near the summit.

We probably will never know with certainty which of the two batteries fired the disputed round. In later years, after conversing with Joe Johnston, Sherman himself had come to believe that the fatal shot came from "the second volley fired from the first battery referred to"—i.e., the 5th Indiana.

Whoever they were, the Yankee artillerists aimed their guns well, for the Confederate officers on the mountain were forced to take cover. Colonel Dilworth remembered, "I had scarcely got down from the parapet when a solid shot from the enemy's battery came over our heads." He urged Johnston, Hardee, and Polk to move back from the crest, or at least to scatter. Hardee and Polk had moved ten paces from the battery when a second enemy shot came down. Hardee and Dilworth walked on; Hardee had said he wanted to inspect the right of the brigade line. Just then Dilworth noticed that Polk was not with them. Something must have distracted him, for he had dropped behind.

Various witnesses have written on what happened next. General Johnston remembered a cannon shot striking a tree as he walked away from the battery, then a second shot passed overhead. Polk was walking with him, but apparently the bishop-general stopped at the crest of the hill for one last look at the enemy position. Then came a third artillery round from the enemy batteries, and someone shouted that General Polk was hit. Hardee, Johnston, and staff officers rushed to him. "Hastening up," Johnston recalled, "I found that the bolt from a field-piece had passed left to right through the middle of his chest." He had been killed instantly. Philip Stephenson, a Confederate artillerist nearby, also remembered the scene:

> General Johnston mounted Beauregard's works and turned his field-glasses to the left, when a shot directed at him came directly from the front. He immediately turned his glasses upon the battery firing, at the same time directing the staff and escort to disperse. General Polk moved off by himself, walking thoughtfully along, his hands folded behind his back, his left side towards the enemy, when a second came, entering his left arm, passing through his body, emerging from his right arm, then struck a tree and exploded.

One of Polk's staff officers recalled the scene, too: "Folding his arms across his breast, he stood intently gazing on the scene below. While he thus stood, a cannon-shot crashed through his breast." Colonel Dilworth repeated Hardee's belief that a 3-inch solid shot had killed Polk, "passing through him, breaking both arms, I think, though I did not examine the wound very closely." "His body, badly torn, was lying upon the ground at full length, with the face upturned," another recounted, "and the arms, though broken, still crossed upon the breast."

A shaken Johnston and Hardee huddled over the corpse. "My dear, dear friend," Hardee lamented; "little did I think this morning that I should be called upon to witness this." Tearfully, Johnston laid his hand on Polk's forehead, murmuring, "I would rather anything but this." Stretcher-bearers came up and the general's body was moved onto a litter. Staff officers escorted it down the mountain; one led the general's horse, "Jerry." Soon Polk's remains were placed in an ambulance and borne to the rail depot at Marietta for transport to Atlanta. General Johnston rode alongside the ambulance, his grief evident. Jerry, riderless, accompanied the procession.

Word of the bishop-general's death spread throughout the army. Hood took the news hard, telling Johnston that he was too sad to come to his headquarters that evening. Johnston wired President Davis in Richmond later that day: "The army and the country this morning had the calamity to lose Lieutenant General Polk, who fell by a cannon-shot directed at one of our batteries." Before long even Sherman knew of Polk's death. His signal officers had decoded the Rebel cipher, so they could read this message sent about noon from Pine Mountain: "Send an ambulance for General Polk's body." Later came another one: "Why don't you send an ambulance for General Polk?" The next day Sherman seemed to take a bit of pride in being able to inform Washington, "We killed Bishop Polk yesterday."

Thus fell Bishop Polk, who was mourned by Southerners more as a moral figure than an able general. "A slow coach—one who cannot be depended on," was how Col. Taylor Beatty unflatteringly phrased it. General Hardee put a pragmatic cast to the event in his letter to Mary on June 17. "I thought the death of our dear old friend General Polk would cause you to feel uneasy about my safety," he wrote. "According to the doctrine of chance, I am in less danger now than before, there being but three Lieut-Generals and one being killed the chances for the safety of the other two are greatly increased." Historians can be equally unsentimental, so the judgment of Polk voiced by one recent writer follows those of earlier students: "He was a mediocre commander whose post would be filled by unremarkable replacements." Until he found a replacement—Johnston wanted Dick Ewell, and so wrote President Davis—Maj. Gen. William W. Loring, division commander, was appointed as interim commander of Polk's corps. Meanwhile, rattled by the day's events, Johnston ordered Bate's division to pull back from Pine Mountain during the night of the 14th–15th.[3]

---

[3] "Abstract from returns of the Army of Tennessee...June 10, 1864," *OR*, vol. 38, pt. 3, 677; "Abstract from returns of the Department of the Cumberland...the Tennessee...the Ohio," showing "Present for duty," *OR*, vol. 38, pt. 4, 373–76; Hardee to "My own dear Mary," June 7, 10, and to "My own dearly loved Mary," June 17, Hardee Letters; Castel, *Decision in the West*, 267–68, 273–75; Hess, *Kennesaw Mountain*, 9–10; McMurry, "'The *Enemy* at Richmond,'" 29; Bragg to Johnston, June 7, 1864, and "[Inclosure.] Hdqrs. Armies of the Confederate States, Richmond, June 4, 1864. His Excellency the President," *OR*, vol. 38, pt. 4, 762; Freeman, *Lee's Dispatches*, 223; Crist, Williams, and Dillard, eds., *Papers of Jefferson Davis*, 10:458; J. B. Jones, *A Rebel War Clerk's Diary* (1958 ed.), 390; S. Davis, "The Death of Bishop Polk," 13; McMurry, "The Battle of Kennesaw Mountain and the Georgia Campaign of 1864," 19, 23, 35; Polk, *Leonidas Polk*, 2:371–72, 374; Dilworth, "A Statement on the Death of Gen. Polk," *Memphis Appeal*, June 29 (reprinted in *Supplement to the Official Records*, pt. 1, vol. 7, 89–90); Sherman, *Memoirs*, 2:52–54; Howard, *Autobiography of Oliver Otis Howard*, 1:563; "How Gen. Polk Was Killed," *Augusta Chronicle & Sentinel*, July 19 (based on the *Louisville Journal* of June 28); Stanley report, undated, *OR*, vol. 38, pt. 1, 223 ("the second shot fired from a rifled section of the Fifth Indiana Battery exploded in a group of rebel generals, killing Lt. Gen. Leonidas Polk"); Whitaker report, undated, *OR*, vol. 38, pt. 1, 243 ("on the 14th a shell from the Fifth Indiana Battery...fired from a 3-inch Rodman gun...killed Lieutenant-General Polk of the rebel army"); Geary report, September 15, 1864, *OR*, vol. 38, pt. 2, 127 ("McGill's battery opened upon the enemy's position on the summit of Pine Hill.... I noticed a group of rebel officers collected near some tents near the summit; calling Captain McGill's attention to it, I directed him to bring his battery to bear on the spot. The shells struck in the midst of and around the group.... Lieutenant General Polk was killed"); "General Polk's Death," 380–81;

After the Confederates withdrew from their salient, early on June 15 Thomas's troops advanced and occupied the abandoned earthworks. There one of General Stanley's men saw a note fastened to a stake in the ground reading, "You damned Yankee sons of bitches has killed our old Gen. Polk." Inspecting the Rebel works, David Conyngham of the *New York Herald* came upon a "mass of blood"; he knew it was Polk's. He and a surgeon picked up some bits of bone; they looked like ribs and *radii*, *ulnae*—arm bones—and pocketed them as ghoulish souvenirs.

Johnston's abandonment of Pine Mountain led Sherman mistakenly to think that the Rebels were withdrawing from their mountain line. On the 15th he ordered all three army commanders to push forward. Schofield on the right was to "threaten" Lost Mountain. McPherson was to do the same on the other flank near Kennesaw Mountain, while Thomas tried to "break the center" of the Rebel line. All tried to do their parts, with varying degrees of success. Schofield's XXIII Corps marched toward Lost Mountain. In the afternoon several brigades made a strong demonstration and captured a line of Rebel works but got no farther. On Sherman's left McPherson moved on Kennesaw. At one point, near Noonday Creek, Hood's troops holding the Confederate right fought back a sally by Brig. Gen. Charles Walcutt's brigade (of Brigadier Harrow's division, XV Corps). The Federals rushed across the creek in "regular storm fashion," according to Capt. Charles Wills of the 103rd Illinois. They overran a line of rifle pits so fast that they captured several hundred prisoners, many from the 40th Alabama, Baker's brigade, Stewart's division. Without orders to do anything more than "threaten" the enemy line, McPherson called Walcutt back to his position south of Big Shanty.

"How General Polk was Killed," 166 (letter of General Johnston, October 9, 1885); C. H. White, *Confederate General Leonidas Polk*, 103 ("the first men to reach Polk's side were staff members Lieutenant Aristide Hopkins and Col. W.D. Gale"); Hay, *Cleburne and His Command*, 223 (Stephenson remembrance; see also Stephenson, *Memoir*, 189–91); "LIEUT. GEN. LEONIDAS J. POLK," *Atlanta Intelligencer*, June 16, 1864 (a shot "tearing off his arm and mashing his body horribly"); "St. Clair," "Special Correspondence of the 'Intelligencer,'" *Atlanta Intelligencer*, June 17 ("a *solid shot* struck him on the arm, and passed through his body, tearing up his ribs and mangling him fearfully"); *Funeral Services at the Burial of the Right Rev. Leonidas Polk*, 5–6; Jefferson Davis to Rt. Rev. Stephen Elliott, July 8, 1864, *OR*, vol. 52, pt. 2, 688 (declining invitation to attend Polk's funeral in Augusta); Govan and Livingood, *Different Valor*, 289; McMurry, *John Bell Hood*, 111 ("too sad to come over"); Johnston to Davis, June 14, *OR*, vol. 38, pt. 4, 775; Van Duzer to Eckert and Beckwith, June 14 ("our signal officers read the enemy's flags") and Sherman to Stanton, June 15, *OR*, vol. 38, pt. 4, 479–80; McWhiney, "A Bishop as General," 9; Bonds, "Pawn Takes Bishop," 58; Pfanz, *Ewell*, 405–406.

Sherman ordered Thomas to throw "a strong well-appointed column of attack" against the left-center of the Rebel line. Major General Howard got Thomas's order at 11 A.M.; it said to start the advance at two o'clock. At 3 P.M. Newton's division moved forward, drove in the Rebel skirmishers but then found the enemy fortifications "so thoroughly constructed and so well manned," as Howard put it, that he called off any further assault. Marching south down the Sandtown Road, Hooker's XX Corps also pushed forward; Butterfield's division approached Gilgal Church, Geary's division to its left and Williams's behind. At 2 P.M. the Federals came within view of the Rebel line. Geary's troops charged across a field marked by *abatis* and *chevaux-de-frise.* Several hours of trying to break the enemy line until darkness ended the fight led to nothing more than casualties, which Geary reported as 82K, 432W and 5M. Butterfield's troops advanced against Confederate entrenchments held by Cleburne's division at Gilgal. The Federals drove in the Rebel skirmishers but faltered under heavy fire from Cleburne's entrenched line. In the area behind the small wooden Baptist chapel Capt. Samuel Foster commanded some dismounted Texas cavalry along with Capt. Richard Goldthwaite's Alabama battery of Napoleons. "Later in the afternoon the enemy drive in our skirmishers," wrote Foster, "and the battery men and men from our Regt. all go to work and tear that church down level with the ground in about 15 minutes. [N]ot one log left on another, even knocked the blocks, and just as the Yanks come in sight about 300 yards away they open on them with grape shot, and canister, and they soon break and run out of sight." "We repulsed him as usual," wrote Maj. Calhoun Benham of Cleburne's staff; "we were in works." In some places the Federals were so pinned down by enemy fire that they could only dig in and wait for nightfall. In the fighting at Gilgal Union casualties were not insignificant. Butterfield's division lost 25–40 dead and 135–175 wounded. Cleburne's loss was not reported but was obviously much lighter.

The next day, June 16, heavy skirmishing took place along the lines. Some of Schofield's troops were able to advance to a position that enfiladed Cleburne's works at Gilgal Church. In the afternoon, the Federals brought up artillery and soon were shelling Cleburne's lines; Brig. Gen. Lucius E. Polk, one of Cleburne's brigade commanders, had his left leg broken by a cannonball. W. H. T. Walker's division, to Cleburne's right, also came under fire. "General Walker was shelled out of his headquarters," observed one Confederate. This was reported to Johnston, who was also aware that an enemy column was working around his left at Lost Mountain. Once again Johnston found himself pressured and maneuvered to give up a defensive position. "I can find no mode of preventing this," he ominously wired General Bragg that evening. Johnston was thus forced to order Hardee during the night to withdraw the left of the line from Lost Mountain and Gilgal, back several miles to a new line west of Mud Creek. Polk's (now Loring's) sector was not moved; Hood still held the right flank running to Brush Mountain.[4]

On the morning of the 17th, pickets in front of the Union right and center discovered that the Rebels ahead of them had withdrawn during the night. Schofield and Hooker ordered their corps forward and, in the center, the IV Corps covered a couple of miles too. By midday the Northerners had come face-to-face with the new enemy position. Historians have called it the "Mud Creek line" because the Confederate left extended from that rivulet northeast to Brush Mountain—Hood's position since June 5. Johnston was concerned, however, that a salient in its center would invite the kind of enfilading artillery fire that had so annoyed Cleburne and Walker the day before. He consequently directed his chief engineer, Lt. Col. Stephen W. Presstman, to lay out a new defensive position a few miles to the south of the Mud Creek-Brush Mountain line, one that would be anchored by two prominent mountains, Big and Little Kennesaw.

In a little over one month, Joseph E. Johnston had abandoned to the enemy, from Dalton to Kennesaw, more than sixty miles of north Georgia. The line his army occupied June 16–17 was its seventh separate defensive position. Newspapers in Atlanta tried to put the best face on all this retreating. "Let there be not even the '*shadow*' of a panic here!" cried the *Intelligencer.* "Atlanta was not, is not and will not be in danger," exhorted the *Confederacy.* Others could not be so sanguine, such as the editor of the *Mobile Advertiser & Register*, who on June 16 derided Johnston as "timorous, mincing, irresolute." A few days later the *Intelligencer* issued a softer criticism. An editorial contrasted Lee's successes in Virginia with Johnston's retreats in Georgia.

---

[4] Hess, *Kennesaw Mountain*, 10–12; Castel, "Death Comes to the Bishop," 160; David P. Conyngham, *Sherman's March through the South* (New York: Sheldon and Co., 1865), 112; Castel, *Decision in the West*, 280; Special Field Orders No. 25, June 14, 1864, *OR*, vol. 38, pt. 4, 479–80; C. W. Wills, *Army Life of an Illinois Soldier*, 262; Luvaas and Nelson, *Guide to the Atlanta Campaign*, 245; Howard report, September 18, 1864, *OR*, vol. 38, pt. 1, 196; Geary report, September 15, *OR*, vol. 38, pt. 2, 127–29; Ward report, September 7, 1864, *OR*, vol. 38, pt. 2, 325; Blount, Jr., *Clash at Kennesaw: June & July 1864*, 48, 52–54; Kerksis, "Action at Gilgal Church June 15–16, 1864," 833–38, 861; Braley, *The Battle of Gilgal Church*, 14–17; Foster, *One of Cleburne's Command*, 95 (diary entry of June 15); Mackall, *Son's Recollections*, 215 (Polk's shell-wound); J. Welsh, *Medical Histories of Confederate Generals*, 175 (June 16 as date of Polk's wound); Sanders, "The Bishop's Nephew," 60 (incorrectly dating Polk's wound as June 27); Johnston to Bragg, June 16, *OR*, vol. 38, pt. 4, 777.

"The great secret of their success (Lee's army)," the paper declared, "seems to be in the fact that General Lee always fights the enemy when he gets opportunity, and never permits his men to become dispirited by the disorganizing influences of retreat—a movement, however, which that that great General"—meaning Johnston—"does not seem to understand." Johnston or someone on his staff must have read the article and complained, for the editor, John H. Steele, rushed to print an apology the very next day. He had not seen the piece before it was printed, Steele explained, adding a compliment to Johnston's "military skill" and expressing confidence in his "final success." But the damage had been done.

Federals spent June 17 testing the enemy position. Sherman in fact called on Thomas to attack if the Rebel works were not too strong, but the Northern troops were content to skirmish and shell. Johnston's worries were confirmed when Walker's and French's sector was subjected that afternoon to an enfilade fire. "Skirmishing during the day and a good deal of artillery firing by the enemy, as usual," Johnston wired Bragg that night. Heavy rains on the 18th suspended operations by either side and gave Presstman more time to construct the new Kennesaw line. But French's division suffered more artillery fire and lost 215 men that day. During the night, amid more rain and on muddy roads, Johnston's army withdrew to Kennesaw Mountain (really a long ridge of Big Kennesaw, Little Kennesaw, and Pigeon Hill). In the seven-mile line laid out by Presstman, Hood's corps occupied the right (east of the mountain), Loring the center (the two Kennesaws), and Hardee the left.

Sherman pushed his lines forward. "Enemy gave way last night in the midst of darkness and storm," he wrote on the 19th. On the Union right, Schofield tried to probe the Rebels' flank but was halted by Noyes Creek, which was high and flooded. Sherman, ordinarily impatient, suspended major movements for several days because of the incessant rain and advised Schofield to wait until the waters dropped. Word of the constant rain made its way all the way to the Richmond papers. The *Sentinel* informed its readers that "both Gen. Johnston and Sherman have given the muddy condition of the roads as the reason for the slowness of military operations in that quarter."

Johnston had time to ponder the Federals' next move. So far in the campaign, Sherman had repeatedly outflanked the Confederates' position. "They out-general us in maneuvering by mere weight of numbers," Lt. Andy Neal wrote to his father on June 20. Johnston had no reason to suspect that the enemy would not again attempt to do so. On June 18 he confessed privately to his wife that he was unable to devise a means of countering Sherman's "Engineering system," and that he envisioned having the army retreat all the way back across the Chattahoochee. For the time being, however, he had to hold his Kennesaw line. On the 19th, to counter Sherman's expected next move, Johnston decided to extend his left—not by shifting his entire line, but by moving Hood's corps from its place on the right, east of the railroad, southwestward to Hardee's extreme left (held by the division of Maj. Gen. B. Franklin Cheatham). Hood would then dig in on a line of works stretching across the Powder Springs-Marietta Road. When Johnston learned that Schofield had started to cross Noyes Creek, he ordered Hood to get his troops marching during the night of June 21–22.

It is worth noting that after Johnston's disappointment with Hood at Cassville, he still called upon him to execute key assignments. More than that, Johnston consulted closely with Hood in his tactical planning. On June 20, Hardee wrote to his wife, Mary, regarding the commanding general, "I don't know what he will do under the circumstances. I found Hood with him when I arrived & I left him there, both seemed in good spirits. Hood, I think, is helping the General to do the strategy, and from what I can see is doing most of it." From these developments we can infer that Johnston gave little notice to Hood's clandestine correspondence with Richmond, if he ever knew of it in the first place.

Meanwhile General Hardee was writing Richmond himself, expressing his disappointment with Johnston. To Mary he confessed, "We get a line of battle established and as soon as the enemy has discovered where it is, he approaches all the weak points of it and commences a siege." To Hardee, Sherman's program of maneuver and continuous pressure seemed to baffle Johnston, at least so far. "When this unpleasant condition of things is to end, Heaven only knows," Hardee lamented. "How to whip Sherman is the problem to be solved, and our General had not yet been able to find a solution." For this reason, Hardee was downright gloomy. "My own convictions are, and have been for some time, that we are drifting to the Chattahoochee and that we shall cross that stream in a week or ten days." Then he added, "Both Hood & myself expressed to Gen'l Johnston our desire to fight before reaching Atlanta, and that we would prefer to go there whipped rather than not to fight at all."

Hardee did not stop with his wife but made the same statement in a letter sent to the president. On June 22, Hardee addressed Davis on several points—mourning Polk's death, suggesting Peter Stewart as the bishop's replacement, and commenting on the course of the "very hard campaign." "I don't see under present circumstances when a collision is likely to occur. The enemy wont attack us in our entrenchments, and we are not disposed to attack him in his," he wrote, adding, "so if the present system continues we may find ourselves at Atlanta before a serious battle is fought." In view of this letter alone, not to mention those being sent to Bragg by other officers in the army, it is unfair of Russell Blount recently to call Hood "the presidential tattler," even as he writes "Johnston...still relies on his one-legged general for advice and continues to expect great feats from him on the battlefields."[5]

Hood started marching his troops to the left during the night of June 21–22. The Confederates' route took them south through Marietta, then to the southwest for another four miles. Hindman's division went into place to the left of Hardee's corps around 1:30 P.M. Carter Stevenson's division marched to Hindman's left astride the Powder Springs-Marietta Road. The Southerners started digging in along their new line. Stewart's division was held back in reserve.

Hood had not arrived a moment too soon, as Federal infantry was probing southeast of Hardee's flank. Confederate cavalry were opposing Schofield's two divisions as they marched toward Olley's Creek, but there were no Southern horsemen in Hood's new front. When two Union regiments (the 14th Kentucky and 123rd New York) advanced up the Powder Springs Road, they ran into the skirmish lines sent out by General Stevenson. The Yankees charged, capturing some thirty Confederates, and pressed closer to Hood's line before halting. Meanwhile the main body of Federal infantry had advanced to near the house of Peter Kolb, a local farmer, about a half-mile in front of Stevenson's line: Williams's division (Hooker's corps) north of the road, Schofield's two divisions (Brigadiers Milo Hascall's and Jacob

---

[5] Castel, *Decision in the West*, 281–82; Hess, *Kennesaw Mountain*, 13–16, 18 (Hardee to his wife, Mary, June 19), 20 (June 21), 54 (map, Kennesaw line); S. Davis, *What the Yankees Did to Us*, 75 (*Intelligencer*, *Confederacy*); Kennett, *Marching through Georgia*, 79 ("timorous, mincing, irresolute"); "Explanation," *Atlanta Intelligencer*, June 22; Johnston to Bragg, June 17, *OR*, vol. 52, pt. 2, 679; McMurry, "The Affair at Kolb's Farm," 21; McMurry, "Kennesaw Mountain," 20; "The Weather," *Richmond Sentinel*, June 29; Strayer and Baumgartner, *Echoes of Battle*, 159 (Neal); W. J. Cooper, *Jefferson Davis, American*, 483 (Johnston to his wife, Lydia, June 18); Ratchford to Hindman and Stevenson, June 19, *OR*, vol. 38, pt. 3, 778 (on proposed march to the left); Hardee to "My own dear Wife," June 20, Hardee Letters; N. C. Hughes, *General William J. Hardee*, 211 (Hardee to wife, June 20, but omitting "both seemed in good spirits"); Little, "General Hardee and the Atlanta Campaign," 14 (Hardee to wife, June 20); Longacre, *Worthy Opponents*, 240 ("Other generals such as Hardee appear to have viewed Hood as Johnston's favorite subordinate, the one on whom the army leader relied for counsel and support"); Hardee to Jefferson Davis, June 22, 1864, MSS 22, box 1, folder 40, Jefferson Davis Collection, (I thank Kathy Shoemaker, Research Services associate archivist, for her help in making this document available to me, as Crist, Williams, and Dillard, eds., *Papers of Jefferson Davis*, offers only excerpts [10:478]); Blount, *Kennesaw Mountain*, 70.

Cox's) south of it. Williams and Hooker met at the Kolb house about 3 P.M.; Hooker told him to expect a Rebel attack and to dig in. Northern troops spent the next two hours building their breastworks, keeping watch to the east.

Hooker was right; Hood intended to attack. Hood's thinking in making the decision can only be assumed, as he did not record his thought process as he determined to send Stevenson's and Hindman's divisions against enemy positions which had not been scouted or reconnoitered. Several possibilities are apparent. Hood may have believed that the approach of the two enemy regiments indicated that the Federals were threatening his lines. Johnston had directed him to prevent further envelopment of his left flank, so Hood could have assumed that the Yankees had to be driven off. Richard McMurry makes this case in his article of 1968 on the battle. Earl Hess has more recently said the same: "Given responsibility for protecting his vulnerable sector, Hood apparently thought the best defense was a quick offense." Because he would thus be carrying out his orders to protect the army's left, Hood "did not inform Johnston, much less seek his commanding officer's permission" to launch his reconnaissance-in-force. Albert Castel has advanced another possibility, a "most likely" one: that Hood thought "that the Federals on the Powder Springs road are in column of march, rather than in line of battle, and that consequently their flank is 'in the air'.... If so, at long last he will be able to do what he has been trying to do since the campaign began—deliver a crushing attack." Castel, however, suggests that if he jumped to this conclusion about the enemy formation, Hood was hasty, as Stevenson's skirmishers would have well told him that the Yankees who drove them back were in no "column of march." Worse, Castel posits that Hood "made no attempt to ascertain it and is simply gambling." Part of that gamble involved Hood not sending scouts out into no man's-land to examine the Yankee works and bring back information about them. In failing to do this, Hess guesses that Hood "did not want to alert the enemy to his planned attack." Dennis Kelly, former historian at Kennesaw Mountain National Battlefield Park, theorizes further: "It is quite possible Hood gambled that his force was extended beyond the Union flank, and by pivoting his direction of attack toward his right he could crash unexpectedly into Sherman's flank and roll it up and win a great victory." Probably the least likely scenario is that which Hood later put forth, that after repulsing an enemy attack—note: by two regiments—he ordered a counterattack—by two divisions. (Sgt. Rice Bull of the 123rd New York was one of the Federal skirmishers approaching the Rebel line; in his memoir he makes it quite clear they were just feeling for the enemy position, to bring word of it back to their main line.) McMurry frames Hood's decision in this light;

"Hood seems to have taken the advance of the two regiments as an attack upon his new position and formed his troops to repel it. When the Northerners fell back, Hood ordered his troops to pursue."[6]

Whatever Hood's motivation was, he ordered an assault on the enemy in his front. "The resulting battle was a simple affair," writes McMurry. Shortly after five o'clock, Stevenson's division led the attack, supported by Hindman's to its right. In Stevenson's, Brig. Gens. John C. Brown's and Alfred Cumming's brigades formed the advance line, Brigadiers Alexander Reynolds's and Edmund Pettus's brigades the second. Heaviest assault was that of Brown's infantry against Brig. Gens. Joseph Knipe's and Thomas Ruger's brigades (Williams's division) north of the Powder Springs Road and on Col. Silas Strickland's brigade (Hascall's) to the south of it. Some of the charging Confederates were not veterans (Cumming's) and under fire melted back after the first enemy volleys. Behind Cumming, Pettus's brigade advanced, and it too was thrown back. James A. Davenport of the 20th Alabama later wrote, "I went Near a half mile under the heavyes fire I ever heard the Shell & Shot & Miney balls flew thick as hale."

Brown's and Reynolds's troops made a better go of it, driving an initial Yankee line before being themselves forced back. Three Federal batteries worked cruelly on the Rebels, breaking up their attack even before they got close to the Union infantry line. Hindman's division moved forth, and it too was stopped very quickly by Union artillery. The Southerners gave way under heavy fire and those who could fell back to their works. "It was a beautiful sight to see their columns shattered and fleeing in confusion," recorded Col. James S. Robinson, whose brigade of Williams's division beat back Hindman's assault. By 6:30 the attack was over; the Federals kept firing at Confederates huddled in a ravine in their front, waiting for darkness to make their escape.

"What will be known as the affair at Kolb's Farm is more a one-sided slaughter than it is a battle," concludes Castel. Confederate casualties in their frontal assault numbered just over a thousand killed, wounded and missing: 807 killed and wounded, 16 missing in Stevenson's division; 215 killed, wounded, and missing in Hindman's. One regiment, the 23rd Alabama, lost 6 officers and men killed, another 34 wounded. The Federals lost some 230

---

[6] McMurry, "Kolb's Farm," 21–22, 25 (map); Hess, *Kennesaw Mountain*, 29–35; Castel, *Decision in the West*, 292–94; Hooker to Whipple, June 22 (4 P.M.), *OR*, vol. 38, pt. 4, 56 ("the whole rebel army lies between my immediate front and Marietta"); Kelly, *Kennesaw Mountain and the Atlanta Campaign*, 28; Bull, *Soldiering*, 131–32; McMurry, *John Bell Hood*, 112.

men, mostly in Williams's division/XX Corps and Strickland's brigade/Hascall's division/XXIII Corps.

The repulse, however, was not very disconcerting to headquarters, as Lieutenant Mackall recorded in his diary:

> 4.00 P.M. demonstration of Eny on Bell Ferry road, and attack on Hood on left East of Manning's near Nose's Ck. just as Hood was arranging line to attack. Stevenson principally engaged, Hindman slightly. Our loss at least 800 killed & wounded in Hd's corps. Hardee lost about 100. Eny driven out of one line of works, but artilly on bald hill in Hardee's front prevent important results.

Two days later General Johnston wired Bragg, "Lieutenant-General Hood, on our left, reports that, being attacked on the afternoon of the 22d, he drove back the enemy, taking one line of his breast-works." This is the story which made its way into Confederate newspapers. "The enemy, about half-past 5 o'clock yesterday evening, attacked our works on our extreme left," reported the *Memphis/Atlanta Appeal* on June 23. "In a spirited action," Stevenson's and Hindman's divisions were said to have turned back the assault and then pursued. "We drove them through three lines of battle, capturing as many breastworks, and sixty prisoners. Some thirteen pieces of artillery were taken, it is reported, but only eight were carried off."

The press may have reported a Confederate victory at Kolb's Farm, but the men who fought there knew better. General Manigault called the battle a "miserable failure." Some blamed Hood for ordering the attack in the first place. Decades after the war, Alabaman J. D. Harwell called Kolb's Farm the place "where Genl Hood exhibited his Generalship." For his decision to order the bloody, futile attack of June 22, Hood has also come under criticism from writers of history. Local historian Eldon B. Richardson sees the Kolb's Farm attack as a "rehearsal for Atlanta's doom": Hood would engage in "bold moves repeatedly taken, regardless of the odds, and with no pause to consider the cost."

On the other hand, some students have tried to take a charitable view of Hood's decision. Wilbur Kurtz credits it to Hood's orders from Johnston "to stop further advances" on the Confederate left. Some, like Castel, think Hood may have jumped at the chance to have his troops catch the enemy in flank. Union General Williams thought that Hood's purpose was to "take advantage of the woods on Knipe's left front to turn that flank." In an article published in 1989, Dennis Kelly reaffirms, "It is quite possible he gambled that his march had put him beyond the Union army's flank, and by wheeling to his right, he would crash into their flank and win a great victory." Even the Atlantan William R. Scaife, known as a harsh critic of Hood, acknowledges that in ordering his "impetuous and ill-advised attack" Hood might have "thought he was striking the enemy's flank or rear." Possibly the most positive comment comes from J. Britt McCarley, who writes, "Though badly managed, Hood's assault had blunted Sherman's flanking maneuver via the Powder Spring Road and bought precious time for the Confederate forces on the Kennesaw line." McCarley goes so far as to call Kolb's Farm a "setback" for Sherman, leading him to his decision to "forgo another flank attack and launch a frontal assault instead" against Johnston at Kennesaw.

Still, Earl Hess is on solid footing when he delivers a "damning criticism" of Hood's decision to attack, for three reasons. First, he could not have possibly mistaken a probe by two regiments for a major enemy attack which needed countering (as he subsequently told Johnston). Second, if he expected the enemy to continue testing his lines, digging into a strong defensive position was the best tactical option. Third, "the worst part of Hood's handling of the affair at Kolb's Farm" Hess concludes, "was that he launched the attack with an appalling ignorance of what lay before his men, even though there was ample time to reconnoiter and learn details of the terrain." On the other hand, a student of the battle, D. Reid Ross, cites the postwar reminiscences of Confederate Brig. Gen. Arthur M. Manigault, who wrote that Hood should have known that the Confederates' field of advance was crossed by an impeding creek, "with a boggy, miry margin on each side." Manigault claimed that General Hindman knew of the creek and had told him of it the day before the battle. This, and Manigault's claim that "the ground had been ridden over by many officers the day before," is nonetheless suspect, given that Hood's corps (including Hindman's division) did not arrive at the Kolb's Farm area until the morning of June 22, after a night march from the army's right flank.

Hood did not mention Kolb's Farm in his memoirs. Given Johnston's eagerness to criticize Hood for Cassville and other miscarriages, in his recollections he treated Hood rather lightly for ordering the assault. Johnston repeated what Hood had reported to him after the battle, "that Hindman's and Stevenson's divisions of his corps had been attacked, and that they had not only repulsed the enemy, but had followed them to a line of light intrenchments and driven them from it" before having themselves been driven back. From all of this, one may infer that by the time they were writing their reminiscences in the 1870s, the painful repulse of June 22, 1864, was just another of those unfortunate events the two generals preferred to forget.

Kolb's Farm figured in a humorous story recounted well after the war by an old Confederate veteran. On the 22nd, after the repulse of his assault, Hood apparently expected a Federal counterattack and set his troops to strengthening their works. According to J. C. Higdon, Hood rode up to General Hindman and, pointing to a hill in his front, said, "Gen. Hindman, when you see the enemy crown that eminence, take your division and charge them off." Hindman reportedly answered rather sharply: "Let me take my division and post them there, and the enemy will not crown that eminence." Hood replied, "Gen. Hindman, why is it that I can never give an order but that you have some suggestion to make?" Hindman: "Because you never give me an order with any sense to it."[7]

On June 23, Sherman took stock of the situation. "We continue to press forward," he wired General Halleck; "we gain ground daily, fighting all the time." (Johnston, for his part, agreed, when on June 26 he lamented to Bragg, "the enemy is gradually pressing us back.") Sherman rather exaggerated the odds against him, though, in adding to Halleck, "the whole country is one vast fort, and Johnston must have full, fifty miles of connected trenches, with abatis and finished batteries." On the 23rd he had Schofield probe farther southward to determine where the enemy's extreme left was (and whether it could be turned). That night Schofield reported it could not; the Rebel line actually extended beyond his, and he (not Hood) was in danger of being turned. Sherman responded cynically to this news in a note written to Thomas: "Schofield reports he can't go ahead...and is far outflanked. I suppose the enemy, with his smaller force, intends to surround us." This development, and the welcome drying-up of the weather, allowed Sherman to

[7] McMurry, "Kolb's Farm," 22–25; McMurry, "Atlanta Campaign" dissertation, 202; Hess, *Kennesaw Mountain*, 35–44; Castel, *Decision in the West*, 295; Rafuse, "Always 'Fighting Joe,'" 240–42; Stevenson report, August 19, 1864, *OR*, vol. 38, pt. 3, 815; S. Davis, *Atlanta Will Fall*, 82; [S. Davis], "The Cost of Kolb's Farm," 24; "Report of casualties in Hindman's division in action of June 22, 1864, near Marietta," *OR*, vol. 38, pt. 3, 769; Williams report, September 12, 1864, *OR*, vol. 38, pt. 2, 32; Strickland report, August 14, *OR*, vol. 38, pt. 2, 647; McMurry, "Mackall Journal," entry for June 22; Johnston to Bragg, June 24, *OR*, vol. 38, pt. 4, 788; O'Connor, *Cavalier General*, 197; Kurtz, "Battle of Kolb's Farm June 22, 1864," in "Locations—Roadside Markers. Atlanta Campaign_of 1864" bound tablet, box 32, folder 1 (also in Scruggs, *Georgia Historical Markers*, 132); Richardson, *Kolb's Farm*, 17; Kelly, "Mountains to Pass," 23; Scaife, *Campaign for Atlanta*, 66; McCarley, *The Atlanta and Savannah Campaigns 1864*, 33; Ross, "Battle of Kolb's Farm," 24; "THE FIGHT YESTERDAY," *Memphis Appeal*, June 23, 1864; Tower, *Carolinian Goes to War*, 192–93; Hood, *A & R*, 124–25 (skipping from New Hope Church to Kennesaw Mountain); J. E. Johnston, *Narrative*, 340; Higdon, "Hindman's Reply to Hood," 69.

begin considering a course of action he had previously said he would not pursue: attack the enemy fortifications head-on. On the 24th he drew up an order calling for an assault at 8 A.M. on the 27th that "will endeavor to break a single point of the enemy line." Sherman's plan called for McPherson, whose army held the Union left near the Kennesaws, and Thomas (center) to select points of attack while Schofield continued maneuvering on the right.

The right of Johnston's line was secured by Big Kennesaw, which soared 691 feet up (held by Maj. Gen. Edward Walthall's division, Loring's corps). Infantry trenches extended up and down the slope; when the Yankees shelled the mountain, its thick forestation prevented any serious casualties among the Southern troops. The top of Big Kennesaw was not very broad, and the Southerners eventually placed only three guns there, arduously dragging them up the slope by hand. Little Kennesaw, on the other hand, was a virtual artillery platform, overlooking the surrounding fields 600 feet down. "Little Kennesaw, being bald and destitute of timber," recorded Maj. Gen. Samuel French in his diary, "affords a commanding view of all the surrounding country as far as the eye can reach." Confederates placed nine pieces of artillery along the mountain crest after Maj. George Storrs, French's artillery chief, found a route by which the guns could be pulled up the hill by rope. (The earthen parapets constructed for them were called Fort McBride after Lt. E. D. McBride, who was mortally wounded there.) Maj. Gens. W. H. T. Walker's and Samuel French's divisions held the Little Kennesaw-Pigeon Hill sectors. To the south, Pat Cleburne's and Frank Cheatham's divisions of Hardee's corps held the Confederate center. There a salient jutted out, held by Cheatham's troops; the point of the salient has in fact come to be called Cheatham's Hill. Farther to the left was Hood's sector. On the 26th Hood reminded his division commanders that the enemy, frustrated in their intent to move around the Confederates' left, could attack on the morrow. He therefore called for a strengthening of his troops' fortifications, particularly "to have the abatis in front of your position made very formidable."

But Hood's works would not be tested. On the other side of the line, McPherson chose as his target the enemy works on Little Kennesaw Mountain and Pigeon Hill. Three brigades of Logan's XV Corps would advance along the Burnt Hickory Road. Thomas chose to attack a point in the Rebel center where the salient brought the opposing lines closer. On June 26 he notified generals Howard and Palmer that a division from each of their corps would make the charge. Thus when the Federal attacks were launched on June 27, Hood's corps was not involved in repulsing them.

And repulsed they were. Fifty-one cannon on McPherson's front opened fire on Kennesaw at 6 A.M. Troops of the XVI and XVII Corps started skirmishing to prevent the Rebels from shifting troops to Little Kennesaw and Pigeon Hill, where three brigades of Brig. Gen. Morgan Smith's and William Harrow's divisions launched their attack at eight o'clock. The Northerners' advance overwhelmed the skirmish lines of Walker's division; 150 Confederates were captured. To Walker's right French's infantry in less than an hour stopped the Union assault. From the top of Little Kennesaw Southerners rolled big rocks down on the Yankees. Few of them even got close to the Confederate position; the Northerners, realizing they would not break the enemy lines, fell back to theirs (if they could). Theodore Upson of the 100th Indiana recorded that after three attempted assaults in the Pigeon Hill sector, "we were in a bad fix." "We could not go ahead and could not get back," so he and a dozen comrades sheltered behind a big boulder while "the Johnnys...yelled and threw stones at us" until nightfall, when they retreated to safety.

Logan would eventually count 586 killed or wounded and 17 missing in the XV Corps; in their demonstrations at Big Kennesaw McPherson's two other corps lost another 257 men. To the Federals' approximately 850 casualties, the Confederates manning the two Kennesaws and Pigeon Hill lost perhaps 200, mainly the skirmishers taken prisoner in the initial Federal advance. General Logan was left with the obvious conclusion that "it soon became evident that the works could not be approached by assault."

The other major Union attack that morning also failed against Cheatham's salient. "The location of the hill was such that Cheatham had to extend his line forward in a V shape to keep the enemy from getting possession and shelling us out," remembered Capt. W. D. Eleazer of the 11th Tennessee. At eight o'clock, after a brief cannonade, Newton's division (3 brigades) from the IV Corps and two brigades from Jefferson C. Davis's division of the XIV launched their charge. The Federals overran the Rebel pickets, capturing a number of them, but some got away to spread the alarm: "Up men, they are coming!" The bluecoats advanced only to falter under heavy musketry. Confederates opened up with cannon whose position they had kept concealed with cut brush. Brig. Gen. Charles Harker, commanding Newton's 3rd Brigade, fell mortally wounded trying to rally his men. So did Col. Dan McCook, commanding one of Davis's. Within two hours the Northern attack had failed as the Federals fell back to their lines. Union casualties for this part of the battle approached 1,500 (Newton 654, Davis 824), while Cheatham's division lost 195 men (Cleburne's, partially engaged, lost only 11). General

Cheatham recognized the one-sidedness of this fight on his hill, commenting, "the loss of the enemy must have been great." On June 27 casualties for both sides, including McPherson's demonstrations on the left and Confederate pickets captured, amounted to 3,000 Federal and 700 Confederate. The bloody day was not without incidents of compassion. On part of Cleburne's front, a brush fire broke out that endangered the wounded Northerners lying in front of the Confederate works. "Boys, this is butchery," cried out Col. Will H. Martin of the 1st Arkansas. He had a truce flag hoisted and called on his men to cross the parapet and help Federal soldiers retrieve their comrades.

The Confederate press reported this act of clemency but did not evince a deep concern for the hapless enemy. A telegraphic report gave details of the battle under the headline, "Confederate Victory Near Marietta! The Yankees Roasting."[8]

---

[8] Castel, *Decision in the West*, 299–301, 303–305, 307–16, 319; McMurry, "Kennesaw Mountain," 24–32 (including excellent map, 26–27); Hess, *Kennesaw Mountain*, 21–22, 51, 71–94, 96, 99–103, 109, 117–21, 129–31, 152–53; Sherman to Halleck, June 23, *OR*, vol. 38, pt. 4, 572–73; Johnston to Bragg, June 26, *OR*, vol. 38, pt. 4, 792; Blount, *Clash at Kennesaw*, 99–103, 125 (casualties); "Effect of the Bombardment of Kennesaw," *Kennesaw Gazette*, March 1, 1887, 1 ("comparatively insignificant" loss from Federal shelling); S. G. French, "Kennesaw Mountain," 506; Raab, *W. W. Loring*, 153 ("hauling the artillery up the steep rocky and wooded slopes was accomplished by tow lines, hooks, toggles, and hundreds of infantrymen"); Storrs, "Kennesaw Mountain," 136; Blythe, Carroll, and Moffson, *Kennesaw Mountain National Battlefield Park Historic Resource Study*, 4–5 (on Fort McBride); George S. Storrs, "The Artillery on Kennesaw," *Kennesaw Gazette*, June 15, 1889, 6–7 (Lieutenant McBride's death from shell-blast); Upson, *With Sherman to the Sea*, 116; Logan report, September 13, 1864, *OR*, vol. 38, pt. 3, 99; Ratchford to division commanders, June 26, *OR*, vol. 38, pt. 4, 794–95; Eleazer, "Fight at Dead Angle," 312; Losson, "The Defender of Cheatham Hill," 51; S. Robinson, "First Tennessee Infantry," 1:164 ("Up men"); Barnes, "An Incident of Kenesaw [*sic*] Mountain," 49 (Colonel Martin's truce); "Confederate Victory Near Marietta!," *Columbia South Carolinian*, June 29, 1864.

Writing for "Battles and Leaders" after the war, Johnston argued that Sherman's purported casualty number at Kennesaw Mountain was way too low. "The Federal dead nearest to Hardee's line lay there two days," he averred, "during which they were frequently counted—at least 1000." Enemy dead counted elsewhere on the field led him to conclude that Sherman's loss on June 27 was "at least 6000" (J. E. Johnston, "Opposing Sherman's Advance," 273; Hess, *Kennesaw Mountain*, 153–54). The *Century* editors tried to tamp down Johnston's claim in a footnote, but on one point the general was correct: the Northern slain lay on the battlefield for two days. On the 29th, Capt. William J. McDill of the 9th Tennessee noted that "the Yankees asked permission to bury their dead under flag of truce." McDill and his comrades were relieved to see the work done, as "their dead had become very offensive, both to us and to them" (Fleming, *Ninth Tennessee*, 162, 164 [McDill to Aunt, July 9]). "Their dead lay on the field until they began to stink," Floridian John Campbell wrote his sister on July 1 (Sheppard, *Noble Daring*, 187). One Confederate remembered that the stench from the unburied enemy dead was so powerful that he and his comrades lost their appetites ("M.," "The Battle of Dead Angle on

Sherman and Johnston, of course, had different reactions to the battle of Kennesaw Mountain. The Federal commander accepted the obvious lesson from his repulse, as he phrased it that night to General Halleck in Washington: "to assault him in position will cost us more lives than we can spare." Thomas had chided him that afternoon ("We have already lost heavily to-day without gaining any material advantage; one or two more such assaults would use up this army"). Sherman chided back, writing Thomas that Schofield's

---

the Kennesaw Line, near Marietta, Georgia," 72). The truce on the 29th was maintained for eight hours (S. Robinson, "Battle of Kennesaw Mountain," 115). Some soldiers used the truce to intermingle with the enemy, exchanging such valued items as newspapers, coffee, and tobacco (Cathey and Waddey, *"Forward My Brave Boys!,"* 178).

Writing from Marietta on June 29, Henry Watterson ("Grape") sent a column to the *Augusta Constitutionalist* about an unusual Federal soldier captured at Kennesaw Mountain: "a girl in the disguise of a soldier of a comely appearance and rather modest mein [*sic*]" ("Grape," "From the Front. Letter from Gen. Johnston's Army," *Augusta Constitutionalist*, July 1, 1864).

On the concealed artillery at Cheatham's Hill, see Hughes and Whitney, *Jefferson Davis in Blue*: "The rebel general opposite—Frank Cheatham—had hidden cannon under some brush and craftily held his fire until Mitchell's men were halfway up the hill" (258). General Cheatham wrote of his batteries, "I directed [them] to be closely masked with brush and under no circumstances to reply to the artillery of the enemy" (Cheatham journal, May 7–July 15, *Supplement to the Official Records*, pt. 1, vol. 7, 144). The journal of Col. Melancthon Smith, Hardee's chief of artillery, gives fuller context:

> The enemy for several days had occasionally opened with artillery, as was their wont, previous to an attack, for the purpose of finding out the position of our batteries, and most of our division commanders would insist upon replying with the batteries on their line, there discovering to the enemy what they required; and General Cheatham was about the only division commander of our Corps who agreed with me to keep them concealed until the assault was made. On this occasion, I think the wisdom of this course was proven (Smith journal, February–September 1864, *Supplement to the Official Records*, Part 1, vol. 7, 77–78).

The "Jefferson Davis in blue" mentioned above was Jefferson Columbus Davis of Indiana. In September 1862 Davis, then a Union brigadier, took offense from remarks by an ex-commanding officer, Maj. Gen. William Nelson. Several days later, in the lobby of the Galt House hotel in Louisville, Davis demanded an apology. He got none; General Nelson called him a "cowardly son of a bitch" and slapped him on the side of the head. Davis shortly thereafter brought back a pistol and shot Nelson in the chest. He died within an hour. General Davis was arrested but two weeks later was released. The army concluded that while Davis had killed a superior officer in an outright act of murder, somehow the crime was justified by Nelson's verbal abuse and contemptuous behavior. A local grand jury indicted Davis a month later, but the indictment was dropped in six months. Despite his rise to corps command during the Georgia Campaign, Davis was not promoted to major general except by brevet (D. A. Clark, *The Notorious "Bull" Nelson*, 144–62; J. Welsh, *Medical Histories of Union Generals*, 241; Warner, *Generals in Blue*, 116).

maneuvering on the right represented "the only advantage of the day." Nonetheless, Sherman sought to justify his attack. "The assault I made was no mistake," he wrote Halleck on July 9; "I had to do it" in order to disabuse his officers and men—much more, the enemy—of any belief "that the assault of lines formed no part of my game." This idea seeped down through the Northern ranks. "It was said that Gen. Sherman wished to show Johnston that we were not afraid to fight him," recalled Col. Benjamin F. Scribner, "and did not depend altogether upon outflanking him." As to the causes of its failure, Sherman hedged: "had the assault been made with one-fourth more vigor"; "had Harker and McCook not been struck down so early," etc. Sherman's worst response was that callously expressed to his wife, Ellen, of all people: "I begin to regard the death & mangling of a couple thousand men as a small affair, a kind of morning dash." Stephen M. Hood is correct in pointing out that historians would have had a field day if John Bell Hood had made such a remark.

The death and mangling of a couple thousand men was the gist of Johnston's telegraphed report that evening to General Bragg in Richmond: the Federals had attacked French, Cheatham and Cleburne and had been repulsed; "their loss is supposed to be great; ours known to be small." Johnston had to know that his "enemy at Richmond" would not be satisfied merely with static defensive victories, so he tried to explain why he and his army had been pushed back the sixty miles from Dalton to Marietta: Sherman possessed numerical superiority and took advantage of topographical features by his effective "engineering operations," chiefly his flanking maneuvers. As a result, Johnston admitted, "I have been unable so far to stop the enemy's progress by gradual approaches on account of his numerous army and the character of the country, which is favorable to that method." To Lydia he was even more despondent. "Such a warfare was never before waged," he lamented. General Hardee was more upbeat when he wrote President Davis on June 29: "On the 27th the enemy attacked Cleburne & Cheatham & were repulsed with loss of about three thousand in killed and wounded our loss less than hundred army in good condition."[9]

---

[9] Castel, *Decision in the West*, 320; Sherman to Halleck, June 27, *OR*, vol. 38, pt. 4, 607; Thomas to Sherman, June 27, *OR*, vol. 38, pt. 4, 610; B. S. Wills, *George Henry Thomas*, 263; Sherman to Thomas, June 27 (4:10 P.M.), *OR*, vol. 38, pt. 4, 610; Vermilya, *The Battle of Kennesaw Mountain*, 142; Sherman to Halleck, July 9, *OR*, vol. 38, pt. 5, 91; Scribner, *How Soldiers Were Made*, 302; Sherman to Ellen Ewing Sherman, June 30, Simpson and Berlin, *Sherman's Civil War*, 660; Hood, *Rise, Fall, and Resurrection*, 220; Johnston to Bragg, June 27, *OR*, vol. 38, pt. 4, 795–96; W. J. Cooper, *Jefferson Davis, American*, 484 (Johnston to Lydia,

Johnston's admission to the administration was a way of telling his superiors that he would continue to retreat. Sherman's flanking pattern, almost always against the Confederates' left, was continuing to be manifested; Lieutenant Mackall recorded as early as June 24, "expect to flank us on our left." Consequently, Johnston ordered his engineers to start building a new fortified line to the rear; both Mackalls rode there on the 25th to inspect it.

On the other hand, the Southern press continued its upbeat reporting. "All looks well here," a correspondent from Johnston's army wrote for the *Memphis/Atlanta Appeal.* "A soldier...said to me this morning: 'General Johnston commands everything here, officers, soldiers, cooks, nurses and herdsmen. Why, sir, he commands those mules yonder, and that accounts for their looking so fat and sleek. He feeds us well, nurses us well, and we love him and believe in him.'" "Such faith, such admiration, such obedience, such cheerfulness," the correspondent concluded, "I have never seen in an army."

Johnston had to hold onto his strong position at Kennesaw for as long as he could. On June 24, according to Lieutenant Mackall, Johnston's friend Louis Wigfall, passing through Atlanta back to his home in Texas, met with Johnston at his Marietta headquarters. The senator brought the dire news that he had heard that President Davis was so disillusioned with Johnston's handling of the campaign that he was thinking of relieving him from command. Wigfall asked the general about his plans and was told bluntly that unless the government ordered a cavalry strike (perhaps by Bedford Forrest, then in Mississippi) against Sherman's line of communications, he would have to retreat back to Atlanta.

President Davis was not the only one alarmed at Johnston's retrogrades. Georgia governor Joseph E. Brown wired Davis on June 28, warning that Johnston needed reinforcements and reminding him that state troops had been sent to the front. Brown had issued proclamations on May 18, May 21, and June 24 calling out the Georgia militia; the governor eventually turned

---

June 30); Hardee to Davis, June 29, Crist, Williams, and Dillard, eds., *Papers of Jefferson Davis*, 10:492.

Cooper concludes that Johnston's gloomy letters to Lydia (e.g., of May 23, June 18, and June 30), together with his pessimistic reports to the War Department, "totally undermine Johnston's postwar claim in his *Narrative*...that he had a definite plan to defeat Sherman" (*Davis, American*, 713 n.30).

Sherman's decision for a frontal attack at Kennesaw Mountain has sometimes been criticized, but not by Liddell Hart, his staunch advocate. Indeed, the English military historian exonerates Sherman for having ordered the assault "in the hope of saving his troops from the strain of a further flank march over rain-swamped roads" (introduction to Sherman, *From Atlanta to the Sea*, 9).

over to Johnston some 3,000 militiamen under Maj. Gen. Gustavus W. Smith. Smith and his troops were posted out to Hood's left, extending the Confederate line farther to the southwest. Brown also put at Johnston's disposal another 1,250 men of the Georgia State Line, two regiments which joined the Georgia brigades of Marcellus Stovall and Alfred Cumming. Still, Brown wrote Davis, "Further re-enforcements are urgently needed.... Is it not in your power to send more troops?" (Actually, Bragg had recently sent Johnston two regiments of Georgia cavalry from Savannah.) The governor was more strident in asking Davis to unleash Forrest's or John Morgan's cavalry upon Sherman's rail lines. "If these are destroyed," Brown argued, "it is impossible for him to subsist his large army, and he must fall back." Atlanta had to be held, Brown lectured; "this place is to the Confederacy almost as important as the heart is to the human body."

On July 1, at his headquarters (the Atkinson house in Marietta) General Johnston received as visitor another Confederate senator, Benjamin H. Hill of Georgia. Governor Brown had asked him to pay a call, converse with Johnston, and develop an argument for a cavalry raid on Sherman's rear that he could then put in a letter to President Davis. Hill agreed but said that he would take the message personally to Richmond—a letter was insufficient medium for a matter so important. In the meeting—significantly, Hood attended it, not Hardee—Hill sat with Johnston and asked the hard questions he knew the Richmond authorities would want answered. If Johnston pinned his hopes on a Confederate cavalry raid behind Sherman's lines, why did he not send his own cavalry, under Wheeler, on such a raid? Johnston answered that he needed all his available horsemen to guard the army's flank.

It is possible that Wheeler and Hood were talking about some infantry-cavalry movement; on June 29 Wheeler wrote Hood, "General Johnston has consented that your proposed expedition be made provided a sufficient force can be left to protect General Smith's right." Hood wrote back, "I am ready to give you all the assistance in my power, and will ride up this afternoon to see General Johnston." There is no further record of what happened with this apparent proposal for action.[10]

---

[10] McMurry, "Mackall Journal," entries of June 24–25 and 30 (Atkinson house); "ENCOURAGING FROM THE FRONT," *Columbia South Carolinian*, June 29; Govan and Livingood, *Different Valor*, 297–98 (on Wigfall's visit, mistakenly dated June 28); King, *Wigfall*, 196–97; McMurry, "*Enemy* at Richmond," 12 (Wigfall's advice in December 1862 that Johnston keep records of his correspondence with the administration); Castel, *Decision in the West*, 302–303, 329; L. B. Hill, *Joseph E. Brown and the Confederacy*, 182; Brown to Davis, June 28, *OR*, vol. 52, pt. 2, 680; G. W. Smith, "The Georgia Militia about Atlanta," 4:331–32; W. H. Bragg, *Joe Brown's Army*, 84–86;

Chapter 5

July 1–17

In truth, Wheeler wanted to be sent on a raid behind enemy lines, as he informed General Bragg in another of his letters to Richmond. Under date of July 1—the very day Senator Hill was meeting with Johnston and Hood—Wheeler wrote,

> I have begged General Johnston to allow me to go to the enemy's rear nearly every day for the last three months and he is anxious that it should be done but states that my presence is necessary upon the flanks, and here I am with one third rations of corn for horses, with my men building and defending rifle pits and the papers abusing me for not being in Sherman's rear. General Johnston tells me not to mind the papers and I try not to do so, but these things are disagreeable. I told General Johnston on yesterday that if he would order Forrest here with his command and would order me to Forrest's place or into Middle Tennessee I would certainly give a good account.

Even the Yankees knew this. On the 1st Kenner Garrard, one of Sherman's cavalry leaders, wrote McPherson, "deserters say the papers have been abusing Wheeler for his inactivity...and that he is now desirous of doing something to redeem his character."

All of this was in the background as Senator Hill met with General Johnston. Hill asked if Sherman also did not recognize the importance of cavalry raids on the enemy's railroads, and if he might be planning a Union raid on the Southern railways then subsisting the Army of Tennessee. Johnston answered that the enemy was not (little did Johnston know that Union Maj. Gen. Lovell H. Rousseau was about to launch a major cavalry raid through east central Alabama to cut the rail line from Atlanta to Montgomery). Given the course of the campaign to date, Hill was entitled to ask Johnston how long he expected he could hold the enemy north of the Chattahoochee. Hill later recalled that the general "did not answer this question with directness."

As he closed the conversation, the senator asked for clarification. "And I understand you to say, General Johnston, that Sherman cannot be defeated except by the proposed attack in his rear, and that this work must be done by Forrest or Morgan or by some such force?" Johnston said yes, and thanked Hill for carrying this message to the government's leaders. "Rarely if ever," Castel observes, "has a military commander striven harder to persuade the head of his government that he is incapable of defeating by his own efforts the army he opposes."[1]

The day after his repulse at Kennesaw, Sherman recommitted himself to flanking Johnston out of position. He ordered Schofield—who already had marched to a point south of Hood's extreme left—to move farther, toward Ruff's Mill on Nickajack Creek. To strengthen the maneuver, Sherman called for McPherson's army to move from the army group's left, behind Thomas and then toward the Chattahoochee on the Sandtown Road. By now, two months into the campaign, Sherman's soldiers were familiar with the maneuvering. "I think unless Sherman can get his flanking machine operating," wrote Edward Carroll of the 14th Ohio on the 1st of July, "there will be home made thunder here before many days."

Confederates on Kennesaw Mountain observed McPherson's march on July 2 and reported it to Johnston, who realized he would soon have to give up his Kennesaw line for the new one already staked out seven miles to the south. Named for the Smyrna Camp Ground at its center, the six-mile line rested on Rottenwood Creek on its right, then ran southwest toward Nickajack Creek, where it refused sharply to the southeast. Confederates marched into these new works during the night of July 2–3. "The extension of the enemy's intrenched line several miles nearer the Chattahoochee than our left," Johnston wired Bragg on the 3rd, "has compelled us to fall back about six miles." As at Kennesaw, Loring held the right, Hardee the center, and Hood the left. The continued retreating of the Confederate army depressed the local

---

[1] S. Davis, *Atlanta Will Fall*, 90; Wheeler to Bragg, July 1, Bragg Papers, box 2, folder 12; Garrard to McPherson, July 1, *OR*, vol. 38, pt. 5, 11–12; D. Evans, *Sherman's Horsemen*, 166 (destruction of the Atlanta-Montgomery railroad), 243; B. H. Hill to Seddon, July 14, *OR*, vol. 52, pt. 2, 704–706.

Senator Wigfall was one of the more influential of Johnston's numerous supporters in Richmond. Alfred P. James enumerates them in "General Joseph E. Johnston, Storm Center of the Confederate Army," 354–59. Because of Davis's well-known antipathy toward Johnston, the general served as a rallying figure for political opponents of the president. In the words of Richard Taylor, "the ancient dwellers in the cave of Adullam gathered themselves behind his shield and shot arrows at President Davis and his admirers, weakening the influence of the head of the cause for which all were struggling" (James, "Storm Center," 354).

population. "During the day general apprehensions existed of the falling back of Johnston's army," recorded William Key, resident of Marietta, "occasioning great anxiety." Key was saddened further when Hardee's infantry passed by his house, "committing many depredations and pretty much destroying all the poultry."

With the Chattahoochee only three to five miles behind the Confederate lines, Sherman thought that Johnston was only holding the Smyrna line long enough for his trains to cross the river, and that McPherson's maneuver would force his retreat. On the 3rd, Thomas's troops marched through Marietta and came up on the Southerners' fortifications. Sherman ordered Thomas the next day to demonstrate against the Rebel works while McPherson and Schofield marched farther around Johnston's left.

On the 4th many on the Northern side celebrated the national holiday. When some Federal bands began playing patriotic tunes—"heard some excellent music," Sgt. Walter Clark of the 63rd Georgia recorded in his diary—Confederate musicians served up a rendition of "Dixie." Lt. George Warren of the 3d Mississippi recorded that this "raised their ire and brought over several shells." The Rebels hurled some shells of their own. "We were shelled all day," remembered Capt. Ira Read of the 101st Ohio. "There was about as much noise as there usually is at home on the Fourth but a little more of the 'shell' than I like."

Around noon two brigades launched Thomas's demonstration against the Confederate center near Smyrna. Col. William Grose's brigade of David Stanley's division and Brigadier James Morgan's of Jefferson Davis's overran the Rebels' skirmishers but failed to break their main line, held by Cleburne's division. The unsuccessful sortie cost the Federals 270 casualties; Cleburne's are not recorded but would customarily have been but a fraction. Among the Confederate wounded was Brig. Gen. Alfred J. Vaughan, commanding a brigade in Cheatham's division, who had his left foot torn off by an enemy shell.

Two and a half miles west of Smyrna, Federals also advanced against the Confederate line held by Hood's corps, but here they enjoyed more success. In the morning Dodge's XVI Corps crossed Nickajack. Brig. Gen. John Fuller's brigade of John Veatch's division moved against Stevenson's division. The Federals drove the Southern skirmishers back to their main line. Later, around 6 P.M., they charged again, forcing the Confederates to fall back and dig in. Union casualties in this brief charge totaled 140 killed and wounded. Among the latter was Col. E. F. Noyes, shot in the ankle. Generals Dodge and McPherson saw him being carried in ambulance to the rear and asked if he were seriously wounded. "General," Noyes answered, "I got their works and they got part of mine. But it's Fourth of July and I don't care a continental." The "battle of Ruff's Mill," as some have termed it, was so slight that Hood in his report does even mention it. Perhaps the most important casualty of the day's activity was Major General Hindman—but not from enemy fire. Hindman was riding through some woods when he struck a tree branch so big that he was thrown from his horse. His eyes took the hit, to the extent that the general could not continue field service. He retired from command for convalescence in Macon; his division was eventually taken over by Brig. Gen. John C. Brown.[2]

The really important event on July 4 was the Federal advance which threatened the Confederate left. By 2:30 P.M. Johnston knew that at least one enemy corps had crossed Nickajack Creek; the Yankees had pushed to within a mile of Hood's headquarters at Alexander Eaton's house. General and Lieutenant Mackall rode to Hood's sector to see for themselves the danger posed by the enemy's encroachments. Then, after four, "Sul" Ross, one of Brig. Gen. William H. ("Red") Jackson's cavalry brigadiers, sent word that the enemy was flanking him; "My position here is no longer tenable.... I am moving back." Then, toward evening, Maj. Gen. G. W. Smith sent in word that his thin line had been pressed hard that afternoon. He suspected that the Federals would attack the next day and he therefore planned to retreat by dawn. Sherman had thus succeeded in flanking the Southerners' line. Johnston had no choice but to order the army to retreat to its next prepared position. That did not stop the croakers in Atlanta from voicing more pessimism over Johnston's prospects. "Speculation was rife yesterday to establish the reason for our retreat," stated the *Intelligencer* on July 5. "To-day it is more eagerly agitated—what will we do next? Our street Generals have it that we will be flanked to the Gulf and then to Richmond."[3]

---

[2] Castel, *Decision in the West*, 322, 324–25, 329, 331–32; Carroll, "There Will Be Home Made Thunder Here," 29; Scaife, *Campaign for Atlanta*, 73–75, f.p. 82 (map); Lane, *Times That Prove People's Principles*, 163 (Key diary entry of July 2); Gottschalk, *In Deadly Earnest*, 373; Johnston to Bragg, July 3, *OR*, vol. 38, pt. 5, 860; W. A. Clark, *Under the Stars and Bars*, 102; Read, "The Campaign from Chattanooga to Atlanta," 271 (Captain Read); Stanley report, undated, *OR*, vol. 38, pt. 1, 225; Grose report, September 5, 1864, *OR*, vol. 38, pt. 1, 260; Davis report, undated, *OR*, vol. 38, pt. 1, 634; Morgan report, August 23, *OR*, vol. 38, pt. 1, 648; McMurry, "Atlanta Campaign" dissertation, 237 (170 casualties in Davis's division); Peterson, *Confederate Combat Commander*, 184; Warner, *Generals in Gray*, 316 (General Vaughan); Kelly, "Mountains to Pass, a River to Cross," 54 (Colonel Noyes); Hood report, undated, *OR*, vol. 38, pt. 3, 762; Neal and Kremm, *Lion of the South*, 196; McMurry, *John Bell Hood*, 115.

[3] McMurry, "Mackall Journal," entry of July 4; "The Alexander Eaton House. Hood's Headquarters," Scruggs, *Georgia Historical Markers*, 146; Ross to Jackson, July 4 (4:10 P.M.),

The line Johnston's troops marched into during the night of July 4–5 had already been constructed. According to the army's chief of artillery, Brig. Gen. Francis A. Shoup, two weeks before, while the Confederates were at Kennesaw, General Johnston had told Shoup that "it was but a question of time, and that a short time" before they would have to retreat across the Chattahoochee. Shoup knew, as did Johnston, that giving up this last river barrier to the enemy would be as fateful as Caesar's crossing of the Rubicon. (Senators Wigfall and Hill would have made the same point.) Thus Shoup offered the idea of building one last defensive line on the north bank of the Chattahoochee around the railroad bridge. Johnston eagerly granted permission.

Thereupon Shoup and Maj. W. B. Foster, Loring's chief engineer, supervised a thousand slaves (as Sherman, in his memoirs, recalled being told) set to work building a new line of fortifications that featured something new: log-and-earth infantry forts which G. W. Smith began calling "Shoupades." There were three dozen of them, spaced anywhere from 60 to 175 yards apart in the four-mile line constructed around the Western & Atlantic bridge. Each was shaped like an arrowhead pointed at the enemy. The outside walls were almost vertical, built with logs laid horizontally up to a height of sixteen feet. Dirt ten to twelve feet thick was packed in between the outer and inner log walls. Inside was a banquette or firing platform for infantry; every fort was intended to be manned by eighty riflemen. Between them were constructed earthen redans, each to hold two cannon. Eight-feet high stockades of logs connected the Shoupades and redans. The key defensive element lay in Shoup's placement: troops in each fortification could pour an enfilade fire toward the next one, all down the line.

General Shoup was proud of his line, unconventional as it was. "When the troops, accustomed to earth works, saw what they were expected to defend, they were greatly amused and made all sorts of ridiculous remarks about them," he recalled. Some, including W. H. T. Walker's men, started tearing down the log revetments to build the earthen parapets they were used to. General Cleburne, however, came to Shoup's defense; eventually Hardee congratulated him for his ingenuity. To assure the troops, who were jittery about

---

OR, vol. 38, pt. 5, 864; G. W. Smith, "The Georgia Militia about Atlanta," 333; S. Davis, *Atlanta Will Fall*, 97 (quoting "The Position," *Atlanta Intelligencer*, July 5).

I wish to thank Dr. Gordon Jones, senior military historian of the Atlanta History Center, for calling my attention to the *Intelligencer* editorial, whose statement about Johnston's being "flanked to the Gulf" echoes President Davis's and other Confederates' talk of Johnston retreating across pontoons to Cuba.

fighting in Shoup's new-fangled stockades, a circular was printed and distributed on the 7th to officers commanding troops in the three dozen Shoupades. Instructions were given on how to strengthen the positions ("line the inside with heavy logs"), weapons to use when attacked (including hand-grenades, slow-match, and port-fire) and how to counter a suspected enemy tunnel. Shoup must have had a hand in preparing this text, for a specific point addressed firing tactics. "The men should be instructed," the circular stressed, "to direct their fire to support the contiguous redoubt rather than against the enemy approaching their own." With Shoup's line dependent upon interlocking fire, each redoubt was integral in defending the line; "in no case must the redoubt be abandoned."

The circular calmed the men down. "Men better pleased," Lieutenant Mackall noted in his journal. One Confederate, Grant Taylor, even wrote his wife on July 9, "we are strongly fortified here and if the Yanks would attack us I believe with God's help we would dress them out nicely." Where the troops had defaced the Shoupades, Johnston ordered them be rebuilt to their original form. Nevertheless, the commanding general asked Shoup to dig standard infantry entrenchments another mile or so down the river to cover Turner's Ferry. When the infantry started to occupy the river line, Hood's corps formed the Confederate left, Hardee's the center, and Loring's the right. For supply, Johnston had six pontoon bridges laid, two for each corps.

On the morning of July 5 Sherman was surprised when his morning pickets reported that Johnston had withdrawn from his Smyrna line and taken position on the north bank of the river. "No general, such as he," Sherman had predicted to Thomas on the 3rd, "would invite battle with the Chattahoochee behind him." (Richard McMurry has wryly remarked that somehow Sherman had forgotten that Johnston had done this very thing at the Oostanaula.) The Federal commander had to see for himself that the Rebels had not retreated across the river as he had expected them to do. On the 5th Sherman was content with Thomas's army drawing up to Johnston's new line. McPherson advanced to the right along Nickajack Creek. Garrard's cavalry fanned out upstream, while McCook's and Stoneman's probed downriver, testing the Chattahoochee's various crossing points.

Johnston's problem was that there were so many crossing points. In the fourteen downstream miles from the Western & Atlantic and wagon bridges were six ferries; upstream there were another seven ferries or fords. Wagon bridges were located at Roswell and at McAfee's, six miles farther upstream. Wheeler's cavalry kept watch on those to the right of the army; Red Jackson's horsemen and Smith's Georgia militia were on the left. Johnston had posted little guard posts with earthworks at some of these places. For instance, on July 3 a hundred state troops, dug in at Aderhold's Ferry ten miles downstream from Shoup's line, drove off some snooping Yankee cavalrymen.

Hood did not think much of the Shoupade line. According to General Shoup, he "began at once to declare his position unsafe." Hood had held the army's left at Kennesaw and Smyrna and knew Sherman's propensity to maneuver around his flank. Now, with a river to his back, Hood felt even more vulnerable. Shoup actually agreed with Hood, stating that his position was indeed weak and that "he ought never to have been put into such a position." Shoup recalled that on the evening of July 7 General Johnston called a council of his corps commanders, who now included Peter Stewart, newly promoted to lieutenant general June 23 and who took command of Polk's/Loring's corps on the 7th (Brig. Gen. John C. Brown took charge of Stewart's division in Hood's corps). At the meeting Hood urged that the army withdraw across the river. Shoup disliked this disregard for his line. "He did not seem to understand the design of the works on his right," he wrote," and I doubt if he ever stopped to think much about them." The council ended without any decision made on a change of position. Johnston would await Sherman's next move.[4]

---

[4] Shoup, "Dalton Campaign," 262–64; Raab, *Lloyd Tilghman and Francis Asbury Shoup*, 193; Scaife and Erquitt, *The Chattahoochee River Line*, 5–6, 14–15, 33; Scaife, "The Chattahoochee River Line," 58 (pointing out that both Francis Shoup's and Andre Maginot's defensive lines were never tested in combat); Dickey, "Johnston's River Line," 25; Scaife, *Campaign for Atlanta*, 77, and map "Federal Approaches to the River Line," f.p. 82; Sherman, *Memoirs*, 2:67 ("this negro explained that he with about a thousand slaves had been at work"); Calhoun Benham, "Maj.-Gen. P.R. Cleburne: A Biography," *Kennesaw Gazette*, August 1, 1889, 5 ("Gen. Walker's men tore down the palisades, not liking them"); Mason, Circular pertaining to construction and fortification of redoubts used by the Army of Tennessee (my thanks to Michael K. Shaffer of Kennesaw State University for providing me copy of this little-known circular); McMurry, "Mackall Journal," entry of July 7; Blomquist and Taylor, *This Cruel War*, 268; Biggs, "The 'Shoupade' Redoubts," 88 ("orders from Johnston restored all the forts"); J. W. DuBose, *General Joseph Wheeler and the Army of Tennessee*, 351 (six pontoon bridges); Elliott, *Soldier of Tennessee*, 194; Sherman to Thomas, July 3, *OR*, vol. 38, pt. 5, 30; McMurry, *Atlanta 1864*, 114; Mitchell, "The Old Ferries and the Ferry Roads," 33–41; Maj. W. Clare to Gen. J. E. Johnston, July 4, *OR*, vol. 38, pt. 5, 562 (on ferries and crossings); "The Upper Chattahoochee," *Columbus Enquirer*, July 16; S. Davis, *Atlanta Will Fall*, 97; Wingfield, *General A. P. Stewart*, 91; General Orders No. 1, July 7, *OR*, vol. 52, pt. 2, 687–88 (Stewart takes corps command); Special Orders No. 86, July 7, *OR*, vol. 38, pt. 5, 868 (General Brown); Castel, *Decision in the West*, 336, 338.

Writing in 2012, James W. Raab notes, "only seven Shoupades still exist of the 36 originally constructed. All are in northwestern Atlanta suburbs, inside Interstate 285" (*Deliverance from Evil*, 143).

As he prepared to force Johnston to retreat across the Chattahoochee, Sherman drew from his bag of tricks a really big one. So far in the campaign he had outflanked the Rebels by maneuvering around their left: at Rocky Face, Resaca, Allatoona, Lost-Brush Mountain, Kennesaw, and Smyrna. Now he would switch up and do the opposite. He had been planning this all along. As early as April 10, he announced to Grant, "Should Johnston fall back behind the Chattahoochee I would feign to the right, but pass to the left, and act on Atlanta or on its eastern communications, according to developed facts."

To keep Johnston guessing about his intended crossing points, Sherman had his forces spread out thirty miles along the Chattahoochee's north bank. To his left, Kenner Garrard's cavalry found that the Rebels had burned the bridge at Roswell on July 5. Schofield's corps was marching to that point from Smyrna. Thomas's army held the center; troops of Howard's IV Corps saw the Rebels cut a pontoon bridge at Pace's Ferry, but they managed to capture it anyway. McPherson marched on the right, sending columns to Howell's Ferry and other points. On the 7th Sherman ordered Mac to "display as much anxiety to cross as possible and as low down, but keep your masses ready to move to the real quarter when required." Union activity downstream from the railroad bridge included random shelling at crossing points such as Turner's Ferry. The *Atlanta Southern Confederacy* reported this on July 8, noting, "The heavy shelling at Turner's Ferry yesterday induced the opinion that the Yankees might attempt to carry that point." George Stoneman's horsemen probed farther downstream still, stretching the Federals from Roswell to Campbellton.

Johnston's soldiers evidently also thought the threat would come beyond their left flank. Stevenson's division of Hood's corps was detached miles down the river to watch developments. One Confederate, writing his wife on July 7 from Green's Ferry, six miles below the Western & Atlantic bridge, predicted that they would probably be sent five miles farther, to Sandtown. "We are only eight miles from Atlanta but by this time tomorrow we may be twenty miles away—we cant tell any thing about it."

Sherman's show was working. Even though Union cavalry found Power's Ferry guarded with artillery, Garrard informed Sherman on the evening of July 7, "My impression is that Johnston will make no attempt on this flank, but that his cavalry has gone to his left." This allowed Sherman to telegraph

General Halleck that night, "Johnston is maneuvering against my right, and I will try and pass the Chattahoochee by my left."[5]

July 8 was the eventful day of the Federals' crossings. Ordering Thomas to make noisy demonstrations downstream to hold Johnston's attention, Sherman directed Schofield's army corps to cross at Roswell or points below. Schofield chose Isham's Ferry, where Sope Creek flows into the Chattahoochee. There a Kentucky regiment paddled across on pontoon boats, landed and drove off the Southern horsemen and their single howitzer, without losing a single man. Soon the pontoon bridge was built, and Cox's division began crossing. A half-mile upstream the Federals found a rock fish dam and walked across on it, unopposed.

The next day, Northern troops planted their third bridgehead at Roswell when a brigade of Garrard's cavalry crossed at Shallow Ford, a mile below the burned bridge. More heralded in the annals was the episode at Cochran's Ford, a half-mile below Sope Creek. Brigadier General McCook ordered Col. James P. Brownlow (son of the East Tennessee Unionist William G. "Parson" Brownlow and commander of the 1st Tennessee [U.S.] Cavalry) to put a detachment across at the ford. Brownlow found the river deep but passable, so his men stripped down to hats, rifles, and cartridge boxes and waded in. On the other side they drove away the Rebels, who were undone by this display of "raw courage." "It was certainly one of the funniest sights of the war," reported General McCook; "citizens on the other side say the enemy were totally unprepared for a crossing on this flank."[6]

---

[5] Sherman to Grant, April 10, 1864, *OR*, vol. 38, pt. 3, 313; D. Evans, *Sherman's Horsemen*, 12, 17–18; Castel, *Decision in the West*, 336; Scaife, *Campaign for Atlanta*, 77–79; Sherman to Schofield, July 7, *OR*, vol. 38, pt. 5, 78; D. B. Connelly, *John M. Schofield and the Politics of Generalship*, 104; Sherman to McPherson, July 7, *OR*, vol. 38, pt. 5, 80; "The Front," *Atlanta Southern Confederacy*, July 8; Candler, "Watch on the Chattahoochee," 427; Garrard to Sherman, July 7, *OR*, vol. 38, pt. 5, 68; Sherman to Halleck, July 7, *OR*, vol. 38, pt. 5, 73.

[6] Sherman to Thomas, July 8, and Thomas to Sherman, July 8, *OR*, vol. 38, pt. 5, 86–87; Maj. J. A. Campbell, SFO No. 46, July 7, *OR*, vol. 38, pt. 5, 85–86; Scaife, *Campaign for Atlanta*, 80; "Raw Courage," 46; McMurry, "More on 'Raw Courage,'" 36–38.

After Garrard's cavalry occupied Roswell on July 5, they burned the town's woolen mill and two cotton mills. Under orders from General Sherman they took the four hundred women working there, with their children and some possessions, by wagon to Marietta, put them on trains, and sent them north to Louisville, where they spent the rest of the war. The incident is described in Bynum, "Sherman's Expulsion of the Roswell Women in 1864," 169–82, and by D. Evans, "Wool, Women, and War," 38–42. Hitt, *Charged With Treason*, is the standard history. Others are Cook, *North Across the River*, and Petite, *"The Women Will Howl."*

Federal cavalry did the same thing at New Manchester, a dozen miles to the west of Atlanta: burned the cotton mill at Sweetwater Factory and deported some 250 women and boys employed there to the north. See Scott, "Total War Comes to New Manchester," 22. One

Confederate officers reported the enemy advances on July 8. Lt. Frank Burkett of the 9th Tennessee Cavalry stated that his pickets were falling back from Isham's Ferry because "the Yankees are supposed to be in considerable force and of all arms." Lt. Col. J. F. Gaines reported enemy scouting parties "as far up as McAfee's bridge," fully six miles east of the Roswell bridge site. The really important news came from cavalryman C. Y. Wilson near Isham's Ford at 9 P.M. on the 8th: the enemy had not only crossed there but advanced more than a mile on the south bank with infantry and artillery. With their pontoon bridge in place more were still to come. "One of my men who was at the point late this evening says he saw about two brigades of infantry this side of the river, with fixed bayonets, marching by fours." This message, passed through ranks to General Wheeler, reached Johnston's headquarters by 1 A.M. on the 9th, when General Mackall added an endorsement calling for more information. Four hours later, still before dawn—headquarters was clearly agitated by the news—Mackall wrote Wheeler, "General Johnston is very anxious to receive your report as to the character of the force which crossed at Isham's Ford." Without any record of Wheeler's reply, Mackall wrote back to him on the 9th that the army would retreat across the river that night.

The Confederates conducted their nocturnal retrograde as quietly as possible, spreading sand and leaves on their bridges to muffle the sounds of transit. When all were across, the bridges were burned and pontoons taken up; those of the latter that could not be hauled in were set afire. Early on the 10th delighted Federal pickets discovered the Rebels' departure and passed the good news through channels. By 6:30 A.M. Sherman was able to telegraph Halleck, "The enemy this morning burned his bridges across the Chattahoochee, the railroad bridge included." Having won the game—or at least this round of it—Sherman, as he had done at the Etowah, planned to rest his forces for the next few days before pushing closer to Atlanta. The Federal commander had good reason to be confident, as reflected in an anecdote carried by Northern newspapers. According to the story, Col. Amos Beckwith, chief commissary officer, informed Sherman that the army faced a possible shortage of beef. "Have you enough to last for two months?" Cump asked. Yes, answered Beckwith. "Well, in less than two months the army will be in

---

of the factory workers, S. H. Causey, who was seventeen at the time and one of the displaced persons, was interviewed decades later: "The Federals…treated us kindly.… None of the women were molested and everyone had plenty of food" (Lewis C. Russell, "Georgia Towns Moved in War," *Atlanta Journal Magazine*, February 28, 1932, 10).

[hell], or in Atlanta; if it goes to the former place, we shall need no beef; if it goes to the latter, we shall find enough."

Sherman had outwitted Johnston, but Johnston had just as much allowed himself to be outwitted. Even Cump, usually kind to Johnston in his postwar writing, recognized this. "I have always thought," he wrote in his memoirs a decade later, "Johnston neglected his opportunity there, for he had lain comparatively idle while we got control of both banks of the river above him." Richard McMurry is harsher, judging Johnston's failure to contest Sherman's crossing of the Chattahoochee to be a blunder as big as his failure to contest Sherman's passing through Snake Creek Gap at the start of the campaign. For his part, Johnston's chief of staff Mackall could only write his wife on the night of July 10, "I suppose they will have a great hurrahing to-day after getting us across the river."[7]

After crossing the Chattahoochee, as it had done before, the Confederate army could withdraw into fortifications already prepared for it. Indeed, as early as May 1863, the Atlanta City Council had called for earthworks to be built at the Chattahoochee to guard bridges against attack by Yankee cavalry. Eventually the project entailed constructing fortifications around the city itself. The army's engineer department in Richmond took over the work, with Maj. Gen. Jeremy F. Gilmer directing Capt. Lemuel P. Grant as commander of the project. From summer 1863 into spring 1864 Grant supervised the digging of a perimeter of infantry trenches studded with a dozen and a half artillery forts—some ten and a half miles of continuous works surrounding the city. Grant's encircling line averaged a mile and a half outside the city's edges. If a Yankee army actually approached Atlanta, this would be far enough to prevent smoothbore artillery (e.g., Napoleon 12-pounders) from shelling downtown, but enemy rifled cannon (e.g., Parrotts) would be within range. Most troubling was that Grant's original entrenchments northwest of the city were closest to Atlanta's exurbs than at any other point—and this would presumably be the direction of an enemy advance. Captain Grant therefore planned a northwest salient to be added to the perimeter, with several more artillery forts; this sector was still being worked on when Sherman approached. The construction was so intense that a thousand shovels and spades

---

[7] Burkett to Col. Robert H. Anderson, *OR*, vol. 38, pt. 5, 869; Gaines to Capt. Terrett, July 8, *OR*, vol. 38, pt. 5, 869–70; C. Y. Wilson to Col. B. J. Hill, July 8, with W. W. Mackall endorsement, *OR*, vol. 38, pt. 5, 870–71; Mackall to Wheeler, July 9, *OR*, vol. 38, pt. 5, 872; Castel, *Decision in the West*, 341; McMurry, *Atlanta 1864*, 117 (pontoons burned); Sherman to Halleck, July 10, *OR*, vol. 38, pt. 5, 102; [De Fontaine], *Marginalia*, 214 (Sherman-Beckwith); Sherman, *Memoirs*, 2:70; McMurry, "A Policy So Disastrous," 244; Mackall, *A Son's Recollections*, 219 (letter of July 10).

were rushed to Atlanta by the Engineer Department in Richmond. After completion, the fortifications surrounding Atlanta extended twelve miles in length. Enslaved African-Americans did most of the work. Captain Grant continuously called for more black laborers; plantation owners were enjoined by the newspapers to give up their slaves for the common defense. Sometimes black men were seized outright in Atlanta as when, according to one news complaint, "negro drivers were taken from carriages, and wood wagons and drays, leaving the vehicles in the street." The number of such workers eventually reached several thousand. One of Sherman's soldiers wrote home on July 7 stating, "a negro that was captord" had reported that 4,000 black men had been toiling for a month and a half, digging trenches and building log stockades. After the crossing of the Chattahoochee, Johnston's engineers set to work on sectors such as the northwest salient, which was still not completed. In his memoirs Johnston affirmed, "as soon as the army passed the Chattahoochee, its engineer-officers joined in the work of strengthening the intrenchments of Atlanta with all the negro laborers that could be collected."

The army, however, did not immediately occupy Grant's entrenched perimeter. With Sherman's army not vigorously crossing the river—it would not do so until the 17th—the Confederates settled into position about three miles south of the Chattahoochee, according to Brig. Gen. Marcus J. Wright, post commandant in Atlanta, who wired Bragg on the 10th. The troops were arrayed south of Peachtree Creek—"high ground south of the creek," Johnston described in his campaign report, "from which to attack the enemy while crossing." Yet the army spent the week after its crossing rather idly. "Camped 4 miles north of the city," entered Arkansas Brig. Gen. Daniel Reynolds in his diary on July 10. "Army resting quietly," was his entry the next day. July 12: "Men washing clothes and resting." "All quiet and no orders to move," Reynolds recorded on July 13; there were apparently no orders even to dig in. Irenus Landingham, a Texan serving in Hood's Corps headquarters staff, wrote his mother on July 14, "we have had an opportunity to rest a little." "12th and 13th. The camp is filled with rumors," recorded General French in his diary; "14th, 15th, 16th. We remained in bivouac." General Johnston stated that the army's position extended along Peachtree Creek to the Chattahoochee, then downstream a bit. "It [the creek] and the river below its mouth were therefore taken as our line," Johnston explained. French's division, on the left of the line, stretched along the Chattahoochee's south bank. His troops were under orders to fire on Yankee cavalrymen when they plunged into the water for a swim.

The bivouacking was apparently easy. Hiram Smith, attached to the army pioneers from the 40th Alabama, would have been busy at work if the army had been set to fortifying in the week after it retreated across the river. But there was no fortifying to be done. After marching into camp on July 11, Smith recorded in his diary, "I judge we will remain here several days, as it will take some time for the enemy to cross the river and until then we may expect a little rest." They got it. "Tuesday, Wednesday, Thursday, Friday and Saturday," Smith wrote, covering the period of July 12–16, "we remained in the same camp, the longest resting space we have had since we left the 'Gap.'"[8]

Johnston had to notify the administration that he had given up the last river barrier before Atlanta. "On the night of the 8th the enemy crossed at Isham's, or Cavalry ford; intrenched," he telegraphed Richmond. "In consequence we crossed at and below the railroad, and are now about two miles from the river, guarding the crossings." The general might have tried to soften the blow of this disturbing news, but in truth during the days and weeks preceding he had pretty clearly signaled his intent to fall back upon Atlanta's fortifications. When he ordered the army into Shoup's river line, he wired the War Department on July 5, informing his superiors that he was now backed up against the Chattahoochee. The next day Bragg acknowledged receiving the message, which he passed on to the president. Davis was naturally agitated. "The announcement that your army has fallen back to the Chattahoochee renders me more apprehensive for the future," he wrote on the 7th. Davis reminded Johnston that "other places have been stripped to re-enforce your army until we are unable to make further additions." Johnston could not accept that he had to contend against Sherman with only the forces at his command. His idea to force the enemy back from Atlanta was the same one he had told Senator Hill in front of Hood on July 1: the administration had to send cavalry from outside his department onto Sherman's railway lines, cut them and thus force the enemy to retreat. In his reply of July 8 Johnston informed the president that he had heard of 16,000 cavalry in Mississippi and

---

[8] [Kurtz,] "Forts to Defend Atlanta," *Atlanta Journal Sunday Magazine*, May 25, 1941, 6; S. Davis, *What the Yankees Did to Us*, 56–63, 86; Fryman, "Fortifying the Landscape," 55; S. Mitchell, "Colonel L. P. Grant and the Defenses of Atlanta," 32–35; Julian, "Atlanta's Defenses," 23–24; Niesen, "'The Consequences of Grandeur,'" 8 (Cpl. Leander S. Davis to wife, July 7); Capt. L. P. Grant to Gilmer, October 30, 1863, Grant letterbook, 1862–1865, box 7, folder 1, Lemuel Grant Papers; Nichols, *Confederate Engineers*, 97; Wright to Bragg, July 10, *OR*, vol. 52, pt. 2, 691; Reynolds, *Worthy of the Cause for Which They Fight*, 135; Irenus Landingham to "Dear Mother," July 14, 1864, Landingham Papers; S. G. French, *Two Wars*, 216; Johnston report, October 20, 1864, *OR*, vol. 38, pt. 3, 618; H. S. Williams, *This War So Horrible*, 103–104.

Alabama. "Might not 4,000 of this cavalry prevent the danger by breaking up the railroad between the enemy and Dalton, thus compelling Sherman to withdraw?"

Davis and Bragg had heard all of this before from Johnston and from others, too. Johnston even sought help from his corps commanders. On June 13—the day before he died—Leonidas Polk wrote the president, asking if Forrest and some of his cavalry could be sent to cut Sherman's rail line below Chattanooga. Hardee added his name in endorsement. (Interestingly, Hood did not, possibly because he knew that Wheeler wanted to launch such a raid himself but had been denied permission.) Governor Brown of Georgia repeatedly called on Davis to send Forrest from Mississippi or bring John Morgan down from Virginia. On the 4th of July the governor of Alabama, Thomas H. Watts, also wrote the president: all Davis had to do was order Forrest to launch the raid. Sen. Richard Walker, also writing on the 4th, informed Assistant Secretary of War John A. Campbell, "No one doubts that Forrest, with 10,000 men, or even less, would effectually destroy the road in Sherman's rear south of Dalton.... Sherman's army would not only be forced to retreat, but...would be destroyed."

The president's patience was wearing thin. In two messages to Johnston on July 11 he spelled out what forces were operating in the Southwest and the enemy forces they faced, all to demonstrate that no further cavalry could be sent to him. "Why not so employ those already sent to you from that department, or others of equal number, for the proposed operation?"[9]

With the president having to ask his army commander such a basic question—and with Sherman's army group poised to cross the Chattahoochee and march on Atlanta—it was obvious that Jefferson Davis needed more information as to General Johnston's plans and intentions than Johnston himself

---

[9] Johnston to Bragg, July 5 and 10, *OR*, vol. 38, pt. 5, 860, 872; Bragg to Johnston, July 6, *OR*, vol. 38, pt. 5, 867; Davis to Johnston, July 7, *OR*, vol. 38, pt. 5, 867; Johnston to Davis, July 8, *OR*, vol. 38, pt. 5, 868–69; Polk to Davis, June 13, *OR*, vol. 38, pt. 4, 774; Johnston to S. D. Lee, June 3, *OR*, vol. 38, pt. 4, 756 ("If you can throw Chalmers and the cavalry brigade...it may produce great results"); Parks, *Joseph E. Brown of Georgia*, 290–92; Brown to Davis, July 5, *OR*, vol. 39, pt. 2, 688; Thomas H. Watts to Davis, July 4, Crist, Williams, and Dillard, eds., *Papers of Jefferson Davis*, 10:499; R. W. Walker to J. A. Campbell, July 4, *OR*, vol. 52, pt. 2, 685–86; Davis to Johnston, July 11, *OR*, vol. 38, pt. 5, 875–76.

Johnston's style of communicating with the Confederate War Department had drawn criticism within the department well before the Georgia campaign. Robert Kean, head of the Bureau of War, had written in his diary on March 22, 1863, when Johnston commanded the Department of the West, "General Johnston has written another of his brief unsatisfactory, almost captious letters." Three weeks later Kean again recorded, "General Johnston has written another sharp captious letter" (*Confederate Government*, 46, 50).

was providing. The year before the president had written General Lee, "I can only learn from him [Johnston] of such vague purposes as were unfolded when he held his army before Richmond," meaning in spring 1862. Davis's remembrance suggests that the president had fixed Joe Johnston as a man who would not communicate, yet he desired information. "At this distance I cant judge of your condition or the best method of averting calamity," he had written Johnston on the 7th. Given John B. Hood's much ballyhooed correspondence with Davis, Bragg and Seddon in the first half of 1864, it is worth noting that the president did not send a surreptitious wire to Hood asking for a report on the general situation in Atlanta. To get a good read of what was going on, Davis turned instead to an informed observer and trusted aide—Braxton Bragg. "You will proceed to Georgia, [and] confer with General Johnston in relation to military affairs there," the president instructed on July 9. Davis wanted Bragg to leave as soon as possible and return as soon as possible, but also to tie into his trip such other matters as he deemed best.[10]

Then came information from another source. Senator Ben Hill had met with General Johnston on July 1 at request of Governor Brown and Senator Wigfall, who hoped that Hill would try to persuade President Davis to send Forrest's or Morgan's cavalry to raid Sherman's railroad supply lines. (Hill was known to be a Congressional ally of the president.) Hill had promised Johnston that he would do this.

Hill arrived in Richmond on July 10. Even though it was a Sunday, he knew that Davis would want to meet him as soon as possible. After checking into a hotel, Hill made his way to the executive mansion. There he met with Davis and Secretary of War Seddon and told them of his discussion with Johnston about sending Forrest or Morgan against Sherman's railroads. The president listened, and then recounted to Hill how S. D. Lee in Mississippi had had to deal with enemy columns from Memphis numbering 15,000 men and how the Federals were marching from New Orleans to take Mobile. As for Morgan, his command had recently been whipped and was incapable of offensive action. Hill grasped that there was thus no further help that could be sent to Johnston. Then, toward the end of their conversation, Davis asked, "How long did you understand General Johnston to say he could hold Sherman north of the Chattahoochee River?" Hill answered that Johnston had implied from 54 to 60 days. Davis then produced the dispatch from Johnston which he had just received stating that the army had crossed the river—ten days after Hill's meeting. The senator from Georgia was stunned. Having

[10] Davis to Lee, July 21, 1863, *OR*, vol. 27, pt. 3, 1030; Davis to Johnston, July 7, 1864, *OR*, vol. 38, pt. 5, 867; Davis to Bragg, July 9, *OR*, vol. 39, pt. 2, 695–96.

originally intended to argue Johnston's case, Hill now became convinced that Johnston could not hold Atlanta.

After their meeting, Secretary Seddon wrote down a summation of what he had heard: that Johnston, with his army then at Kennesaw, had retreated to that point because Sherman had used his masses to flank him successively; that he had found no favorable opportunity to launch an attacking battle; "that with the army and means at his command he was unable to repel the enemy from Georgia or prevent his advance, and that the only mode by which, in his opinion, such results could be obtained was to have the enemy's communications with his supply base cut, and thus to compel him to withdraw; that this could only be done by forces other than those under his command," and that he could not spare any of his own cavalry for such a raiding operation. Seddon wanted to be clear on a salient point: "on your emphatic inquiries, varied in form several times in order to leave no doubt on your mind or on his as to the full import of the statement, he stated distinctly and positively that with his army and resources he could not repel the enemy nor protect the State, and that aids from other commands to cut communications were indispensable." There were two other points Seddon wanted to be sure of. First, "at no other time nor in any other view did he seem to contemplate the probability of his fighting." Second, "at the time of your conference, General Johnston, being at Kennesaw Mountain, expected to be compelled to fall back again, but believed....that at least a month or more must elapse before the retreat, if it continued to be made, would reach that river."

Seddon took such care in writing his narrative because Hill's was the most up-to-date and complete report on Johnston's plans and frame of mind yet received by the administration. He therefore asked Hill if it was correct. According to Robert Kean in the War Department, "Mr. Hill said it was, except that it was not strong enough and he wrote it out himself." Hill claimed that it was still "fresh in my mind," even though the meeting had taken place a week and a half before. In his memorandum, the senator laid out the details clearly. When Governor Brown and others asked him to write President Davis, urging him to order Forrest and his cavalry to raid into Sherman's rear, he had offered to meet with General Johnston, carry the import of their conversation to Richmond and personally communicate it to President Davis. On July 1 he had met with Johnston, who eagerly accepted Hill's offer to explain to the president how things stood in Georgia. From their "free conversation" Hill took away five points:

1. Johnston hoped Sherman would attack his troops in their entrenchments, but with the exception of a few such assaults (as at Resaca, New Hope, and Kennesaw Mountain), Sherman had not done so;
2. Sherman offered no opportunity to be attacked in the open field, as he always approached cautiously, entrenching as he did so;
3. Johnston had not the strength to attack Sherman in his fortified lines;
4. Sherman used his superior numbers and "the character of the country" to outflank Johnston successively from his positions; and
5. "General Johnston distinctly stated that the only way to get Sherman's army out of the country was by an attack on his rear—by cutting the railroad and destroying his supplies, &c.... His desire was that Forrest or Morgan, one or both, should be at once ordered to do this work." There were 15,000 cavalry in Mississippi and Alabama under General S. D. Lee; a third of them under Forrest's command could accomplish the feat.

There were a couple more points Hill wished to make. The senator told Johnston that when he retreated to the Chattahoochee, the Federals would be able to pass down the river and burn the bridge on which ran the railroad to Montgomery, and that his army's supply line to Alabama would be cut. Johnston countered that the railroad bridge at West Point was well guarded. "No," Hill remembered saying, "I am well informed that there is not a spade of dirt amiss, nor a pile driven, in the way of fortification, and the militia called there to defend it, I understand, had been sent home." Johnston looked surprised. When a senator had to inform an army commander about the state of his logistical support system, Hill seemed to imply, things were not good. He noted that only once did Johnston predict that the two armies would eventually clash in a "bloody fight." He also asked why Johnston could not send Wheeler out on a raid; the general replied that his cavalry was needed to guard the army's flanks.

Hill repeated that three times he had said, "And I understand you to say, General Johnston, that Sherman cannot be defeated except by the proposed attack in his rear, and that this work must be done by Forrest or Morgan or by some such force?" When Johnston agreed, Hill told him he would go to Richmond and "do all in my power to have the order issued for the re-enforcements to the rear of Sherman," but he worried that there was not enough time. Johnston disagreed. Hill called upon General Hood, who apparently had been silent during the conversation. "General Hood thought the time was passing, if it had not already passed. We all agreed that no time was to be lost."

Hill asked Johnston how long it would take the enemy to force him back to the Chattahoochee. Rather evasively, Johnston said that Hill could calculate that, based on the fact that it had taken over a month for Sherman to advance the few miles from New Hope Church. (He ignored the fact that thirteen days of rain during June had slowed Sherman's progress.) Apparently Hood at that point chimed in, saying that it was premature to think of giving up the Kennesaw Mountain line. "General Hood said our present line was the strongest in that country; that the Kennesaw was the great military base in that region, and that when we left that line we should go back much more rapidly to the Chattahoochee River." At the end of the conversation Johnston had thanked him.

Such was Ben Hill's report. He told President Davis he would be in Richmond a few more days and would help in any further way.[11]

Even though Johnston had been contemplating a retreat across the Chattahoochee, when the Confederate army did so, Atlanta's civilian population reacted with demoralization and even panic. A newspaper correspondent in the city wrote,

> I can give you no idea of the excitement in Atlanta. Everybody seems to be hurrying off, and especially the women. Wagons loaded with household furniture, and everything else that can be packed upon them, crowd every street.... Every train of cars is loaded to its utmost capacity.... The excitement beats anything I ever saw.

In truth Johnston had been signaling his intentions for days. On July 5 he had ordered the invaluable munitions machinery in the city, especially that of the rolling mill, to be moved to Augusta, Macon, and other places. The percussion cap factory, for instance, went to Macon. The army's hospitals were to pack up and move south to Macon and nearby towns. After a ride into the city on the 7th, Lieutenant Mackall recorded that the evacuation orders had created obvious alarm among the citizenry. His uncle, Brigadier General Mackall, related this news that night in a letter to his wife: "I suppose that there is great excitement in Atlanta, as I have ordered the hospitals removed." But he grumbled, "people will never think that it would be well to

---

[11] T. L. Connelly, *Autumn of Glory*, 362, 407–408; Castel, *Decision in the West*, 328; J. Davis, *Rise and Fall*, 2:557–61 (Hill to W. T. Walthall, October 12, 1878); Hill to Davis, October 12, 1878, Rowland, *Jefferson Davis Constitutionalist*, 8:284–85; Seddon to Hill, July 13, *OR*, vol. 52, pt. 2, 693–95; Kean, *Confederate Government*, 166; Hill to Seddon, July 14, *OR*, vol. 52, pt. 2, 704–707.

get the sick and wounded out of the way if it was decided to fight near the town." There were other signs that the Chattahoochee line would be given up. The army's trains rolled to the south bank on the 6th. That day Johnston moved his headquarters to the Delegall Campbell house south of the river. On the 8th Stevenson's division of Hood's corps was ordered to the south bank. That day the editor of the main Atlanta newspaper, the *Intelligencer*, warned the people that the army would soon have to retreat back into the fortifications ringing the city. Felix de Fontaine, a newspaper correspondent in the city, wrote on the 8th, "one gloomy fact is patent above all others, namely, that the citizens of Atlanta are deeply, darkly, desperately blue. I have yet to hear a cheerful word." About the same time, Georgian Celathiel Helms wrote his wife from camp. "The men is all out of heart and say that Georgia will soon have to go under," he despondently acknowledged on July 6, "and they are going to the Yankees by the tens and twentys and hundreds most every night." "Johnston's army is very much demoralized," he concluded.

And that was before Johnston's army retreated across the river during the night of July 9–10. "Alarm in Atlanta. Citizens fleeing," Lieutenant Mackall recorded on the 10th. "There is much discontent arising among the people around Atlanta," noted his cousin General Mackall, "and they are beginning to abuse the General." Irenus W. Landingham, a member of the 5th Texas serving as a clerk on Hood's staff, wrote his mother on the 14th, "our presence on the south side of the Chatahoochee created considerable alarm, and the good people of [Atlanta] have been getting to the rear in a hurry." The retreat across the river was also a psychological blow to a number of Johnston's officers and men. "I have grown very wearied of this eternal retrograde," Maj. Gen. William H. T. Walker wrote his wife on July 12. "We feel much dejected and low spirited at our prospects," Capt. W. L. Trask, on Hardee's staff, recorded in his journal on July 10. That same day Pvt. Robert D. Patrick, Co. A, 4th Louisiana, entered in his diary, "I don't believe Johnston can hold Atlanta." "We have gone as far as we ought to go unless we intend to give up all our country and not fight anymore," Texan Benjamin Seaton wrote in his diary on July 9. For his part, General Shoup was disappointed that his unique fortified line had not been tested in combat. Col. Charles Olmstead of the 1st Georgia saw an officer of the 63rd Georgia, Capt. Wallace Howard, watching the troops crossing the Chattahoochee "with melancholy earnestness." Olmstead commented that the enemy's retirement upon Atlanta was actually a positive event, as it would likely bring on the decisive battle that everyone wanted. "I don't know," Howard replied. "I don't like giving up so much territory, it looks to me like the beginning of the end and as though we were going right straight down to the Gulf of Mexico." Writing his wife on July 6, Pvt. Mathew Dunn of the 31st Mississippi tried to be upbeat, but he looked ahead to things turning worse: "our defeat will have a great bearing on the minds of the northern People in the approaching Elections."

Soldiers who earlier had expressed confidence and given Johnston glowing compliments now felt differently. Example is Capt. Sidney S. Champion of the 18th Mississippi. On June 9 he had written his wife, "Our beloved Genl Johnston has the confidence of almost every soldier...based upon skills as a strategist." The day after the army's river crossing, though, Champion felt "not so sanguine as I have been." On July 14 he was downright negative about his commander: "I have been speculating as to the movements of the army because no one can comprehend Joseph E. Johnston but his corps generals." "We are very much disappointed in having to give up so much, such magnificent country, without fighting for it," wrote Lt. Col. C. Irvine Walker of the 10th South Carolina to his fiancée, Ada, on July 9. "I should not be at all surprised if we lost Atlanta," he lamented. "I can't help thinking that Genl. Bragg would have conducted the campaign very differently and perhaps with other results," Walker added, leading one to infer that if some officers pined for Braxton Bragg over Joe Johnston, things were troubled indeed. That same day Capt. S. H. Dent wrote "My Dear Darling" about his "fear that Atlanta will be given up. If Johnston can be flanked out of every position from Dalton [to] here I do not see why he cannot be flanked out of Atlanta." Another Confederate was just as despondent. In his letter to Frances on the 12th, James W. Watkins wrote, "I think our confedricy is about to go up the spout."

Still, some Southerners maintained faith in their Uncle Joe. Although he began to ponder the consequence of losing Atlanta, Capt. George K. Miller confided to his wife on July 12, "I don't remember a time within the last year that I felt more confident of our final success than of present." "The army has still the most unbounded confidence in Genl. J. and is not at all demoralized by having to fall back so far," wrote Joseph B. O'Bryan to his sisters on July 13. That same day Lt. Robert Mitchell of the 29th Georgia wrote his wife, Amaretta, "we are able to carry on this war fifty years and I know they cannot whip us." On July 15 Capt. Peter Marchant of the 47th Tennessee wrote home to Susan, "as for Gen Johnston I can't tell what he will do whether he will still fall back before Sherman or not." Still, he added that "evry one has the gratest confidence in Gen Johnston." The day after that, Tennessee surgeon U. G. Owen affirmed to his wife that the army was confidently waiting for the Yankees to cross the river and launch their attack: "We are ready for them."[12]

Senator Wigfall on June 24 had already warned Johnston that tongues were wagging in Richmond about his possible relief from command, but Johnston had evinced no concern. Two weeks before that, Lt. Richard Manning had written that Johnston was aware of all the "criticizing—blaming—abusing & undermining" of himself, adding, "What does our General think of all this? He thinks little of it." Besides, there were still many who supported Old Joe. On the 4th of July, Alabama Senator Richard Walker had stated to Assistant Secretary Campbell that in his recent travel through Georgia he had seen "no indication whatever that the popular confidence in General Johnston had been at all shaken by his retreating policy."

But to many Southerners, Johnston's giving up of any further defense north of the Chattahoochee was the last straw. An Athens lawyer, M. C. Fulton, was only one Georgian not afraid to express his opinion, and he did so in a letter written July 11, sent directly to President Jefferson Davis. "His infernal Fabian policy will take us all to ruin and that very rapidly"; Fulton advised the president to either fire Johnston or issue "positive orders to *fight* and not to *fall back* another foot." Davis was getting such advice from all quarters. Maj. Gen. Samuel French recalled that Davis told him that he was beset "by deputations, committees, individuals, officials, and the press demanding to know if Atlanta and the State of Georgia were to be given up without a battle." Henry Brewster told Mary Chesnut that when he met with the president,

---

[12] "Atlanta Correspondent of the Mobile News," *New York Times*, August 7, 1864; "Getting Out of Atlanta," *New York Times*, August 7, 1864; S. Davis, *Atlanta Will Fall*, 97, 100, 102, 107; Vandiver, "General Hood as Logistician," 3 (reprinted in Hewitt and Bergeron, *Confederate Generals in the Western Theater, Volume 1*, 177–91); Mackall, *Son's Recollections*, 218–19 (letters of July 7, 11); Mackall circular, July 6, *OR*, vol. 38, pt. 5, 867; McMurry, "Mackall Journal," entries of July 7, 10; J. W. Ratchford to Stevenson, July 8, *OR*, vol. 38, pt. 5, 871; McNeely, "Felix Gregory de Fontaine," 86; D. Williams, *Georgia's Civil War*, 194 (Helms); Hess, *The Battle of Peach Tree Creek*, 9; S. Davis, "A Georgia Firebrand," 456; W. L. Trask war journal, entry of July 10 (typescript at Kennesaw Mountain National Battlefield Park, copy courtesy of Dennis Kelly); Strayer and Baumgartner, *Echoes of Battle*, 201 (Patrick); Shoup, "Dalton Campaign," 264; Castel, *Decision in the West*, 364 (Seaton entry, July 9); S. Walker, *Hell's Broke Loose in Georgia*, 145 (Colonel Olmstead); Dunn, "Notes and Documents," 119; Longacre, *Cavalry of the Heartland*, 278, 280 (Captain Champion); C. I. Walker, *Great Things Are Expected of Us*, 128; Mathis, *In the Land of the Living*, 103 (S. H. Dent); Watkins to wife, July 12, box 1, folder 2, Watkins Papers; Phillips, *Diehard Rebels*, 106 (Miller); Clampitt, *Confederate Heartland*, 74; Bobbie Mitchell to Nettie Mitchell, July 13, L. J. Gordon, "Courting Nationalism," 201; Peter Marchant to "Dear Susan," July 15, Marchant Correspondence; Owen, "Letters of a Confederate Surgeon," 172.

carrying Hood's verbal message (sometime in late May), even then Davis had told him that Johnston would have to be removed if Sherman were ever to be "blocked." About this time Mrs. Chesnut entered in her diary that she had heard Buck Preston tell her father that General Hood and Doctor Darby "are backing down into the Gulf of Mexico with Joe Johnston, for aught we know."

To be sure, the Confederate press fulfilled its role as upbeat public propagandist, even in these perilous days. Or at least it tried to. The *Milledgeville Southern Recorder*, publishing weekly in Georgia's capital, reviewed "the military prospect" in its issue of July 12, with the editor asking more questions than providing answers.

> Can the work be accomplished? Can Sherman pass through Johnston's lines and take Atlanta? Will not Johnston maintain his present strong position near the river? Can the flanking policy further avail Sherman? Is not his rear in peril? Has he not already advanced too far from his base, to regain it, if driven back? Would not a retreat be fatal to his army; even with a force inferior in numbers to that of the Yankees, cannot Johnston with his veterans whip the whole in a general fight? Has not Johnston evinced qualities as a strategist which ought to assure us of the wisdom of his policy? Does he not possess the unbounded confidence of his officers and men? All these questions can be answered favorably to our cause, as the crisis will soon demonstrate.

"Personne"—Felix de Fontaine—contributed a cheerier column from the Chattahoochee on July 14. "The retreat of Gen. Johnston is one of the most skillful and wonderful achievements which the history of war records," he announced. "He has saved an army from annihilation, inflicted tremendous damage on his antagonist, and at this hour confronts him with the original disparity between the two forces so much reduced, that in a fair and open fight, there is not a question as to our success."

On the other hand, Johnston's retreat across the river allowed correspondent Henry Watterson to get the last laugh, albeit in a depressing way. He had been harshly criticized by other newspapers for predicting in late May that Johnston would be defeated. "When I wrote you a month ago," he addressed the *Advertiser & Register* editors, "that Atlanta would fall, yourself, and numbers of my friends of the Atlanta press, were somewhat violent in your dissent, and yet the latter have packed their traps and are now branching out in search of new places of refuge." "I still adhere to my original opinion," he affirmed.

Amidst these forebodings, some in Davis's own cabinet now wanted Johnston relieved. After the war Senator Hill remembered that on July 11, after he gave Secretary Seddon his narrative of the meeting with Johnston, Seddon told him that he favored Johnston's removal, but that the president was opposed. Davis himself acknowledged that he had been slow to make a decision on removing Johnston from command. "I did not anticipate the abandonment of the mountain region of Georgia," he wrote Herschel Johnson, Georgia's other senator two months later, "and if General Johnston had informed me that he would retreat to Atlanta, he would have been sooner relieved."[13]

The situation in Georgia was deteriorating quickly. On the 11th Johnston wired Bragg—not knowing that he had just left on the long train trip to Atlanta—that the Federal prisoners held at Andersonville should be sent elsewhere to places of greater safety. Johnston did not give his reasons; his telegram simply read, "I strongly recommend the distribution of the U. S. prisoners, now at Andersonville, immediately." This was an unsettling message indeed, with its implication that enemy forces could raid 110 miles south of Atlanta. Davis telegraphed Johnston that day, saying Senator Hill had arrived and had met with him. As for Johnston's reference to 16,000 cavalry in Alabama and Mississippi (in his telegram of July 8), Davis added, "I have called for exact statements from the War Department." (General Mackall wrote his wife on the 12th, "Mr. Davis sent Johnston a very sneering telegram yesterday.") Davis seems to have begun building a case for relieving Johnston. On

---

[13] McMurry, "*Enemy* at Richmond," 29; Walker to Campbell, July 4, *OR*, vol. 52, pt. 2, 685; Fulton to Davis, July 11, Crist, Williams, and Dillard, eds., *Papers of Jefferson Davis*, 10:515; S. G. French, *Two Wars*, 218; Woodward, *Mary Chesnut's Civil War*, 620, 635; "The Military Prospect," *Milledgeville (GA) Southern Recorder*, July 12, 1864; "Personne," "Behind the Chattahoochee," *Columbia South Carolinian*, July 19; "Shadow," "Our Army Correspondence. Letters from Gen. Johnston's Army," *Mobile Advertiser & Register*, July 17; Hill to W. T. Walthall, October 12, 1878, J. Davis, *Rise and Fall*, 560–61; Davis to Herschel V. Johnson, September 18, 1864, Rowland, *Jefferson Davis Constitutionalist*, 6:336.

Fulton's mention of Johnston's "Fabian policy" refers to Quintus Fabius Maximus (c. 280–203 B.C.), the Roman general who in the Second Punic War earned the name Cunctator ("The Delayer") for his tactical policy of retreating before Hannibal's army, giving up northern Italy in the interest of gaining time and falling back on the capital. It is said that Johnston did not like being thought of as a modern Fabius, as he believed the Roman general had a choice of tactics facing Hannibal, whereas facing Sherman he believed that he did not (Govan and Livingood, *Different Valor*, 393).

The Civil War community remains divided on Johnston's tactics. Writing in 1959, Frank Vandiver observed, "Despite the brilliance of his Fabian campaign," Johnston was removed from command (Vandiver, introduction to Johnston's *Narrative*, xxviii). More recently Earl Hess offers a charitable definition of Fabian cunctation: "Johnston sought to maintain the integrity of his army as long as possible while waiting for an opportunity to catch a cautious opponent at a disadvantage" (Hess, *The Civil War in the West*, 222).

the 11th, Burton Harrison, Davis's secretary, wrote General Cooper asking what reinforcements had been sent to Johnston and when, since the start of the campaign. Then in a wire on July 12—as if he needed to say it again—Johnston repeated to the president that he was greatly outnumbered, he could not send any of his cavalry on a raid to Sherman's rear, and that he needed help from S. D. Lee's department.

As he wrestled with what to do before Atlanta—whether to fire Joe Johnston or not—President Davis logically sought the advice of his most trusted adviser and most successful general, Robert E. Lee. News of Johnston's retreat across the Chattahoochee on July 9–10, Senator Hill's disturbing report of the 10th, Johnston's Andersonville telegram of July 11, and Johnston's admission (again) of his weakness the next day compelled Davis to confide in General Lee on the 12th. His telegram was to-the-point: "General Johnston has failed, and there are strong indications that he will abandon Atlanta. He urges that prisoners should be removed immediately from Andersonville. It seems necessary to relieve him at once. Who should succeed him? What think you of Hood for the position?"

Davis's suggestion of Hood may not have been a surprise to Lee, and it was certainly logical. Davis respected Hood's military service and was personally fond of the young lieutenant general. Moreover, with his reputation as a bold attacker in Lee's army, Hood offered the alternative to Johnston's passive "strategy" of retreat. His correspondence with Richmond that spring and sending Colonel Brewster to converse with the president further assured the administration of his desire to take the initiative. Yet it is possible that even Hood feared it was too late to defeat Sherman. Probably about this time he wrote Buck Preston, who read the letter to Mrs. Chesnut. "They had left all the good fighting ground behind them," Mary recorded in her diary. "An army by constant retreating loses confidence in itself. There was a talk of some new man to be put in command. He hoped not. God help the new commander, whoever he should be, for he had a rough road to travel."

Davis on July 12 was far from resolved upon Hood as a possible successor to Johnston, especially after he received General Lee's reply, "It is a bad time to release the commander of an army situated as that of Tenne." "We may lose Atlanta and the army too," Lee wired back. As for Hood, he added, "Hood is a bold fighter. I am doubtful as to other qualities necessary." Davis's and Lee's wires encoded certain key words; Lt. Col. Walter Taylor, Lee's A. A. G., had the responsibility of rendering these with cipher. Thirty years later, writing his memoirs, Taylor remembered the momentous dispatches. "Their contents were more positively and permanently impressed upon my memory," he recalled, than the usual messages he dealt with.

That night Lee wrote a longer response to the president, repeating, "It is a grievous thing to change commander of any army situated as is that of the Tennessee." But here he added, "still if necessary it ought to be done." Johnston would have been pleased with Lee's recommendation, "It would be better to concentrate all the Cav^{y} in Miss^{pi} & Tenn: on Shermans communications." Then in further answer to the president's question, Lee wrote, "Hood is a good fighter very industrious on the battle field, careless off & I have had no opportunity of judging of his action, when the whole responsibility rested upon him. I have a high opinion of his gallantry, earnestness & zeal. Genl Hardee has more experience in managing an army." Historian Thomas Robson Hay paraphrases Lee's obvious meaning: "his regard for Hood, as the commander of the army, was not high. He preferred Hardee." Dr. Freeman's assessment is predictably more nuanced. Lee was aware of Davis's low estimation of Johnston; because of this, the general's telegram and letter "seem all the more remarkable in their frankness." "The whole tone of the letter," Freeman professes, "seems...to caution President Davis against a hasty change....Lee, it is needless to say, had a very high opinion of General Johnston's military qualities."

Such a measured response was so characteristic of Robert E. Lee. Focused on the Virginia front, he chose not to assert his opinion strongly regarding other theaters. He left to the nation's chief executive the decision which would be his alone ("May God give you wisdom to decide in this momentous matter"). More telling was his half-hearted endorsement of Hood. The president had clearly stated a tentative preference of Hood over, say, Hardee, but General Lee seemed to favor the latter to succeed Johnston. It is profoundly significant that a full-throated and hearty endorsement of John Bell Hood could not be given by the commander of the Army of Northern Virginia, in which Hood had earlier in the war won his spurs and earned his reputation.

Davis received Lee's telegram at 8:45 P.M. on the 12th. The next day he wired back, informing Lee that Bragg had reached Atlanta and would report on the situation. "It is a sad alternative," he wrote, meaning the proposed relief of Johnston from command, "but the case seems hopeless in present hands." The feeling was so desperate in official circles about the situation in Georgia that on July 13, J. B. Jones, clerk in the War Department, entered into his diary, "Some think Lee is going hither."[14]

At this point, General Bragg's arrival in Atlanta was most opportune. After a long train ride from Richmond, the president's military adviser reached his destination on the morning of July 13. Reporters soon learned that he would meet with Johnston and his generals, about which there was much speculation. "Ora," a reporter for the *Montgomery Advertiser* (Samuel Chester Reid, Jr.), wrote, "it is not to be supposed for a moment that either the President or Gen. Bragg would attempt to interfere with the plans of Gen. Johnston." But others were not so sure. "He looks as grim and wired as ever," wrote Henry Watterson ("Grape") for the *Augusta Constitutionalist*; "I have just passed him in a carriage going out to army headquarters." Bragg was not the only one "grim and wired," as General Mackall noted in a letter of the 13th: "Joe looks uneasy this morning. I am sorry to see him so fretted." Still, Mackall put the best cast upon Bragg's visit: "I don't think he has any special mission," yet Mackall could not avoid speculating that it had to do with "the design of Mr D…to make a display of his distrust of Johnston & if he finds he can infuse it into the army, to relieve him." Halsey Wigfall, the Texas senator's son serving on General Hood's staff, wrote his father on July 15, "What the object of his visit is I don't know." General French visited Johnston's headquarters on the 13th and found Bragg there. "What is his mission?" he asked in his diary; "Who knows?" Some soldiers were suspicious. "That bird of evil omen, Gen. Bragg, has arrived," recorded Robert Patrick in his diary. Even Sherman got word of the visitor from the Rebel capital: "Bragg's arrival from Richmond must be to consult." For his part, Bragg later wrote that Johnston greeted him "kindly and courteously."

Even before he met with Johnston, Bragg wired the president that "indications seem to favor an entire evacuation of this place." Then, after meeting

[14] Johnston to Bragg, July 11, *OR*, vol. 38, pt. 5, 876; Davis to Johnston, July 11, *OR*, vol. 38, pt. 5, 876; Mackall, *Son's Recollections*, 219 (letter of July 12); Harrison to Cooper, July 11, *OR*, vol. 38, pt. 5, 874; Johnston to Davis, July 12, *OR*, vol. 52, pt. 2, 692; Davis to Lee, July 12, *OR*, vol. 52, pt. 2, 692; Woodward, *Mary Chesnut's Civil War*, 630 (entry of August 3, 1864); Lee to Davis, July 12, in Freeman and McWhiney, *Lee's Dispatches*, 282–85; W. H. Taylor, *Four Years With General Lee*, 139; Hay, "Davis, Bragg, and Johnston in the Atlanta Campaign," 42; Davis to Lee, July 13, *OR*, vol. 52, pt. 2, 692; J. B. Jones, *A Rebel War Clerk's Diary* [2015], 2:223.

Steven E. Woodworth believes President Davis took too long to decide upon Johnston's removal ("Davis, true to form, hesitated") (Woodworth, "How to Lose a City," 31). My interpretation instead credits Davis with taking the time he felt he needed to make his decision, knowing how fateful it would be to the fortunes of the country.

with Johnston at his headquarters (the Dexter Niles house on the Marietta road), at 1 P.M. Bragg sent a telegraphic report of what he had learned. Wheeler reported at least two enemy infantry corps had crossed the Chattahoochee —Johnston, entrenched south of Peachtree Creek, was giving Sherman all the time he wanted. Then there were the numbers: "our army is sadly depleted, and now reports 10,000 less than the return of 10th of June." For all of these reasons, Bragg added, "I find but little encouraging." Moreover, Bragg related the enemy position as more threatening than Johnston had led Davis to believe. In his wire to Richmond of July 12, Johnston had stated rather calmly, "The enemy...has troops intrenched on this side. Elsewhere everything quiet, except a little occasional skirmishing at long range across the river." Bragg was en route to Atlanta by then, so Davis replied for him on the 12th, addressing Johnston's warning that the Andersonville prisoners be removed. "You have all the force that can be employed to distribute or guard prisoners; know the condition of the country and prospects of military operations."

Most important of all, Bragg could not report any plans by Johnston to assume the initiative. Sherman's army was then divided—three corps south of the Chattahoochee, four north of it—yet Johnston evidently chose to remain idle. If he were planning an attacking battle, Johnston certainly should have told Bragg about it, and Bragg would have reported this to Davis. Indeed, only after the war did Johnston claim to have a plan for attacking Sherman—but this was purportedly while the Federal forces were crossing Peachtree Creek, not while they were crossing the Chattahoochee. Connelly notes this, pointedly: "Curiously, he gave Sherman an entire week to cross his army and concentrate....Here Johnston lost probably the greatest opportunity of the campaign."

On the 14th Bragg sent no telegrams to Richmond, but Johnston did. Aware that Bragg was now probably feeding the administration information he had been reluctant to divulge, Johnston on July 14 wired General Cooper that three Federal infantry corps were across the river, according to reports from General Wheeler. Bragg started the day at Johnston's headquarters. "Genl B at hq 9 am meets Corps Commanders by invitation of Genl. J & afterward visits them," Lieutenant Mackall recorded. (Bragg may have visited Hardee, but the latter complained to President Davis two months later that Bragg "never said a word to me on the subject of the campaign.") A lot was happening at the Niles house that day; Maj. Gen. G. W. Smith dropped by, according to Mackall; so did Robert Toombs of Washington, Georgia, in retirement since his resignation as brigadier in the army, March 1863. In the early afternoon Bragg visited the headquarters of corps commander Peter Stewart. Brigade leader Daniel Reynolds recorded in his diary for the 14th, "At 1 P.M., I went in company with generals Walthall and Quarles to Gen. Stewart's headquarters to meet Gen. Bragg. He seems to be in fine health and spirits." Bragg then spent a good part of the 14th with generals Hood and Wheeler, whom he would by then have known were critical of Johnston's tactical policies. Wheeler would have told him how he had wanted to launch a raid on Sherman's rearward rail line (just the day before, General Johnston had informed General Cooper in Richmond that Red Jackson's cavalry division—half the army's horsemen—numbered 5,370 troopers present).

Thus far Hood and his infantry had been the army's main offensive arm. It had carried out, at Johnston's orders, assaults at Resaca May 14–15; it had been instructed to do so at Cassville May 19 and Pickett's Mill May 28 (but had not); and it had done so without orders at Kolb's Farm June 22. Thus it is curious that Hood apparently told Bragg that during the campaign he had been disappointed that the army had not struck the enemy an offensive blow. Apparently Bragg asked Hood to put his thoughts into writing, for Hood wrote Bragg a long letter under date of July 14, which emphasized Hood's frustration with Johnston's passive and retrogressive tactics and urged strong, aggressive action against the enemy. "During the campaign from Dalton to the Chattahoochee River it is natural to suppose that we have had several chances to strike a decisive blow," Hood wrote; "we have failed to take advantage of such opportunities." "Our loss cannot be less than 20,000," he added, "without having fought a decisive battle." Here, either Bragg was coaxing him to be selective of his facts, or Hood was deliberately being disingenuous in not mentioning that it had been he, not Johnston, who had called off the attacks at Cassville and Pickett's Mill, and that it had been he and General Polk, not Johnston, who had argued for a retreat from Cassville, rather than stand ready for defensive battle on May 20. (For these reasons, Hess has called Hood's letter of July 14 "an outright lie.") Yet Hood knew that Bragg (and Davis) wanted an end to the retreating policy of Johnston, and in a very blatantly self-serving passage offered to give his superiors what they wanted (in hopes, we may well assume, that he would be promoted to command the Army of Tennessee).

> Our present position is a very difficult one, and we should not, under any circumstances, allow the enemy to gain possession of Atlanta, and [I] deem it excessively important...that we should attack him, even if we should have to recross the river to do so. I have, general, so often urged that we should force the enemy to give us battle as to almost be regarded reckless

> by the officers high in rank in this army.... Please say to the President that I shall continue to do my duty cheerfully and faithfully.[15]

Far more than his earlier missives to Lee, Davis, Bragg, or Seddon, Hood's letter of July 14 was perhaps the most controversial of the entire war. "Hood and Bragg met more as conspirators than friends," write Govan and Livingood, who judge that Hood prepared the document to bring about the firing of General Johnston and as "a bid for the command for himself." Symonds concludes that in writing his letter, whether Bragg asked for it or not, "Hood stepped over the line from unprofessional to outright subversive." Hood does not mention the letter in his memoirs; neither does O'Connor in his biography. Dyer wonders even if it had any effect on Bragg when Hood handed it to him. McMurry, as usual, is on the mark: "No motive other than ambition can be ascribed to the writing of this letter." Connelly challenges Hood's claim that he had constantly advocated giving battle, concluding that "at best, Hood was a chronic liar." *Chronic* may be debatable, but it is undeniable that in this letter "Hood was making his last bid for the army's generalship." Castel agrees, noting that with his references to "officers high in rank in this army," he was referring to Hardee, who was senior corps commander and stood over him in line to succeed Johnston. Hood doubtless knew also that Hardee had wanted to attack at Cassville and had opposed the army's retreat from there. But Hardee had to be discredited. "The truth was that as often as not it had been Hood who advised retreat and Hardee who had dissented," writes Woodworth, "but Bragg and Davis were ignorant of that, and the ambitious young general was not about to enlighten them." "Hood," Castel agrees, "is motivated by ambition."

Stephen M. Hood, recent apologist for Hood, observes that Hood's correspondence with Richmond leads some authors to "accuse the general of secretly scheming to replace Gen. Joseph E. Johnston," a charge for which

---

[15] "Telegraphic," *Augusta Chronicle & Sentinel*, July 13, 1864 ("Gen. Bragg arrived here this morning"); "Letter from Ora," *Montgomery Advertiser*, July 16; "Grape," "Letter from the Georgia Front," *Augusta Constitutionalist*, July 16; Andrews, *South Reports*, 550 (Reid as "Ora") and 551 (Watterson as "Grape"); S. Davis, *Atlanta Will Fall*, 105–108; Symonds, *Johnston*, 323; T. L. Connelly, *Autumn of Glory*, 397, 415–17; S. G. French, *Two Wars*, 216; Hallock, *Braxton Bragg*, 192–93 (quoting Patrick and Sherman); Bragg letter to Davis, July 15, *OR*, vol. 39, pt. 2, 712; Bragg to Davis, July 13, *OR*, vol. 38, pt. 5, 878; Kurtz, "At the Dexter Niles House," 5, 20; Johnston to Bragg, July 12, *OR*, vol. 38, pt. 5, 877; Davis to Johnston, July 12, *OR*, vol. 38, pt. 5, 877; J. E. Johnston, *Narrative*, 350; Johnston to Cooper, July 13 and 14, *OR*, vol. 38, pt. 5, 878–79; McMurry, "Mackall Journal," entry of July 14; Crist, Williams, and Dillard, eds., *Papers of Jefferson Davis*, 11:67 (Hardee to Davis, September 26, 1864); Reynolds, *Worthy of the Cause*, 135; Hood to Bragg, July 14, *OR*, vol. 38, pt. 5, 879–80; Hess, *Bragg*, 232.

Hood argues "there is no conclusive evidence." I believe that General Hood's letter of July14 to Bragg is more than circumstantial evidence to support such a charge.[16]

Thus the leader of the Confederate States of America faced in mid-July perhaps the most profound crisis of his presidency. The general of arguably the country's largest army claimed he could not defeat the enemy forces in front of him without outside help—which could not be summoned. A numerically superior enemy army was lodged just a few miles outside of the country's second-most important city. His personal military adviser, at the scene, was attesting that its fall seemed imminent. Davis later told an audience, "I was not one of those who considered Atlanta lost when our army crossed the Chattahoochee"; he was resolved that the city would not fall with "an honest and manly blow." Johnston seemed incapable of striking such a blow. In truth, such action was wanted not only by the Confederate government, but by the Southern people. "Atlanta will not be surrendered without fierce and bloody struggle," Felix G. de Fontaine, correspondent for the *Savannah Republican*, wrote on July 13. Citizens, including Georgians (according to General French) were clamoring for Johnston's removal; after the war Postmaster General John H. Reagan remembered that Davis had been "urged by letters and telegrams and by deputations sent to Richmond, in the most earnest manner to make the change." "Everybody has at last come to the conclusion that Johnston has retreated far enough," wrote Col. Josiah Gorgas, the Confederate army's chief of ordnance on July 13. Even Senator Ben Hill, who had been one of Johnston's supporters, had turned on him, advocating his removal. Hill, according to Gorgas, "came on here after a full interview with Gen J. & urged a change of commanders, having ascertained to his entire satisfaction that Johnston would not fight even for Atlanta but would con-

---

[16] Govan and Livingood, *Different Valor*, 310; Symonds, *Johnston*, 324; Dyer, *Gallant Hood*, 245; McMurry, *John Bell Hood*, 118; T. L. Connelly, *Autumn of Glory*, 417; Castel, *Decision in the West*, 357; Woodworth, "How to Lose a City," 33; Hood, *Rise, Fall, and Resurrection*, 19. Earl Hess has noticed the disingenuousness of Hood's claim that he had "so often urged that we should force the enemy to give us battle." Hess observes that "William J. Hardee, and sometimes Hood, expressed a strong desire to fight a major, deciding battle with the enemy before giving up Atlanta" (Hess, *Civil War in the West*, 222). Yet Hess is off the mark with his recent observation that Hood's letter to Bragg of July 14 reflects "the game he had learned to play from Polk"—meaning, intrigue behind the back of one's superior, which Polk had engaged in under Bragg (Hess, *Peach Tree Creek*, 15). As we have seen, Hood was sending self-serving letters to President Davis as early as March 1864, two months before Leonidas Polk even joined Johnston's army.

tinue the retrograde under the flanking movements of Sherman." Yet the senior officer in the army most capable of delivering that "honest and manly blow" to save Atlanta (Hood) did not have the full confidence of the nation's most successful general (Lee), who suggested instead replacing Johnston with the quavering corps commander (Hardee), who seven months before had declined the opportunity to command that same army.

Such was Jefferson Davis's quandary when he turned to his last body of counsel, his presidential cabinet. From their postwar correspondence, we know the cabinet members met at least once during this fateful time, but without minutes from the highly confidential meeting we can only guess its date. A good argument can be raised that it occurred on July 14. That was the date on which Robert G. H. Kean, head of the War Bureau, entered into his diary, "the present excitements are...(with the Cabinet) the retreats of Johnston." More significantly it was also the day on which Senator Hill wrote his long narrative of his meeting with General Johnston. Seddon had called for it and had written one of his own from Hill's meeting of July 10 with Davis, "in order to have the basis on which to found an official recommendation to the President." Hill already knew that Seddon favored Johnston's removal; after the war Hill recalled Seddon had told him that appointing Johnston to replace Bragg had been "a great mistake" and at this juncture he wanted to undo it. Seddon told Hill he would use his document in the cabinet meeting and invited the senator to sit in on it. By this time Hill too had come to the conclusion that Johnston had to go. After the cabinet meeting, Hill dined with James Lyons, the prominent Richmond attorney and congressman from Virginia, at his home outside the city and told him of the meeting. Lyons after the war recalled how Hill had joined in the call for Johnston's removal. Still, Hill remained sympathetic to Johnston's plight. On July 14, probably right after the cabinet meeting, Hill sent an urgent telegram to Johnston in Atlanta: "You must do the work with your present force. For God's sake do it."[17]

---

[17] "President Davis in Macon," *Macon Telegraph and Confederate*, September 24, 1864; S. G. French, *Two Wars*, 218; "F. G. de F." (Felix G. de Fontaine), "From the Army of Tennessee," *Savannah Republican*, July 13, 1864; Reagan to Davis, February 7, 1878, and Lyons to W. T. Walthall, June 10, 1878, in Rowland, *Jefferson Davis Constitutionalist*, 8:79, 215–16; Gorgas, *The Journals of Josiah Gorgas*, 121, 123 (July 13 and 22, 1864, entries); Kean, *Confederate Government*, 165 (diary entry, July 14); Hill to Seddon, July 14, *OR*, vol. 52, pt. 2, 704–707; Seddon to Hill, July 13, *OR*, vol. 52, pt. 2, 693; Hill to W. T. Walthall, October 12, 1878, J. Davis, *Rise and Fall*, 2:561; Dodd, *Jefferson Davis*, 332; Hill to Johnston, July 14, *OR*, vol. 38, pt. 5, 879.

After the war Hill denied that he had advocated Johnston's removal. "I have heard it said that I advised that removal," he told an audience in Atlanta. "This is not true. I gave no advice

Indeed, it is only from their postwar correspondence that we can reconstruct what the several cabinet officers said. Seddon remembered how "one of the gravest objections existing to him [Johnston] was, that he would not give satisfactory information or assurances" on "his plans and purposes." Just as grave were the indications Johnston had given that he intended to evacuate Atlanta without a fight. "Up to that time," Seddon recollected, "frank communication of his plans or assurances to the President that he would not evacuate, without making decisive battle, would have kept him commander."

Secretary of State Judah P. Benjamin also wanted Johnston removed. He had formed an opinion of Johnston in late 1861 when he had then headed the War Department. As he later explained to Davis, "From a close observation of his career I became persuaded that his nervous dread of *losing a battle* would prevent at all times his ability to cope with an enemy of nearly equal strength, and that opportunities would thus constantly be lost which under other commanders would open a plain path to victory." For this reason, by mid-July, Benjamin was "most anxious and urgent that he shd. be replaced by some other commander." Naval Secretary Stephen R. Mallory recorded in his diary that Benjamin exclaimed, "Johnston is determined not to fight, it is of no use to re-enforce him, he is not going to fight." Thus secretaries Seddon and Benjamin, with possible help from Senator Hill, led the arguments for relieving Johnston. Virginia Congressman Lyons tells the rest.

> That evening the President rode out to my farm and took Tea, as he often did, when visiting the Batteries on my farm and told us of the removal—I expressed my regret at it and asked him why he did it—when he replied, I could not help it—Hill urged it on behalf of the people of Georgia and Benjamin and Seddon were so violent that they would listen to nothing....

Everyone recalled that the president sat back and allowed consensus to form. "I also remember that when you were appealed to afterwards to remove Gen. Johnston from the command of that army, by letters and by personal solicitation," remembered Postmaster General Reagan, "you hesitated to do

---

on the subject, because I was not a military man." On the other hand, Secretary Seddon would not have invited Hill to attend the cabinet meeting if he did not intend for him to speak, at least about his conversation with Johnston. This has led some to accuse Hill of "betraying" the general. His biographer Haywood Pearce has a more balanced view: Hill had started out to help Johnston, but had been swayed by Davis and Seddon's arguments that Johnston could not be saved except by Forrest's cavalry, &c., &c. ("Address Delivered Before the Southern Historical Society, at Atlanta, Ga., February 18, 1874," in B. H. Hill, *Senator Benjamin H. Hill of Georgia*, 101; Haywood Pearce, "Removal of Gen. Johnston," *Atlanta Journal Sunday Magazine*, April 1, 1928, 13, 23.)

so, and called attention to the danger of doing so in the midst of a campaign." Eventually a vote was taken, "and the Cabinet was unanimous," Benjamin remembered. So did Seddon: "the whole Cabinet concurred." What President Davis did at that point was remembered by Congressman Lyons, from the night of the cabinet meeting. Senator Hill had visited him at his home at Laburnum and related that after the vote Davis said he was left with an important question: "Gentlemen it is very easy to remove the Genl. But when he is removed his place must be filled and where will you find the man to fill it?"[18]

President Davis's question to his cabinet members was a real one. Confederate military policy was quite clear: full generals alone could command armies. "By 1864," writes Connelly, "the South was running out of full generals." Or, to put it another way, those whom it had were in fixed positions. The venerable Samuel Cooper was behind his desk in Richmond, as was Braxton Bragg. Lee indispensably commanded the army in Virginia; Kirby Smith was managing the Trans-Mississippi. Only Beauregard had been moved around so much (Virginia—Mississippi—South Carolina—Petersburg) that he could presumably be moved again, but Davis did not trust his abilities for army command. According to Robert Kean, writing in his diary on July 14, "The real trouble is to find a successor. The only solution is to send Beauregard, but the President thinks as ill of him as of Johnston."

Johnston's replacement would have to be a promotable lieutenant general, which logically meant either Hardee or Hood (R. S. Ewell had been disgraced at Spotsylvania, and S. D. Lee in Mississippi was too new to the command). Fortunately, just that spring the Confederate Congress had passed a law empowering the president to appoint generals of all four grades to temporary rank in military emergencies "for such time as the temporary exigency may require." In the choice between Hardee and Hood, the cabinet

[18] Seddon to Davis, June 17, 1872, Seddon to W. T. Walthall, February 10, 1879, Benjamin to Davis, February 15, 1879, Lyons to W. T. Walthall, July 31, 1878 and John H. Reagan to Davis, February 7, 1878, Rowland, *Jefferson Davis Constitutionalist*, 7:320, 8:78, 216, 353, 356; Catton, *Never Call Retreat*, 330 (quoting from Mallory's diary).

Harry Hansen concludes that "Davis was just as reluctant as Lee to see a change in the top command at Atlanta" (*The Civil War*, 432). Given his reluctance to concur with the cabinet vote, it is unfair for Govan and Livingood to state regarding Davis's communications with Lee, July 12–13, "Davis wished advice only if it affirmed the decision he had already made" (*Different Valor*, 315).

I place much emphasis on the cabinet meeting, even if Hood's biographers do not. O'Connor (*Cavalier General*, 199) and Dyer (*Gallant Hood*, 243) mention it once, without citing the secretaries' voluminous postwar correspondence regarding the meeting; McMurry does not mention it at all.

might have favored Hardee, except for the fact that he had already declined the appointment six months before. Secretary Seddon recalled that "Genl. Hardee, from his connection with and repute in the Army, seemed a natural appointment, and the Cabinet and President would have preferred him, had he not on a previous occasion when temporarily succeeding Gen. Bragg, declared conviction of his own incompetency to such command." Postmaster Reagan also remembered that Hardee's refusal of army command was a factor in Hood's favor.[19]

The cabinet had voted to fire Joe Johnston, but President Davis was not ready to act until he heard further from Bragg at Atlanta. And he also continued to agonize over the opinions expressed on the 12th by General Lee, who obviously had to be informed of the cabinet vote. So the next morning, July 15, Secretary Seddon took the 5 A.M. train from Richmond to Petersburg. The visit, Davis benignly remembered in his memoir, "was designed merely to secure General Lee's estimate of qualifications in the selection of a successor for the command." In his biography Dr. Freeman posits that "Secretary Seddon visited him, told him that the removal of Johnston had been decided upon, and asked his advice, as Davis had, concerning a successor."

"There is no record of Lee's reply," Freeman adds, but there is. Seddon conferred with Lee at his headquarters, then promptly boarded the train and by 11 A.M. was back at the capital, where he met with the president for at least an hour. According to Wade Hampton, writing a decade after the war, Lee urged the president not to relieve Johnston, for if he could not command an army "we had no one who could."

There is more. In his memoir John B. Gordon recalls a conversation he had with General Lee about the transfer of command. "He said enough to indicate his opinions," Gordon writes. "He said 'General Johnston is a patriot and an able soldier. He is upon the ground, and knows his army and its surroundings and how to use it better than any of us.'"

If this indeed were the information brought back by Seddon, then President Davis's quandary was that much greater.[20]

---

[19] T. L. Connelly, *Autumn of Glory*, 406, 412; Symonds, *Johnston*, 321; Kean, *Confederate Government*, 166; M. J. Wright, *General Officers*, 13; Vandiver, "Proceedings of the Second Confederate Congress First Session," 108–109, 167; S. Davis, *Atlanta Will Fall*, 111; Seddon to Walthall, February 10, 1879, and Reagan to Davis, February 7, 1878, in Rowland, *Jefferson Davis Constitutionalist*, 8:79, 351; Patrick, *Jefferson Davis and His Cabinet*, 138 ("the secretaries unanimously agreed that Johnston should be removed").

[20] Kean, *Confederate Government*, 166–67; Hampton to Johnston, undated, J. E. Johnston, "Opposing Sherman's Advance," *Battles and Leaders*, 4:277; Freeman, *R. E. Lee*, 3:462; J.

Davis, *Rise and Fall*, 2:561n.; J. B. Gordon, *Reminiscences of the Civil War*, 131–32; R. M. Hughes, *General Johnston*, 252 ("beyond question...General Lee advised against it").

After the war, former Confederate general Bradley Johnson claimed that Lee actually *wrote* to Seddon, protesting Johnston's removal. There is absolutely no documentary support for this contention (B. T. Johnson, *A Memoir of the Life and Public Service of Joseph E. Johnston*, 117).

Symonds believes the Confederate cabinet meeting occurred on July 17 (*Johnston*, 326), but such an assumption does not square with Seddon's visit to General Lee on the morning of the 15th, Davis's challenging telegram to Johnston of the 16th, and the several secretaries' postwar recollections, which place their meeting before Davis had finally made up his mind.

Ben Hill's speech in Atlanta, February 1874, was so widely reprinted that General Johnston read it and was disturbed by Hill's apparent implication that Lee had acquiesced to Johnston's removal ("I know he [Davis] consulted with General Lee fully, earnestly, and anxiously before his removal" [B. H. Hill, *Senator Benjamin H. Hill*, 409]). A decade after the war, Joe Johnston was still sensitive about his relief from command. Johnston wrote Alfred H. Colquitt, president of the Georgia branch, Southern Historical Society, "As Mr. Hill's allegations were published by you, and are, I believe, a part of your archives, I respectfully ask that this paper be published also, and may be kept in the juxtaposition to the oration." Johnston inserted text of a letter written to him by Wade Hampton, dated June 6, 1874, in which Hampton replied to correspondence sent him by Johnston on May 19. "Your letter of the 19th May reached me," Hampton wrote.

> Col. Venable is perfectly correct in saying that Gen. Lee had spoken freely to me in regard to your removal from command of the army of Tennessee. On that occasion, he expressed great regret that you had been removed, and that he had done all in his power to prevent it. The Secretary of War had recently been at the General's headquarters, near Petersburg, to consult as to this matter, and Gen. Lee assured me that he had urged Mr. Seddon not to remove you from command, and had said to him that if you could not command the army we had no one who could. He was earnest in expressing not only his regret at your removal, but his entire confidence in yourself. ("Important Correspondence. General Wade Hampton—Gen. Joseph E. Johnston—Gen. Robert E. Lee," *New Orleans Times*, August 14, 1874)

A few months later Hampton wrote again about this correspondence.

> Genl Johnston wrote to me saying that Mr. Hill of Geo. in a public speech had stated, or intimated, that Gen. Lee had advised his removal from command and that upon the publication of this speech he, Genl. Johnston, had written to Col. Venable asking if he could give any information on this point. He went on to say that Col. Venable had replied that Gen. Lee had not spoken to his staff on this subject, but that he had done so to me. Upon this Gen. Johnston requested me to let him know what Gen. Lee had said. I felt that it was due to Gen. Johnston and to truth to give so much of Gen. Lee's conversation on that occasion as would do away with the charge that he had advised the removal of Johnston. I could no less in justice to a brother officer and a friend. In the conversation with Gen. Lee, he referred only to his interview with the Sec. of War and whilst giving his estimate of Genl. Johnston as a soldier, expressed regret at his removal. (Hampton to Davis, September 5, 1874, in Rowland, *Jefferson Davis Constitutionalist*, 7:399).

At least Braxton Bragg wanted Johnston gone. He made that very clear in a long letter he wrote to Jefferson Davis on July 15 after he had conducted a review of the troops ("greeted with loud & prolonged cheers," according to a soldier in the 24th Alabama). Lieutenant Mackall noted that Bragg did not visit Johnston's headquarters on the 15th; Bragg had gotten as much information as he had expected. "I have made General Johnston two visits, and been received courteously and kindly," he wrote. "He has not sought my advice, and it was not volunteered. I cannot learn that he has any more plan for the future than he has had in the past." As it turned out, Bragg's letter and that written by Hood on July 14 would have no effect on the course of events; they would be carried by a staff officer, Lt. Col. Harvey W. Walter, to Richmond and hand-delivered to the president. Walter's train trip from Atlanta to the capital would take at least two days. Having left Atlanta late on the 15th, he would not arrive at the capital at least by late on the 17th, maybe the 18th.

Davis would act well before Bragg's letter arrived. Nevertheless, it is instructive as to his thinking. Bragg suggested that "there is but one remedy" in the current crisis: "offensive action." Yet Johnston would not likely bring it on: "the commanding general has ever been opposed to seeking battle, though willing to receive it on his own terms in his chosen position.... General Hood has been in favor of giving battle," he added, while "on the contrary, General Hardee generally favored the retiring policy." These statements were not quite true for either officer's conduct so far in the campaign, but Bragg knew the president's dissatisfaction with Joe Johnston. "The removal of the commander, should such a measure be considered," he acknowledged, "would be attended with some serious evils," including denunciations from the administration's critics. "I do not believe the second in rank"—meaning Lieutenant General Hardee—"has the confidence of the army to the extent of the chief." Then he offered his assessment of Hood. "General Hood would give unlimited satisfaction, and my estimate of him, always high, has been raised by his conduct in this campaign. Do not understand me as proposing him as a man of genius, or a great general, but as far better in the present emergency than any one we have available."

Bragg wired Richmond several times on the 15th, indicating that the tactical situation was worsening. Another enemy infantry corps had crossed, and reports indicated that the Federals were extending their bridgehead several miles eastward, inching toward the Augusta railroad. One of Bragg's telegrams sent to Richmond on July 15, the most arcane, has for decades provoked discussion among scholars: "I am decidedly opposed, as it would perpetuate the past and present policy which he has advised and now sustains. Any change will be attended with some objections. This one could produce no good." The assumption by most is that Davis had instructed Bragg, if events had reached crisis, that he was to relieve Johnston and appoint Hardee to succeed him. Bragg's wire is taken to mean, however, that he did not favor Hardee. Such an assumption is made in context with another telegram which Davis sent that day to Bragg: "The selection of a place must depend upon military considerations so mainly that I can only say that if C. is thus indicated adopt advice and execute as proposed."

Many scholars have assumed that "C." was Davis's cryptic reference to *change* of army commanders. Richard McMurry (1980) and Steven Woodworth (1990) have surmised that Bragg may have taken a written order relieving Johnston, which he could use if circumstances warranted. More recently Woodworth has written that "on July 14 Davis telegraphed Bragg that he should use his own discretion but that if it really seemed necessary to replace Johnston it might be best to go with Hardee." On the other hand, David Evans reasons that Davis's "C. telegram" was actually part of a protracted exchange about moving Union prisoners from Andersonville to either Cahaba or Silver Run, Alabama. By Evans's reading, Davis was mindful that Johnston had just three days before recommended the removal of inmates from Camp Sumter, the Andersonville prison, that he was wiring Bragg about a new place for the Federal prisoners, and that "C." was *Cahaba*. Evans's case is strengthened by a little-cited telegram Bragg sent General Cooper on July 15: "No more prisoners should be sent this side of South Carolina for the present. Have ordered the new depot at Cahaba and the transfer immediately." Indeed, leaving Atlanta late on the 15th, Bragg boarded the train for Montgomery and a tour of inspection in Alabama. The most important reason to reject essentially a generation of historians' speculations on the "C. telegram" is that the leading participants neither during the war nor after it ever mentioned the possibility of Bragg firing Johnston on the spot, July 13–15. Indeed, Seddon, Benjamin, and Davis all exchanged *post bellum* recollections that it was not until at least the morning of July 17 that the president finally made up his mind to relieve Johnston. The event that precipitated Johnston's demise was of his own doing, as we shall see.[21]

---

[21] Hess, *Bragg*, 234 (review of the troops); S. Davis, *Atlanta Will Fall*, 113; McMurry, "Mackall Journal," entry of July 15; Bragg letter to Davis, July 15, *OR*, vol. 39, pt. 2, 712–14; Bragg telegram to Davis, July 15, *OR*, vol. 38, pt. 5, 881; Bragg telegram to J. B. Sale, July 15, *OR*, vol. 52, pt. 2, 707; Bragg telegram to Davis, July 15, *OR*, vol. 52, pt. 2, 707; McMurry, *John Bell Hood*, 119–20 ("Bragg enclosed Hood's letter of July 14), 220 n.20 (Walter's arrival

During these several critical days in mid-July there could have occurred an event not usually seen in the campaign literature. According to Bragg's first biographer, writing in the 1920s, several members of Congress sympathetic to Johnston visited the general in Atlanta to warn him of President Davis's growing displeasure. One of them, George G. Vest, congressman from Missouri, after the war recounted the meeting in which Johnston grew indignant when he was told that the administration wanted him to attack Sherman. "You may tell Mr. Davis," Vest recalled Johnston saying, "that it would be folly for me under the circumstances to risk a decisive engagement. My plan is to draw Sherman further and further from his base in the hope of weakening him and by cutting his army in two. That is my only hope of defeating him." Vest mentioned that the president had said that if he were in command, Sherman would be whipped. "Yes, I know Mr. Davis thinks that he can do a great many things that other men would hesitate to attempt," Johnston retorted. "For instance, he tried to do what God had failed to do. He tried to make a soldier of Braxton Bragg and you know the result. It couldn't be done." Vest remembered that at that point a courier came in with news that Sherman's forces were moving south of the Chattahoochee and that a division of infantry

in Richmond "at the earliest, the evening of the seventeenth"); Woodworth, *Jefferson Davis and His Generals*, 283; Woodworth, *Decision in the Heartland*, 113; Davis, *Man and His Hour*, 561 ("Davis may even have authorized him [Bragg] to remove Johnston. Certainly he gave Bragg some authority"); Hallock, *Braxton Bragg*, 193; Castel, *Decision in the West*, 355; Davis telegram to Bragg, July 15, *OR*, vol. 52, pt. 2, 704; D. Evans, *Sherman's Horsemen*, 502 n.36; Crist, Williams, and Dillard, eds., *Papers of Jefferson Davis*, 10:520 (accepting Evans's point that "C." means Cahaba); Bragg telegram to Cooper, July 15, *OR*, ser. 2, vol. 7, 467.

Judith Lee Hallock has found Bragg's letter to the president of July 15 in the Civil War Collection at the Henry E. Huntington Library in San Marino, California, and has thus read that portion of it which was deleted from its published version in the *OR* (vol. 39, pt. 2, 714). In it Bragg refers to Georgia governor Brown's proclamation of July 9, calling out volunteers to reinforce Johnston's army. Brown used the document to criticize Davis's refusal to send Forrest or Morgan against Sherman's railroad lines: "If the Confederate Government will not send the large Cavalry force...to destroy the long line of Railroad...the people of Georgia...must, at all hazards and at any sacrifice, rush to the front" (Candler, *The Confederate Records of the State of Georgia*, 3:715; also in *OR*, vol. 52, pt. 2, 688–91). Bragg stated that he was enclosing a copy of Brown's proclamation and noted that General Johnston had appended a note on it, probably expressing agreement with the governor's call for a cavalry strike. "It has been most unfortunate for Genl. Johnston that his name has been associated with this weak but treasonable document," Bragg wrote, adding that "his influence in the Army is injured by it" (Hallock, *Braxton Bragg*, 194). The excision of this paragraph, critical of Johnston, may be another example of Johnston's friends among the staff of the *Official Records* project deliberately deleting text that would have been damaging to the general (see note 17, chapter three above).

had crossed Peachtree Creek. "Gentlemen, the time has come to strike," Johnston announced, suddenly rising. "Sherman has cut his army into three pieces and I believe now, by rapid movements, I can whip him in detail."

Indeed, an artillery sergeant in Johnston's army claimed well after the war that on the afternoon of July 16, "a circular battle order was promulgated among the troops from General Johnston, which stated that the favorable opportunity had arrived for striking the enemy." Sherman had divided his armies, wrote T. G. Dabney, and "we would advance at daylight next morning and beat him." No such attack, of course, was made on the morning of the 17th. (Had Johnston done so he would have doubtless saved his job.) Sergeant Dabney's recollection must thus be judged as awry, as no circular to the army on July 16 appears in the *Official Records.*

With Sherman's forces on the advance, as reported by both Johnston and Bragg, President Davis had to act. Even before Bragg's packet of correspondence arrived, the president took one more step on the 16th. Perhaps thinking of Lee's advice to Seddon the previous day, Davis effectively gave Johnston one final chance to save his job. On July 16 he telegraphed a very direct message to the general: "I wish to hear from you as to present situation, and your plan of operations so specifically as will enable me to anticipate events." Already the cabinet had voted unanimously to fire him. Seddon after the war recalled the prevalent opinion in a letter to Davis.

> I can only say Genl Johnston had been strangely reticent with the Government as to his plans and purposes, and one of the gravest objections existing to him was, that he would not give satisfactory information or assurances on this very point. From such communications as the Government could extract from him, and the best information it could obtain from other accredited sources, it was the unanimous opinion of your Cabinet and yourself,—certainly my firm conviction, taken against prior provocations I recommended removal,—that he did not intend to fight a battle for the relief of Atlanta, but was already arranging, by another disastrous retreat, to abandon his position there and elude the threatening enemy.

With this as the mindset of the Confederate government's leaders, Johnston's answer, telegraphed later the same day, was stunning. After confirming the movements of the enemy, which Bragg had reported the day before, Johnston reminded the president, "as the enemy has double our number we must be on the defensive." Then came the bombshell: "My plan of operations must, therefore, depend upon that of the enemy. It is mainly to watch for an opportunity to fight to advantage. We are trying to put Atlanta in condition to be held for a day or two by the Georgia militia, that army movements may be freer and wider."

"Freer and wider"? For a general already known for his captious communications, this telegram was by far his most captious indeed, suggesting that he would leave Atlanta in the hands of Smith's militia and march his army out of the works in some undefined movement. With this vague answer to the president's very direct question, Joseph E. Johnston had essentially penned his death-wish. "That was the clincher," writes Woodworth.

Davis and his advisers after the war remembered that Joe Johnston's telegram of July 16 had forced the conclusion. A decade later to Seddon, Davis remembered that he had "persisted in objection to his removal from a command," but that it was "his telegram to you [actually, it was to Davis himself] which overcame my objections." Judah Benjamin also recalled, "the telegrams from Mr. Seddon were urgent and finally amounted to a direct order to risk a battle at all hazards," but that Johnston had refused. "There was still hesitation" to relieve the general, Benjamin affirmed, "until his purpose was made to continue the retreat of his army and to abandon Atlanta to the defence of the militia; there was then an end of all doubt."

Years afterward, Seddon was also recalling the events to Lt. Col. William T. Walthall, explaining how the administration had effectively asked Johnston whether he would fight to hold Atlanta. "The answer was deemed evasive and unsatisfactory, and then and not till then, under the belief that Genl. J. really meant to abandon Atlanta without decisive engagement did the President finally decide and authorize his removal." In February 1878 John Reagan also remembered that Davis's reluctance to relieve Johnston melted away with the general's telegram of July 16: "the change was not finally made until, as I understood, General Johnston would not give the Secretary of War to understand certainly whether he would give battle rather than uncover Atlanta."

Two weeks after his final decision, in an interview with William D. Gale, the president explained how it had all happened. In their conversation, held in the executive mansion on July 31, Gale recounted to Davis how Johnston had not committed to decisive battle. Later that day Gale wrote,

> He seemed to writhe in his chair in torture, as with a countenance expressive of agony he replied "Yes I had hoped & trusted from day to day that I should get news of a general battle, in [which] I felt sure we should succeed, but day after day came and went, and there was no fight, while the army was still falling back, back, back. I remonstrated with Genl Johnston, urging him to give battle, telling him that if the enemy were ever permitted to cross below a certain distance, that nothing could prevent his (Sherman) sending raids through a country (Ala./& Ga/) already drained of troops

and cutting his communications. At last he fell back to Kennesaw Mountain & then to the Chattahoochie. *There I lost all hope of a battle.*"[22]

Now that Johnston had indicated that he planned to "uncover" Atlanta, the reaction from Richmond came swiftly.

Richmond, July 17, 1864

General J. E. Johnston:

Lieut. Gen. J. B. Hood has been commissioned to the temporary rank of general under the late law of Congress. I am directed by the Secretary of War to inform you that as you have failed to arrest the advance of the enemy to the vicinity of Atlanta, far in the interior of Georgia, and express no confidence that you can defeat or repel him, you are hereby relieved from the command of the Army and Department of Tennessee, which you will immediately turn over to General Hood.

S. Cooper,

Adjutant and Inspector General.

---

[22] Seitz, *Braxton Bragg*, 445; Dabney, "Gen. A. P. Stewart on Strong Topics," 31; Dabney, "When Hood Superseded Johnston," 406 (repeating the claim of "a circular 'battle order'"); T. L. Connelly, *Autumn of Glory*, 420; McMurry, *John Bell Hood*, 220 n.20 (Walter's estimated date of arrival); Hampton to Johnston, *Battles and Leaders*, 4:277; Seddon to Davis, June 17, 1872, Davis to Seddon, August 18, 1874, Benjamin to Davis, February 15, 1879, Seddon to Walthall, February 10, 1879, Reagan to Davis, February 7, 1878, in Rowland, *Jefferson Davis Constitutionalist*, 7:320, 396, 8:78, 353, 356; Davis to Johnston, July 16, *OR*, vol. 38, pt. 5, 882; Johnston to Davis, July 16, *OR*, vol. 38, pt. 5, 883; Woodworth, "How to Lose a City," 33; William D. Gale to Katherine Gale, July 31–August 1, 1864, Crist, Williams, and Dillard, eds., *Papers of Jefferson Davis*, 10:570.

Given the cabinet officers' recollections that the president held out until the 17th, upon receiving Johnston's "death-wish" telegram, Schott is off the mark in claiming that by the time Bragg left Atlanta on July 15, "Davis had already decided to sack Johnston" (Schott, "Lieutenant General William J. Hardee," 2:183 n.37).

Craig L. Symonds agrees with Davis's cabinet officers that Johnston's "unwillingness to accommodate himself to the president's intense desire for information" was an important cause of his removal from command (Symonds, "Johnston's Toughest Fight," 61). Gallagher has also weighed in on the Johnston controversy: "Typically overlooked in analysis of Johnston's activities is the degree to which his style of warfare failed to meet Confederate/civilian expectations," he writes. "The Confederate people wanted aggressive officers who supplied unequivocal battlefield victories, and Johnston's failure to meet this standard prompted considerable criticism" ("The Battle for Joe Johnston," 6). More recently, Gary Ecelbarger, historian of the Atlanta Campaign, has chimed in about Johnston's demise: "he never concretely declared that he would wage battle for the most vital city of the Confederacy outside of Richmond." "This," he adds, "more than anything else, is what induced Davis to dismiss him" (*Slaughter at the Chapel*, 33).

Assuming that Davis, Seddon, and Cooper met and discussed their course for several hours on the morning of the 17th—including the selection of Hood over Hardee as new commander of the army—Cooper's fateful telegram was sent probably that afternoon. (The president was meanwhile preparing to receive two Northerners, James Gilmore and James Jaquess, who wanted to explore possibilities of a negotiated peace; that night Davis would tell them that he expected Richmond and Atlanta to be held.)

Cooper's telegram was received in Atlanta at the army telegraph office near the city's Express Office in the Gate City Hotel, at Alabama and Pryor streets. Maj. Charles W. Hubner, chief of the telegraph service, oversaw the deciphering of the key words and phrases (coded in dots and dashes punched on the paper tape). Hubner himself carried the message out to the Niles house, three miles from downtown. Lieutenant Mackall recorded that it arrived about 9 P.M.

Probably before Major Hubner reached the Niles house, word of the surprise news had gotten out. Excited staffers at the telegraph office rushed out to tell friends. Samuel Richards, bookseller and newspaper printer who lived at Washington and Peters streets, three blocks from the telegraph office, heard it in time to include it in his diary entry for that Sunday night. To Richards, the news meant that there would be "a fight before Atlanta is given up," as Hood was a fighter, even "if he *has* only one leg."

General Johnston and Lieutenant Colonel Presstman, his chief engineer, were looking at maps, discussing a line of entrenchments to be dug south of Peachtree Creek. Johnston did not record his feelings upon reading Cooper's telegram, but he could not have been very shocked or surprised, given the "death-wish" character of his own telegram the previous day. Therefore it took no great deal of time for Johnston to have dictated and his staff to have written out the general order that announced his removal and bade farewell to his troops. "I cannot leave this noble army without expressing my admiration of the high military qualities it has displayed.... No longer your leader, I will watch your career, and will rejoice in your victories."

This order was written before midnight, as was a message to General Hood congratulating him on the assumption of his new command. Hood, probably by nine or ten o'clock, received General Cooper's telegram informing him that he had been promoted to temporary rank of full general and would take command of the Army of Tennessee. To Johnston's congratulatory message, Hood replied that he would come to army headquarters in the morning for discussion.

In another wire to Hood, Secretary Seddon took the opportunity to lay out the government's expectations of him: "It may yet be practicable to cut the communication of the enemy or find or make an opportunity of equal encounter whether he moves east or west." Thus the administration signaled its expectations: send out Wheeler's cavalry to cut Sherman's railway line (something Johnston had refused to do) and launch a major attacking battle if he saw favorable opportunity (which Johnston also had not done). Unstated, but obvious, was that even though "position, numbers, and morale are now with the enemy" (Bragg's words in his letter of July 15), the Confederate government hoped that Hood would at least hold Atlanta as long as he could. There was a final expectation, too, which Seddon carefully phrased in his telegram: "You are charged with a great trust. You will, I know, test to the utmost your abilities to discharge it. Be wary no less than bold."[23]

---

[23] Cooper to Johnston, July 17, *OR*, vol. 38, pt. 5, 885; Nelson, "'Independence or Fight,'" 11, 13; S. Davis, *Atlanta Will Fall*, 89, 116–17; S. Davis, *What the Yankees Did to Us*, 89–90; Hubner, "Some Recollections of Atlanta during 1864," 6 ("our headquarters were in the old American Hotel, which stood on the corner of Alabama and Pryor Streets" [the Gate City changed its name to American after the war]); M. H. Walker, *Charles W. Hubner*, 27, 30–31; Richards, *Sam Richards's Civil War Diary*, 227; Keller, "On the Road to Atlanta," 35 (stating erroneously that "on the night of July 17 as Johnston was preparing orders in his tent for the next day's moves against the Union army"); McMurry, "Mackall Journal," entry of July 17 ("9 P.M Order recd relivg Genl J"); General Order No. 4, July 17, *OR*, vol. 38, pt. 5, 887; Hood to Johnston, July 18 (1 A.M.), *OR*, vol. 38, pt. 5, 889; Govan and Livingood, *Different Valor*, 317; Symonds, *Johnston*, 329–30; Hood to Cooper, July 18, *OR*, vol. 38, pt. 5, 888; Seddon to Hood, July 17, *OR*, vol. 38, pt. 5, 885; Bragg letter to Davis, July 15, *OR*, vol. 39, pt. 2, 713.

A half-century ago many in Atlanta's Civil War remembrance community castigated Davis for firing General Johnston. Local newspaper columnist Bill Key captured the mood in terming Johnston's removal the result of "political skullduggery" (*Battle of Atlanta*, 46).

Richard McMurry summarizes the administration's thought process as to Johnston's replacement.

> Lt. Gen. William J. Hardee, the senior corps commander, was not selected because he had declined command of the army the winter before. In addition, Hardee was a bitter enemy of Gen. Braxton Bragg, President Davis's military adviser. Therefore, Hardee probably could not have worked well with the Confederate government. The third corps commander of the army was Lt. Gen. P. Stewart, who had been made a corps commander only ten days prior to the removal of Johnston. The delay that would follow if anyone outside the army were brought in to replace Johnston dictated that the new commander had to come from the army's ranks. Hood was the only alternative. ("Rise to Glory," 51.)

Recently a previously unknown letter from G. W. Smith to Hood has been found and published. "I wonder if old Joe did intend to leave my little band in charge of Atlanta whilst the three corps and the cavalry were hunting for Sherman's right or left flank," Smith wrote his

Hood must have reflected much that night about his new responsibilities. In *Advance and Retreat* he claimed that after receiving Cooper's and Seddon's telegrams, "I remained deep in thought throughout the night." On this point we can take Hood at his word. Halsey Wigfall, serving on Hood's staff, later wrote home, "the first time I saw him that day [July 18], I shook hands with him and congratulated him. He spoke very sadly and said he hardly knew if it were a subject of congratulation." But on another point we must judge Hood as disingenuous. In his memoir he claims, regarding the government's telegram, "This totally unexpected order so astounded me, and overwhelmed with sense of responsibility thereto attached." Hood was surprised? He had worked hard positioning himself for this very promotion, telling his superiors what he believed they wanted to hear, and finally writing that self-serving, deceitful letter to Bragg on July 14. Now that he had gotten what he obviously wanted, it is hard to believe he was either "astounded" or "overwhelmed." Moreover, in *Advance and Retreat*, Hood deleted a long passage of the letter Peter Stewart had sent him after the war concerning the change of command. Stewart did not like the idea of relieving Johnston; he learned of it on the night of July 17 when he visited Johnston's headquarters. Hood left out this part of Stewart's letter:

---

friend and relative on January 23, 1879. "Carrumba! Wouldn't that have been a kettle of fish?" (S. Hood, *Lost Papers*, 71).

John Bell Hood became the youngest of the eight Confederate officers to be promoted to the rank of full general; closest to him was Kirby Smith, born in 1824 and promoted to general five months before Hood (T. H. Williams, "The Military Leadership of North and South," 36).

Hood's rise from first lieutenant in May 1861 to full general thirty-eight months later has been termed "the most spectacular rise in rank of any officer North or South" (H. B. Simpson, *Hood's Texas Brigade in Reunion and Memory*, 16 n.14). In making this statement, Simpson probably has the facts on his side. But some would argue that Nathan Bedford Forrest's rise in rank was at least as spectacular: from private in June 1861 to lieutenant general in February 1865 (B. S. Wills, *A Battle from the Start*, 45, 300). John B. Gordon's rise from captain in May 1861 to major general three years later deserves mention also (Tankersley, *John B. Gordon*, 1, 153). Edmund Kirby Smith's ascent from lieutenant colonel in April 1861 to full general in February 1864 could be seen as less stellar. Actually, among Confederate officers' fast promotional tracks, the surprise comes from John C. Pemberton, who rose from lieutenant colonel in April 1861 to lieutenant general in less than a year and a half (Ballard, *Pemberton*, 87, 120).

Among Federal officers, honors for fastest meteor could go to Phil Sheridan, who rose from captain in May 1861 to major general seventeen months later (Dawson, "Philip Henry Sheridan," 471). Williams, cited above, notes that among high-ranking officers of both sides Hood and Sheridan, born less than three months apart, "represent the lowest age brackets" ("Military Leadership," 36). Both achieved the rank of major general in their respective armies at the age of 31.

> I was aware that Genl. Johnston had correctly anticipated the movements of the federal army; and had made his dispositions with a view to giving battle as the enemy should cross Peach Tree creek in his advance upon Atlanta. It seemed to me therefore, that Sunday night, that battle on Peach Tree creek the following day was inevitable. I felt that the removal of Genl. Johnston at such a time would be unjust to him and, to say the least, extremely embarrassing to yourself; and that *any* change of commanders in the very face of the enemy was hazardous. I therefore urged Genl. Johnston not to act upon the order, nor turn over command until the fate of Atlanta should be decided. He declined to do this, stating that he could not disobey the President's order. On leaving his quarters at a late hour, I rode to Hardee's quarters for the purpose of consulting him and inducing him to join with me in urging the same course upon Genl. Johnston, which he agreed to do. You will remember that I addressed a note to you requesting you to meet us next morning at an early hour as practicable. You replied that you would meet us at Johnston's headquarters at sunrise.

As he had promised, at dawn Hood rode to the Niles house. There he joined Stewart and Hardee in urging Johnston to retain command until the decisive battle for Atlanta had been fought. Johnston declined, saying that the president's order was to be obeyed. (According to one recollection, Johnston said, "Gentlemen, I am a soldier. A soldier's first duty is to obey.") Afterward, Hood sent a telegram to General Cooper, asking that the relief of Johnston be suspended. "I deem it dangerous to change command of this army at this particular time," Hood wrote, "and to be to the interest of the service that no change should be made until the fate of Atlanta is decided." Davis's reply arrived back in Atlanta that afternoon (5:20, according to Lieutenant Mackall). He admitted great hesitation at changing command of the army, adding, "reluctance to make the change induced me to send a telegram of inquiry to the commanding general on the 16th instant. His reply but confirmed previous apprehensions." "The order has been executed," Davis closed, "and I cannot suspend it without making the case worse than it was before the order was executed."

Hood, Hardee, and Stewart took the president's answer in stride. So did Johnston—sort of. On the morning of July 18, he sent a final telegram to Richmond. After having been relieved of duty essentially for its dereliction, Johnston could not resist taking one last poke at Jefferson Davis, and he did so by jabbing the president's favorite general, Robert E. Lee. "As to the alleged cause of my removal," he wrote, "I assert that Sherman's army is much stronger compared with that of Tennessee than Grant's compared with that of Northern Virginia." (Wrong: Johnston's and Lee's armies were roughly on par against their respective enemies.) "Yet the enemy has been compelled to advance much more slowly to the vicinity of Atlanta than to that of Richmond and Petersburg, and has penetrated much deeper into Virginia than into Georgia." (Wrong again: Grant and Sherman, starting at the same time in early May had covered roughly the same distance, about 85 miles—the Rapidan to Petersburg; Tunnel Hill to the Chattahoochee.) Moreover, Johnston certainly could not have claimed to have punished the enemy in his front as harshly as Lee had punished his. Finally, he took a swipe at General Cooper's phrasing ("express no confidence") when he closed by asserting, "confident language by a military commander is not usually regarded as evidence of competency."

Having fired his last figurative shot, the ex-commanding general settled into his new status. On the evening of the 18th, Halsey Wigfall visited Johnston to say goodbye and found him to be quite composed, even serene. "No one could ever have told from his countenance or manner that anything unusual had occurred," he wrote his sister Louise; "indeed he seemed in rather better spirits than usual." The general and his wife then boarded the train for Macon. By one account (Johnston's *Narrative*), Johnston stayed in Atlanta long enough to explain his plan of attacking Sherman's forces as they crossed Peachtree Creek. By another (Hood's *Advance and Retreat*), he left so quickly that he was in Macon by the evening of the 18th. Lieutenant Mackall, however, is to be believed when he entered into his journal on the 19th, "Genl J. takes cars for Macon 6½ PM." Corroboration comes from the *Macon Telegraph*, which on the 21st noted that the general and Mrs. Johnston arrived in town on the afternoon of July 20, and were staying at the home of Maj. Gen. Howell Cobb.[24]

---

[24] Hood to Johnston, July 18 (1 A.M.), *OR*, vol. 38, pt. 5, 889 ("Much to my surprise I received the appointment you refer to"); Hood, *A & R*, 126–27 ("this totally unexpected order so astounded me," and excerpted Alex P. Stewart to Hood, August 7, 1872), 162; S. Hood, *Lost Papers*, 58–60 (Stewart to Hood, August 7, 1872); Hess, *Bragg*, 234, 239 (Hood "never really was angling for Johnston's position"; I disagree); Hess, *Battle of Peach Tree Creek*, 23 (Wigfall to mother, August 7); Milner, "The Protest Sent to President Davis," 507; Hughes, *Hardee*, 217; Castel, *Decision in the West*, 362–63; Hood to Cooper, July 18, *OR*, vol. 38, pt. 5, 888; Davis to Hood, July 18, "(Same to Generals Hardee and Stewart)," *OR*, vol. 38, pt. 5, 888; Johnston to Cooper, July 18, *OR*, vol. 38, pt. 5, 888; Johnston's army returns of July 10, *OR*, vol. 38, pt. 3, 679 (59,196 officers and men present); Sherman's returns of June 30, *OR*, vol. 38, pt. 1, 115–16 (106,070 present); Lee's returns, June 30, OR, vol. 40, pt. 2, 707 (with Early's corps detached, 52,225 officers and men present); Grant's returns, *OR*, vol. 40, pt. 1, 177 (92,989); Wright, *Southern Girl in '61*, 181–82; J. E. Johnston, *Narrative*, 350; Hood, *A & R*, 128 ("left that evening for Macon, Georgia"); McMurry, "Mackall Journal," entry of July 19; "General Johnston in Town," *Macon Telegraph*, July 21.

In Richmond, the president dealt with public reaction to his decision. Ordnance chief Josiah Gorgas summarized the view of the administration in his diary entry of July 19: "People are, I think, generally satisfied with the removal of General Johnston. They have praised him and waited for him to fight until he has lost all Georgia, and they have got tired of him." The thinking was apparently different in the Army of Tennessee. "The news that Johnston is out and Hood is in as commander makes a strong impact on the army, and most of it is negative," Castel judges. Hood's own staff officer, Taylor Betty, entered in his diary for the 18th regarding the command change, "I am afraid it may do harm at this moment, because the army still has confidence in Johnston & does not know Hood." One division commander in Hardee's corps expressed the opinion of probably many officers. "Hood has 'gone up like a rocket,'" Maj. Gen. W. H. T. Walker wrote to this wife, Mary, on the morning of July 18; "it is to be hoped...that he will not come down like a stick." "He is brave," Walker conceded, but then added, "whether he has the capacity to Command armies (for it requires a high order of talent) time will develop." The wait-and-see attitude held by Walker was not shared by General Hardee. "It is well known that I felt unwilling to serve under General Hood," he later wrote, "because I believed him, though a tried and gallant officer, to be unequal in both experience and natural ability to so important a command." A month after Johnston's relief, Hardee hinted to General Mackall that he would have liked to have been offered the position. "The President is endeavoring to create the impression," he confided, "that in declining the command at Dalton, I declined it for all future time."

General Frank Cheatham, one of Hardee's division commanders, did not record his immediate response to Hood's accession, but his men did. In his diary for the 18th, Alfred T. Fielder of the 12th Tennessee wrote, "Johnston was the most popular Gen. that was ever in command of this department and his removal gives great dissatisfaction amongst the army." Van Buren Oldham

---

In excerpting parts of Stewart's letter of August 1872 in *A & R*, Hood also left out Stewart's statement, "I learned that Genl. Johnston went into Atlanta Monday evening," which would have implied that Johnston did not leave on the 18th (S. Hood, *Lost Papers*, 62). Regardless of whether he left on July 18 or 19, Connelly disapproves of Johnston's departure from Atlanta: "Johnston could have put aside his anger and injured pride and remained to assist Hood through the Atlanta crisis," he writes. "Even Bragg, in December of 1863, had offered to stay on as his chief of staff under similar circumstances" (*Autumn of Glory*, 425).

Several days after leaving Atlanta late on July 15, General Bragg learned that the government had relieved Johnston. Bragg later confessed surprise—not because he had not seen the removal coming, as Earl Hess surmises, but because President Davis had acted without further consultation with him (*Bragg*, 234).

entered into his journal that day that he and his comrades were "completely dumbfounded" by the change. Other soldiers were also critical. "The change is not well received by the soldiers," Alabaman Joel Murphree informed his wife, Jo, by letter of the 19th.

Writing to his wife, Bettie, on the 18th, Lt. Robert M. Gill of Mississippi was even harsher: "Hood in command. I fear the result. Hood is the most unpopular Gen'l in the Army and some of the troops are swearing that they will not fight under him. Brig. Gens., Cols. and Company officers have been called together to forestall anything like an outbreak. All regret that Johnston is gone. As Hardee once declined command it was given to Hood the butcher." Kentuckian Johnny Green, in the Orphan Brigade, affirmed the general melancholy in the ranks with his journal entry for the 18th: "the removal has cast a gloom over the army." He was a bit softer on Hood than others in writing, "the boys have no objection to him but they don't think there is another General in the world eaqual to Genl Joseph E.—except Gen R. E. Lee." "The change does not suit the boys at all," commented James M. Morey of the 32nd Tennessee. "Gen Johnston is universally loved by the troops and they all feel as if they had lost their best friend some are even moved to tears. All believe that Gen Hood is too rash." "I don't think it was good policy to remove Gen Johnston at this time," wrote Missourian Albert Allen to a friend on July 18. "Officers and men had great confidence in him. Much dissatisfaction is shown by the troops." "Quite an unexpected and startling announcement this morning," observed Robert Patrick in his diary, adding disapprovingly, "I have no doubt old Bragg has had something to do with this." Capt. Samuel Foster of the 24th Texas recorded in his diary for July 18 that he heard men talking about going home, so deep was their resentment and anger. "All over camp, (not only among Texas troops) can be seen this demoralization—and at all hours in the afternoon can be heard Hurrah for Joe Johnson and God D—n Jeff Davis." In view of these contemporaneous expressions—really only a sampling in the literature—one must conclude that in the Army of Tennessee there were many more soldiers who disapproved of the change in command than who liked it.

Lt. Hamilton Branch of the 54th Georgia was very outspoken about his affection for Old Joe.

> We learned yesterday to our great surprise and sorrow that our beloved gallant commander had been relieved from the command of this army. I never have seen or heard of an army so wrapped up in a commander as this army proved itself to be on yesterday. When it was announced everyone seemed to feel as if they had lost their best friend and the general remark

> was, well this army is lost, and everyone seemed to be whipped. As for myself I have never felt so downhearted in my life as I did on yesterday, and if we had not been ordered off, I know that I could not have helped from crying. Genl Johnston had the love and confidence of every man in his army, and not one doubted but that he would annihilate Sherman before he had finished the campaign. Genl Hood the present commander of this army is a fighting man and no doubt a fine officer and under him we will gain the victory, but he is not Genl Johnston.

Less eloquent was Sgt. John W. Hagan, 29th Georgia, writing on July 19: "Sad news yesterday the grate Genl J. E. Johnston who has led this army to so many glorious victories & he who was loved by all who ever Sirved under him sent to us his farwell address...all moan the loss of our grate Leader." Another commentator was Lt. W. L. Trask, on General Hardee's staff, who made an acute entry into his journal.

> Tuesday, July 19th—Yesterday morning everyone in the Army was astounded at the intelligence that General Johnston had been superseded by an order of Secretary of War and that General Hood had been placed in command of the Army. General Hardee and Cheatham were as much surprised as anyone, and expressed unfeigned regret that Johnston should be relieved at this juncture. Everyone from the highest officer to the lowest private is ignorant of the motives which could have influenced the President to make the change which has been felt with such startling effect by the entire Army—and no doubt the country. General Johnston had the unbounding confidence and esteem of almost every man in thc army and he had inspired it with the utmost enthusiasm. With the troops, nothing he did was wrong—every move he made was strategic. He was beloved far beyond any commander that ever controlled any fraction of the army, and the soldiers will part with him with much regret. His farewell was announced to the army in a brief and eloquent address early yesterday morning. This morning he leaves for the rear—many rumors and conjectures were current yesterday as to the probable cause of this action. The most likely of which is that it is because he has not fought a general battle before now.

Col. Ellison Capers of the 24th South Carolina, in Gist's brigade, recorded his men's unhappiness in his campaign report under date of September 10.

> [T]he 18th...a courier handed me a circular order from General Hood, announcing General Johnston's removal and assuming command. Shortly after the farewell address of General Johnston was received and read to the

> regiment. It is due to truth to say that the reception of these orders produced the most despondent feelings in my command. The loss of the commanding general was felt to be irreparable. Continuing the march and passing by his headquarters Walker's division passed at the shoulder, the officers saluting, and most of the latter and hundreds of the men taking off their hats. It had been proposed to halt and cheer, but General Johnston hearing our intention requested that the troops march by in silence.

Brig. Gen. Clement Stevens, brigade leader in Walker's division, went so far as to write Johnston on the 18th, relating that the order announcing his removal "was received by officers and men in silence and deep sorrow." He judged that the loss of Johnston as commanding general "is irreparable," adding, "we would hail with joy your return to command us."

Henry Watterson, the veteran journalist writing as "Grape" for the *Augusta Constitutionalist*, was with the army at the time of the command change and recorded what he saw and heard on July 18. News of Johnston's removal had produced a "shock upon the nerves of the army this morning"; "the strange tidings," he observed, "may be plainly read in every face and every group gathered to discuss the news." The removal, "Grape" attested, "has, to use a mild term, introduced dissatisfaction." Despite his retreats, General Johnston was much beloved by the men, so that the announcement of his relief was greeted by "surprise, and not unfounded discontent." Watterson made a point of describing the new army commander. Hood looked like "a tall bony man, with light blue eyes, and brown hair, a heavy suit of beard of the same color falling upon his bosom and an expression of imperturbable benevolence all over his open face. He stoops a little from the use of his crutch, but moves about with ease."[25]

---

[25] Strode, *Jefferson Davis, Tragic Hero*, 83; Castel, *Decision in the West*, 363; Hess, *Battle of Peach Tree Creek*, 29 (Beatty), 55 (Whitehead); S. Davis, "A Georgia Firebrand," 456; R. K. Brown, *To the Manner Born*, 257–58; Hughes, *Hardee*, 243–44; Hardee to Mackall, August 24, 1864, *OR*, vol. 38, pt. 5, 987; Losson, *Tennessee's Forgotten Warriors*, 171; Murphree, "Autobiography and Civil War Letters of Joel Murphree of Troy, Alabama," 184; Wiley, "A Story of 3 Southern Officers," 33; Green, *Johnny Green of the Orphan Brigade*, 142; Clampitt, *Confederate Heartland*, 75, 77; Gottschalk, *In Deadly Earnest*, 379; Hallock, *Braxton Bragg*, 195 (quoting Patrick); Foster, *One of Cleburne's Command*, 106–107; Joslyn, *Charlotte's Boys*, 268; Hagan, *Confederate Letters of John W. Hagan*, 51; Trask diary, entry of July 19; Capers report, September 10, 1864, *OR*, vol. 38, pt. 3, 717; Stevens to Johnston, July 18, *OR*, vol. 38, pt. 5, 890–91; "Grape" (Henry Watterson), "From the Army of Tennessee," *Augusta Constitutionalist*, July 22; Hess, *Battle of Peach Tree Creek*, 27.

Patrick's suspicion of Bragg's involvement in Johnston's relief from command was a logical inference, one that Bragg himself objected to. In his memoir titled "Notes of Gen'l Bragg's

Other soldiers, however, seemed to approve of the change of command. J. P. Douglas of Texas wrote his wife on the 18th, "I suppose the President has good reasons for the change. One thing certain Gen'l Hood will fight the enemy, and I believe he can whip them." The next day a surgeon in Loring's division, P. F. Whitehead, wrote his friend Irene, "To Hood no objection is made, every one thinks that he will *fight here*." Similarly, artillery lieutenant Andy Neal noted to his father that "the change was very unexpected in the army and deeply regreted but I cannot regard it as a calamity." Two weeks before, Neal had written, "There is not an officer or man in this Army ever dreamed of Johnston falling back this far.... I think he has been woefully outgeneraled." Jack King, another soldier in Johnston's army, wrote his wife on July 19 that he understood the commanding general had been relieved "for not fighting and allowing the Yanks to penetrate so far into Georgia." Probably for this reason, Lt. Col. Barrington S. King (serving in Virginia) wrote

---

services in the cause of the Confederacy after leaving the Army of Tenn. in Dec. 1863," Bragg recounted his visit to Atlanta. There is more than one fib in the following excerpt:

> Gen'l Bragg asked for and obtained the order to visit Atlanta and see Gen'l Johnston, to whom he had ever been most friendly, with a view, if possible of obviating his removal. Upon arriving at Atlanta he telegraphed the President the exact situation of the two armies, without comment, and advised a suspension of all action till his written report could be received. Without waiting, however, the President promptly ordered the removal, and said to the Sec'y who handed him the dispatch, "I regret now that Gen'l Bragg was permitted to go to Atlanta, as he will no doubt be charged with complicity in this act when it is wholy my own."(Series I, container 2, folder 13, Bragg Papers)

In his recent work, Robert D. Jenkins presents additional contemporaneous writings by Confederates in the Army of Tennessee which show the men's unhappiness over the change of command: "there was great indignation among the rank and file" (Sgt. William P. Chambers, 43d Mississippi, journal entry c. July 18); "regretted by the whole army as Gen. Johnson had won the respect and admiration of all" (Mumford Dixon, diary entry of July 17); "the troops are dissatisfied with the change" (J. N. Wyatt, August 10); "the army was astonished yesterday" (Lt. Col. Columbus Sykes, 43d Mississippi); "this is a great blow to our cause and has cast a gloom over the whole army" (J. P. Cannon, diary entry of July 18); "I am sorrow of it" (Pvt. Angus McDermid, 29th Georgia, July 21); "great gloom in the army" (Pvt. John T. Kern, 45th Mississippi, journal entry of July 18) (Jenkins, *To the Gates of Atlanta*, 171–78).

Gary Ecelbarger puts these expressions of discontent in context: "although the anti-Hood voices in the army left the impression that they spoke for the vast majority of the Confederates defending Atlanta, no evidence exists that they represented even half, but their opinions easily outweighed those of Hood's muted supporters" (*Slaughter at the Chapel*, 33).

to his mother in Roswell, "I rejoice that Johnston has been superceded by Hood."[26]

Newspapers across the South weighed in on the dramatic events, and just as men in the army were divided in their remarks, "the Southern press was divided in its editorial comments," as J. Cutler Andrews puts it. Papers that tended to favor the Davis administration sided with the president; the *Richmond Sentinel*, for instance, observed that "the President is a good judge of military merit and ability," and that the removal of Johnston had been compelled by "the most urgent State necessity." It added, "The people have seen enough of Johnston's retreats to satisfy them that, if let alone by the Government, he had no idea of making a stand this side of the Gulf of Mexico." The *Richmond Dispatch*, which also usually sided with the president (an anti-Davis paper called it the "yelper in the administration kennel"), hailed Johnston's successor as "a man who knows how to fight," "the very man to infuse spirit into an army." A correspondent of the *Dispatch* on July 21 commented that "the President was reduced to the alternative of retaining Johnston and losing Atlanta, or losing Johnston and the possibility of saving Atlanta." Likewise, the *Richmond Enquirer* asserted that Hood "justified the confidence of the President" and had earned his new appointment.

On the other hand, those journals opposed to the administration criticized the decision. John M. Daniel of the *Richmond Examiner* in his editorial of July 19 protested that "the whole army and the whole people of Georgia had unshaken confidence in Johnston...and this event cannot fail to produce unhappy effects." Johnston's removal was not unexpected, he asserted; "for some time the wretched supplejacks of the Government have been busy blackening him." When Daniel stated, "they have talked about a letter from a young lieutenant saying how he would not have retreated, disapproving the campaign, etc., etc.," the editor could only have been talking about Hood. Worse, Daniel complained that Hood was "notoriously incapable of managing anything larger than a division." The *Richmond Whig* was a bit kinder, claiming that "Hood's abilities may be adequate to the task assigned him," but the situation in Georgia was too critical for the testing of new generals. "The case does not admit experiments," it argued; "too much is at stake." For this reason, it leveled harsh criticism at the Davis administration: "Our authorities are diseased in mind, and the craziest of their crazes is the fancied possession of an intuitive knowledge of men, especially military men." "Hardee is surely

[26] Clampitt, *Confederate Heartland*, 77; Neal to "Dear Pa," July 20, Neal Papers; McMurry, "Atlanta Campaign" dissertation, 311; Gallagher, *The Confederate War*, 133 (Jack King); Galloway, *Dear Old Roswell*, 80–81.

entitled both by seniority and by greater experience to the promotion," it added. "When and where has Hood displayed the capacity to command one of the largest armies of the Confederacy and to conduct a campaign on which the salvation of the cause to a great measure depends?"

The complexity of the recent events, and the shared understanding of the importance of Atlanta, meant that Southern newspapers offered a wide range of editorial opinions. The *Charleston Courier* disliked the need for a change in the army's command but approved of Hood as a "fighting general." The *Charlottesville (Virginia) Chronicle* criticized the president for experimenting with a "young, inexperienced and not very remarkable officer." The *Chattanooga Rebel* praised Hood as "one of the most successful officers in the Confederacy." The *Mobile Evening News* offered the rather reserved comment that "the change of men was made for a change of policy." The *Columbia South Carolinian* hinted the same point in its commendation of General Hood: "The new leader has passed in rapid succession through every grade from Captain to Lieutenant-General, and has never found himself in any position to which he was not more than equal." "The public may feel assured that in every event," the editors exhorted, "he will prove himself a brave and prudent, but not too prudent, General." That last dig was to Joseph E. Johnston, whose excessive "prudence" had cost him his job.

Given the range of journalistic comments, one must question Mary Boykin Chesnut's diary entry, "All the papers were down on the president for this change of commanders—except the Georgia papers." The Georgia papers were just as divided as those throughout the rest of the country. In Atlanta, the *Appeal* voiced optimism that Hood would infuse his "fiery, resolute spirit" into the army. Poking obviously at General Johnston, the *Appeal* editors intoned, "It is time perhaps that Stonewall Jacksonism had usurped the place of caution and strategy. There is a limit to prudence.... This war must end and the final battle be fought. Why not here, and even now, as well as at a later period?" Another relocated paper, the *Knoxville/Griffin Rebel*, praised General Hood as "one of the most gallant and successful officers in the Confederacy," opining that "the Army of Tennessee will be safe in the hands of such a man." In similar vein, the *Macon Telegraph* observed, "we cannot censure the Administration for relieving [Johnston] from command." Moreover, the editorial asserted, the president should not be thought to have succumbed to bitter personal feelings against Johnston. "It will be unjust to charge President Davis with relieving General Johnston on account of any personal ill-feeling that may exist between them. We would not imagine for an instant that our chief magistrate would allow private enmity to trample over the public good, and with that opinion we shall support him in his last action."

"We cannot doubt the President has well considered so important a step," the *Telegraph* concluded; "let us keep cool, wait and watch." In similarly measured tone, the *Georgia Journal and Messenger*, also of Macon, opined that "such changes are much to be regretted." Yet it added, "the army and the public generally seem to have entire confidence in the military capacity and management of Gen. Johnston during the time he has been in command." In purpler prose, the *Macon Confederate* criticized Hood as a "reckless rattle-brain."

Across the state in Augusta, the city's two main dailies split on this issue. In a column titled "Who Is to Blame?" the *Chronicle & Sentinel* accused the administration of not doing enough to help Johnston; specifically, in not permitting Forrest to raid into Tennessee and cut Sherman's rail lines. "It was apparent to Gen. Johnston's friends...that not much would be done at Richmond, if anything, to assist him in achieving a victory." For this reason, the paper concluded that Johnston should not be blamed for his retreats: "the fault of all our disasters was attributable to action of other parties than those connected with the Army of Tennessee." On the other hand, the *Augusta Constitutionalist* rather welcomed the removal of Johnston, "the Fabius of Confederate Generals," and the appointment of Hood, "*the fighting general*." "If it means anything it means this and this alone: Atlanta will not be given up without a fight," the paper concluded. Indeed, the editors hailed "our fighting General, whose maimed body and wooden leg gives evidence of his bravery and impetuosity—whose glorious form will tower amidst the thickest of the fight, and whose trumpet call will excite and urge on his men to the deadly-breach to the last."[27]

---

[27] Andrews, *South Reports*, 453 (Richmond press divided); "Removal of Johnston and Appointment of Hood," *Richmond Sentinel*, July 21, 1864; Govan and Livingood, *Different Valor*, 325–26 (*Enquirer*; *Dispatch* as "yelper"; *Examiner* on "supplejacks"; *Whig*, "too much is at stake"); Patterson, "John Bell Hood," 20 (*Richmond Whig*); Clauss, "The Atlanta Campaign, 18 July–2 September 1864," 41–44; Hay, "The Davis-Hood-Johnston Controversy of 1864," 71–72; T. L. Connelly, *Autumn of Glory*, 405; Dyer, *Gallant Hood*, 251; untitled editorial, *Columbia South Carolinian*, July 20 ("It will be sometime"); Woodward, *Mary Chesnut's Civil War*, 623; "The War in Georgia"; editorial, *Griffin Rebel*, quoted in *Augusta Constitutionalist*, July 27; "General Johnston Relieved," *Macon Telegraph*, July 20, and "Defense of Atlanta," *Macon Telegraph*, July 20; *Georgia Journal and Messenger* (Macon), July 20, quoted in Hudson, *Odyssey*, 165; "Who Is to Blame?," *Augusta Chronicle & Sentinel*, July 21; "The Army of Tennessee," *Augusta Constitutionalist*, July 20.

With a troubled, divided army behind him, and a strong, dangerous foe in his front, John Bell Hood assumed command of the Army of Tennessee. In an address written to his officers and men after the president's answer had been received, Hood pledged to "bend all my energies and employ all my skill" in his new post. "I look with confidence to your patriotism to stand by me, and rely upon your prowess to wrest your country from the grasp of the invader."[28]

Southerners continued to weigh in on the Army of Tennessee's change of command. On August 10, Sen. Herschel Johnson of Georgia wrote Jefferson Davis on various matters. He pronounced the removal of Johnston a wise move although probably too late to save Atlanta. Nevertheless, he added, the decision "does not meet universal approval. It took the army & Country by surprise, & produced momentary alarm...nothing short of the success of Genl. Hood in closing the campaign, will procure a final verdict of approval."

As usual, Mrs. Chesnut adds color to the story. "Brewster says Joe Johnston is like the snakes, so blinded by his own venom.... Says he was afraid to risk a battle, for fear the Davis generals would betray him." She also wrote that Brewster told her that in an earlier visit to Richmond, "Mr. Davis said Johnston would have to be removed and Sherman blocked. He could not make Hardee full general because when he had command of that army he was always importuning the War department for a general in chief to be sent there—over him. Polk would not do...but Mr. Davis did not once intimate to him who it was that he intended to promote to the head of the western army."[29]

Hood's three biographers side with Hood and the administration in the transfer of command. O'Connor states that "the President, the Cabinet and the War Department had lost faith in Johnston's ability to fight aggressively." Dyer emphasizes the days-long agonizing process by which the president made his decision to replace Johnston: "Davis did not act hastily." After reviewing the information and options that the president had before him in mid-July 1864, McMurry concludes, "Davis's decision to remove Johnston was wise."[30]

Jefferson Davis's removal of Johnston and promotion of Hood have brought comment from generations of Davis biographers. Alfriend (1868) stresses the president's deliberate delay in making his decision until he was

---

[28] Hood's address to the army, July 18, *OR*, vol. 38, pt. 5, 889.

[29] Crist, Williams, and Dillard, eds., *Papers of Jefferson Davis*, 10:599; Woodward, *Mary Chesnut's Civil War*, 624, 635.

[30] O'Connor, *Cavalier General*, 198; Dyer, *Gallant Hood*, 245; McMurry, *Hood*, 123.

absolutely certain of Johnston's lack of an aggressive plan. Dodd (1907) describes Davis's writing to Johnston, asking for his plan to stop Sherman, "but he received no assurance." Dodd refers to "delegations of prominent Georgians" and Senator Hill, all advocating Johnston's removal. Gordon (1918) simply states that Davis bowed before "a public clamor" calling for the general to be fired. Schaff (1922) disparages Johnston's performance in the campaign, opining that "if Lee, Grant, Stuart, Stonewall Jackson, or Hampton had been in Johnston's place we think Sherman would have had to fight for his life long before the Etowah and Chattahoochee." Tate (1929) rather overstates Bragg's role in the administration's decision. Cutting (1930) reiterates that Davis announced his opposition to the cabinet's decision at its meeting, July 14. McElroy (1937) seems to question Davis's decision by quoting General Grant: "When I heard your Government had removed Johnston from command I was as happy as if I had reinforced Sherman with a large army corps." Strode (1964) notes that Johnston had many friends and admirers who would criticize the president for relieving him, but "a majority of thinking Southerners then believed that Davis had been too lenient and long-suffering with Johnston's retreats, delays, and muddy replies." Eaton (1977) judges, "Today Hood's appointment is controversial among historians, but the consensus regards it as a serious mistake" without citing sources other than recommending Connelly's *Autumn of Glory* on the campaign as a whole. William C. Davis (1991) takes a broad view in asserting that "Davis's only mistake with Johnston was assigning him to the command in the first place and then retaining him while he gradually lost most of north Georgia." Davis also suggests that the Confederate president would have done better to replace Johnston with someone other than Hood—Beauregard or, best of all, Robert E. Lee. Allen (1999) emphasizes Johnston's telegram of July 17 ("army movements may be freer and wider") as the last straw, forcing an anxious, hesitant president to fire his general. Cooper (2000) justifies Davis's decision by contrasting Johnston's performance against Sherman in north Georgia with Lee's against Grant in Virginia. Lee had battered and bled Grant's army from the Wilderness to Petersburg and there held him in a stalemate. Johnston, on the other hand, "never slowed down or weakened Sherman, who reached Atlanta relatively more powerful than Grant at Petersburg." Finally, Hattaway and Beringer (2002) agree with McMurry's conclusion that Johnston needed to be relieved and that Hood was probably the administration's best pick as his successor.[31]

---

[31] Alfriend, *The Life of Jefferson Davis*, 568–69; Dodd, *Davis*, 332; A. C. Gordon, *Jefferson Davis*, 228; Schaff, *Jefferson Davis His Life and Personality*, 203; Tate, *Jefferson Davis*, 257;

Confederate general Dabney Maury recorded an incident that occurred long after the war. He, General and Mrs. Johnston were enjoying dinner when outside a scream interrupted the conversation. Johnston looked away, tried to resume talking, but was interrupted again, and again. A little girl outside was being attacked by a turkey-gobbler, a house pet. "Why don't you run away? Why don't you run away?" Johnston shouted. "Well that's fine advice for a general to give," Maury commented. Angered, Johnston spat out, "If she will not fight, sir, is not the best thing for her to do to run away, sir?" Mrs. Johnston broke the tension by laughing, "That used to be your plan always, I know, sir."[32]

Davis laid out very thoroughly his criticisms of Johnston's generalship in a memorandum dated February 18, 1865. Congress the month before had passed a resolution calling on the reassignment of Johnston to command the Army of Tennessee. Davis objected, and wrote a long response outlining the deficiencies he had observed in Johnston throughout the whole war. But he never sent the paper to Congress, acquiescing instead to Robert E. Lee's suggestion that Johnston be brought out of retirement for field command. Nonetheless, Davis's "unsent message" is a cogent summary of the president's reproaches, which number more than a dozen and a half. One of them concerns the general's failure to inform the administration of his plan of operations. In Virginia, Mississippi and Georgia, Johnston had not provided "any communications of his plans or purposes beyond vague statements of an intention to counteract the enemy as their plans might be developed."

In mid-August 1864 Senator Wigfall offered to Mary Chesnut a reason for Johnston's reluctance to convey his ideas or intentions. "Wigfall said to me," she entered in her diary, "Joe Johnston does not exactly say that Jeff Davis betrays them to the enemy, but he says he dares not let the president know his plans, as there is a spy in the war office who invariably warns the Yankees in time." If Mrs. Chesnut and Wigfall are to be believed, that General Johnston really thought there was a Yankee spy in the Davis administration, this

---

Cutting, *Jefferson Davis, Political Soldier*, 230; McElroy, *Jefferson Davis*, 416; Strode, *Tragic Hero*, 83; Eaton, *Jefferson Davis*, 254; Davis, *Man and His Hour*, 563; Allen, *Jefferson Davis*, 390; W. J. Cooper, *Jefferson Davis, American*, 485; Hattaway and Beringer, *Jefferson Davis, Confederate President*, 323

[32] B. T. Johnson, *A Memoir of the Life and Public Service of Joseph E. Johnston*, 299; Kelly, "Epilogue: 'Gen'l J. Is Not the Man We Thought Him," 60 (telling the same story).

is certainly one of the stranger rationales given for Johnston's routine failures to communicate with his government.[33]

Johnston's apologists strain to justify the general's reticence. Comparing the regularity with which Sherman telegraphed Washington (every night during the campaign), Philip L. Secrist has to admit that Johnston's dispatches "leave much to be desired." Secrist even posits as a reason, "there is the likelihood that many of Johnston's messages were lost or destroyed in the capture and burning of Atlanta and Richmond before the war's end." Secrist should not have strained so hard; the *Official Records* show more than forty telegrams or letters sent by Johnston to Davis, Bragg, and Seddon from the time the campaign commenced (May 5) to the eve of the general's removal from command (July 17). True, not all of these are progress reports on the campaign; some involve administrative matters (such as officers' appointments), and some Johnston used to argue his case for the president to send Forrest's cavalry against the enemy's railroad supply lines. Nonetheless, the mere number of messages, averaging better than one every two days, suggests that Joe Johnston was at least talking to the administration, even if he was not telling it what it wanted to hear.[34]

The question of "why Joseph E. Johnston would not fight" continues to this day in Civil War scholarship. "I am blamed for not fighting," he acknowledged in a letter to his friend Dabney Maury on September 1, 1864 (the day Hood gave up Atlanta). Jay Luvaas has offered the opinion that "if Johnston could not in good conscience carry out the policies of his government [i.e., fight] then he ought to have resigned." The title of Newton's more recent article says it all: "Why Wouldn't Joe Johnston Fight?" Here he reverses Luvaas's statement: "a government that understood his limitations could have employed Johnston quite profitably." One recalls, for instance, that in fall 1862, as Johnston was recovering from his Seven Pines wound, Jefferson Davis gave him some consideration to be secretary of war. Postmaster General Reagan recalled that at a cabinet meeting when George W. Randolph was about to leave the War Department, "the President, in speaking to the members of the Cabinet about the appointment of a successor, said that he was considering Mr. Seddon, Gen. Joseph E. Johnston and Gen. G. W. Smith in

---

[33] Symonds, *Johnston*, 341–42; Davis to James Phelan, March 1, 1865, *OR*, vol. 47, pt. 2, 1303, with "Inclosure" (the memorandum of February 18, 1865), 1304–11; Woodward, *Mary Chesnut's Civil War*, 635 (entry of August 14, 1864).

[34] Secrist, "Jefferson Davis and the Atlanta Campaign," 11–12; S. Davis, "Issues Affecting Jefferson Davis' Decision to Fire Joe Johnston," 12.

connection with that appointment, and preferred them in that order." (James A Seddon, of course, got the job.)[35]

Military historian Bevin Alexander changes the issue from why Johnston wouldn't fight to why Johnston fought that way in the first place. "He was certain Sherman would attack him directly," Alexander explains; hence he took one defensive position after another, awaiting an assault that rarely came (except, for instance, Federal attacking battles at New Hope Church and Kennesaw). "Johnston was incapable of conducting a war of maneuver," Alexander adds.

> Building entrenchments was no solution. But the unimaginative Johnston stuck to the same approach of defending the railroad with fortifications—around which Sherman could continue to move. Instead of relying entirely on static defense, Johnston could have sent part of his army under cover of the mountains and forests to descend unexpectedly on Sherman's most vulnerable target—the railroad leading back to Chattanooga.[36]

---

[35] "Two Letters," 395; Luvaas, "An Appraisal of Joseph E. Johnston," 31; Newton, "Why Wouldn't Joe Johnston Fight?," 53–54; Reagan, *Memoirs*, 161.

[36] Alexander, *How the South Could Have Won the Civil War*, 252–53.

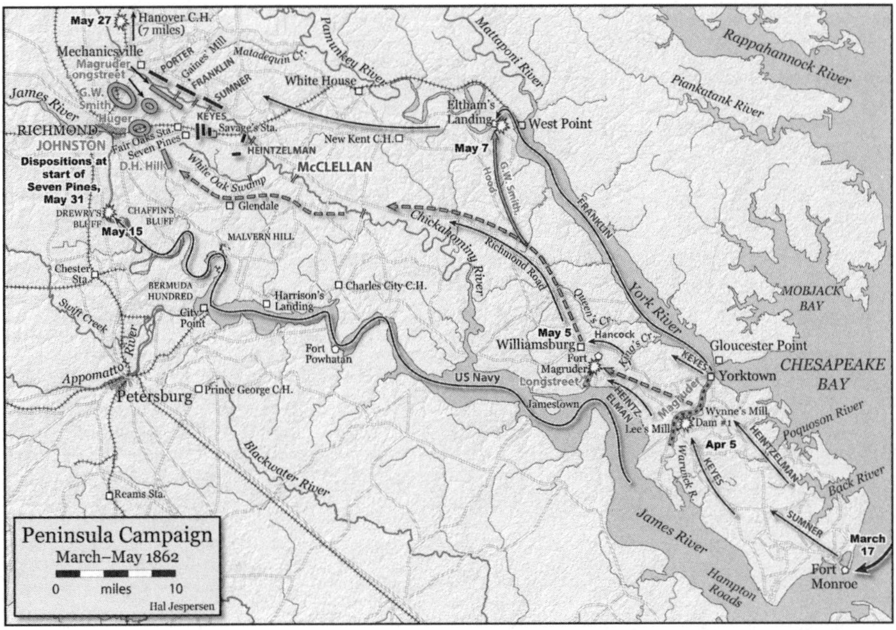

PENINSULA CAMPAIGN

March–May 1862

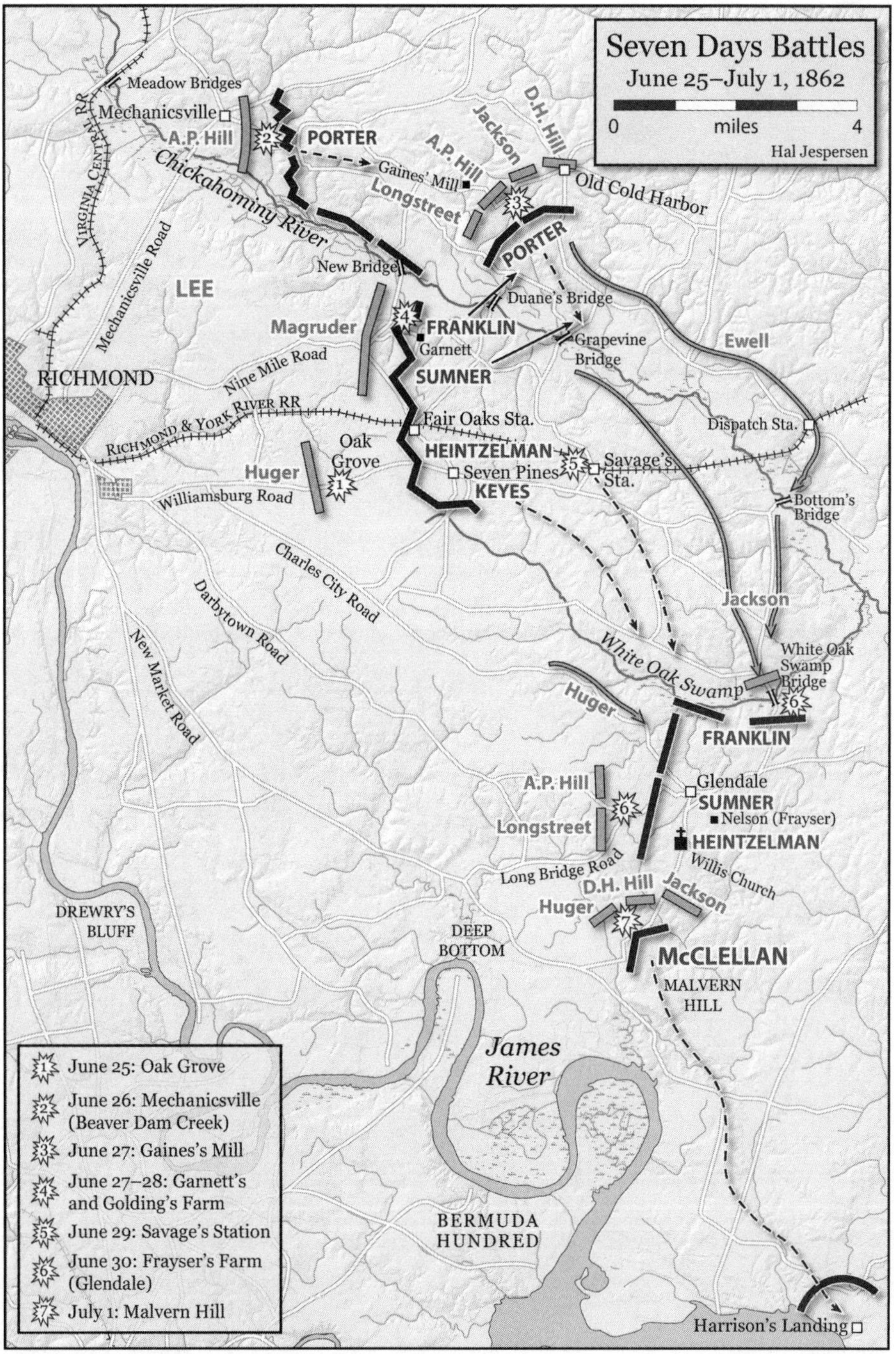

SEVEN DAYS BATTLES

June 25–July 1, 1862

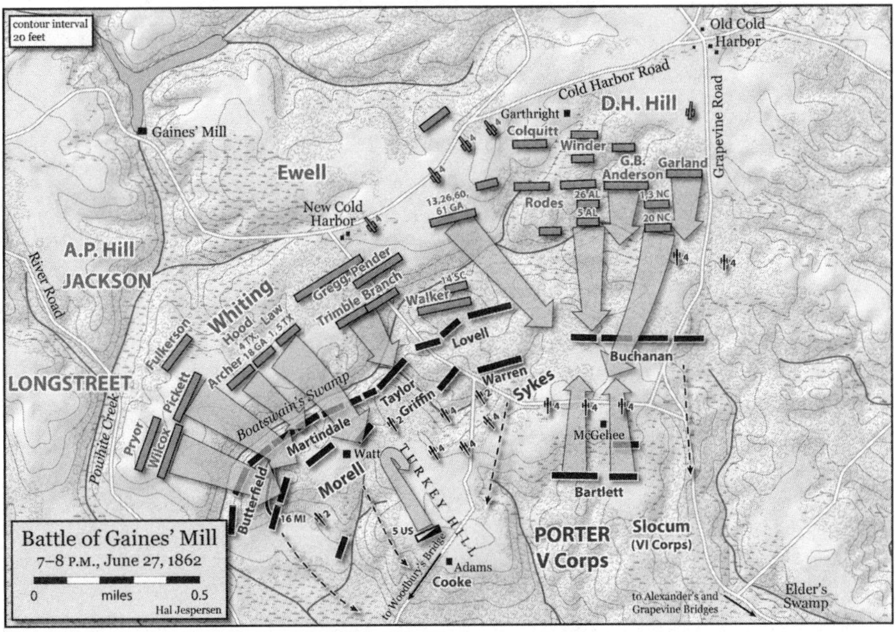

BATTLE OF GAINES' MILL

June 27, 1862

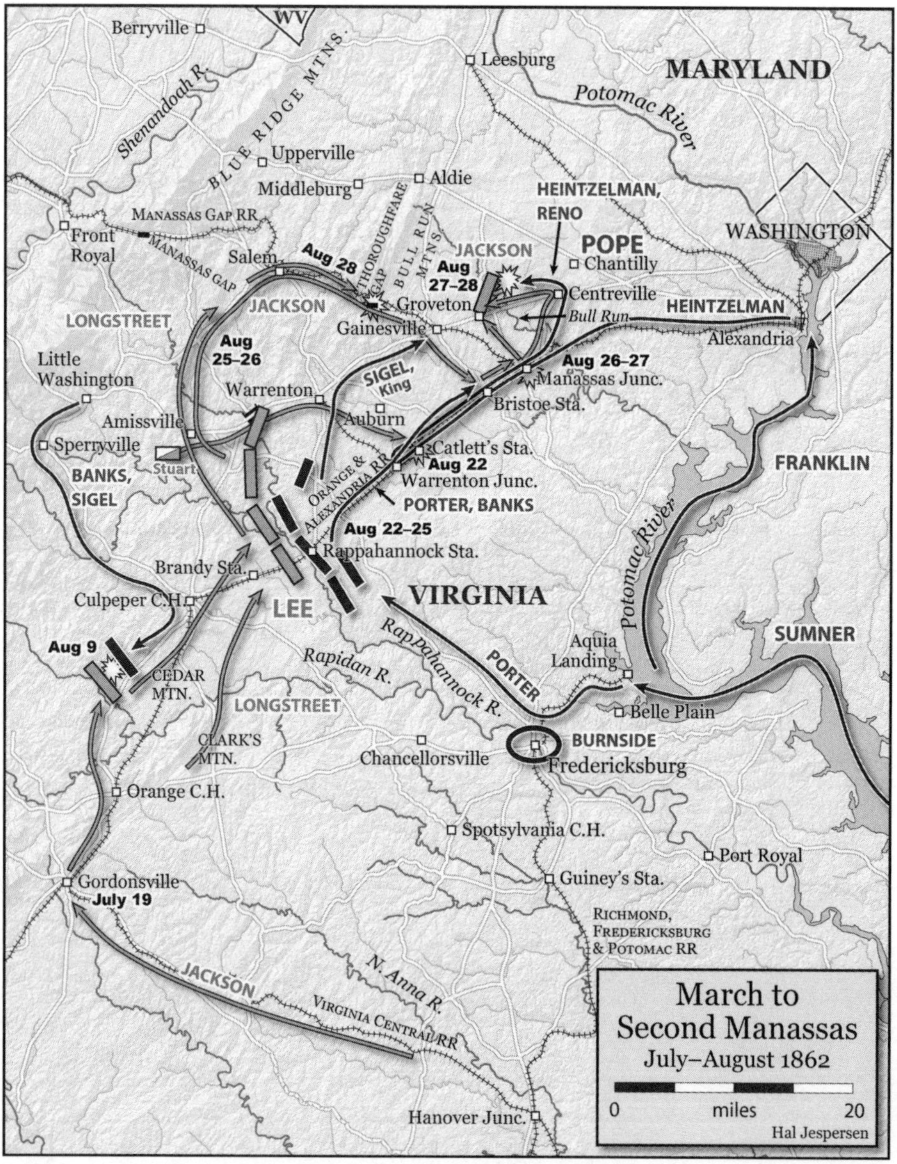

MARCH TO SECOND MANASSAS

July–August 1862

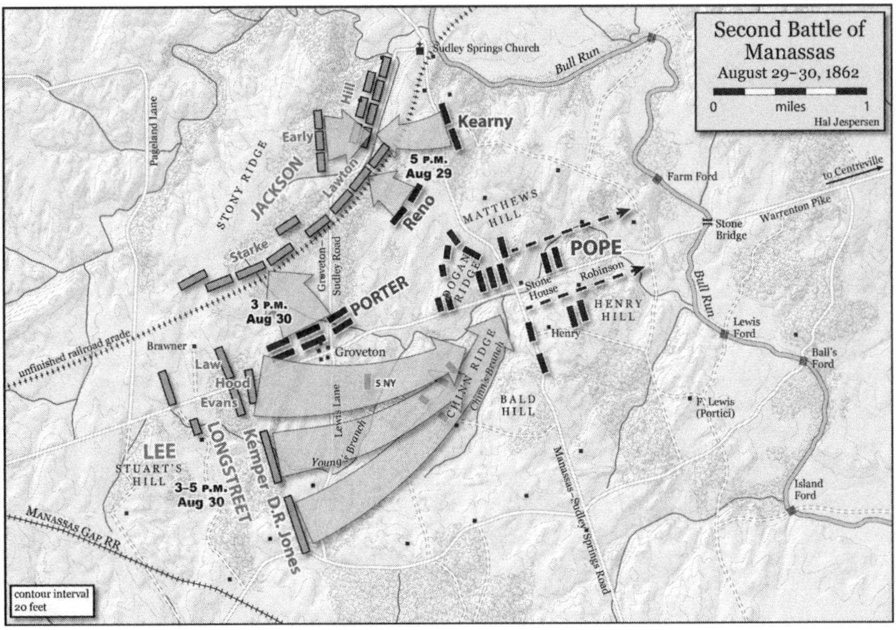

SECOND BATTLE OF MANASSAS

August 29–30, 1862

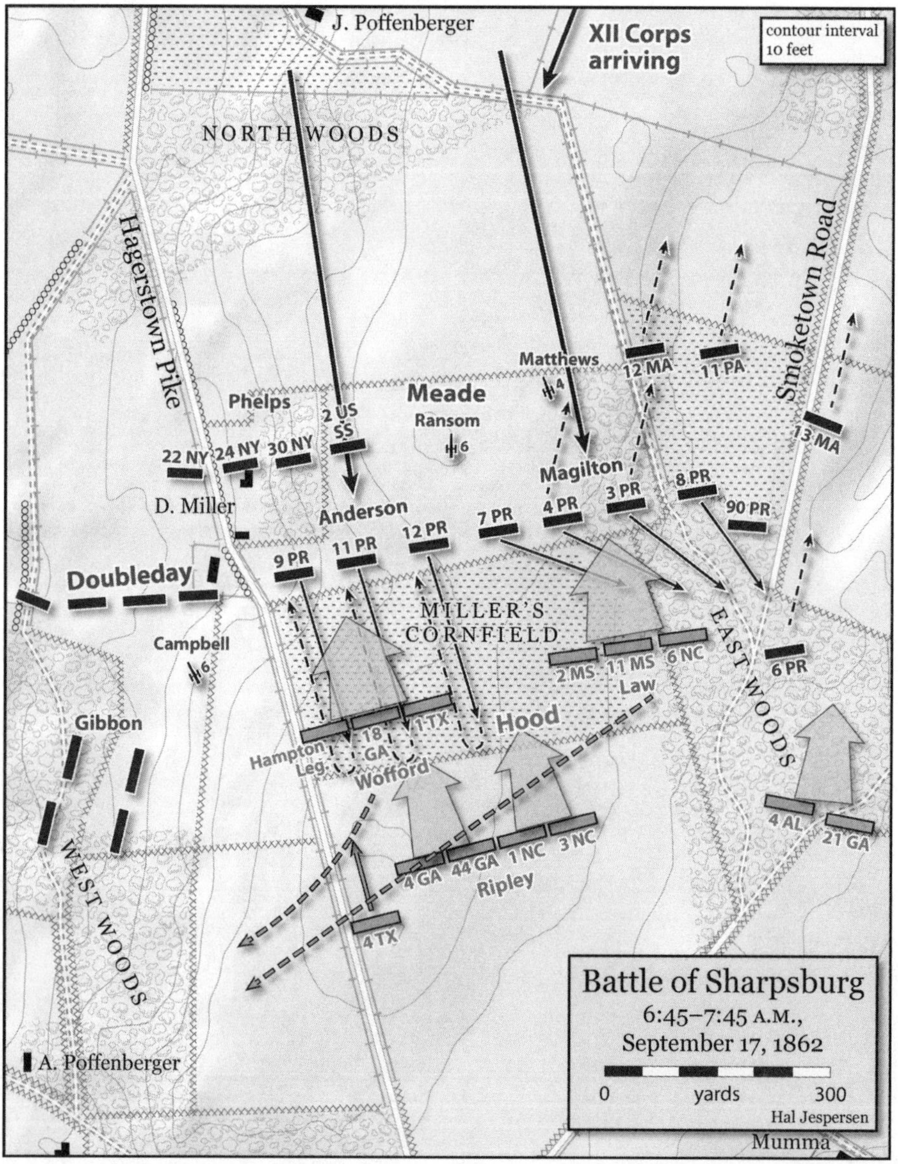

BATTLE OF SHARPSBURG

September 17, 1862

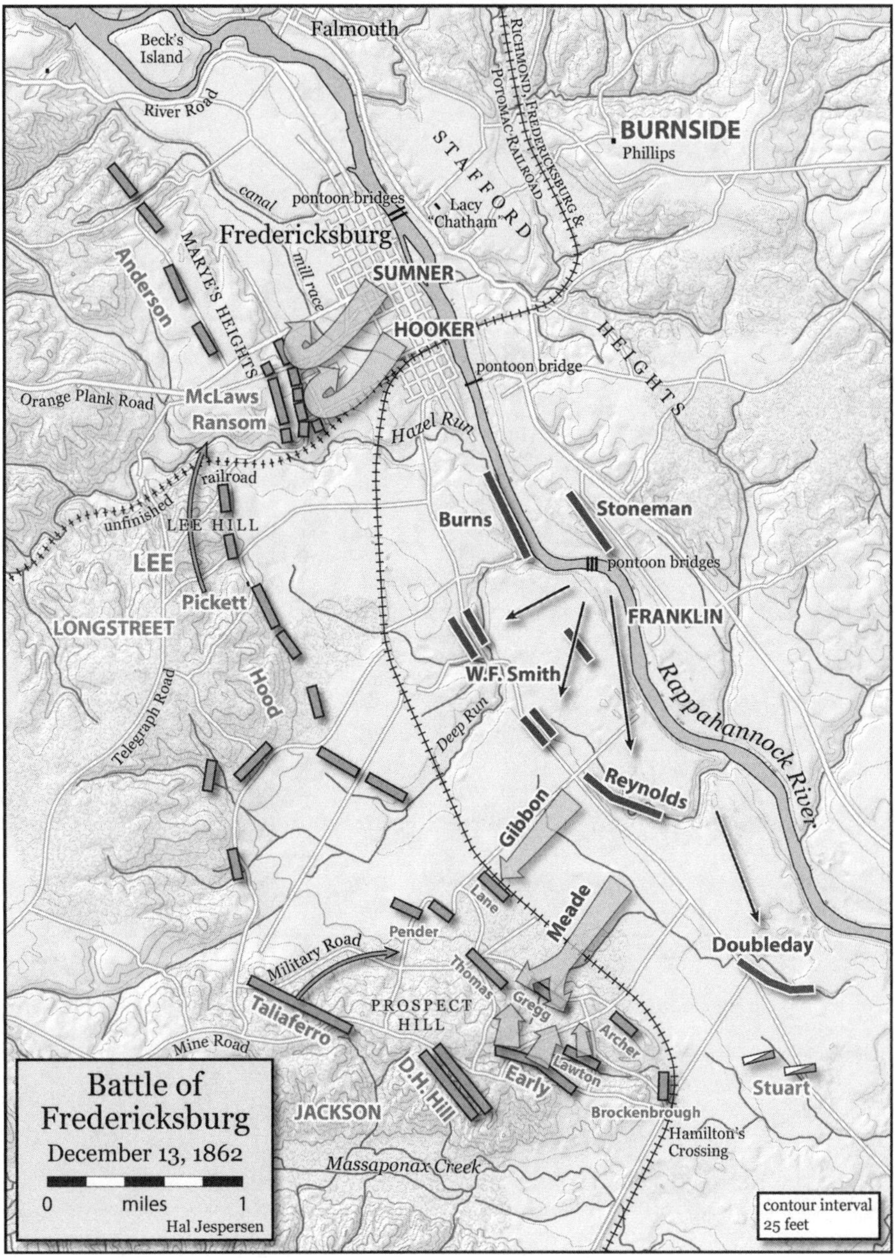

BATTLE OF FREDERICKSBURG

December 13, 1862

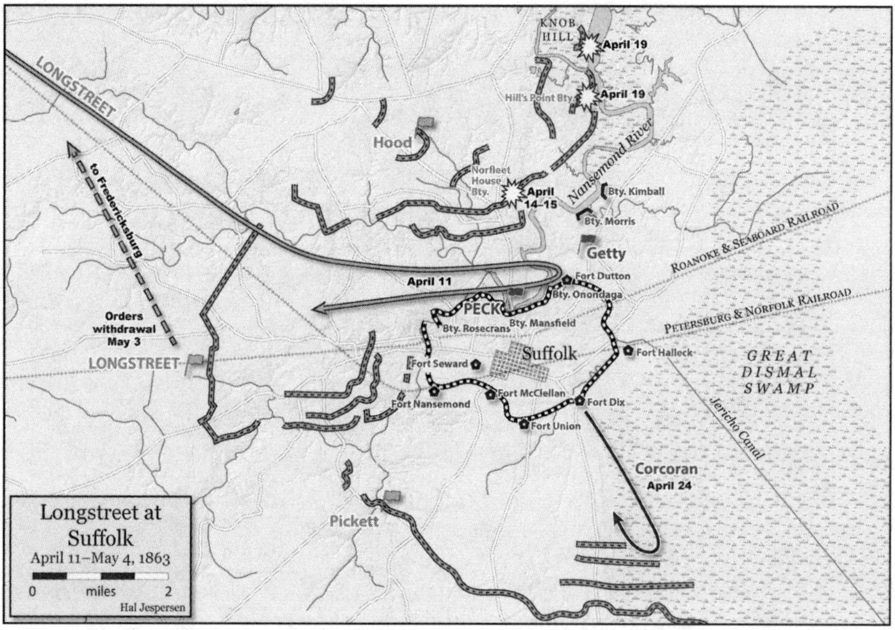

LONGSTREET AT SUFFOLK

April 11–May 4, 1863

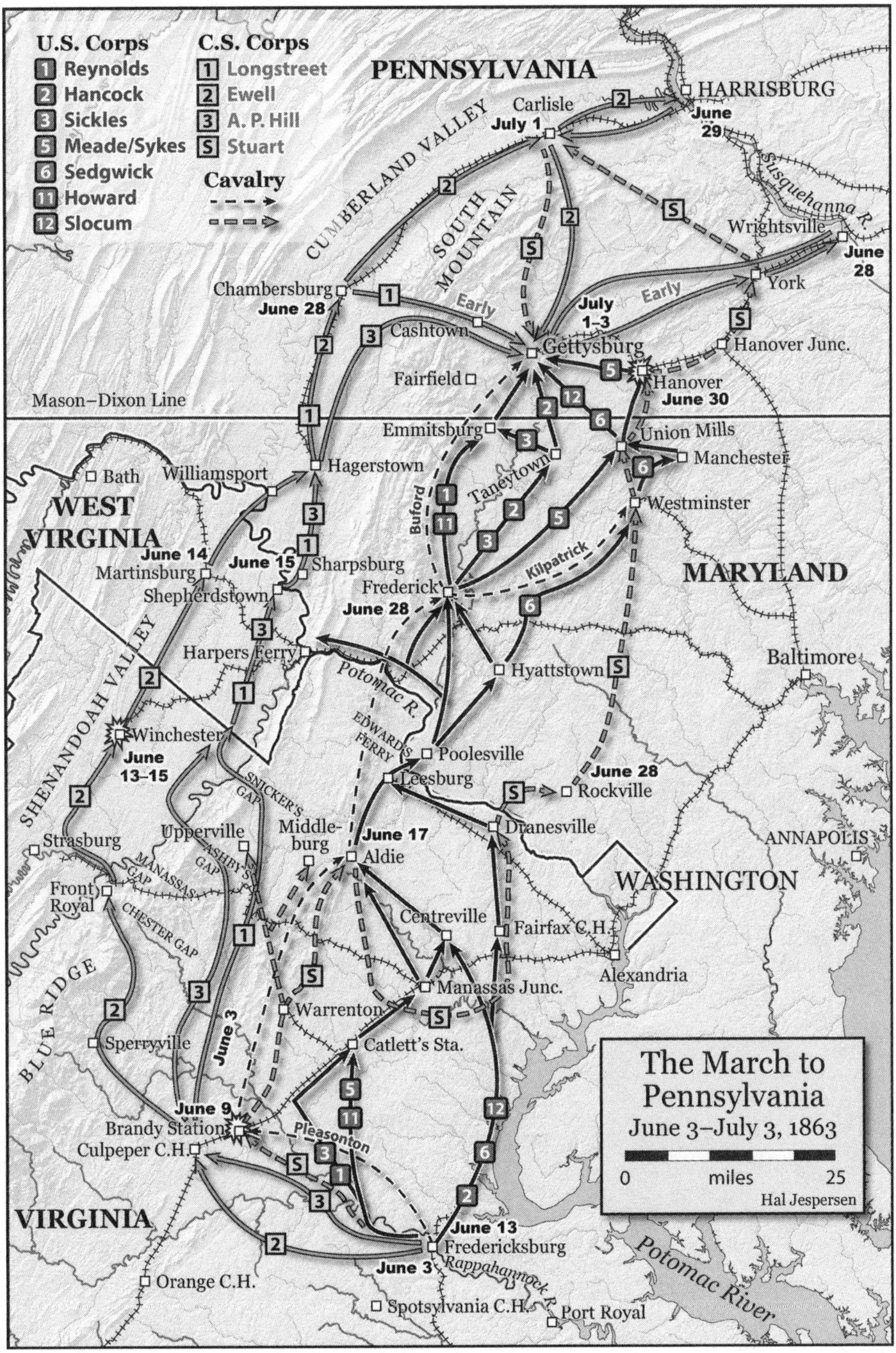

MARCH TO PENNSYLVANIA & GETTYSBURG

June 3–July 3, 1863

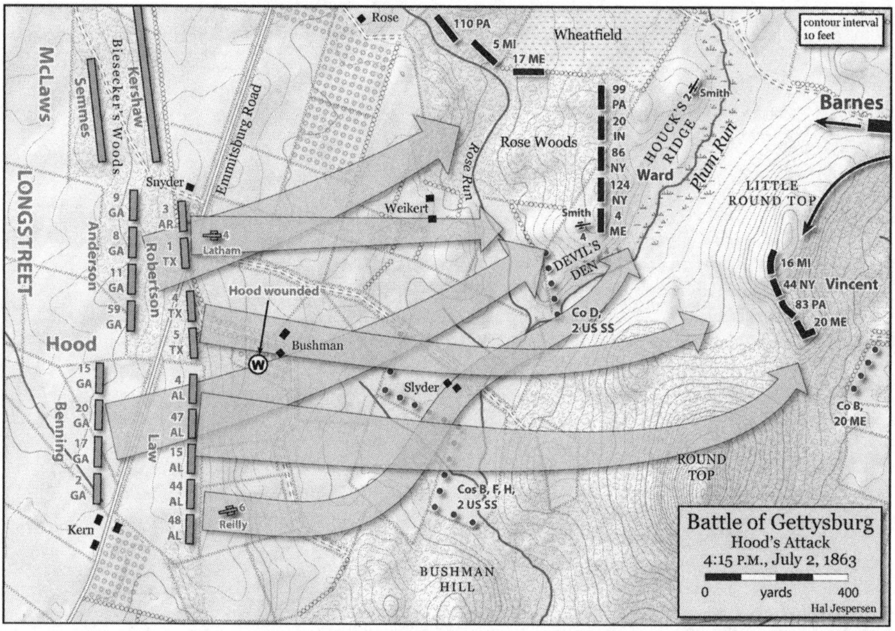

BATTLE OF GETTYSBURG & HOOD'S ATTACK

July 2, 1863

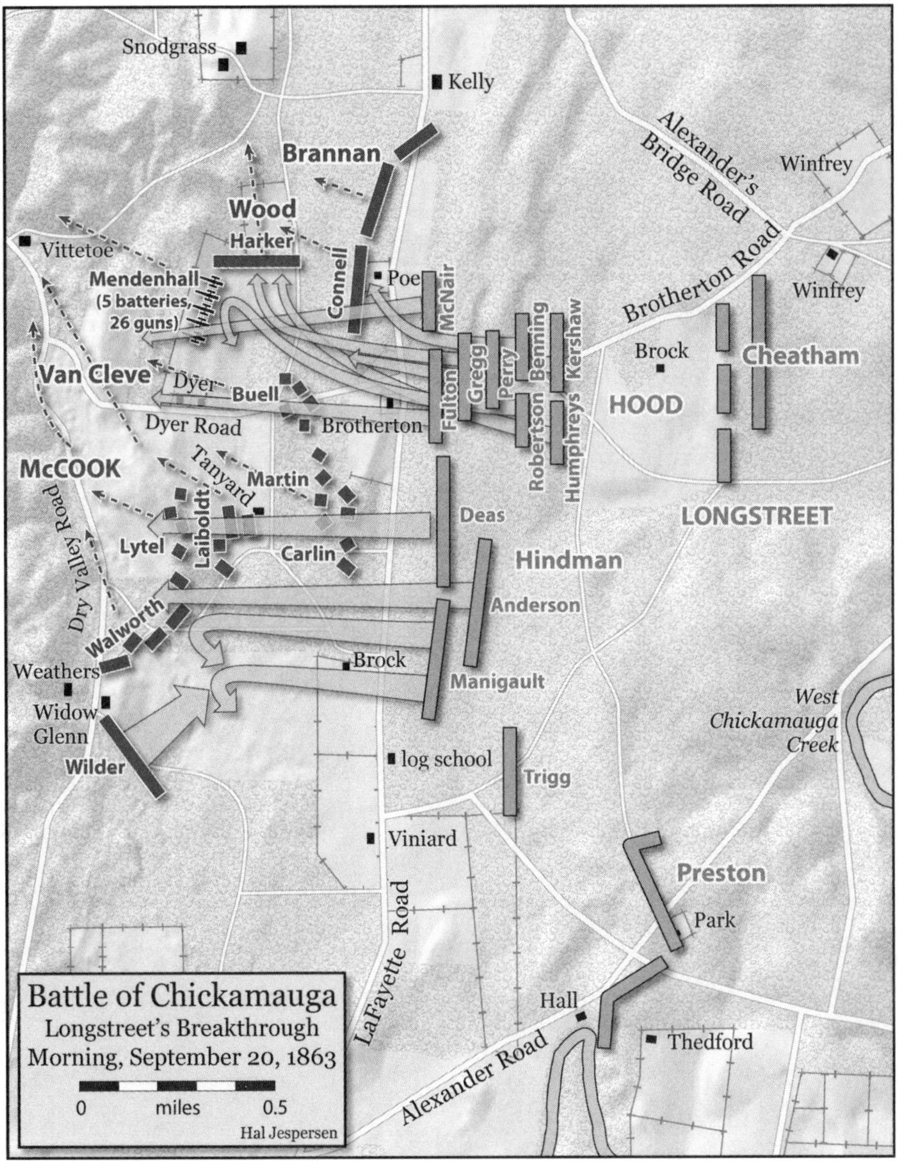

BATTLE OF CHICKAMAUGA

September 20, 1863

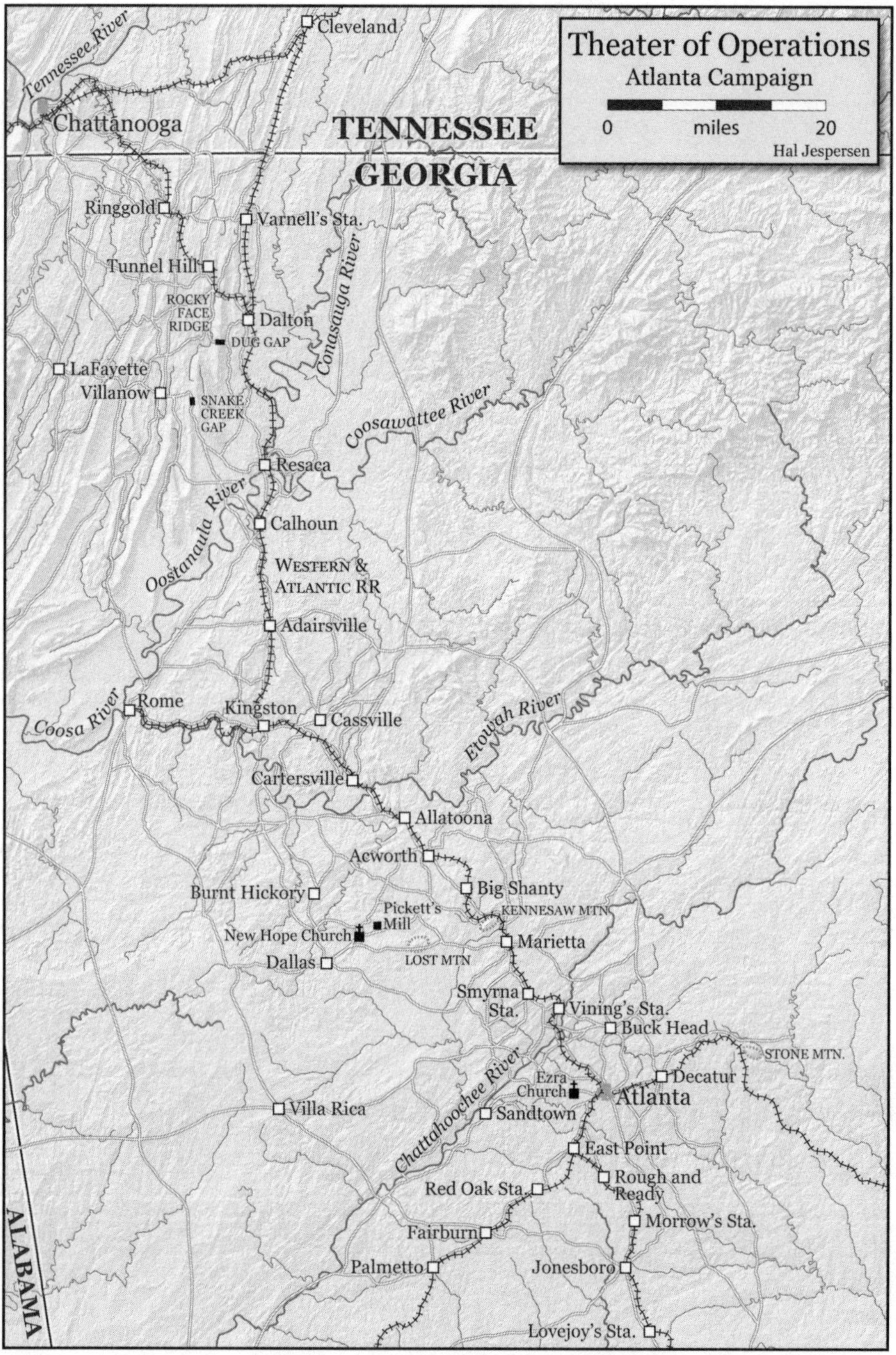

ATLANTA CAMPAIGN

Theater of Operations

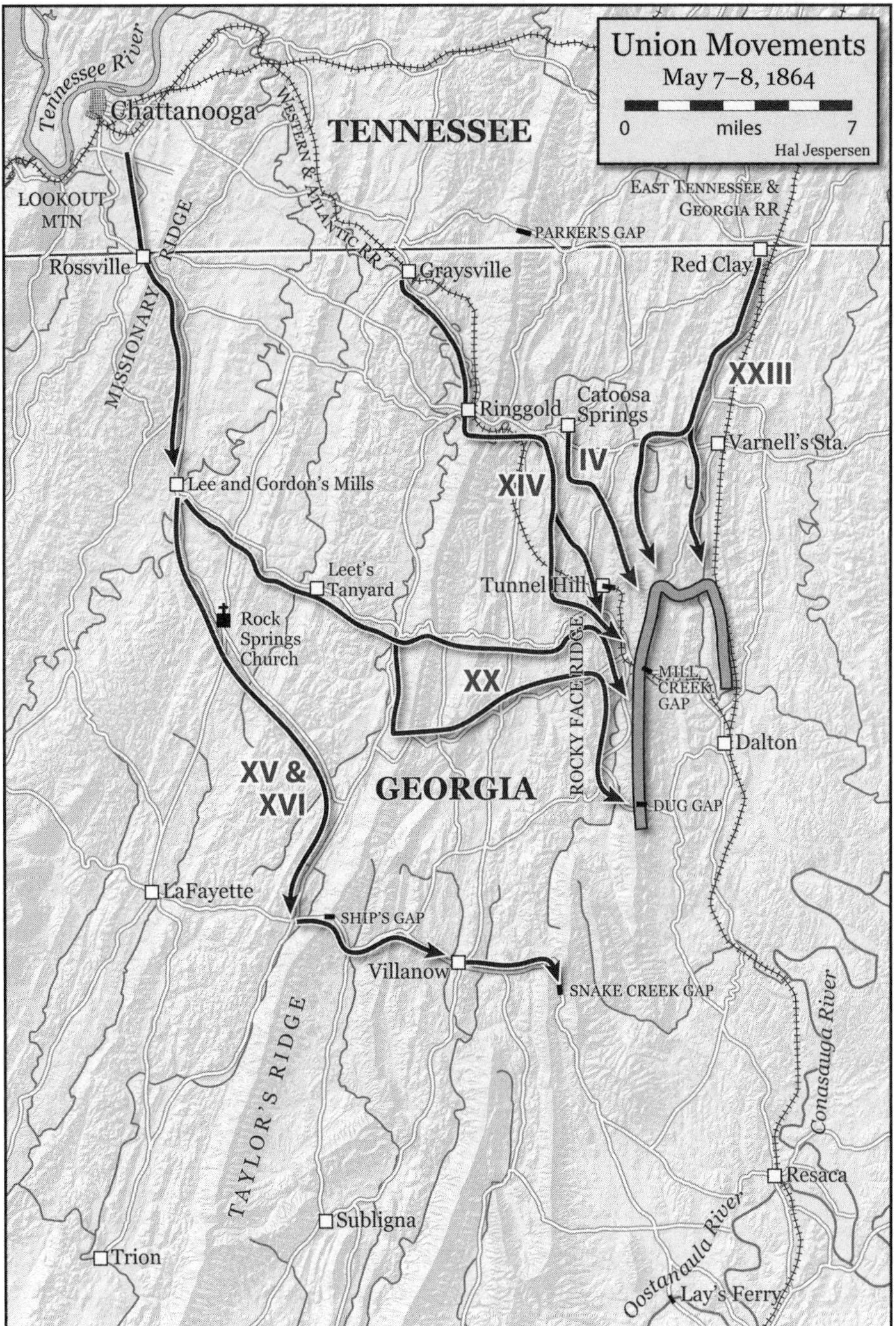

ATLANTA CAMPAIGN

Union Movements, May 7–8, 1864

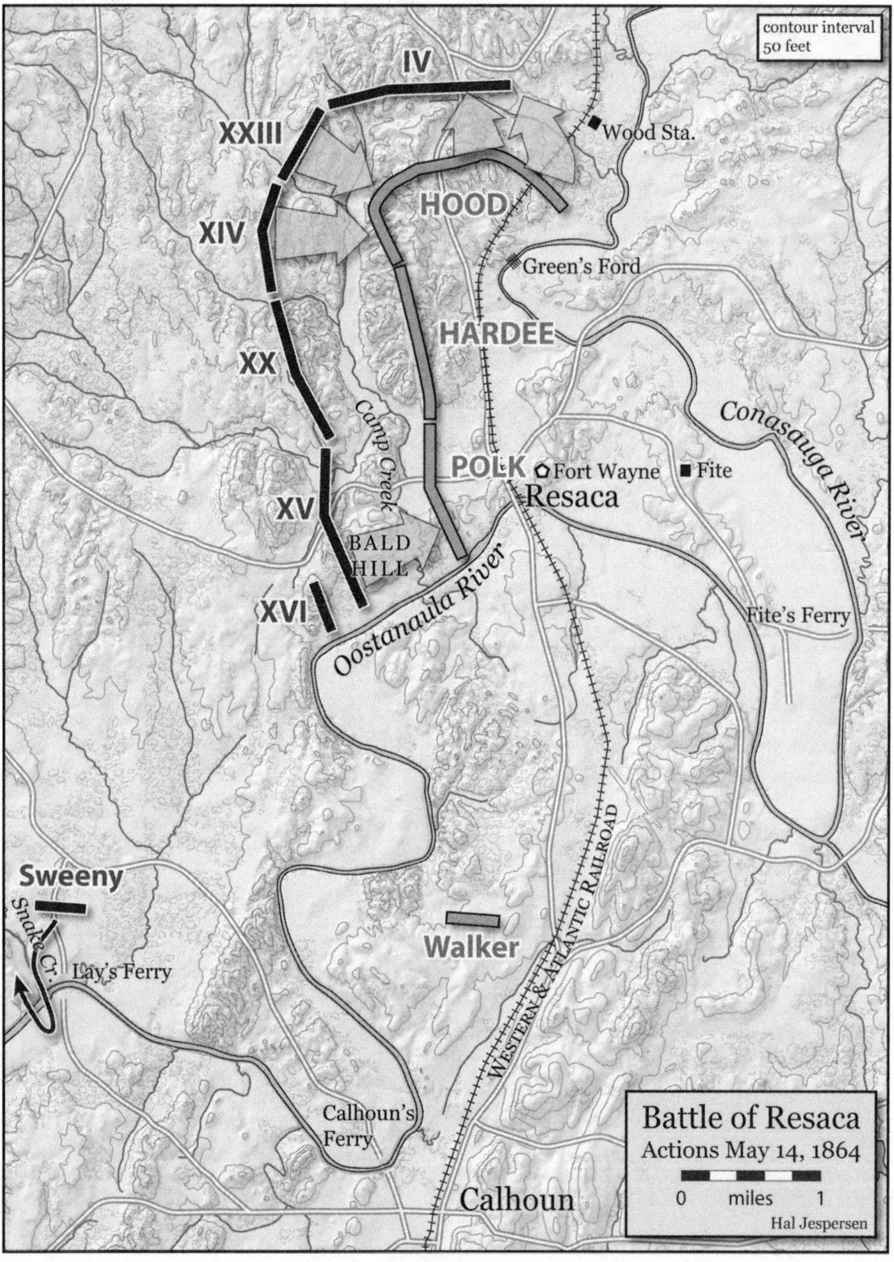

BATTLE OF RESACA

May 14, 1864

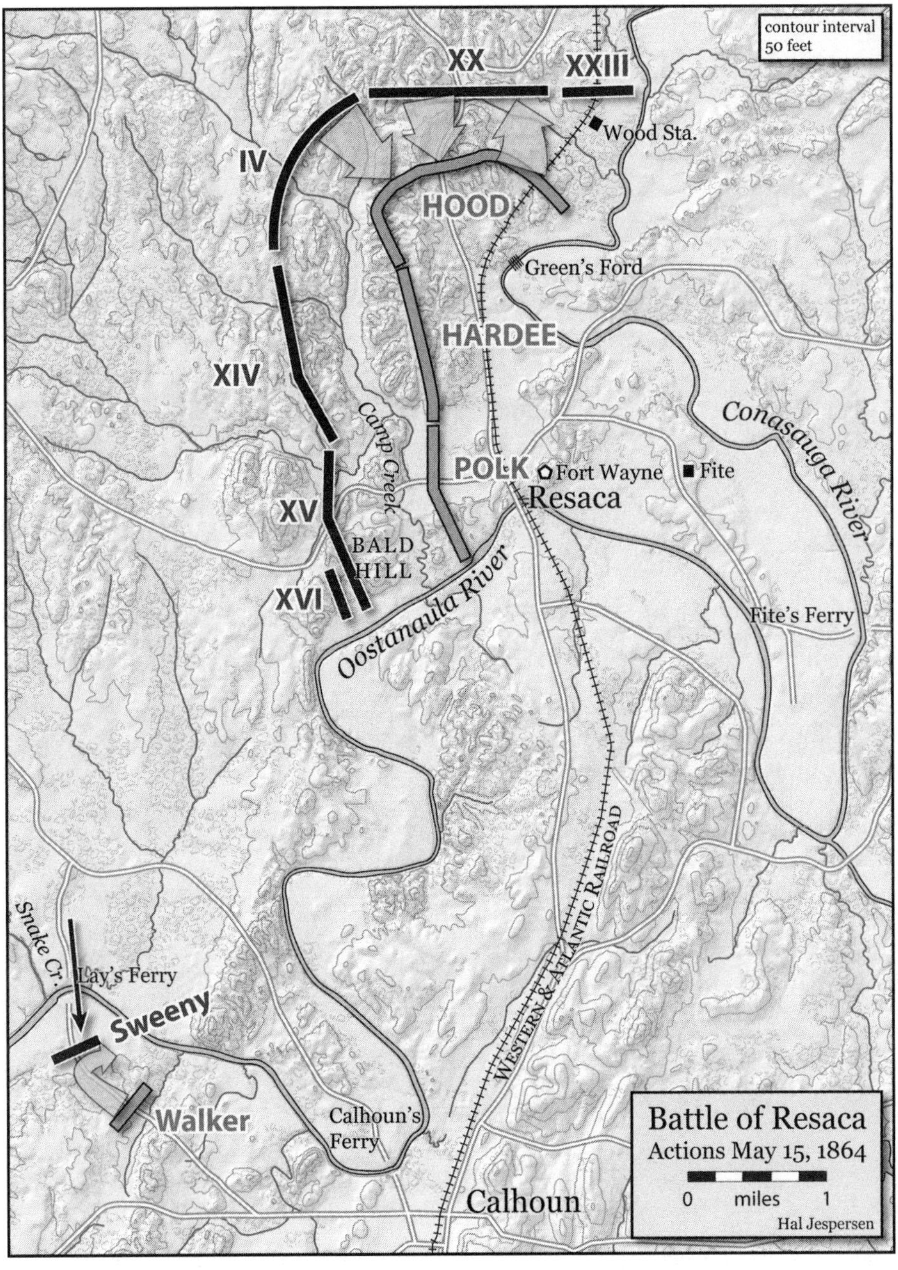

BATTLE OF RESACA

May 15, 1864

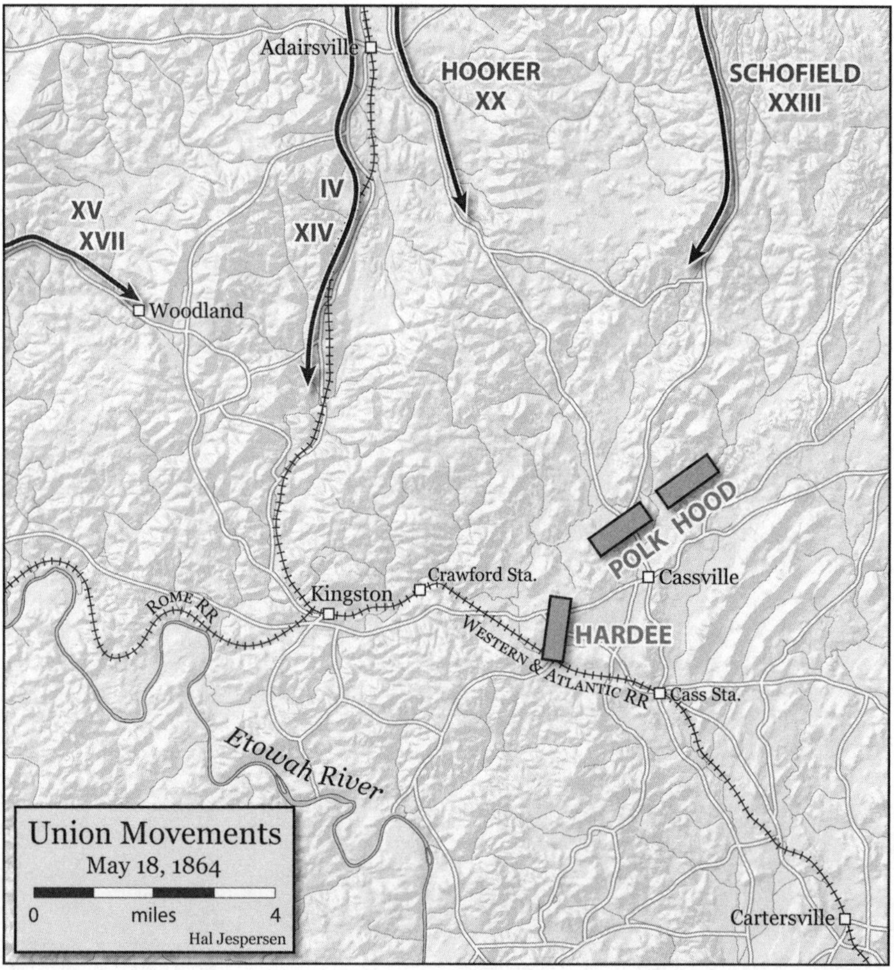

ATLANTA CAMPAIGN

Union Movements, May 18, 1864

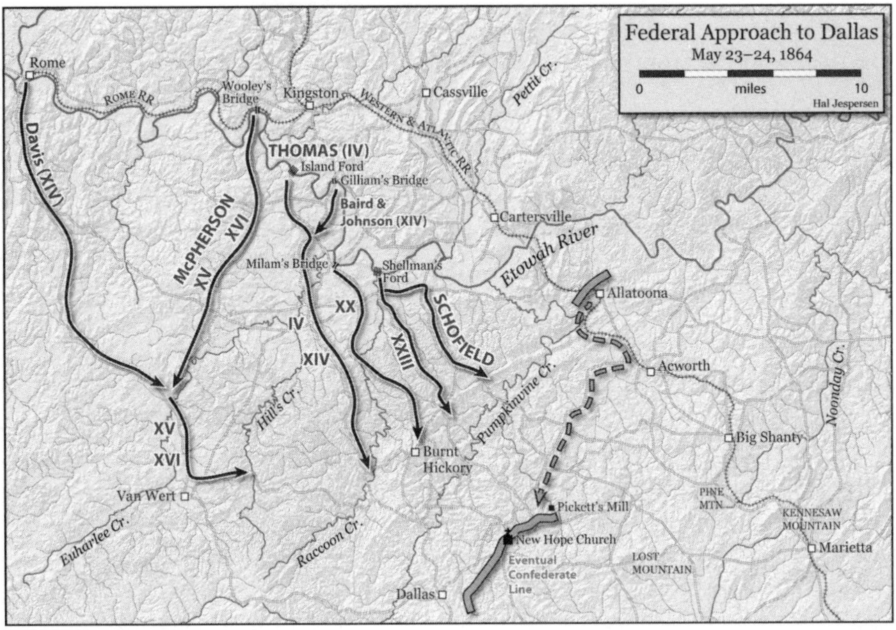

ATLANTA CAMPAIGN

Federal Approach to Dallas, May 23–24, 1864

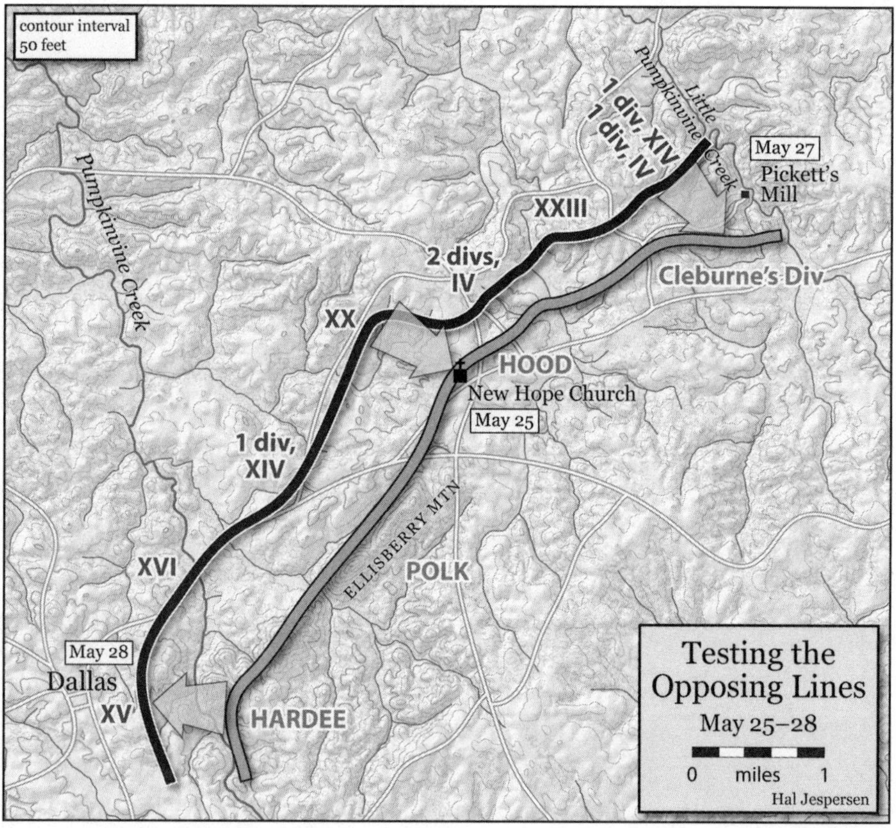

ATLANTA CAMPAIGN

Testing the Opposing Lines, May 25–28, 1864

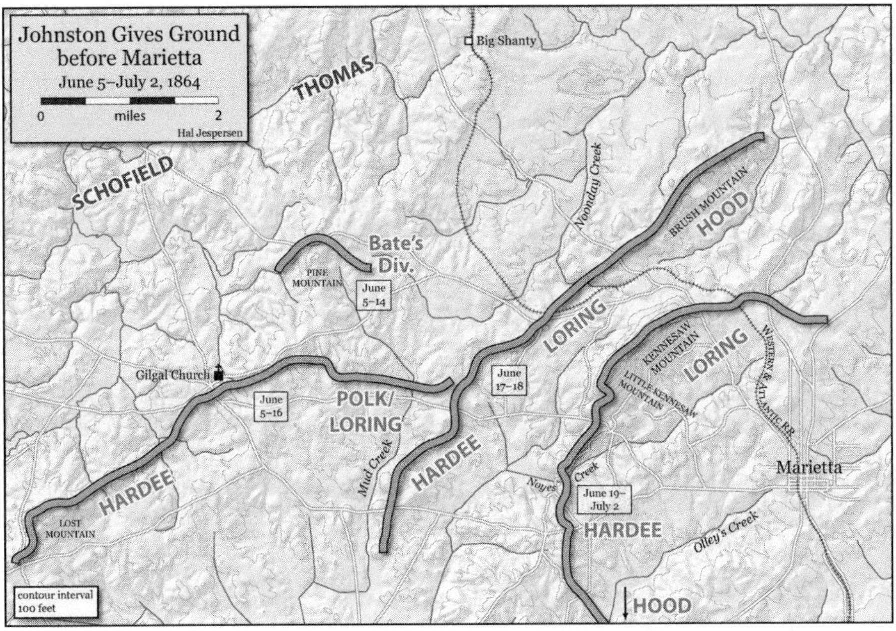

JOHNSTON GIVES GROUND BEFORE MARIETTA

June – July, 1864

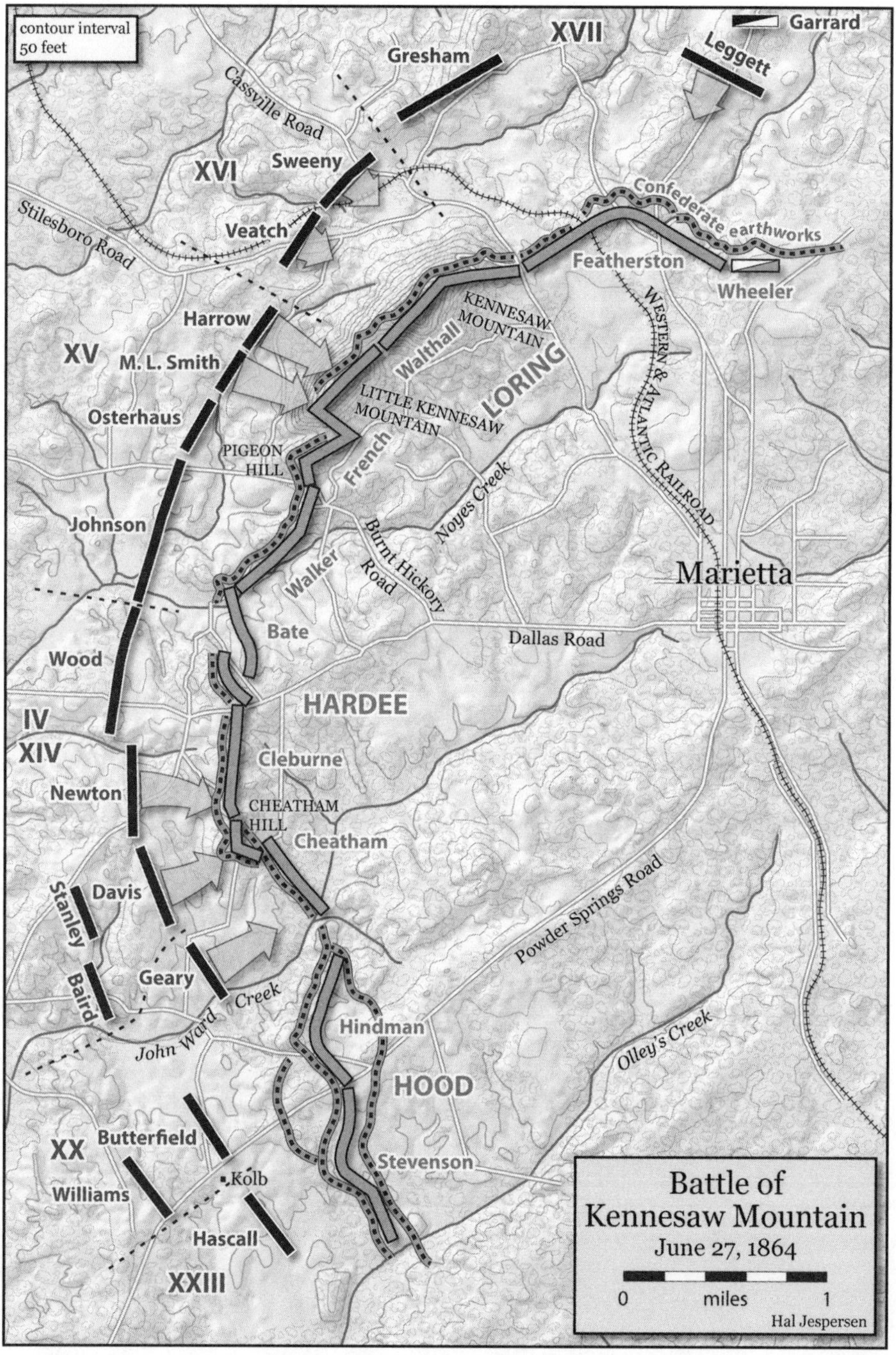

BATTLE OF KENNESAW MOUNTAIN

June 27, 1864

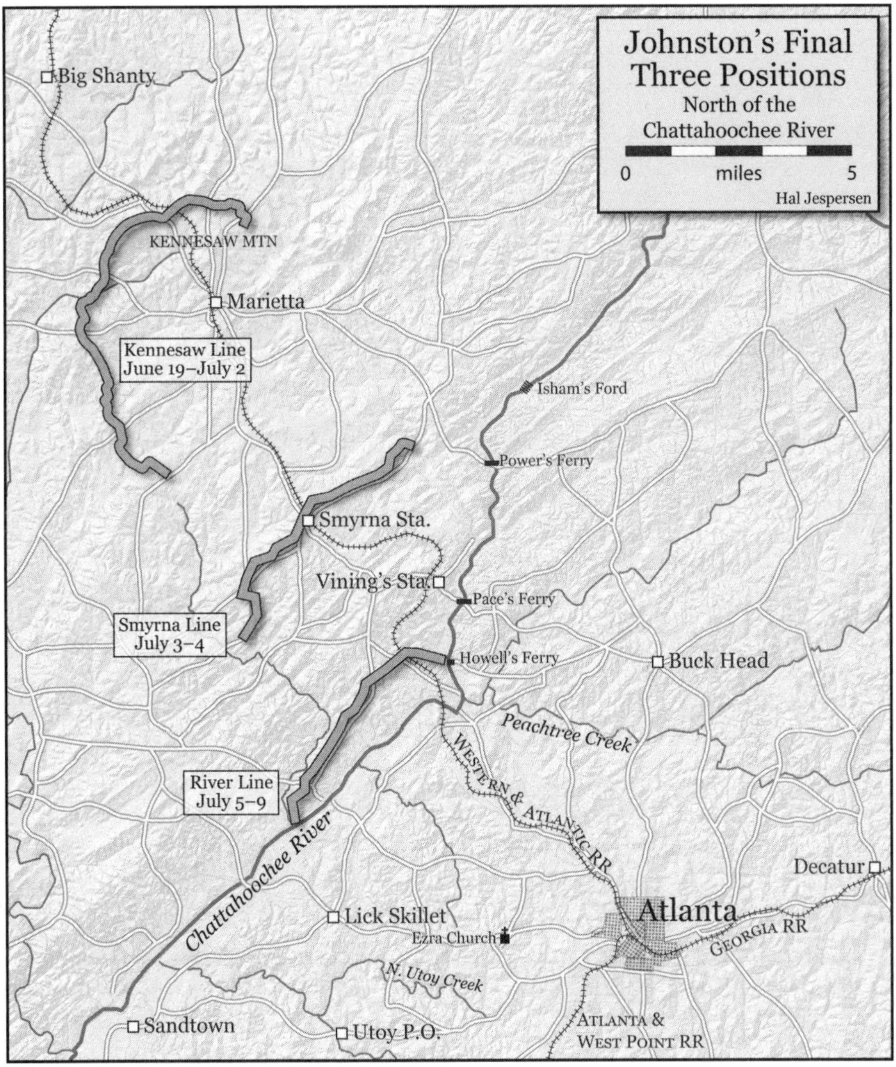

JOHNSTON'S FINAL THREE POSITIONS

North of the Chattahoochee River: June – July, 1864

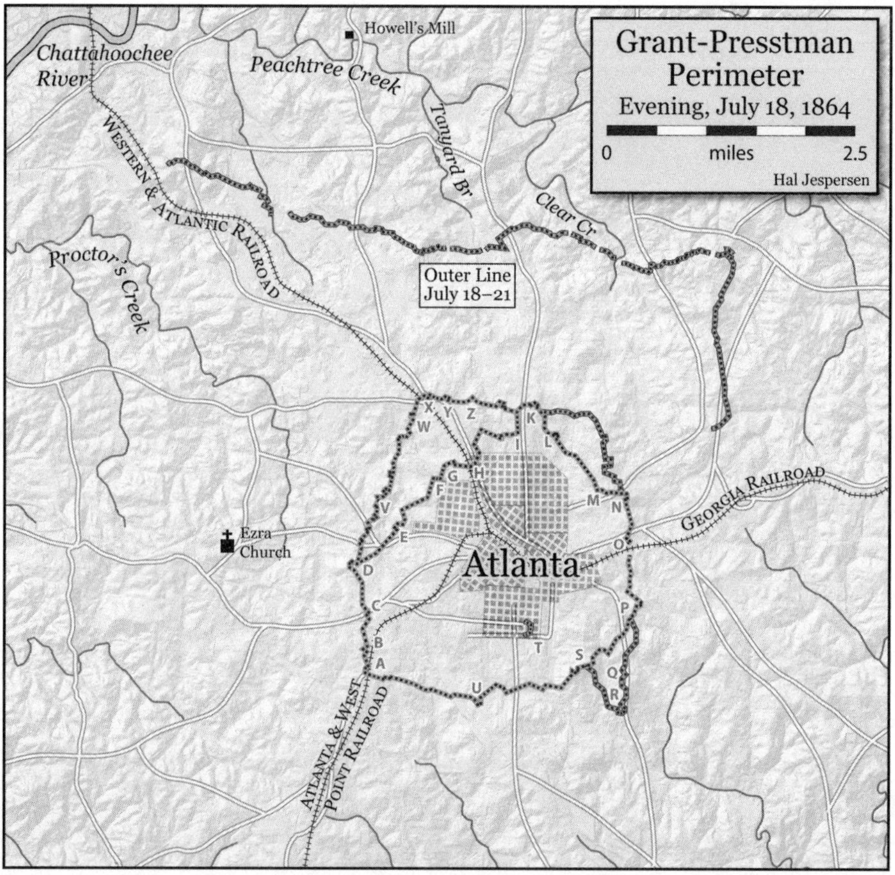

GRANT-PRESSTMAN PERIMETER

July 18, 1864

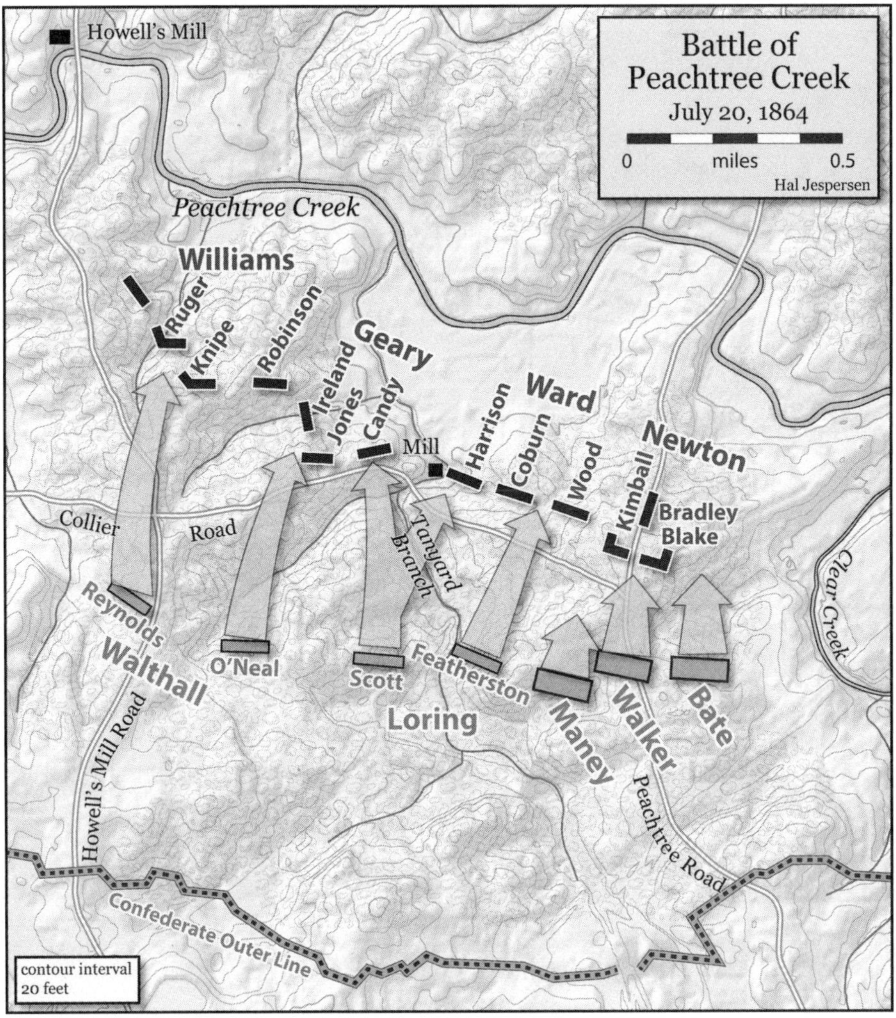

BATTLE OF PEACHTREE CREEK

July 20, 1864

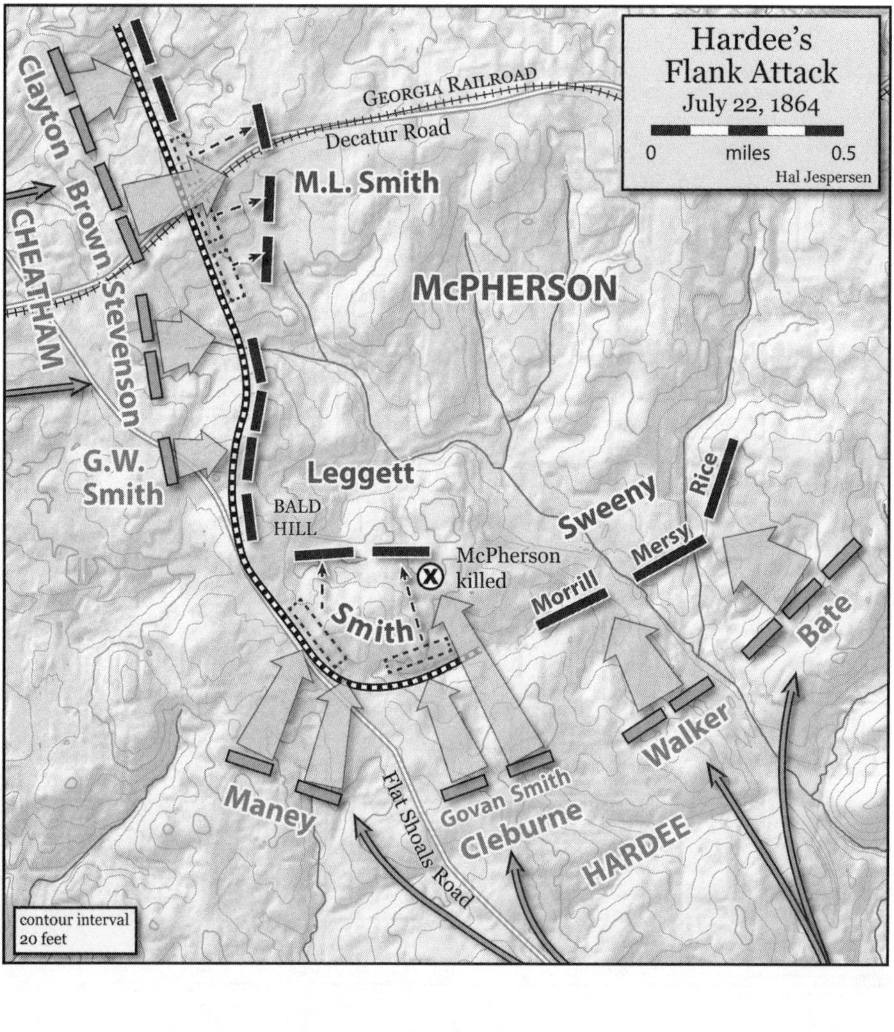

HARDEE'S FLANK ATTACK

July 22, 1864

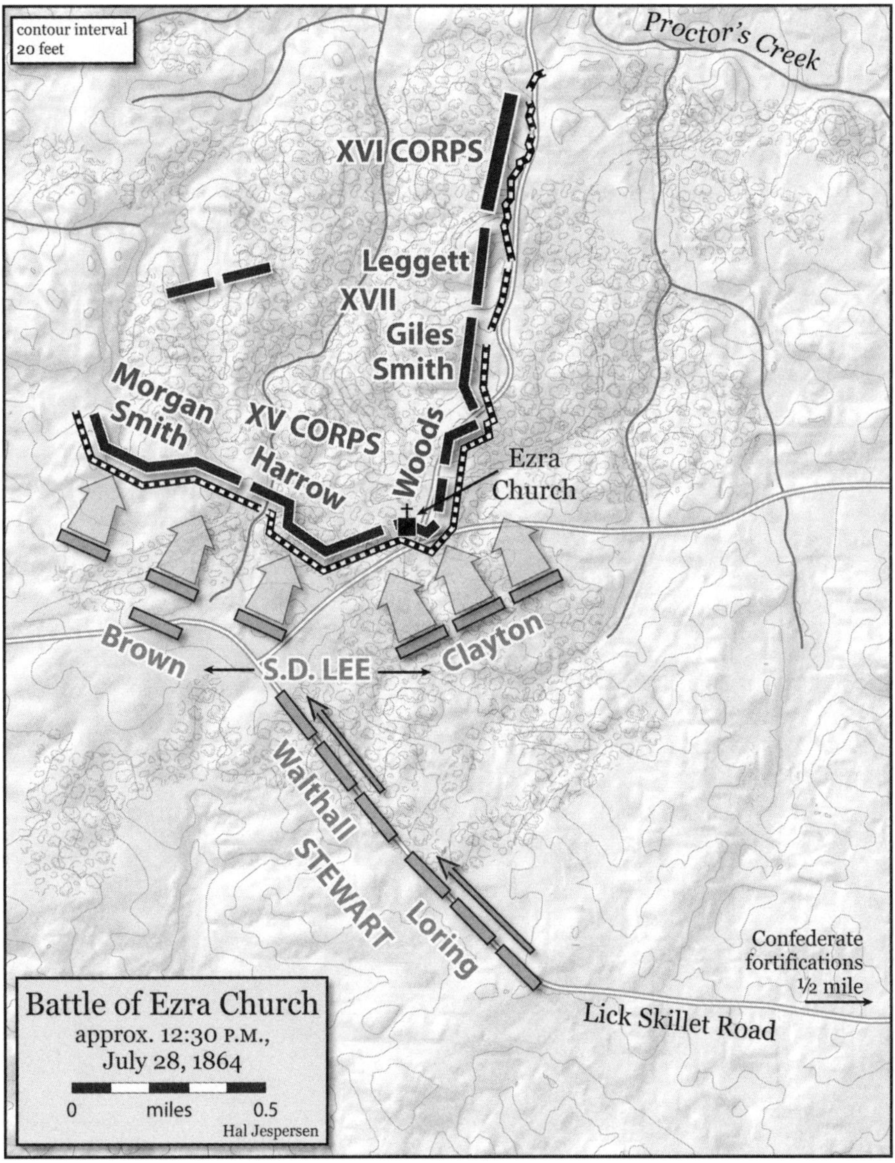

BATTLE OF EZRA CHURCH

July 28, 1864

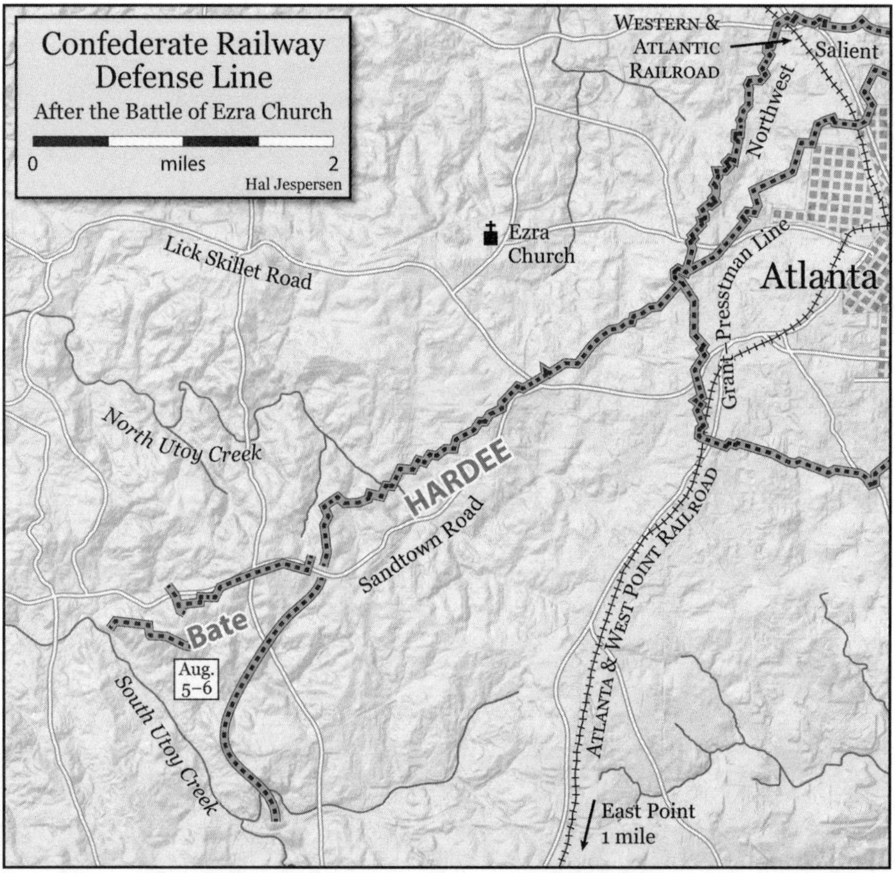

AFTER THE BATTLE OF EZRA CHURCH

Confederate Railway Defense Line

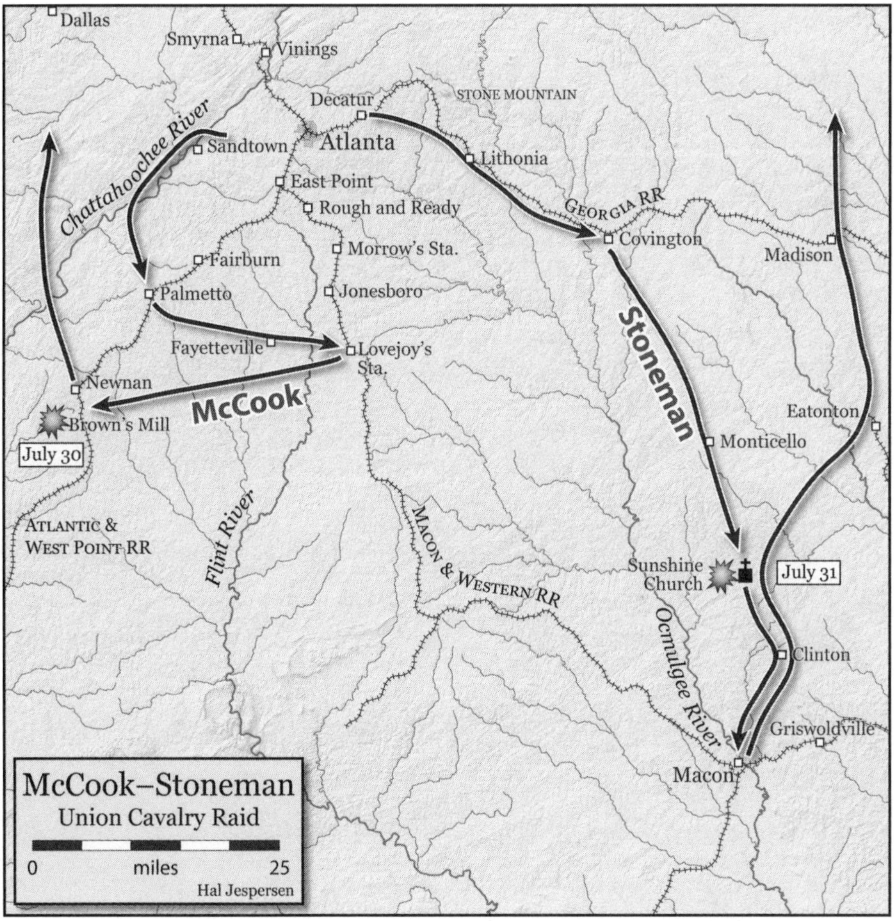

UNION CAVALRY RAID

McCook & Stoneman

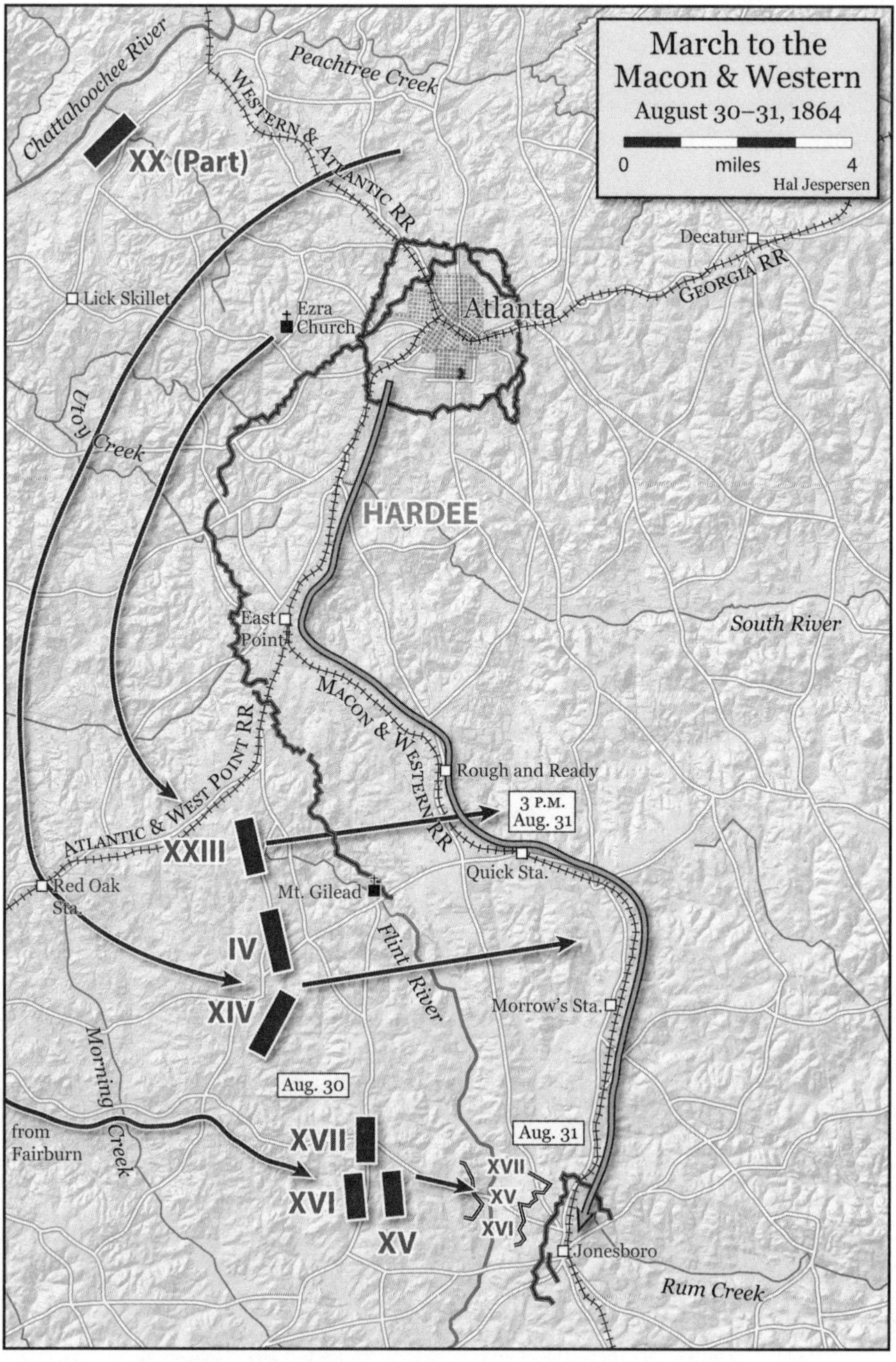

MARCH TO THE MACON & WESTERN

August 30–31, 1864

Part 2

Hood as Army Commander at Atlanta, July 18–September 2, 1864

Chapter 6

July 18–31

So John Bell Hood, who had graduated from the service academy at the age of twenty-two and entered the army of the Confederate States as first lieutenant at twenty-nine, had at the start of his thirty-third year become the youngest of the eight full generals rising to that rank in the Confederate army.[1]

When Hood on July 18, 1864, sent the telegram to President Davis, "I have assumed command of the Army and Department of Tennessee," his ambition—that lust for fame and popular adulation which had long marked his military career—now reached its zenith.

Hood's headquarters issued to the troops a special announcement on the occasion.

> SOLDIERS: In obedience to orders from the War Department I assume command of this army and department. I feel the weight of the responsibility so suddenly and unexpectedly devolved upon me by this position, and shall bend all my energies and employ all my skill to meet its requirements. I look with confidence to your patriotism to stand by me, and rely upon your prowess to wrest your country from the grasp of the invader, entitling yourselves to the proud distinction of being called the deliverers of an oppressed people.[2]

Moreover, Hood was taking command of the largest army the Southern nation had in the field at that time. Robert E. Lee's Army of Northern Vir-

---

[1] McMurry, *John Bell Hood*, ix, 123. The others were Cooper, A. S. Johnston, Lee, J. E. Johnston, Beauregard, Bragg, and Kirby Smith (M. J. Wright, *General Officers*, 9–13).

[2] Hood to "His Ex^y The President," July 18, in Hood, Telegrams Book I (identical to Gen. Cooper is in *OR*, vol. 38, pt. 5, 889); proclamation, July 18, *OR*, vol. 38, pt. 5, 889; Hood to Brig. Gen. William H. Jackson, July 18, Hood Service Record (with same text as announcement to the army). Hood's note to Jackson suggests he probably sent a personal copy of the announcement to his senior officers.

ginia, once the Confederacy's biggest, counted 52,225 officers and men present as of June 30 (its Second Corps, under Jubal Early, was detached in service to the Valley, not soon to return). In spring 1864, the Confederate States army had eight other departments, each of which had troops assigned to and positioned within it, ranging from West Virginia to the Trans-Mississippi. Largest of these forces was the Department and Army of Tennessee. At the end of April, just before the start of the campaign, it numbered 52,216 officers and men present for duty. Then came all the reinforcements sent to Johnston during May when the War Department "drained" the southwest and transferred coastal garrisons. By one authoritative chronicler, these additional troops totaled 24,646, although most authorities settle on around 20,000. Most of them had come with Bishop Polk, who wired Johnston on May 5, "I shall bring you about 10,000 infantry and 4,000 cavalry." This brought the army's strength to 76,862 by the end of May. (Characteristically, Sherman knew his enemy's numbers, as he wrote his wife on June 12, "Johnston has 60,000 Infantry, 15,000 Cavalry and a good deal of militia.") But by then had come the casualties. Five weeks into the campaign, on June 13, Lieutenant Mackall recorded, "Cas since May 7—kill & w in infy 5700 in cavy about 800." Dr. Andrew Jackson Foard, medical director for the Army of Tennessee, tabulated killed and wounded in the first weeks, May 7–20, as 3,384; then from the Etowah to Lost Mountain, May 21 to about June 4, another 2,005. But this was for infantry only; Wheeler reported losses in his cavalry corps, May 6–31 as 73 killed, 341 wounded, 53 captured, and 81 missing for a total of 548. Doctor Foard also counted 4,583 infantrymen killed and wounded from Lost Mountain to the Chattahoochee (to July 9), for a total loss of 9,972 in the first two months of the campaign. A year after the war, apparently at Johnston's request, Foard went through his papers and compiled another tally of infantrymen killed and wounded May 7–July 4 and came up with a slightly lower figure, 9,450. Johnston published the report in his memoirs. Most interestingly, given their bitter disagreements about the Georgia Campaign, so did Hood in his.[3]

[3] Ecelbarger, *The Day Dixie Died*, 25 ("largest Confederate army on the continent"); "Abstract from field return of the Army of Northern Virginia, General Robert E. Lee commanding, for June 30, 1864," *OR*, vol. 40, pt. 2, 707; Newton, *Lost for the Cause*, 185–326 *passim*; "Abstract from returns of the Army of Tennessee," April 30, 1864, *OR*, vol. 38, pt. 3, 676; Dawes, "The Confederate Strength in the Atlanta Campaign," 4:281 (24,646); Polk to Johnston, May 6, 1864, *OR*, vol. 38, pt. 4, 670; Chapman, "Comparing Two Great Armies," 101 (21,600 reinforcements); Fullerton, *Armies in Gray*, 699–701 (21,477 in Polk's three infantry divisions and Jackson's cavalry division); Sherman, *Home Letters of General Sherman*, 296 (Sherman to Ellen, June 12); McMurry, "Mackall Journal," entry of June 13; Foard, "Casualties

The army returns for July 10 also reported the men who had been captured and were AWOL: 6,994. Johnston evidently had this report when he was writing his reminiscences, as he was able to break this number down: 2,159 in Hardee's corps and 4,835 in Hood's. POWs in Stewart's corps are not given but would have added at least a couple thousand more. The total of men missing is thus even greater than that recorded by the Federals, who usually liked to inflate such figures. Union Col. J. G. Parkhurst, provost marshal general, reported for the month of May 1,829 Rebel officers and men captured and another 610 received into the lines as deserters. Another Federal officer stated that 1,308 Rebels were taken prisoner during June and 531 more as deserters, figures McMurry accepts.

Despite the fact that in Confederate army hospitals in summer 1864 there were about twice as many soldiers sick as there were wounded, figures for the number of men away from the ranks because of disease are hard to verify. In late May Dr. Samuel H. Stout, hospital director for the Army of Tennessee, called for expansion of facilities in Atlanta to treat the sick and wounded. Especially during the rainy month of June, hundreds of men were sickened by pneumonia, rheumatism, and other ailments. Diarrhea, dysentery, and malaria also hospitalized many soldiers. On June 4 Dr. Stout reported that fully 18,000 men were in hospitals under his direction. Johnston claimed to Bragg on June 26 that "we are losing from the ranks by sickness some 300 men a day since the heavy rains." Fortunately, only 2 or 3 percent of Confederates who entered the hospital died there, but those sick and convalescing were still away from the ranks. By these calculations—10,000 *hors de combat*; at least 7,000 missing; and an uncountable several thousand soldiers in hospital—one comes to terms with Johnston's return of July 10 showing 59,196 officers and men present for duty. Just days before, General Bragg had observed that the army's strength was reduced by 10,000 since the returns of June 10, which showed 69,946 present.

Such was the strength of the Army of Tennessee—about 59,000— when Hood took charge on July 18. Besides its strength, the new commander had to tend to administrative matters in the army as well. Hood's old corps needed

---

in the Confederate forces May 6–July 9," *OR*, vol. 38, pt. 3, 686–87; "Report of casualties in Wheeler's cavalry corps from May 6 to 31, 1864," *OR*, vol. 38, pt. 3, 949; Foard, "Consolidated Summaries in the Armies of Tennessee and Mississippi…furnished for the Information of General Joseph E. Johnston," in J. E. Johnston, *Narrative*, 576–77, and Hood, *A & R*, 218–20; J. Jones, "The Medical History of the Confederate States Army and Navy," 131 (accepting Foard's figures); Stevenson report, August 19, 1864, *OR*, vol. 38, pt. 3, 815 (689 men lost from June 22 to July 20 "in skirmishing, &c." as opposed to big battles, the usual causes of an army's casualties).

a leader now, and for the moment Carter Stevenson, senior major general in it, rose to the post. But Hood was not satisfied with this and on Hardee's recommendation named Maj. Gen. Frank Cheatham as temporary commander, "although he did not desire it," as Hood told Richmond. Instead, Hood asked for Maj. Gen. Mansfield Lovell (idle since having lost New Orleans), Lt. Gen. Stephen D. Lee (commanding the Department of Alabama, Mississippi and East Louisiana after Polk), or—surprisingly—Maj. Gen. Wade Hampton, his old friend from their Virginia days together, but now in charge of Lee's cavalry. Cheatham would serve about two weeks until his eventual replacement, General Lee, arrived in Atlanta from Alabama on July 26.[4]

Hood knew that he had to assume the offensive if he were to blunt the enemy's advance on Atlanta. What he did not know was how the enemy planned to capture the city. Sherman would not assault the Confederate fortifications, which through spy reports he knew to be strong. "I was willing to meet the enemy in the open country," he later wrote, "but not behind well-constructed parapets." Nor would he seek to surround the city's ten- to twelve-mile perimeter of entrenchments in a siege, as he had insufficient numerical strength to do so. What ensued was a "semi-siege," with Sherman's armies digging in around large parts of Atlanta's defenses, skirmishing with and shelling the Rebels, but never completely encircling their works. Instead, he

---

[4] "Abstract from returns of the Army of Tennessee...July 10, 1864," *OR*, vol. 38, pt. 3, 679; J. E. Johnston, *Narrative*, 575; Parkhurst to Brig. Gen. William D. Whipple, June 4 and Lt. H. M. Duffield report, *OR*, vol. 38, pt. 1, 147, 153; McMurry, "Atlanta Campaign" dissertation, 308; McMurry, "Confederate Morale," 229; Daniel, *Soldiering*, 78; Eastwood, "Confederate Medical Problems in the Atlanta Campaign," 285; S. L. Clark and Riley, "Outline of the Organization of the Medical Department of the Confederate Army and Department of Tennessee," 81; Johnston to Bragg, June 26, *OR*, vol. 38, pt. 4, 792; Breeden, "Medical History of the Later Stages of the Atlanta Campaign," 56; "Abstract from returns of the Army of Tennessee," July 10, *OR*, vol. 38, pt. 5, 679; McMurry, *Atlanta 1864*, 143, 195 (explaining Confederates' "present for duty" and "effective strength," and stating that the former number is better to use); Bragg to Cooper, July 13, *OR*, vol. 38, pt. 5, 878; General Orders 57, July 18, *OR*, vol. 38, pt. 5, 892 (Stevenson); Hood to Seddon, July 19, *OR*, vol. 38, pt. 5, 892 (Cheatham, Lovell, Hampton, and Lee); Losson, *Tennessee's Forgotten Warriors*, 173; Hattaway, *General Stephen D. Lee*, 126 (giving July 25 as Lee's date of arrival at Atlanta); Bragg to Davis, July 27, 1864 (from Atlanta), *OR*, vol. 52, pt. 2, 713 (giving date of Lee's arrival as July 26, which I accept).

Lee received his order to report to Hood on July 22, when he was in Montgomery (*OR*, vol. 52, pt. 2, 711–12). He assumed command on the 26th (52, 2, 713).

Edward Longacre claims that after assuming army command, "Hood wanted [Maj. Gen. Wade] Hampton to command the cavalry of his army with the rank of lieutenant general," but offers no documentation for his statement (*Gentleman and Soldier*, 205).

would pursue what J. Britt McCarley terms a "logistic strategy": he would swing his armies around Atlanta to cut the three railroads which ran into it from the east (Georgia Railroad, to Augusta), south (Macon & Western, to Macon), and southwest (Atlanta & West Point to West Point, Georgia, and from there to Montgomery).

Sherman's first tactical objective was to cut the railroad to Augusta. On July 14 he laid out his plan for his three armies' movements. Once across the Chattahoochee, Thomas would advance southward toward Buckhead, Stoneman's cavalry to his right. McPherson would swing wide to the southeast, aiming to strike the Georgia Railroad at Stone Mountain; Garrard's horsemen would guard his flank. Schofield would march in between. Sherman did not expect the enemy to sally forth against him, at least not for the next few days. A Rebel advance, he told his generals, was unlikely "until we reach below Peach Tree Creek." Sherman remained supremely confident. After General Halleck in Washington telegraphed him that Johnston might soon receive reinforcements, Sherman shrugged off the warning. "I do not fear Johnston with re-enforcements of 20,000 if he will take the offensive," he replied.

July 17 was the first day of active movement by Sherman's forces in a week. By nightfall that day the XVII Corps had only crossed the river at Roswell, but the XIV and XX Corps of Thomas's army had reached Nancy Creek, which flows into Peachtree Creek a mile from the Chattahoochee. The IV Corps was behind them, east of Power's Ferry. The XV, XVI, and XXIII Corps were also to Nancy Creek well upstream; the XVI was actually across it. Confederate cavalrymen contested the Yankees' advance as best they could. On the 18th, the Kentucky brigade of Brigadier John S. Williams and two guns fired at the Federals until forced back. Southern infantry north of Peachtree Creek also had to withdraw; Stewart had at least one brigade of Loring's division in the angle between the river and creek which had to fall back before the enemy advance. Sherman's columns kept moving on the 18th, with Thomas's infantry marching toward Buckhead.

At the Niles house, which, after taking over from Johnston, Hood kept as his headquarters, the new army commander worked hard to better grasp the tactical situation: the harsh reality was that enemy forces were now five miles from Atlanta. Worse, on July 18 McPherson's infantry and Garrard's cavalry struck the Georgia Railroad near Stone Mountain, thirteen miles east of Atlanta. One of the Southerners' three railroads supplying them in Atlanta had been cut. That night Sherman could report that Garrard's cavalry and Morgan Smith's infantry had wrecked five miles of track between Stone Mountain and Decatur.

On the night of July 17, Johnston and Lieutenant Colonel Presstman had been discussing a line of fortifications south of Peachtree Creek. Fortifying began on the 18th, Johnston's last day in command, when the Confederate army took up new positions. Douglas West, A. A. G. for Stewart's corps, issued orders that the troops "will without delay be placed in line of battle on the line adopted by the engineers.... The engineer officers will indicate the line to division commanders, and also the positions selected for artillery." Just after noon, Col. Thomas B. Roy, Hardee's A. A. G., instructed Cleburne to "put your division in motion at once." In his diary Major General French wrote, "18th. I moved into a position where my left rested on the Marietta road, and commenced intrenching at night." Presstman's "outer line" extended west-east from near the Chattahoochee and across the W. & A. almost six miles before turning south for another mile and three-quarters. When the Confederate infantry occupied it, the three corps deployed in the usual manner: Stewart on the left, Hardee in the center, and Hood/Stevenson on the right. All four of Hardee's division commanders were to throw a hundred pickets out toward the creek to watch for the enemy's possible approach. Wheeler was out in front; around noon General Mackall instructed him to burn every bridge over Peachtree Creek. That night Hood wired Richmond, "The enemy advanced to-day...and established his line on Peach Tree Creek."

Sherman knew that the Rebels were busy. "The enemy are at work on their forts and entrenching around Atlanta," he wrote McPherson on the 18th. General Dodge sent information brought in by a "scout" who had just left Atlanta. "Says the enemy were moving troops all night last night there." In one of the Rebels' abandoned camps, someone found a document that gave round figures for every infantry brigade, division, and corps in Hood's army. It thus contained more detail than Hood's own official abstracts. Moreover, it was surprisingly accurate, conforming roughly to the numbers of "Present for duty" in Hood's return of July 10. It appears in the *Official Records*, from the records of the Army of the Cumberland:

HARDEE'S CORPS	
First Division, Cheatham's:	
Maney's brigade	1,200
Strahl's brigade	1,000
Wright's brigade	1,400
Vaughan's brigade	1,200
	4,800

Second Division, Cleburne's	
Polk's brigade	1,000
Govan's brigade	1,200
Lowrey's brigade	1,500
Granbury's brigade	1,200
	4,900
Third Division, Walker's	
Mercer's brigade	2,000
Stevens's brigade	1,500
Jackson's brigade	1,000
Gist's brigade	1,200
	5,700
Fourth Division, Bate's	
Lewis's brigade	700
Finley's brigade	1,000
Tyler's brigade	1,300
	3,000
	18,400

HOOD'S CORPS

First Division, Hindman's	
Deas's brigade	1,200
Manigault's brigade	1,200
Tucker's brigade	1,000
Walthall's brigade	1,200
	4,600
Second Division, Stevenson's	
Cumming's brigade	1,500
Brown's brigade	800
Baker's brigade	1,000
Reynolds's brigade	1,200
	4,500
Third Division, Stewart's	
Gibson's brigade	800
Stovall's brigade	1,200
Clayton's brigade	1,500
Moore's brigade	1,000
	4,500
	13,600

STEWART'S CORPS	
First Division, Loring's	
Featherston's brigade	1,500
Adams's brigade	1,500
Scott's brigade	1,200
	4,200
Second Division, French's	
Cockrell's brigade	2,000
McNair's brigade	1,000
Sears's brigade	1,200
	4,200
Third Division, Walthall's	
Cantey's brigade	2,000
Ector's brigade	1,000
Quarles's brigade	1,000
	4,000
	12,400
Hardee	18,400
Hood	13,600
Stewart	12,400
	44,400[5]

[5] S. Davis, *What the Yankees Did to Us*, f.p. xxii (Kurtz map of 1938 showing entrenched line), 86 (Milton Glass's spy report on fortifications); Sherman, "Grand Strategy," 253; McCarley, "'The Great Question of the Campaign Was One of Supplies,'" 38 ("logistic strategy"); 60 n.21 ("siege in name only"); Federal SFO 35, July 14, *OR*, vol. 38, pt. 5, 142–43; Halleck to Sherman, July 16, 4:30 P.M. and Sherman to Halleck, July 16, 11 A.M., *OR*, vol. 38, pt. 5, 150; Kurtz map, "Military Operations June 22–July 17, 1864," Garrett, *Atlanta and Environs*, 1:598, 604 ("Colonel Presstman, the chief engineer, was at headquarters, having been summoned by Johnston to receive instructions about the new line of fortification to be erected the next day"); Elliott, *Soldier of Tennessee*, 200; Kurtz, "Williams' Ky. Cavalry at Nancy's Creek" (July 18), handwritten marker text, box 35, folder 1 (Scruggs, *Georgia Historical Markers*, 208); Logan report, September 13, 1864, *OR*, vol. 38, pt. 3, 101; Garrard report, September 4, *OR*, vol. 38, pt. 2, 802; McMurry, *John Bell Hood*, 126; Sherman to Halleck, July 18 (7 P.M.), *OR*, vol. 38, pt. 5, 169; Kurtz, "At the Dexter Niles House," 20 ("Orders had been issued prior to the 18th for the three corps to take position north and northeast of Atlanta, and this morning they began to move to their assigned oppositions"); West, Gen. Orders No. 4, July 18, *OR*, vol. 38, pt. 5, 891; Roy to Cleburne and Roy circular, July 18, *OR*, vol. 38, pt. 5, 890, 892; S. G. French, *Two Wars*, 216–17; Scruggs, *Georgia Historical Markers*, 217–18 ("Outer Defense Line" ["News of Federal crossings of the Chattahoochee July 17, 1864, prompted the building of a defense line N & E. of Atlanta"] Kurtz, historical marker texts, box 35, folder 1); Mackall

As if he did not have enough to handle already, Hood learned on July 18 that Union cavalry raiding through east central Alabama had cut the railroad connecting Atlanta with Montgomery. "Last night R.R. between WPt & Montg cut," was Lieutenant Mackall's terse journal entry. Indeed, cutting the railroad into Alabama had been part of Sherman's campaign plan since he laid it out to Grant in April, although he believed he would not order the strike until Johnston had crossed the Chattahoochee. But when Maj. Gen. Lovell Rousseau in mid-June proposed a cavalry raid into Alabama, Sherman eventually agreed, stipulating that once Rousseau's column reached the Montgomery railway, it should go to work "breaking up the road and twisting the bars of iron." Rousseau and 2700 horsemen rode out of Decatur, Alabama, on July 10. One week and 210 miles later, the raiders struck the Montgomery & West Point Railroad west of Auburn. Over the next several days, the Federals wrecked 26 miles of track before setting off to the northeast to unite with Sherman's forces at Marietta. Southerners knew of the raid but had no troops to chase the enemy. "There is no cavalry here to pursue them.... The situation is most unsatisfactory," General Bragg telegraphed Richmond from Montgomery July 18. After Rousseau rode away, the Confederates were eventually able to repair the M. & W. P., but not until August 26 (even Hood in late July sent an officer looking for iron). To be sure, wagon traffic across the wrecked length brought some supplies from Montgomery to Atlanta, but for the rest of the campaign, the rail line essentially ceased to function as a major railway supply route for Hood's army. "Good for Rousseau!" was Sherman's delighted response upon hearing the news.[6]

---

to Wheeler, July 18 (11:45), *OR*, vol. 38, pt. 5, 889–90; Hood to Seddon, July 18, *OR*, vol. 38, pt. 5, 889; Sherman to McPherson, July 18, *OR*, vol. 38, pt. 5, 177; Dodge to McPherson, July 18, *OR*, vol. 38, pt. 5, 177; "Estimated strength of Hood's army," c. July 18, *OR*, vol. 38, pt. 5, 178–79.

[6] McMurry, "Mackall Journal," entry of July 18; Sherman to Grant, Apr. 10, 1864, *OR*, vol. 32, pt. 3, 313 ("break up the road between Montgomery and Georgia"); D. Evans, *Sherman's Horsemen*, 29, 31, 99 (map), 121 (map), 166 (26 miles of track destroyed), 476 (wagon traffic); Bragg to Sale, July 18, *OR*, vol. 38, pt. 5, 888; [Hood] to Maj. J. M. Hottel, Opelika, July 28; [Hood] to "Commanding Officer Opelika," August 22 (5:10 P.M., "How much of the Montgomery road remains yet unfinished?") in Hood, Telegram Book III; Hood to Bragg, August 1 (12 m.), *OR*, vol. 38, pt. 5, 937 ("I have done all in my power to send iron"); Sherman to Thomas, July 19, *OR*, vol. 38, pt. 5, 185 ("Good for Rousseau!"); McMurry, "Riding through Alabama, Part I," 16 (map, "The route of Rousseau's raiders"); McMurry, "Riding through Alabama, Part II," 40 ("Not until late August did the Southerners have the Montgomery & West Point back in operation").

After Rousseau's break of the Montgomery line, Confederate officials worried how to repair or replace it. General Bragg suggested building an alternate rail route: laying a line from

Having reached Peachtree Creek on July 18, Sherman's columns advanced farther the next day, when Thomas's infantry began to cross it. Meanwhile, creating a front of his own, McPherson marched toward Decatur, a town six miles east of Atlanta. On the 19th, troops of Logan's XV Corps reached the Georgia Railroad at Decatur and started tearing it up. Sherman's orders were quite specific as to how it should be done. "Officers should be instructed that bars simply bent may be used again, but if when red hot they are twisted out of line they cannot be used again," read SFO 37, dated July 18. "Pile the ties into shape for a bonfire, put the rails across, and when red hot in the middle, let a man at each end twist the bar so that its surface become spiral." Farther to the east Garrard's cavalry also tore up track. As result, after July 19, the link to Augusta was knocked out and the Confederate army in Atlanta was down to one serviceable railroad, that to Macon. But Hood's immediate worry was the advance of McPherson's army. Confederate cavalry contested the Yankees' march, yet Wheeler, with 2,500 men, could not hold back McPherson's 25,000. The enemy advance east of the city worried Hood's headquarters because of its speed. "Send your best scouts close in," Wheeler

---

Montgomery to Union Springs, Alabama, which was connected by a railway to Columbus, Georgia, and then to Macon. He estimated that the length of track to be laid would be about forty miles; rails could be taken, he suggested, "from unimportant roads" in the area. Hood, apparently at Bragg's request, added his voice in support of this proposal in a telegram to Seddon on July 25. But scarcity of iron, as always, plagued the Confederacy. For this reason, Major General Gilmer of the Engineer Bureau, to whom the matter was referred, determined that the Montgomery-West Point line should be repaired, as it would involve less track to be used. Gilmer accordingly dispatched an officer to oversee the work, and ordered Captain Grant in Atlanta to assist. Eventually Maj. George Whitfield of the Railroad Bureau began reconstruction of the Montgomery & West Point in mid-August. The work was completed in less than three weeks, but by then Hood had abandoned Atlanta and the campaign was over (Black, *Railroads of the Confederacy*, 251–52; Bragg to Col. John B. Sale, July 23, Hood to Seddon, July 25 and Gilmer endorsements, July 26, *OR*, vol. 38, pt. 5, 904–905, 908).

General Hood paid close attention to the repair of the railroad to Montgomery. In early August, he and Shoup were wiring Capt. J. C. Shoup in Opelika both about finding material ("cannot use the iron from the Thomaston road") and labor ("how many negroes on break?"). After Maj. Gen. Smith arrived as the army's chief engineer, Hood assigned him to further supervise the reconstruction ([Hood] to J. C. Shoup, August 4 [11:20 A.M.]; to Shoup, August 5 [2 P.M.], in Hood, Telegram Book III).

In his reminiscences Sherman is harsh on Rousseau, stating that in his Alabama raid "as usual, the cavalry did not work hard, and their destruction of the railroad was soon repaired" (*Memoirs*, 2:69). Elizabeth Schafer is closer to the mark. "The railroad was indeed reconstructed," she writes, "but the repairs were hastily completed, and accidents destroyed several trains" ("Jaded Mules, Twisted Rails and Razed Depots," 46).

was told; "hold the enemy in check as much as possible." Events were unfolding so rapidly that Wheeler was asked to put the hour at which he sent in each of his reports.

Hood's troops during all this time continued working on their entrenched line south of Peachtree Creek in anticipation of the enemy forcing a crossing. Hardee's division commanders were instructed to "cause the banks of Peach Tree Creek to be thoroughly examined…and cause the best defenses that can be made to be placed there. The object is to enable a small force to resist the enemy's crossing for some time." On the afternoon of the 19th, a Union brigade (Caleb Dilworth's of Davis's division) crossed the creek and advanced against Confederate pickets. When Daniel Reynolds's Arkansas brigade came up, a sharp several hours' skirmish broke out. In this, the only hard fighting of the day, Federals suffered 245 casualties. Reynolds claimed that his loss was 8 killed, 48 wounded, 3 missing. In his evening telegram, Hood reported to Seddon that Reynolds's brigade charged "handsomely," and took almost 150 prisoners. But the Federals were successfully across the deep-gorged creek. Pioneers of Geary's division threw across a wooden footbridge and added two more during the night. Meanwhile Schofield marched toward Atlanta from the northeast, but at a distance from Thomas's left. Sherman ordered two of Thomas's divisions to close toward Schofield in an effort to reduce the gap.[7]

It was on the morning of July 19 that Sherman learned he now faced a different opponent. General Grenville Dodge, head of the XVI Corps, was riding with his troops toward Decatur when "one of my scouts joined me, he having left Atlanta that morning, bringing the intelligence of the supersedure of Johnston by Hood." The "scout" was a soldier of the 2nd Iowa who brought with him the morning *Memphis/Atlanta Appeal*, which carried the news of the command transfer. Dodge took his newspaper to generals Schofield and Sherman, who were conferring and who had already heard rumors of the switch. Sherman asked Schofield, who had been Hood's classmate at West Point,

---

[7] Castel, *Decision in the West*, 366–67; Logan report, September 13, 1864, *OR*, vol. 38, pt. 3, 101; SFO 37, July 18 (orders for July 19), *OR*, vol. 38, pt. 5, 179; D. Evans, *Sherman's Horsemen*, 80–81, 83; Mackall to Wheeler and Mason to Wheeler, July 19, *OR*, vol. 38, pt. 5, 893; Roy to Cleburne, July 18, *OR*, vol. 38, pt. 5, 890 ("commence fortifying"); Roy circular, July 19, *OR*, vol. 38, pt. 5, 894 ("best defenses"); Reynolds report, July 23, *OR*, vol. 38, pt. 3, 935–37; Hood to Seddon, July 20, *OR*, vol. 38, pt. 5, 894 ("late yesterday evening a force of the enemy, supposed to be a brigade, crossed Peach Tree Creek and attacked Reynolds' brigade"); Kurtz, historical marker texts, "Federal Crossings," "Green Bone Creek," and "Geary's Three Bridges," box 35, folder 1 (Scruggs, *Georgia Historical Markers*, 218–19); Daniel, *Days of Glory*, 412 (two-mile gap).

what it meant; Dodge after the war remembered him as saying, "This means a fight; Hood will attack you within twenty-four hours." Other Federal generals sized up the Rebels' new commanding general. O. O. Howard had been a year behind Hood at West Point and knew him well enough to offer comment to his wife: "He is a stupid fellow, but a hard fighter, does very unexpected things." Further corroboration of the change in command came when Thomas sent to Sherman another Atlanta paper that had been obtained by Brig. Gen. Thomas J. Wood, division commander.[8]

Meanwhile Wheeler's dismounted horsemen resisted McPherson's march from Stone Mountain and Decatur along the Georgia Railroad. The enemy was known to be moving westward steadily, if not speedily. Maj. A. P. Mason, Hood's A. A. G., instructed Wheeler on July 20 at 10:20 A.M., "You must retard the enemy as much as possible." To McPherson's north, Schofield's XXIII Corps on the morning of July 20 was marching southwestward from the area of today's Emory University. The Confederates' outer defense line, held in that sector by Stevenson's division of Cheatham's corps, would contain Schofield's advance. But the line ended less than a mile north of the Georgia Railroad (in the area known after the war as "Copenhill," and which today is the grounds of the President Jimmy Carter Center and Library). South of that point only Wheeler's cavalry stood in front of McPherson's advance, which had pushed to two and a half miles east of Atlanta along the railroad.[9]

The Federals had gotten close enough to the city to begin lobbing shells into it. On the 19th Sherman issued orders that if any of his artillerymen got within range, "the place must be cannonaded without the formality of a demand." The Federal commander's motives in ordering a bombardment of Atlanta and of its civilian population (by then reduced to perhaps 5,000, given the panicked exodus of the previous few weeks) are unclear. But by this stage of the war, the shelling of cities had become accepted as common military

---

[8] Dodge report, November 25, 1864, *OR*, vol. 38, pt. 3, 383; Dodge, "The Late J. M. Schofield," 461; Hirshson, *The White Tecumseh*, 227 ("Dodge kept two spies in Atlanta"); Howard to Sherman and Sherman to Howard, July 19, *OR*, vol. 38, pt. 5, 188; Sherman to Thomas, July 19, *OR*, vol. 38, pt. 5, 183; Sherman, *Memoirs*, 2:72; D. B. Connelly, *John M. Schofield and the Politics of Generalship*, 104 (Howard to Elizabeth, July 23); Thomas to Sherman, July 19, *OR*, vol. 38, pt. 5, 184.

[9] Wheeler report, October 9, 1864, *OR*, vol. 38, pt. 3, 951–52; Mason to Wheeler, *OR*, vol. 38, pt. 3, 892 (July 19, but which McMurry and Castel recognize is misdated, and was actually July 20 [McMurry, *John Bell Hood*, 220; Castel, *Decision in the West*, 607 n.53]); William Key, "War Comes to Atlanta," *Atlanta Constitution*, July 20, 1930 (accompanied by Wilbur G. Kurtz map, "Embattled Atlanta"); Castel, *Decision in the West*, 378.

procedure. Indeed, the year before, after the fall of Vicksburg when he marched on Jackson, Mississippi, to drive away Joe Johnston's small army, Sherman had shelled the city as a complement to his cavalry raids and infantry sorties against the Rebels. Now, when Capt. Francis DeGress's Battery H, 1st Illinois Light Artillery reached a hill east of Atlanta around 1 P.M., DeGress gave orders for his 20-pounder Parrotts to unlimber and throw a few rounds into the city. They fired three. Reports differ as to where they fell: the Western & Atlantic depot, three blocks northwest of the Car Shed; the Female Institute five blocks northeast of it; maybe near City Hall (according to the Rt. Rev. Henry C. Lay, Episcopal bishop of Arkansas, staying at the home of William Solomon across the street). The Yankee shells created great public consternation, but no casualties.[10]

As DeGress's guns opened fire, McPherson formed line of battle. His right (Dodge's XVI Corps) reached for Schofield's left north of the railroad; Logan's XV advanced in the center. South of the railroad Frank Blair's XVII Corps pushed westward. Brig. Gen. Walter Gresham's division came up on Wheeler's troopers aligned on a ridge. The Confederates opened with artillery and musketry; Gresham was shot in the leg by a sharpshooter on a bald hill upon which Wheeler's right rested. "It's too bad, Gresham," General Blair consoled. "I was racing John Logan to get into Atlanta and you go and get shot." Blair ordered Brig. Gen. Mortimer Leggett to charge and take the bald hill. Leggett's infantry moved forward around 4 P.M., according to Wheeler, whose men were forced back at some points. Leggett, however, failed to push hard enough to take the hill. As he had done at Snake Creek Gap on May 9, McPherson on July 20 east of Atlanta grew cautious, apprehensive of a Rebel attack. Federal assault on the bald hill would await morning of the 21st. Wheeler clung to his line about two and a quarter miles east of the city.[11]

General Hood moved his headquarters on the afternoon of July 19 to Alexander Luckie's house on Peachtree Road just outside of the city limits.

---

[10] Sherman's Special Field Order 39, July 19, *OR*, vol. 38, pt. 5, 193; S. Davis, *What the Yankees Did to Us*, 94–97, 106–110 (debunking the postwar legend that the first Yankee shell killed a little girl).

[11] Castel, *Decision in the West*, 378–79; Kurtz, "Walter Q. Gresham at Atlanta," *The Constitution Magazine*, August 24, 1930, 9; Wheeler report, October 9, 1864, *OR*, vol. 38, pt. 3, 952; Dyer, *Gallant Hood*, 257–58; Howell and Purdue, *Pat Cleburne Confederate General*, 347.

The bald hill eventually taken by Leggett's division came to be called "Leggett's hill" by the Federals who took and later defended it (Ecelbarger, *Day Dixie Died*, 205, 211). Among some writers today there is a movement afoot to refer to it without using the Yankee general's surname (e.g., "the battle of bald hill").

There he planned a battle. The government had charged him to seek "an opportunity of equal encounter," and he now thought he saw one developing. Thomas's army was advancing on Atlanta from the north and had to get across Peachtree Creek, whose deep gorge would offer obstacle. Attacking the enemy while crossing, or just after they reached the south bank before they had a chance to dig in, offered prospects for a successful attack as sanguine as any handed to the Confederates thus far in the campaign. Moreover, the gap (of some two miles) between Thomas and Schofield meant that the Army of the Cumberland, while Sherman's largest, would likely have to fight without help from Schofield, much less McPherson. At least that is what he wrote in his memoirs:

> I perceived at once that the Federal commander had committed a serious blunder in separating his corps, or Armies by such distance as to allow me to concentrate the main body of our Army upon his right wing, whilst his left was so far removed as to be incapable of rendering timely assistance.... Although a portion of the enemy would undoubtedly be found under cover of temporary breastworks, it was equally certain a larger portion would be caught in the act of throwing up such works, and just in that state of confusion to enable our forces to rout them by a bold and persistent attack.

In another of their spats over the facts, Hood's memory was different from Johnston's as to who laid the initial plan for the upcoming attack on Thomas's army at Peachtree Creek. The documentation surrounding Hood's battle plan for July 20 is so skimpy as to allow Joseph E. Johnston to claim that it was his idea and Hood to claim that it was his. Three months after his removal, Johnston stated that before he left on July 18, he explained his plan to attack the Federals as they were crossing Peachtree Creek. After the war, in his reminiscences, Johnston asserted that on the morning of the 18th, "in transferring the command to General Hood I explained my plans to him. First, I expected an opportunity to engage the enemy on terms of advantage while they were divided in crossing Peach-Tree Creek." Hood, on the other hand, claimed that it was Wheeler's intelligence (that came to him on the 19th) about the three enemy armies' positions that led him to develop his battle plan. Hood further states that he had no recollection of Johnston's advising him about an attack at Peachtree Creek.[12]

---

[12] McMurry, "Mackall Journal," entry of July 19 (Hood's HQ); [Garrett and Kurtz], "Key to Map," 28 (Site 151, "Site of the old Alexander F. Luckie home. Used by Gen. Hood as his headquarters on July 19, 20 and 21, 1864, during the Battle of Peachtree Creek"); Johnston report, October 20, 1864, *OR*, vol. 38, pt. 3, 618; Hood, *A & R*, 141, 162–65, 168; J. E. Johnston, *Narrative*, 350.

Not surprisingly, Johnston's biographers side with him: "He did explain his plans to Hood in detail"; "Johnston had two conversations with Hood in which he explained his plans for an attack as the Federals crossed Peachtree Creek."[13] Hood's chroniclers take a different view. "Actually, there is no evidence to show that Johnston intended to defend Atlanta," declares O'Connor. Dyer notes that when Hood began contemplating an attacking battle, "Johnston wrote in later years he planned to do" the same thing. McMurry mentions Johnston's contention without comment, but more recently, in his *Atlanta 1864*, he does not even refer to Johnston's claim of ownership of the Peachtree Creek battle plan: "After studying the scouts' reports, Hood realized that he had an opportunity to strike a heavy—perhaps a mortal—blow at a part of Sherman's forces." To McMurry, the Confederates' real opportunity lay in the gap between Thomas's left and Schofield's right, allowing Hood to send Hardee's infantry into the gap and against the flank of the Army of the Cumberland.[14]

Other historians have taken both sides in the generals' dispute. Stanley Horn wonders why Johnston did not tell the administration of his attack plan when he had the chance a few days earlier. "There is strong suspicion that the plans of which he wrote so confidently in 1874 were the product of reflection after the event," he reasons. Woodworth credits Hood for devising his plan of attack without mention of Johnston. Castel does the same in his text but cites Johnston's claim to the battle plan in an endnote. Connelly judges that "Johnston probably did describe his Peachtree plan to Hood." His source is the redoubtable Mrs. Chesnut, who under date of August 14, 1864, entered into her diary a statement made by Henry Brewster: "The day Hood took command he and General Johnston talked all day. The new commander asked the old one for all his views and plans. And they were given freely—apparently." Mary Boykin Chesnut would have had no reason to fabricate this point about Joe Johnston, just as Henry Brewster, Hood's friend, would have had no reason to say good things about Johnston unless they were based on what he had heard. What Connelly cannot conclude with certainty, however, is whether all of Johnston's "views and plans" included an attacking battle south of Peachtree Creek.[15]

---

[13] Govan and Livingood, *Different Valor*, 318; Symonds, *Johnston*, 330.

[14] O'Connor, *Cavalier General*, 204; Dyer, *Gallant Hood*, 252; McMurry, *John Bell Hood*, 126–27; McMurry, *Atlanta 1864*, 146–47.

[15] Horn, *Army of Tennessee*, 348; Woodworth, *Davis and His Generals*, 286; Castel, *Decision in the West*, 366, 606 n.48; T. L. Connelly, *Autumn of Glory*, 424.

Ex-Confederates well after the war were still talking about Johnston's "plan" to attack at Peachtree Creek. In a speech given in May 1891, Leigh Robinson claimed that Johnston had told Gen. Peter Stewart to "find the best position, on our side of Peach Tree Creek, for our army to occupy.... He did not say that he would attack Sherman on the crossing of Peach Tree, 'but,' says Stewart, 'his dispositions were evidently made with a view to so attack, and were inconsistent with any other purpose'" (L. Robinson, "General Joseph E. Johnston," 361). More far-fetched is the statement by a Tennessean, John Bright, that around July 8–9, when Sherman was shoving troops across the Chattahoochee, he had conversation with General Cheatham, whom he quoted forty-five years after the fact. "He replied," Bright wrote, "'Johnston keeps his own counsel; but I have heard that Sherman has divided his army and one part has crossed the Chattahoochee at Sandtown and the other part is crossing at Roswell's Factory, some thirty miles apart, and General Johnston is moving two corps of his army to attack Sherman's force at Roswell at daylight while they are disorganized, and that General Stewart's Corps will resist the advance of Sherman's forces at Sandtown" (Bright, "The States in the Confederate War," 395).

Example of a recent historian who credits Johnston with planning to attack at Peachtree Creek (but who cannot cite primary evidence) is Christopher Losson, who writes, "Johnston had planned to attack the Federals in the vicinity of Peachtree Creek, and Hood adopted the strategy as well." For this statement, Losson's endnote cites only Hood's *A & R*, which gives no credit to General Johnston, and T. L. Connelly, *Autumn of Glory*, 439–40, in which the author assumes that "the opportunity to use Johnston's attack plan was at hand" (Losson, *Tennessee's Forgotten Warriors*, 174, 315 n.22).

The most thorough chronicler of the battle of Peachtree Creek, Robert Jenkins, sides with Johnston in the argument over battle plan authorship (Jenkins, *The Battle of Peach Tree Creek*, 12–17). Of the evidence he presents the most interesting is the statement of Col. Ellison Capers, 34th South Carolina, that on July 17 Johnston "published an address to the army and announced that he would attack General Sherman's army so soon as it should cross the Chattahoochee" (Capers report, September 10, 1864, *OR*, vol. 38, pt. 3, 717). By July 17 Sherman's army *had* crossed the Chattahoochee, and, while Johnston liked to issue martial proclamations before he planned to launch an attack (as at Cassville), no such announcement of July 17 appears in the *Official Records* or elsewhere.

Writing in Atlanta around 1900, Thomas H. Martin gives a rather bizarre account of the planning for Peachtree Creek, constructing it around an alleged recent conversation with a "General B.," said to have been a former Confederate officer in the Army of Tennessee. "The facts were given me by Gen. Jos. E. Johnston, in person, directly after he was relieved from command of the Confederate forces," "General B." claimed, recalling Johnston's battle plan quite clearly after thirty-five years. Johnston purportedly planned to take a strong defensive position north of Atlanta, wait for Thomas's Army of the Cumberland to attack him, then repel and crush the enemy's forces. "It was Johnston's belief that he could destroy, or capture, the army of the Cumberland," and after that, turn on the separate forces of Schofield and McPherson. "Johnston had absolute confidence in his plan of battle," "General B." reportedly told Martin. "For three long months it was on his mind constantly. Every detail had been carefully gone over and arranged." Of course, Johnston was removed from command before he could execute this masterpiece ([T. H. Martin], "The Story of the Battle of Peachtree Creek," 13, 16, 21–22). My thanks to Dr. Gordon Jones of the Atlanta History Center for putting me onto this pamphlet.

Martin is quite cryptic about his source, "General B." He implies that the alleged conversation occurred in fall 1899 ("this October day," 14; "thirty-five years ago," 13). Martin's

After Hood decided to launch his attacking battle, on the night of July 19–20 he met at his headquarters with generals Hardee, Cheatham, Stewart, Wheeler, and G. W. Smith to discuss their roles in it. Hardee's and Stewart's corps would assault Thomas's troops beginning at 1 P.M. The attack would begin on Hardee's right and move to the left: each division advancing *en echelon*, meaning in sequence one after the other (similar to Lee's plan of battle on the second day of Gettysburg). The order would start with Bate's division leading off, then (east to west) Walker's and Maney's (with Cleburne in reserve). Stewart's corps would then take up the attack with Loring's, Walthall's, and French's divisions. Hood made certain that his corps commanders knew the objective: "everything on our side of the creek to be taken at all hazards." The Confederates, if successful, would push the Yankees back against Peachtree Creek and the Chattahoochee, trapping them against the stream's steep banks. Cheatham would guard the right and keep Schofield and McPherson from bringing aid. East of the city Wheeler would continue to contest McPherson's approach from Decatur. Smith's militia (some 700 men in the fortifications near the Georgia Railroad) would help if needed. In his memoirs Hood states that at the end of the meeting each officer was asked whether he understood his orders. All answered yes. General Stewart summed it up in his report: "The plan was...to attack the enemy, drive him back to the creek, and then press down the creek to the left. Should the enemy be found intrenched his works were to be carried, everything on our side of the creek was to be taken." It was clear to all that this would be the big battle the officers and men had been expecting since the opening days of the campaign. "The fight was to be a general one along our lines," as Brig. Gen. Winfield S. Featherston (in Loring's division) put it, "and the victory to be made as decisive as possible."[16]

---

veteran "had commanded a brigade in General Stewart's division" (14), but of Stewart's brigadiers in 1864 there was only Alpheus S. Baker, who died in 1891. It is plausible that Martin gave the general—if there were even such an actual individual in the first place—a misleading initial in order to disguise the fact that there may have in fact been no "General B." at all, in which case there would have been no conversation about Johnston's "plan" at Peachtree Creek. The cynical conclusion is that Martin resorted to much disguising indeed because Johnston *had no plan* at Peachtree Creek.

[16] Davis, "Reappraisal," 71, 73; Smith to Wheeler, July 20, *OR*, vol. 38, pt. 5, 895 (700 men in "Georgia Railroad Fortifications"); Stewart report, January 12, 1865, *OR*, vol. 38, pt. 3, 871 Featherston report, July 23, 1864, *OR*, vol. 38, pt. 3, 882.

T. L. Connelly claims that Hood did not hold the planning meeting with his corps commanders until the morning of July 20, criticizing him for "not allowing enough time for preparation" of the attack (*Autumn of Glory*, 440). Gen. Stewart months later recalled that the meeting was indeed held "on the morning of the 20th" (Stewart report, January 12, 1865, *OR*, vol.

Hood was eager to fight his first battle as army commander. That morning a reporter for a Savannah newspaper, Felix De Fontaine, caught him at his headquarters, "his eyes flashing with a strange indescribable light, which gleams in them only in the hour of battle." Before mounting and riding to the front, Hood said, "At one I attack the enemy. He has pressed our lines until he is within a short distance of Atlanta, and I must fight or evacuate. I am going to fight. The odds are against us, but I leave the issue with the God of Battles."

Things started going wrong for Hood on the morning of the battle. Lieutenant Mackall recorded in his journal for July 20 that Hardee's and Stewart's corps were "to attack in echelon of divns at 1 P.M." But around ten o'clock that morning Hood's plan began to unhinge. Wheeler notified headquarters that the enemy (Logan's XV Corps in the advance) was pressing his 2,500 troopers and was marching on the city. Hood responded at 10:20 by ordering to him another thousand cavalrymen (from Red Jackson). He also ordered Cheatham, whose right flank in the entrenched outer defense line did not cover the railroad, to move a mile or so southward to block the enemy's route (in his report, Hood called it "a division front to the right"). Hardee and Stewart were in turn directed to sidle a half-division front. Cheatham, however, mismanaged the maneuver and marched too far to the south. As a result, several hours were spent in the shift of the infantry line and the start of the attack had to be delayed. Hood was partly to blame as well. That afternoon he was at Stewart's headquarters on the left of the line, not Hardee's, despite the fact that the battle was to begin with Hardee's troops on the right. (General Stewart later believed that Joe Johnston would have managed events better.)[17]

By noon of July 20, all of Thomas's infantry had crossed Peachtree Creek. The Federals had had to build bridges across the stream, which Brig. Gen. Richard Johnson described as "deep with very abrupt and high banks." On Thomas's right, near the confluence of the creek and the Chattahoochee, the three divisions of Maj. Gen. John Palmer's XIV Corps were aligned and, in the words of Johnson, had constructed "temporary breast-works" against a

---

38, pt. 3, 871). McMurry, however, sides with Hood and his memoirs that the meeting occurred late on the 19th, reasoning that if it was held after midnight, "both accounts [Stewart's and Hood's] would be correct" (McMurry, *John Bell Hood*, 127, 220 n.32).

[17] "Personne" [Felix De Fontaine], "From the Army of Tennessee," *Savannah Republican*, July 25, 1864; McMurry, "Mackall Journal," entry of July 20; Castel, *Decision in the West*, 371; Hood report, February 15, 1865, *OR*, vol. 38, pt. 3, 630; Losson, *Tennessee's Forgotten Warriors*, 174–75; A. P. Mason to Wheeler, July 20 (10:20 A.M.), *OR*, vol. 38, pt. 5, 895; Elliott, *Soldier of Tennessee*, 203.

Rebel attack Palmer thought was likely that day. Hooker's XX Corps infantry encountered rough terrain all along the south bank of the creek. Lt. Col. Harvey Chatfield, in the 3rd Brigade of Geary's division, reported that his regiment had moved forward, "but owing to the dense undergrowth and rough nature of the ground, the advance was rendered very difficult and the line irregular, besides an almost utter inability to keep in view the other regiments of the brigade." Once through the brush, Hooker's line extended eastward toward the Peachtree road: Williams's, Geary's, then Ward's divisions. On the Federal far left was Brig. Gen. John Newton's division of Howard's IV Corps. The divisions were only roughly aligned along a ridge (marked by today's Collier Road). More important, Newton's left was in the air. Thomas had no cavalry on his flank, and even after he sent his two divisions off toward Schofield, the two-mile gap existed between Newton and the dispatched troops. "We must not mind the gap," advised Thomas when he sent Howard off with the two divisions. "We must act independently."

Hooker's three divisions had not even entrenched, as many Federals (despite the news of Hood's accession) did not expect the Rebels to attack. To his division commander Brig. Gen. William T. Ward, Col. John Coburn expressed concern about the Federals' vulnerability, with Peachtree Creek behind them. "That is true," Ward replied, "but I don't believe the rebels are coming." Sgt. Rice C. Bull of the 123rd New York (Knipe's brigade, Williams's division) described the situation:

> No orders were given to put our Brigade in position to defend itself in case of attack....On this day, everyone from our Corps Commander to private soldier seemed certain that the enemy would make no stand until behind the entrenchments at Atlanta; so there was no need for placing our Brigade in line of battle. Our officers must have thought it unnecessary to use even the most ordinary precautions for the skirmish line was not advanced as far in front as usual.[18]

Hardee did not begin shifting his troops rightward until almost 1 P.M. He had posted a staff officer, Maj. S. L. Black, at the point at which the left of his corps was to halt after moving a half division's length. But when the

---

[18] Daniel, *Days of Glory*, 412; Johnson to Capt. A. C. McClurg, July 18, *OR*, vol. 38, pt. 5, 172; "Itinerary of the Fourteenth Army Corps, May 6-September 8," *OR*, vol. 38, pt. 1, 507; Brig. Gen. Richard W. Johnson report, August [undated], 1864, *OR*, vol. 38, pt. 1, 524; Castel, *Decision in the West*, 369–71; McKinney, *Education in Violence*, 345; Kurtz, historical marker text, "Geary's Division," box 35, folder 2 (Scruggs, *Georgia Historical Markers*, 215) ("this commanding ridge...was the critical point of the Federal position"); D. Evans, "The Atlanta Campaign," 33; Bull, *Soldiering*, 146.

troops kept marching—Cheatham had gone more than a mile, maybe two—Major Black rode off to General Hardee, who told him that because Cheatham had kept moving, he had been ordered to do so as well. At 3:15 Hardee sent a note to Hood, complaining that "the movement to the right has caused great delay, and I fear some disorder." Black after the war recalled that "General Hardee expressed his impatience at the delay, and his annoyance at the repeated movements to the right." Stewart was also anxious to get the battle started, and sent a staff officer to Hardee, asking if the rightward march could be halted (it could not). At 2:30 Hardee sent word to Stewart that he expected to begin his attack soon, but it was 3:15 before he finally launched his advance.[19]

When Hardee's assault began, between three and four o'clock, it almost immediately went awry. Advancing into "an almost impenetrable thicket" (Hardee's words), Bate's troops wandered without knowledge of what lay ahead. It turned out they never made direct contact with the enemy, having stepped forth into the gap existing between Newton's and the other two divisions of the IV Corps. As it neared Peachtree Creek, General Bate ordered a section of the Washington Artillery to press forward and fire at any enemy crossing the stream. After a few shots, the Southern gunners themselves came under fire from Newton's artillery on the other side of the creek. Bate's division eventually fell back after coming under the artillery shelling.

Walker's division was next to advance *en echelon*; the Yankees ahead of it were Newton's, occupying high ground across Peachtree Road (today's Piedmont Hospital). By now they had begun to dig in and erect a log barricade to cover them; some had cut down trees in their front as a killing ground. Walker's three brigades were more or less in line: Brig. Gen. Clement H. Stevens's on the left, Brig. Gen. States R. Gist's in center, and Brig. Gen. Hugh W. Mercer's on the right. The Southerners began their attack "with a rapidity and an absence of confusion I have never seen equaled," according to a Northern officer. Two of Stevens's regiments, the 1st and 66th Georgia, overran a portion of the enemy works, but had to withdraw under heavy fire. Unfortunately for the Southerners, Walker's brigades launched uncoordinated attacks. Though they outnumbered Newton's troops, after driving in the enemy skirmishers, the Confederates failed to break their line. General Stevens was mortally wounded, shot in the head; Col. James Cooper Nisbet of the 66th remembered that captains Briggs Napier and Charles Williamson

---

[19] Hughes, *Hardee*, 221; Stewart report, January 12, 1865, *OR*, vol. 38, pt. 3, 871 ("At 1 o'clock I found the left of Hardee's corps just beginning to shift to the right"); Roy, "General Hardee and the Military Operations Around Atlanta," 348; Elliott, *Soldier of Tennessee*, 202.

were both wounded carrying Stevens from the field. (The general died five days later.) The Northerners' musket fire was destructive. "We began firing by volleys on the first appearance of the Johnnies," recorded Lt. Ralsa C. Rice of the 125th Ohio; "the shock of our broadside was terrific." Rice added, "We were evening up matters with June 27 (Kennesaw) in mind." Much of Gist's brigade fell back before getting into close rifle range. Artillery was also effective. General Thomas observed the fighting and directed the fire of nearby batteries. "Load after load as fast as the artillerymen could hand their pieces followed—a continuous shower of murderous iron," observed Ohio Cpl. John Raper. "No troops on earth could stand that long." Because Mercer's brigade overlapped Newton's left, it never became seriously engaged. The brevity of the Confederates' assault—twenty minutes before Stevens and Gist were driven back, ten minutes for Mercer, according to Robert Jenkins—minimized Walker's casualties that afternoon. As a result, division losses on July 20 did not exceed 300.[20]

The assault of Maney's division, next to go in, indicates the difficulties encountered by the Confederates that day at Peachtree Creek. The afternoon of July 20 was very hot, and Hardee's men had already marched a good distance before they began their charge. "Some of the Southern troops advanced at the double-quick, for perhaps a half-mile," writes Lee Kennett, "so that when they finally reached the Northern lines they were completely winded." Henry McIntosh, of the 24th Wisconsin (Kimball's brigade, Newton's division) observed, "they were so tired and it was so hot they could only walk."

Tired, hot men took every opportunity to stop and take a canteen swig. Those behind them did the same. The rough terrain in front of them was another deterrent: uphill, underbrush, ravines and a streambed (Tanyard Branch). Once the men stopped, it was hard to get them going again, especially when close to the enemy and under hot fire. In the case of Robert Jarman's Mississippi command, it was a fence in front of the enemy line that brought the charge to a stop: "Our lines halted and it was fatal to them, for never did they advance again."

Hardee had been unable to send scouts ahead to recon the enemy position, so Maney's troops had no idea of what lay ahead of them. It came as a

---

[20] Hardee report, April 5, 1865, *OR*, vol. 38, pt. 3, 697; Jenkins, *Peach Tree Creek*, xxxviii–xxxix (maps), 77, 112, 117, 122 ("commenced a line of works"), 129, 133 (Raper), 135–36, 138, 140 ("much if not all of Gist's brigade failed to get farther than rifle range of the Federal works"), 142, 379; N. C. Hughes, *The Pride of the Confederate Artillery*, 197; Castel, *Decision in the West*, 376; R. K. Brown, *To the Manner Born*, 260; S. Davis, "Atlanta Campaign," 14; Nisbet, *4 Years on the Firing Line*, 209.

hard surprise to find the Yankees dug in, "posted behind breastworks," as one Confederate put it, too strong to "make it pay" for them to launch a determined assault. So they did not. It did not help that Maney's men had been able to see the rough handling of Walker's division to their right (the usual problem with echelon assaults). Col. Michael Magevney's and Col. John C. Carter's brigades got no farther than sixty yards from the enemy lines—Newton's right and the left of the next Union division, Brig. Gen. William T. Ward's—before halting. Brig. Gen. Otho Strahl's and Col. Francis Walker's brigades, behind them and in support, did not even get that far. The Confederates exchanged fire for the next couple of hours, but the attack of Hardee's corps was over. General Maney's report for the battle of July 20 is not to be found in the *Official Records*, so Atlantan Russell S. Bonds's explanation of Maney's failure is as good as anyone else's:

> Maney's force could and should be considered the heart of Hardee's attack—four brigades, perhaps 4,500 men.... But rather than striking the hardest blow along the line that day, Maney's Division crept forward and barely fought at all. The Tennesseans stepped off on the advance but soon encountered thick, screening woods in their front. Unwilling to fight blind, not knowing what was in front of them, Maney's front line called a halt and sent out skirmishers to see what was ahead as the afternoon wore on and the daylight failed. No one really knows why Maney's men appeared to flinch and failed to press their attack, and the contrast between their timid advance and the ferocious charges by the Confederates to the right and left of them is striking.[21]

When General Loring heard the sound of firing to the right—the commencement of Hardee's attack—he knew it was time to send in his two brigades, those of Brigadiers W. S. Featherston and Thomas Scott. Pvt. J. P. Cannon, an Alabaman in Scott's, remembered that General Stewart addressed his troops beforehand, announcing that "we were going to assault the enemy in his works, and we must carry everything, allowing no obstacle to stop us; that the fate of Atlanta probably depended on the result of this battle." Stewart's words had an impact upon the troops as they put forward. Union Brig. Gen. John Geary, whose division would face Scott's brigade, remembered the sight of the Rebels once they crossed Tanyard Branch, got out of the woods and marched out into an open field.

---

[21] Kennett, *Marching through Georgia*, 187; Fowler, *Mountaineers in Gray*, 152; Jenkins, *Peach Tree Creek*, 143–46; Castel, *Decision in the West*, 370 (map), 376–77; Bonds, *War Like the Thunderbolt*, 95–96.

> The appearance of the enemy as they charged upon our front across the cleared field was magnificent. Rarely has such a sight been presented in battle. Pouring out from the woods they advanced in immense brown and gray masses (not lines), with flags and banners, many of them new and beautiful, while their general and staff officers were in plain view, with drawn sabers flashing in the light, galloping here and there as they urged their troops on to the charge. The rebel troops also seemed to rush forward with more than customary nerve and heartiness in the attack.

Featherston's troops charged up the Collier road ridge, atop which were dug in two regiments as skirmishers, the 136th New York and 22nd Wisconsin. The Confederates drove them back to the main Union force, Ward's division. Ward's troops were aligned with Wood's brigade on the left, Col. John Coburn's in the center, and Col. Benjamin Harrison's on the right. The Southerners fired a volley, overshooting as they descended the ridge. The Federals were unfortified, but the sounds of battle to the east had alerted them, and they were ready. "The brigade poured deadly fire into the enemy, who charged in large numbers down the slope," Coburn reported. Featherston's troops took heavy casualties and began to fall back. As one soldier in the 40th Mississippi put it, after taking the first enemy line on the ridge, they "crossed an open field in which the brigade was shot to pieces. Nothing remained but to retreat across the field, yielding the toll of sacrifice at every step." After checking the Rebel assault, the Northerners charged back up the Collier ridge and there, from behind fence railings, kept firing at the retreating Rebels. Along the way the Confederates lost two regimental battle flags, those of the 31st and 33rd Mississippi. The Mississippians halted in the woods south of the road and kept up musketry until darkness, when they withdrew to their original lines. General Featherston reported fully half of his men killed, wounded or missing, 616 out of 1,200 carried into the fight. The Northerners' numerical superiority—Coburn's brigade alone had 1,315 officers and men—even without entrenchments had brought them success. To Coburn were reported 216 killed, wounded, or missing. Wood reported only that his brigade's loss was "slight in comparison with the loss and havoc that were inflicted upon the rebels."[22]

Scott's brigade fared better. Marching at first through woods and briers, the Southerners moved into a field and began to take fire from their front

[22] Jenkins, *Peach Tree Creek*, 81, 94–96, 161–63, 189, 199–200, 233, 239, 255–59; Geary report, September 15, *OR*, vol. 38, pt. 2, 140; Coburn report, July 28, *OR*, vol. 38, pt. 2, 390; Castel, *Decision in the West*, 377; Featherston report, July 23, *OR*, vol. 38, pt. 3, 883; Wood report, September 23, *OR*, vol. 38, pt. 2, 443.

(Harrison's brigade), but also from artillery posted on high ground across Tanyard Branch (Geary's front). The branch and flanking fire from the left caused some of Scott's troops to veer right, crossing the stream, then the Collier road on the ridge top, passing Collier's mill house and charging the lines of Colonel Harrison, "portion of which they took," Scott reported. But his troops came under heavy fire (some of Harrison's men were armed with Spencer repeating rifles). The Southerners were also winded. "The long distance the enemy charged on the double-quick," Harrison reported, "halted their advance." "I could observe many of his men lying down and a few even turning back, while the officers, with their drawn swords, were trying to steady their lines and push them forward." Just then Harrison ordered a counterattack, which pushed the Southerners back up and across the Collier ridge. In their advance, soldiers of the 105th Illinois captured the flag of the 12th Louisiana.[23]

The attack of Scott's brigade, Loring's division, at the Collier ridge on July 20 illustrates the military historian's troubles in writing about Hood's first battle as army commander. General Scott does not identify the enemy his troops fought; neither does Colonel Harrison. As a result, standard histories state that Scott's brigade did not attack Harrison at all, but advanced to the west of Tanyard Branch, against the left of Geary's division line on the other side of the creek. Details of this part of the battle are important, if only, as William R. Scaife writes, because Scott's brigade "made the most determined and successful Confederate penetration of the day." But Scaife does not mention Scott versus Harrison; neither does Castel; both only pitch Scott against Geary. Maps accompanying these authors' texts follow suit (that for Castel even has Scott veering to the *right* of Geary's division). Jenkins observes that the 55th Alabama, 57th Alabama, and three companies of the 12th Louisiana charged Geary's artillery and two of Harrison's regiments west of Tanyard Branch. But he also states that the right (or east) section of Scott's brigade hit the left part of Geary's division and posits that the heavy casualties suffered by the 55th, 57th, and 12th came from charging Geary's artillery, not from being pushed back by Harrison's counterattack. The map Jenkins drew for his book shows Scott's Alabamans and Louisianans charging Geary's left west of Tanyard Branch. Indeed, the author so emphasizes Scott's fight against Geary that he discredits evidence that the battle flag of the 12th Louisiana was captured by the 105th Illinois (Harrison's brigade). Instead, Jenkins states that

---

[23] Scott report, July 23, *OR*, vol. 38, pt. 3, 895; Harrison report, August 12, *OR*, vol. 38, pt. 2, 345; Lt. Col. Everell F. Dutton report, August 5, *OR*, vol. 38, pt. 2, 364 (12th Louisiana colors); Rollins, *The Returned Battle Flags*, 21 (12th Louisiana flag returned March 25, 1905).

the Confederate colors "most likely" were those of the 40th Mississippi, one of Featherston's regiments.

On the other hand is the sturdy work of the locally heralded Wilbur G. Kurtz. Kurtz was an indefatigable researcher of the Civil War, writing articles for the Atlanta newspapers and other publications. He also meticulously rendered fact into picture with his many ink illustrations and color paintings. In the early 1950s, Kurtz was hired by the Georgia Historical Commission to select locations and to write the text for more than four hundred roadside historical markers pertaining to the Atlanta Campaign and to choose the sites of their placement. Kurtz's voluminous notebooks reflect decades of research into the campaign of 1864, and for several years in the mid-1950s he poured his knowledge into his carefully written GHC marker texts. "This extraordinary educational accomplishment," writes David O'Connell in a recent assessment, "has not been fully appreciated by posterity."

Kurtz is firm and clear in his several marker texts detailing this part of the battle of Peachtree Creek and describes the split of Scott's attack, its right going against Harrison and its left against Geary.

Gap in Federal Line (front yard of 423 Collier Road, N.W.)

The 129th, 105th Ill. and 70th Ind., the left of Harrison's brigade, together with the rest of Ward's div., 20th A.C. (F), were posted in the low ground 350 yds. N., having just crossed Peachtree Creek. No immediate attack was expected.

Scott's (C), dashed across the rd. & down the slope met the alerted 129th Ill.—the sole guardian of the left flank of Geary's div. for a brief period. Assailed from front & flank, the 129th held on until the 105th Ill. & 70th Ind. came up.

The gap was closed with the alignment E. of Coburn's brigade. Scott's troops were driven S. of the road & in a left wheel, the 129th Ill. moved E. against the left of Featherston.

Scott's Brigade (Collier Road at Tanyard Branch)

Pressing forward, the left of Scott's line (C) moved toward Geary's batteries at Collier Rd., while the right of it, diverted by gun-fire, crossed Tanyard Branch & the road and continued down the slope N.

Harrison's Brigade (Collier Road just east of the bridge over Tanyard Branch at Redland Road)

Scott's (C) assaulting line was broken by Geary's artillery fire; the left of the brigade attempting to seize the guns while the right was diverted to the east of Tanyard Branch, where its threat to outflank both Harrison & Geary failed for lack of numbers.

On Geary's Front (Collier Road at Walthall Drive)

The left of Scott's brigade was driven from Geary's front; the right was diverted E. of the branch by Geary's oblique-fire.

My point in elaborating these details is not to engage a local historian against other esteemed scholars of the Georgia Campaign. It is only to suggest that after 150 years, if students are still working out the tactical fine-points of John B. Hood's battles, then the larger question of the general's historical reputation is at least equally open for debate.[24]

The attack of the left of Scott's brigade is easier to document. General Geary had his three brigades north of the Collier road near today's Northside Drive: Col. Charles Candy's in front, Patrick Jones's to its right, and David Ireland's in reserve to the rear. Around noon Geary ordered a regiment, the 33rd New Jersey, several hundred yards out in front of his line to hold a hill and prepare it for an artillery battery. "With no opposition we advanced to the knoll and were about to stack arms preparatory to making a place ready for the battery," recalled Lt. Stephen Pierson. "The woods were dense," such that both sides were surprised when the Confederates came on, threatening to overlap and surround the Jerseymen. "The enemy rushed upon us with his dense masses and pressed us so close," reported Lt. Col. Enos Fourat, commander of the 33rd, that "to stand longer was madness, and I reluctantly gave the order to retire fighting." "Retire fighting" is always a hard order to pull off; Lieutenant Pierson remembered that in actuality "we were all driven in great confusion and disorder." In their charge, the Confederates dropped the enemy's color bearer and captured the flag of the 33rd New Jersey along with prisoners.[25]

---

[24] Scott report, July 23, *OR*, vol. 38, pt. 3, 895; Harrison report, August 12, *OR*, vol. 38, pt. 3, 345; Scaife, *Campaign for Atlanta*, 89, f.p. 91; Castel, *Decision in the West*, 370, 379; Jenkins, *Peach Tree Creek*, 181 ("the right…Geary's division"), 183, 190 (55th Alabama, 57th Alabama and 12th Louisiana against Sloan's battery, Geary's division), 238 (105th Illinois allegedly capturing flag of 40th Mississippi); D. O'Connell, *The Art & Life of Atlanta Artist Wilbur G. Kurtz*, 117; A. P. Kurtz, *Atlanta and the Old South*, 35 (Kurtz's watercolor of fighting at Collier's Mill); Kurtz, historical marker texts, box 35, folder 2; Scruggs, *Georgia Historical Markers*, 214–16; "Georgia Tourists Should Be Told that Historical Signs Are Ahead," *Atlanta Journal*, November 18, 1954; "State Making Preparations to Erect More Historical Markers," *Atlanta Journal*, November 26, 1954.

[25] Scaife, *Campaign for Atlanta*, f.p. 91 (map); Geary report, September 15, *OR*, vol. 38, pt. 2, 139 (map, "Position on commencement of the Battle"); Strayer and Baumgartner, *Echoes of Battle*, 209 (Lt. Pierson); Fourat report, July 23, *OR*, vol. 38, pt. 2, 225; Jenkins, *Peach Tree*

Scott's left—the consolidated 27th-35th-49th Alabama plus nine companies of the 12th Louisiana, kept advancing up to Candy's breastworks and charged so fiercely that they overran them, drove the enemy back and captured three, maybe four cannon (Napoleon smoothbores of Knap's Battery, Pennsylvania Light Artillery). Col. S. S. Ives, in charge of the Alabamans, reported that the Yankees offered only "feeble resistance," and that "in their precipitous flight threw away their knapsacks, guns, and accouterments." Though Colonel Ives later complained that "we had no support on our right or left," Scott's attack received support from the next brigade to its left, that of Col. Edward A. O'Neal (Maj. Gen. Edward C. Walthall's division). Just as Scott was assaulting Candy's left, O'Neal struck Jones's right. The result was near-disaster for the Federals, as Colonel Candy himself had to admit: "About 3 P.M. the enemy made his appearance in my immediate front, and while battling with him in that direction the right flank and rear of this brigade was enveloped by the enemy in solid column, pouring their murderous fire in the rear of my line." The Rebels' charge was "so sudden and impetuous" that three of Candy's regiments broke and ran; so did the 154th New York and other regiments on Jones's right. Union artillery helped slow the enemy attack; Capt. Henry Bundy's 13th New York Light Artillery (Napoleon smoothbores) was especially effective against O'Neal's infantry. "Enveloped on three sides," Candy wrote, "it required desperate fighting to hold the ground, and if the artillery had not rendered the assistance it did the entire brigade would have been compelled to fall back and change front." General Hooker himself helped rally the troops; Ireland's brigade came up and the Federals began pushing the Rebels back. Eventually Scott gave the order for his troops to retire, and the Northerners retook their lost guns. Scott reported significant casualties in his brigade: 55 killed, 249 wounded, 86 missing for a total of 390 out of around 1,450 men carried in. In Geary's division were counted 82 killed, 229 wounded, and 165 missing, totaling 476 casualties.[26]

Colonel O'Neal was very precise in his reporting of the action of July 20. His brigade (37th Mississippi, 29th Alabama, 17th Alabama and 26th Alabama) had been posted in the outer defensive line to the right of Pace's Ferry road when the order came to advance. It marched a half-mile, halted "near

---

Creek, 175; Castel, *Decision in the West*, 370 (map), 377 ("capturing many of the 33rd New Jersey and its state flag"); Zinn, *The Mutinous Regiment*, 129–30.

[26] Scott report, July 23, *OR*, vol. 38, pt. 3, 895; Ives report, July 24, *OR*, vol. 38, pt. 3, 896 ("3 pieces of artillery"); Candy report, August 1, *OR*, vol. 38, pt. 2, 158; Jenkins, *Peach Tree Creek*, 182–83, 288, 300 (Scott's 1450); Dunkelman, "'The Worst Sight I Ever Saw,'" 73–74; Castel, *Decision in the West*, 377 (Hooker); Geary report, September 15, *OR*, vol. 38, pt. 2, 141.

the church" (Mt. Zion Baptist, according to Wilbur Kurtz, today on Howell Mill Road just north of Interstate 75). There the troops formed into battle line and stepped forth. When O'Neal's pickets exchanged fire with the Federals', "I gave the command to charge the enemy, and continue forward and drive every obstacle before them, which order was obeyed with a cheer, driving in a heavy line of skirmishers and one line of battle." O'Neal's assault struck the right of Jones's brigade and overran it. "The men seemed to be panic-stricken," a Northern officer lamented, calling the whole scene "truly deplorable." Men of the 5th Ohio saved two guns of Bundy's battery, pulling them back so they could resume firing at the charging Rebels. The Southerners tried to take the Yankees' cannon. "At one time they charged right up to the batteries of our division and a hand-to-hand conflict for the possession of the guns lasted for some minutes," wrote a Union officer, "when the enemy were forced to retire, with only the contents of the guns, instead of the guns themselves, in their possession." The Federals' tenacity and the terrain combined to stop O'Neal's assault. As they overlapped Jones's right, the Confederates poured over Collier ridge into a deep ravine, taking fire from their right (Geary's division) as well as their left (Williams's). "In danger of being flanked and captured," Colonel O'Neal reported, "the brigade fell back, not in very good order," adding that it incurred 279 casualties.[27]

To O'Neal's left, Brig. Gen. Daniel H. Reynolds led his brigade out from the Confederate outer line possibly as early as 2:30. His Arkansans drove in the Yankee skirmishers and on their left pushed on to even take a section of works held by Col. Anson McCook's brigade (Richard Johnson's division, XIV Corps). "The enemy advanced upon us in great force," wrote Union Brig. Gen. Alpheus Williams, "driving in our skirmishers with his line of battle, and, under cover of the thickets and undergrowth, coming close upon our lines before being seen." Col. Douglas Hapeman of the 104th Illinois rallied his men for a counterattack. Enfilading fire from Williams's infantry, plus Capt. John D. Woodbury's Battery M, 1st New York Light, forced the Rebels to withdraw seventy-five to one hundred yards. With O'Neal no longer supporting, and with no help on his left, Reynolds ordered his troops to retire. In the fighting, Buck Preston's brother, Maj. William Preston, commanding the artillery battalion attached to Walthall's division, was killed in action. "Willie Preston killed—his heart literally shot away—as he was getting his battery in

[27] O'Neal report, July 31, *OR*, vol. 38, pt. 3, 941; Kurtz, historical marker texts, "Old Mt. Zion Church," "O'Neal's Brigade" and "O'Neal's Brigade at the Ravine," box 35, folder 2 (Scruggs, *Georgia Historical Markers*, 216–17); S. Davis, "Atlanta Campaign," 17 ("truly deplorable"); Col. Patrick Jones report, August 1, *OR*, vol. 38, pt. 2, 209.

position," was the entry Mary Chesnut made in her diary on July 25. The next day Mrs. Chesnut remembered when, a year before, Willie had visited her. "He was worn out.... For once there was no laugh in his bonny blue eyes." Major Preston was only one casualty in the attack of Reynolds's brigade. Of 540 effectives, Reynolds counted his loss as 6 killed, 52 wounded, and 9 missing.[28]

Brig. Gen. William Quarles's brigade, Walthall's reserve, was not called up until after O'Neal's repulse. General Quarles reported moving his men forward, but by the time they got into position it was sunset and too late to launch an attack. His brigade came under incidental fire nonetheless, and lost 7 killed, 17 wounded that afternoon. French's division, farther to the left still, also saw the battle petering out before its time came to advance. "I did not engage the enemy except with the skirmishers," French later wrote. Hood's battle south of Peachtree Creek was over.[29]

What went wrong for the Confederates at Peachtree Creek? Battle is complicated, and the usual factors include numbers engaged, but here neither side held the advantage. Livermore fixes the two sides' present for duty as 21,655 U.S. versus 20,250 C.S. McMurry counts approximately 23,000 Confederates "available to fight some twenty thousand Yankees." Jenkins agrees: 20,000–24,000 Southerners engaged almost 20,000 Northerners. Numbers aside, the Federals held advantages that cumulatively decided the outcome. The rough terrain and dense undergrowth that the Southerners had to cover, plus the considerable distance to march from their outer defensive line, tired at least some of the troops before they got to their jump-off points. Lack of coordination among the Confederate divisions was also a contributor. Echelon attacks are difficult to launch properly, and several authors have noted that Stewart's troops began their advance ahead of Maney. Then, too, at some places the Federals had some form of fortification; Sherman recorded Newton's loss as slight, "as his men were partially covered by a rail barricade." Northern artillery, as noted by several Union officers, also helped in the re-

---

[28] Reynolds report, July 25, *OR*, vol. 38, pt. 3, 938; Williams report, September 12, *OR*, vol. 38, pt. 2, 34; Jenkins, *Peach Tree Creek*, 335–42; Adolphson, "An Incident of Valor in the Battle of Peach Tree Creek," 411–16 (Hapeman and the 104th Illinois); Woodbury report, September 13, *OR*, vol. 38, pt. 2, 479 ("the artillery assisted materially in repulsing three impetuous charges of the enemy"); Kurtz, historical marker text, "Maj. William C. Preston C.S.A.," box 35, folder 2 (Scruggs, *Georgia Historical Markers*, 217); Woodward, *Mary Chesnut's Civil War*, 623, 625.

[29] Quarles report, July 25, *OR*, vol. 38, pt. 3, 931; Jenkins, *Peach Tree Creek*, 69, 329; French report, July 25, *OR*, vol. 38, pt. 3, 902.

pulse. Sometimes the battle literature adds in the two or three hours lost because of the Confederates' rightward shift, but the timing probably did not matter. Hood judged that the three hours' delay in starting the battle was "unfortunate," but not a significant cause of the Confederates' repulse. Castel believes the Southerners were actually in better position at 4 P.M. than they would have been at one o'clock. McMurry agrees: "By sheer luck, the Secessionists were ideally positioned to deliver exactly the type of assault Hood envisioned." Bate's overlap of Wood's left due to the gap in Thomas's line allows some to credit the Confederates with a plan for a flanking attack at Peachtree Creek. "He intended to envelop the Union left in the vicinity of Peachtree Creek," Britt McCarley has recently written.

What must also be noted is that the Southern soldiers had charged, at least at some points, with an admirable elan. Union surgeon William Grinsted, attached to Ward's division, XX Corps, reported that 110 Rebel wounded were brought to his field hospital. They "were very severely wounded," he wrote, "many having from three to five wounds; a single wound being exceptional." Only men charging with grit will become such casualties.

Then there is General Hardee. The battle had ended just a few hours before Lieutenant Mackall recorded in his journal, "Hardee does not think proper to attack eny's strong works," a statement that could only have reflected the talk at headquarters. Hood blamed his senior corps commander for having failed to push the attack vigorously. Two months after the battle, on September 13, Hood wired President Davis, "in the battle of July 20 we failed on account of General Hardee." Then, five months later in his campaign report, he charged Hardee on two counts: first, that he had failed to post a staff officer to oversee the troops' rightward march in the hours before the battle; and second, that "Hardee failed to push the attack, as ordered." (Hood had ordered each corps commander to hold a division in reserve. What we do not see in Hood's critique is why Hardee chose his best combat division, Cleburne's, as his reserve during the battle.) Thus a battle plan calling for the enemy line to be rolled up from east to west failed at the outset because Hardee's attack had not dislodged and driven back the Yankees. The result was that "the enemy, remaining in possession of his works on Stewart's right, compelled Stewart by an enfilade fire to abandon the position he had carried." Stewart felt the same way; he blamed Loring's repulse on the failure of "the left division of the other corps" (Maney's division of Hardee's corps) to have attacked vigorously, which allowed the enemy to enfilade Loring's troops from the right. Stewart concluded, "had the plan of battle, as I understood it, been carried out fully, we would have achieved a great success." Word began to circulate that Hardee had lacked the enthusiasm needed for what was essentially the Army of Tennessee's first major attacking battle of the campaign. Allegation of Hardee's overcaution even made its way into the Southern press, as when the *Augusta Constitutionalist* claimed that Hardee's role in the battle of July 20 turned "a partial success out of what would have been a complete victory." Confederate General French sided with Hood when, based on his diary, he wrote after the war that "the failure of Hardee deranged the plan of battle."

Newspaper publication of Hood's report in February 1865, with its contention that "Hardee failed to push the attack," compelled Hardee to reply, which he addressed to General Cooper in April, just weeks before the Confederate collapse. Hardee vehemently denied having disobeyed orders. He argued that he had posted a staff officer, Maj. Samuel L. Black, "to prevent my command from moving more than half a division length to the right." Hardee in turn blamed Hood for not being on the battlefield and not being available to help direct troop movements (he claimed Hood was "in Atlanta" during the early afternoon of the 20th). As for the allegation that he had "failed to push the attack," Hardee wrote that late in the afternoon, after Walker's repulse, he was about to send in Cleburne's division, but "I received information from General Hood that the enemy were passing and overlapping the extreme right of the army, accompanied by an imperative order to send him [Wheeler] a division at once." In other words, it was Hood, not Hardee, who had ended the attack before its possibly decisive punch could be delivered.

Hardee died in 1873, unable to parry the further allegations Hood made in his memoir, published in 1880. In it Hood piled on even more, repeating "Hardee's failure to obey my specific instructions in regard to the extension of the one-half division front to the right." He also contrasted Hardee to Peter Stewart, who "carried out his instructions to the letter." Stewart had personally exhorted his troops before they went into battle on the 20th, with the result that they achieved the greatest results of the day. "An address from a corps commander to his troops, upon the eve of battle, is always productive either of great good or evil, according to the spirit in which it is spoken." Hood accused Hardee, on the other hand, of cautioning his troops about enemy earthworks ahead of them. As a result, "the troops of Hardee—as their losses on that day indicate—did nothing more than skirmish with the enemy." Hood went so far as to reconstruct a conversation he claims to have had after the battle with Pat Cleburne, in which he referred to his "peremptory orders to halt at nothing on our side of Peach Tree Creek." "Cleburne seemed surprised, and thereupon informed me that as his Division was about to move forward to the attack, on the 20th, General Hardee rode along the line, and, in the presence of those around him, cautioned him to be on the lookout for breastworks." As a result, Hood characterized Hardee as a commander who was competent on the defensive and in retreat, but one who "was wanting in that boldness requisite for offensive warfare."

Defense of Hardee's reputation against these allegations fell to Col. Thomas B. Roy, the general's A. A. G. (and son-in-law), who composed a long and meticulous essay published in the *Southern Historical Society Papers* in fall 1880. Roy objected to the fact that Hood's wartime accusations were now "reproduced, amplified and reinforced." He reprinted Hardee's full report to Cooper, adding his own points of emphasis. He maintained that it was not Hardee's fault that Cheatham's troops marched farther than they should have and delayed the attack set for 1 P.M. Moreover, Roy had gotten in touch with several officers, asking their recollection of events. He excerpted the letter Major Black, the officer overseeing the march, had written him May 31, 1880, which affirmed that Black had protested Cheatham's errant march but was powerless to stop it. He also produced Capt. Irving A. Buck's letter to him of March 27, 1880, repeating that Hood had ordered a division (Cleburne's) late on the 20th to Wheeler's assistance east of the city. Roy points out that Hood fails to mention this important fact in his memoir.

Writing late in their lives, some of Hood's veterans sided with him on such issues as Peachtree Creek. "Hardee did not press the fight in front of him," was the terse judgement of William J. Worsham of the 19th Tennessee, writing at the turn of the century.

In the dispute between Hood and Hardee, historians, as in most questions of fact, take both sides. In an early article, Thomas R. Hay concludes that Hardee "does not seem to have made the best use of his forces at the point of superiority, *viz.*, in front of Newton's small division." As a result, while Hardee cannot entirely be blamed for the Confederates' failure on July 20, Hay judges that "he was, in a measure, responsible for the lack of a *complete success.*" Robert D. Little, writing on "General Hardee and the Atlanta Campaign" in 1945, approves of Colonel Roy's essay in *SHSP*, calling it "an overwhelming answer to Hood." Stanley Horn (1941) accepts the judgment of Union general Jacob Cox that in making his accusations against Hardee, Hood "was neither just nor generous." Indeed, Horn asserts, "Hardee seems to have done all that could have been expected of him in the circumstances." Hardee's biographer Hughes asserts that the general was overcautious in his rightward troop movement before the battle and thus took too much valuable time in making it. But he calls Hood's (and Stewart's) allegations of Hardee's lack of aggressiveness "unfair." Connelly charges that Hardee "handled the attack badly" and that "he did not have control of his divisions," yet he sides with Hardee about Hood's non-presence: "Hood did not remain on the field, but retired to Atlanta, four miles distant. In his absence his design disintegrated along Peachtree Creek." McMurry alleges that Hardee "sulked through the day and did not give his best efforts to making the attack a success." Castel seems agree with Hood in stating that "Stewart, in bitter words, has told him that it [the attack] would have succeeded if Hardee had done his part." On the other hand, Castel does not accept Hood's postwar claim that Cleburne told him of Hardee's warning about breastworks. Another writer, David Chollet, quotes a letter in which well after the war a Confederate attached to Hardee's headquarters recalled that after Peachtree Creek one of Maney's brigade commanders "told me, positively, that his brigade had orders not to attack breastworks."

McMurry faults both Hood and Hardee for not having sent staff or engineer officers ahead to examine the ground to be covered in the attack. As we have seen earlier, General Lee took that prudent precaution before his attack on the second day of Gettysburg. Jenkins too criticizes Hardee for no "proper reconnaissance," but he takes Hardee's side in the matter of Hood failing to supervise the battle. More important, Jenkins accepts Hood's contention that Hardee failed to convey the desperateness of the attack ("everything on our side of the creek was to be taken," in General Stewart's words). "As there is a complete absence of any record that such an order was ever given to any of Hardee's divisions, brigades, or regiments," Jenkins contends, "Hood's order for such a charge must have stopped with Hardee." For these reasons, he concludes, "it is fair to lay the loss at Peach Tree Creek at the feet of Hardee and two of his subordinates, Bate and Maney."

Hardee continues to have his defenders. At Peachtree Creek, Thomas E. Schott concludes, "Hardee did his duty, followed his orders, and rendered critical decisions when he had to." Moreover, a recent student's charge that "Hardee allowed his troops to attack piecemeal" overlooks the inherent danger of echelon attacks becoming "piecemeal."

As usual, Castel finds a pithy way to summarize the Southerners' side of the battle: "where the Confederates had the advantage in strength [Hardee's corps], they did not fight well; and where they fought well [Stewart's corps], they were too weak."[30]

---

[30] Livermore, *Numbers and Losses*, 122; McMurry, *Atlanta 1864*, 150–52; Jenkins, *Peach Tree Creek*, 395, 397, 405, 407–408; Davis, "Hood Fights Desperately," 17; Sherman to Halleck, July 21, *OR*, vol. 38, pt. 5, 211 ("rail barricade"); Hughes, *Hardee*, 221 ("Some overeager

Still, Hardee's point about Hood's absence on the battlefield of July 20 needs to be considered. Thomas's biographers make much of their general's presence on the field and the impact it had on the battle's outcome. When, late in the fight, Bate's division finally seemed to be advancing, Thomas rode to Peachtree Creek and up the ridge with several artillery batteries. He even thwacked the rumps of the battery horses with his sword to speed them along, then he personally directed the cannonfire. "Here he sat on his horse and directed their fire," writes Van Horne, who admiringly concludes, "by his conduct and his swiftness in moving and firing his artillery—his only resource—he saved his flank, and defeated the enemy. Seldom has an army commander done so much by direct act, to defeat an enemy and win a victory." Another early biographer praises Thomas's "quickness of thought and splendid valor in action." Freeman Cleaves refers to General Newton's seeing Thomas "calm and resolute" as he directed his guns "and rescued his flank." There were other observers. An officer in the 70th Indiana saw Thomas tugging at his whiskers and knew that he was worried; "he had his whiskers all out of shape." "But when he saw the Rebels at last retreating Thomas threw his hat on the ground. 'Hurrah! Look at the Third Division! They're driving them!'" The officer noted, "His whiskers were soon in good shape again."[31]

Francis McKinney credits Thomas for having "held the left" of his line by bringing up the batteries to repel the enemy assault. Wilbur Thomas, in *General George H. Thomas*, also hails Thomas's artillery direction, "at which he was a master," having taught artillery at West Point before the war. The general's trio of recent biographers hit the same note. "Thomas led the guns forward himself"; "at one pivotal moment, he helped bring up a battery.... It

---

units in his [Stewart's] command had attacked as early as 2:30 P.M."), 224–25 (including *Constitutionalist*); Castel, *Decision in the West*, 375, 379, 383; McCarley, "'Great Question,'" 39; Clauss, "Atlanta Campaign" dissertation, 97 (Dr. Grinsted); McMurry, "Mackall Journal," entry of July 20; Hood to Davis, September 13, *OR*, vol. 39, pt. 2, 832; Hood report, February 15, 1865, *OR*, vol. 38, pt. 3, 630–31; Stewart report, January 12, 1865, *OR*, vol. 38, pt. 3, 871; S. G. French, *Two Wars*, 219; Hardee report, April 5, 1865, *OR*, vol. 38, pt. 3, 698; Hood, *A & R*, 168–69, 171, 185–86; Roy, "General Hardee," 338–40, 345–50; Worsham, *The Old Nineteenth Tennessee Regiment*, C.S.A., 128; Hay, "The Atlanta Campaign" (Part One), 38; Little, "Hardee," 18; Horn, *Army of Tennessee*, 353; J. D. Cox, *Atlanta*, 162; T. L. Connelly, *Autumn of Glory*, 440–41, 443–44; McMurry, *John Bell Hood*, 130; Chollet, "Advance and Retreat: Rage or Reason?," 62; Cisco, *States Rights Gist*, 127 ("Hardee allowed his troops to attack piecemeal"); Schott, "Lieutenant General Hardee," 160.

[31] Jenkins, *Peach Tree Creek*, 381–82; Van Horne, *The Life of Major-General George H. Thomas*, 244; Coppee, *General Thomas*, 219; Cleaves, *Rock of Chickamauga*, 230; O'Connor, *Thomas*, 282.

was the turning point of the battle"; "Thomas stood his ground and blasted the danger into oblivion."[32]

Against this rich documentation of Thomas's presence on the battlefield at Peachtree Creek we have no such record of Hood's, because the Confederate commander was no closer to the battlefield than General Stewart's headquarters at the Ira Foster house, a mile and a half away from the fighting. To be sure, one may say that an army commander's proper place, once the battle has begun, is behind the lines, at his or a corps commander's HQ (think of Lee at Gettysburg, or Rosecrans at Chickamauga). For Hood, too, riding horseback was a more trying exercise than it was for George Thomas. Whether Hood on the field could have better coordinated his divisions in their echelon attack is an academic diversion. Still, given the flatteries by his biographers, Thomas's role in the battle of July 20 stands in sharp contrast to Hood's. If Hood criticized Hardee for mismanaging the troop movement to the right, Hardee's defenders are right in asking where the commanding general was as his first plan of battle was being delayed, if not derailed.[33]

Casualties for the battle of Peachtree Creek began to be counted soon after the firing ended. Sherman reported to General Halleck a day after the battle a loss of some 1,600 of his soldiers. In contrast, he stated that Newton had buried 200 Rebels and was satisfied that "at least 1,200" of the enemy had been wounded. Hooker buried "about 400 of the rebel dead," and "thinks the

---

[32] McKinney, *Education in Violence*, 350–51; W. Thomas, *General George H. Thomas*, 488; Einolf, *George Thomas*, 252; Bobrick, *Master of War*, 253; Wills, *True as Steel*, 270.

[33] McMurry, "Mackall Journal," entry of July 20 ("Gen Hd stays at Stewarts Hdqrs on Pace Ferry road"); Kurtz, historical marker texts, "Gen. Stewart's H'dq'rs," box 36, folder 1 (Scruggs, *Georgia Historical Markers*, 227); Kurtz, "Map of Atlanta as of 1938" (showing site #29, "Gen. A. P. Stewart's H'q'rs—July 20").

Sherman's response to Thomas's victory at Peachtree Creek shows his coarser, petty side. For hours he was unaware that the battle had even begun. At 3:25 P.M. he sent to Thomas instructions that "all your troops should push hard for Atlanta sweeping everything before them." As Castel has acidly written, this message was delivered when Thomas's troops were in the very thick of their fight. Castel further notes that at 8 P.M. Sherman wrote McPherson that because of "Thomas's slowness and excessive caution, 'our only good chance of entering Atlanta by a quick move if possible is lost.'" In his nighttime message to Thomas Sherman did not even mention the battle, and, in a telegram to Halleck at 9 P.M., he merely reported, "we have taken several hundred prisoners, and had some short severe encounters but they were partial." This dismissive, callous behavior causes Ethan Rafuse to add "the way he sneered at the character of the men in the Army of the Cumberland at Peachtree Creek" to his list of reasons why Sherman is among "the generals we love to hate" (Castel, *Decision in the West*, 379; Castel, *Winning and Losing*, 98; Sherman to Thomas, July 20 [3:25 and 8 P.M.], Sherman to McPherson, July 20 [8 P.M.], Sherman to Halleck, July 20 [9 P.M.], *OR*, vol. 38, pt. 5, 195–98; Rafuse, "Generals We Love to Hate," 35).

rebel wounded in his front equal to 4,000." These numbers would have added up to Sherman as 5,800 Confederate casualties, but he commented, "I don't like to make guesses in such matters." A few weeks later, General Thomas confirmed his "severe" losses as 1,600 killed and wounded for Hooker's XX Corps and Newton's division of the IV. Although General Sherman affirmed that he didn't like guessing at numbers, in his campaign report he did just that, hazarding a conjecture that Hood's loss at Peachtree Creek "could not have fallen short of 5,000, whereas ours was covered by 1,500 killed, wounded and missing." A decade later, in his memoirs, Sherman repeated Hooker's estimate of 400 Rebel dead and 4000 wounded, noting that "this was conjectural." Ten years after that, writing for *Battles and Leaders*, Sherman refined the casualty figures as 4,796 Confederate and 1,710 Federal. This latter number reflected the losses reported by the four division commanders engaged at Peachtree Creek (Newton, Ward, Geary, and Williams). But Col. Anson McCook's brigade of Johnson's division, XIV Corps, also got into the fight; its loss of some 70 men would bring the Federal casualty count up to around close to 1,800.[34]

Hours after the battle, at about 11 P.M., Hood sent his telegraphic update to Secretary Seddon: "At 3 o'clock to-day a portion of Hardee's and Stewart's corps drove the enemy into his breast-works, but did not gain possession of them." This was an accurate summation, to which he added, "our loss slight." In his campaign report, written months later, Hood made no mention of casualty figures for Peachtree Creek. The first calculation of Confederate losses was apparently made by Livermore, who put them at "about 2500" killed and wounded. Writing at the turn of the century, the Englishman made scrupulous use of the various Confederate officers' reports in the *Official Records*, and thus would have compiled his number as follows: Loring's division 1,062, Walthall's three brigades (Reynolds, O'Neal and Quarles) 370, and French's division a mere 19, for a total of 1,451. He would have done some informed guesswork for Hardee's loss, based on numbers of troops engaged, ventured about 1,000, and thus concluded that Confederate killed and

[34] Sherman to Halleck, July 21, *OR*, vol. 38, pt. 5, 211; Thomas report, August 17, *OR*, vol. 38, pt. 1, 156; Sherman report, September 15, *OR*, vol. 38, pt. 1, 71; Sherman, "Grand Strategy of the Last Year," 4:253; Newton report, September [undated], 1864, *OR*, vol. 38, pt. 1, 298; Ward report, September 7, *OR*, vol. 38, pt. 2, 329; Geary report, September 15, *OR*, vol. 38, pt. 2, 141; Williams report, September 12, *OR*, vol. 38, pt. 2, 34; Brig. Gen. Richard Johnson report, August [undated], *OR*, vol. 38, pt. 1, 524 ("slight loss"); Jenkins, *Peach Tree Creek*, 343 (71 casualties in McCook's brigade). Thomas's loss on July 20 represented less than half of his army's casualties for the month. Breakdown of KWM by divisions is in Palumbo, *George Henry Thomas*, 250, from *OR*, vol. 38, pt. 1, 158–59.

wounded on July 20 totaled approximately 2,500. The most recent and authoritative chronicler of Peachtree Creek, Robert Jenkins, has scoured the sources and comes up with an even lower figure: 2,316 Confederates killed, wounded, and missing.[35]

General Sherman may be forgiven exaggerating Rebel casualties at Peachtree Creek, but when Atlantans accept Cump's number it is altogether a different matter. At the time of the Centennial, Atlanta's Civil War community largely sided with Joe Johnston in the Johnston-Hood-Davis controversy. In such a frame of mind, some Atlantans preferred to accept Sherman's estimate of Hood's battle losses as a way of criticizing Hood's generalship. William Key, a writer for the local papers, is an example: "The Confederate losses on July 20, 1864, at Peachtree Creek stood at 4,796 as compared to only 1,710 for the Federals." Allen Julian, head of the Atlanta Historical Society, wrote the text for several metal plaques placed in 1964 by the Georgia Civil War Centennial Commission in an Atlanta park at the site of the Peachtree Creek battlefield, and repeated the 4,796 number (the plaques are still there today). In his article for *Civil War Times Illustrated* in summer 1964, Wilbur Kurtz followed suit by repeating Sherman's tally of 4,796 Confederate killed, wounded and missing. Writing in 1972, Dr. McMurry also accepted the local "wisdom" that "it seems now that 4,700 is a reliable estimate of Confederate casualties." Richard repeated this a few years later in his biography of General Hood when he wrote, "the Confederate casualties were perhaps 5,000." Atlantan William R. Scaife maintained the tradition in his narrative, stating that "the attacking Confederates had approximately 18,450 engaged and losses of 4,800."[36]

In 1989 Jim Miles, another Georgian, accepted 4,800 as Hood's loss at Peachtree Creek, noting Sherman's 1,780. So did Atlanta suburbanite Wiley Sword, writing in 1992 that Hood at Peachtree Creek lost "nearly 5,000 men." A writer for the local newspapers, Webb Garrison, in 1995 wrote his summation of the campaign, and posted the 4,796 Confederate number against the 1,710 Federal. Writing for the *Encyclopedia of the American Civil War*, Atlanta

---

[35] Hood to Seddon, July 20, *OR*, vol. 38, pt. 5, 894; Livermore, *Numbers and Losses*, 122; Jenkins, *Peach Tree Creek*, 397 (2,316 Confederate casualties).

[36] William Key, "Hood Replaces Johnston," *Atlanta Journal*, July 30, 1964, 6; [Allen P. Julian], Atlanta Historical Society plaque at Tanyard Branch Park, Peachtree Creek battlefield, Atlanta; B. DuBose, "Georgia's Civil War Centennial Activities," 63 (on Col. Julian as author of the plaque texts; my thanks go to Helen Matthews of the Atlanta History Center for finding this source); Kurtz, "Sherman Versus Hood in the Fighting at Atlanta," 10; McMurry, *The Road Past Kennesaw*, 44 (reprinted in *Battlefields of the Civil War*, 452–521); McMurry, *John Bell Hood*, 128; Scaife, *Campaign for Atlanta*, 57.

suburbanite John E. McKay repeated that casualties for the opposing forces numbered 1,780 Union and 4,800 Confederate. Philip Secrist, staunch apologist for Johnston, used the "more than 5,000" toll as recently as 2006.[37]

As one might suspect, it was Connelly who broke this logjam of mistakes, writing in *Autumn of Glory* (perhaps with an eye back to Livermore) that Hood's losses at Peachtree Creek were "probably not exceeding twenty-five hundred." Castel, working through the *Official Records*, came to the same conclusion, noting with some exasperation that "most historians have followed him"—that is, Sherman and his inflated number of 4,796 Rebel casualties. At last, McMurry has also come around in his campaign study. Observing this historiographical ebb and flow, some years back for Savas and Woodbury, I observed that one of Vietnam's lessons to historians is that one should very warily accept the United States Army's enemy "body counts." Now, fifty years after the Centennial, it appears the lesson is at last sinking in, as evidenced by David Coffey's, Russ Bonds's, and Gary Ecelbarger's more recent work. Nevertheless, the redoubtable John Keegan trots out the old 4,796 vs. 1,710 numbers in his general history of the war.[38]

---

[37] Miles, *Fields of Glory*, 137; Miles, "The Campaign for Atlanta," 55 (repeating "Hood's army had suffered 4,800 casualties"); Sword, *Embrace an Angry Wind*, 33; Garrison, *Atlanta and the War*, 151; McKay, "Atlanta Campaign," 141; Secrist, *Sherman's 1864 Trail*, 143.

[38] T. L. Connelly, *Autumn of Glory*, 444; Castel, *Decision in the West*, 381 ("most historians have followed him" [Sherman and his 4,796]); McMurry, *Atlanta 1864*, 152; Davis, "Reappraisal," 81; Coffey, *John Bell Hood and the Struggle for Atlanta*, 71; Bonds, *War Like the Thunderbolt*, 106; Ecelbarger, *Day Dixie Died*, 26; Keegan, *The American Civil War*, 265.

E. B. Long and Barbara Long give the 4,796 casualty count in *The Civil War Day by Day*, 542. Some historians have glibly and inexplicably exceeded even Sherman's exaggeration of Confederate casualties at Peachtree Creek. See Linderman, *Embattled Courage*, 137: "At Peachtree Creek Hood lost between 5,000 and 6,000 in killed, wounded, and missing to Sherman's 1,800."

Larry Daniel agrees that Confederate casualties at Peachtree Creek totaled 2,500, but he errs in stating that this number is "a staggering 25 percent of those engaged." This does not conform to the figures authoritatively set by Robert Jenkins, that Hood's loss on July 20 in terms of percentages was well lower than that, about 10 percent of the 20,000–24,000 Confederates engaged (Daniel, *Days of Glory*, 414; Jenkins, *Peach Tree Creek*, 397).

Daniel also relates an incident which he places in his narrative of Peachtree Creek but consequently misdates: "The next day, July 23 [*sic*], a Federal yelled across the line, 'Johnny, how many men have you got left over there?' Came the reply, 'About two more killings.'" The story seems to have been popular among Sherman's troops because it was told in other settings and given other dates. General Cox relates the yarn ("Oh! About enough for another killing"), but attaches it to *all three* of Hood's attacking battles, July 20, 22, and 28, not to one solely. But in another iteration, the sardonic exchange occurred after Bate's repulse at Dallas, May 28. Russell Blount tells the story ("Well I guess there's enough for another killin'"), based on John W. Clemson's article in *National Tribune*, September 1897, in which Clemson tells how the

Hood does not mention it in his memoirs, and we do not see it in the *Official Records*, but there is some evidence that the general's battle blood was up and that he resolved to continue attacking the Federal lines on July 21. McMurry mentions this, citing Lieutenant Mackall's journal, which he found at William and Mary. In this important document, under date of July 20, is the entry, "At night determined to 'pierce eny's centre.'" The next day, however, Mackall wrote, "Order of renewal attack countermanded." There is no reason to doubt Mackall's veracity, and his comments shed light on John Bell Hood's inclinations as commander. Stymied and frustrated by a lost or unfulfilled tactical opportunity (that of rolling up Thomas's troops south of Peachtree Creek), according to Mackall, Hood's instinct was to attack them again the next day, under far less favorable circumstances. Unfortunately, we will see this behavior again, at Franklin.[39]

From Peachtree Creek, Hood's eyes turned toward Decatur. The Federals' advance against him on July 20 had spurred Wheeler to call for help. At 1:10 P.M. General Mackall at headquarters sent an urgent message: "Are you driven back? Hold at all hazards." Mackall informed Wheeler that a brigade was being sent from Cheatham's division (Brown's brigade), which was ordered to extend to the right, reaching the railroad. But Wheeler at 5:45 sent further word that the enemy was pressing him. Hood was at Stewart's headquarters when he got this message. He at once ordered Hardee to send Cleburne's division to march to Wheeler's help. Maj. A. P. Mason, Hood's

---

Federals at Dallas got the Rebels to raise their heads above their trench parapets by cheering and blowing bugles as if in preparation for a charge. The curious "Johnnies" were the ones who got shot by Federal sharpshooters. Afterward came the yelled question, "'Say Johnny, how many of you are over there?'" Bill Scaife gives a version as well, but this time the exchange between Yank and Reb takes place after Ezra Church, July 28 (although his source is General Cox). Albert Castel has found the anecdote in the letter of Union Maj. Thomas Taylor to his wife, written July 30 after Ezra Church. Taylor relates that he had asked a captured Confederate officer how many men he had left, and was told, "Hood has about enough left to make two killings." Earl Hess cites three additional sources placing the incident after the battle of July 28 (Hess, *Ezra Church*, 163).

Peachtree Creek, Dallas, Ezra Church: there are as many "enough for another killing" (or two killings) stories as there are "Lee to the rear" episodes in the apocrypha of the war (Commager, *The Blue and the Gray*, 996). Maybe one day writers of history will reach an agreement about this alleged incident that really fits it into the chronology of the Georgia Campaign (Daniel, *Days of Glory*, 414; J. D. Cox, *Atlanta*, 186; Clemson, "Surprised the Johnnies," *National Tribune*, September 30, 1897, cited in Blount, *New Hope Church*, 150; Scaife, *Campaign for Atlanta*, 108; T. T. Taylor, "A Buckeye Warrior at Atlanta's Gates," 81; Castel, *Tom Taylor's Civil War*, 158; Hess, *Ezra Church*, 163).

[39] McMurry, "Mackall Journal," entries of July 20 and 21; McMurry, *John Bell Hood*, 130.

A. A. G., sent word at 7:15 that evening: "Cleburne's division is moving to your support."

By twilight, with the battle of Peachtree Creek sputtering out, Cleburne had his division in motion. The march route entailed three miles to the Car Shed downtown, then a couple more out along the railroad. At the edge of the city limits Cleburne allowed his men to rest for two hours before resuming their march. They reached Wheeler's line before dawn only to find that the cavalrymen were not well fortified, having only dug rifle pits. This meant more work for troops who were already exhausted. Cleburne posted one brigade north of the railway, and the rest of the division moved into line south to the bald hill.

At daylight on July 21, the Yankees started a fierce bombardment that killed forty Confederates and wounded a hundred more in less than five minutes. "I have never before witnessed such accurate and destructive cannonading," recorded Brig. Gen. James A. Smith, who had taken over Granbury's brigade. About 8 A.M. came the Federal attack against the hill. Leggett's infantry quickly took it; Brigadier Samuel Ferguson's cavalrymen "gave way in some confusion," as Wheeler put it. Then Blair ordered Brig. Gen. Giles Smith (who that day took over for the wounded Gresham) to charge north of the hill. There, too, the cavalrymen gave way. A counterattack by Smith's Texas brigade drove the Federals back in that sector, but they still held the hill, which was the real prize; Leggett placed artillery on it. "By 11:00 A.M. most of the killing was done for the day," writes Gary Ecelbarger, the recent chronicler of the battle of July 22. In those three hours the Federals lost 700 men taking and defending the hill; Cleburne lost 300 men defending it against Leggett's first attack, then trying to retake it (J. A. Smith counted in his brigade 47 killed, 120 wounded, and 19 captured). Wheeler did not provide casualty figures for July 21, but given the celerity of the cavalrymen's departure, they would have been light (Ecelbarger estimates from fifty to a hundred).

A number of Southerners had been captured, and these were marched to Blair's headquarters where the general, standing beneath the Stars and Stripes, proceeded to lecture the Rebels on the principles of patriotism and love of country. "He made us a speech and told us what bad boys we were and that we ought to honor the flag," wrote Pvt. Alfred Gibbons of the 1st Georgia Cavalry, "but I felt more like cutting his throat than listening to his speech."

With Union artillery now placed on the hill, Cleburne made no further effort to retake it. After 11 the fighting turned into long-range (at 150 yards) skirmishing and artillery firing. Cleburne still needed help. At around 9 A.M., Hood directed Hardee to send Maney's division marching to extend Cleburne's right; Maney's troops began arriving after noon. A couple of hours later Hood's headquarters ordered Wheeler to "hold the gap between Generals Cleburne and Maney," and informed him that infantry would be sent that night to fill the gap. The fighting that morning had been hard; Cleburne called it "the bitterest" of his life. But the Southerners had accomplished their larger objective, slowing McPherson's advance on Atlanta. Colonel Roy later described the achievement as "a combination of good luck, audacity, and hard fighting."[40]

Aware that his troops would at some point have to give up their outer line and retire into the main Grant-Presstman fortifications ringing the city, Hood had his engineers directing work to strengthen them. Johnston had begun some of this after crossing the Chattahoochee on July 10. Because Captain Grant's original line impinged too closely upon the city, in mid-April a salient was being laid out to throw works farther northwest of town. But as of July 21, these positions had not been completed. On the afternoon of July 13, a spy named Moore had slipped out of the city on the Peachtree Road, then filed a report with Union Brig. Gen. Thomas J. Wood. Along their fortified line a mile north of town, Moore observed "the rebels were busily engaged in constructing four separate forts on the separate hills." But at the time of Hood's assumption of command, this labor was not completed.

"The works already constructed for the defense of the place," Hood later wrote, were "wholly useless from their position." A big fort had been started at the tip of the northwest salient, but entrenchments connecting it with other positions had not been finished. When Hood at last brought his army within the entrenched perimeter, during the night of July 21–22, further digging ensued with earnest, including at the big northwestern fort. "Not until...the night of July 21 did this isolated battery emplacement become an integral part of the defense line of Atlanta," writes Kurtz. "This fort, and a few others eastward, were then in feverish haste, strengthened and connected by rifle pits,

---

[40] Mackall to Wheeler and Cheatham to Wheeler, July 20, *OR*, vol. 38, pt. 5, 895–96; Mason to Wheeler, July 19 [*sic:* 20], 7:15 P.M., *OR*, vol. 38, pt. 5, 893; Symonds, *Stonewall of the West*, 224; Purdue and Purdue, *Pat Cleburne*, 347–48; Castel, *Decision in the West*, 383–86; Wheeler report, October 9, 1864, *OR*, vol. 38, pt. 3, 952; Smith report, August 5, *OR*, vol. 38, pt. 3, 746; Blair report, undated, *OR*, vol. 38, pt. 3, 544 (728 casualties in Leggett's and Smith's divisions); Ecelbarger, *Day Dixie Died*, 34, 45, 252 n.32 (Wheeler's loss); Strayer and Baumgartner, *Echoes of Battle*, 219; Maj. A. P. Mason to Cleburne, July 21, 9 A.M., *OR*, vol. 38, pt. 5, 898 (Maney's division); John S. Smith to Wheeler, July 21 [no time given] and 2:30 P.M., *OR*, vol. 38, pt. 5, 899; Hay, *Cleburne and His Command*, 233; Roy, "General Hardee," 353.

obstructions and palisades, the negro laborers and the soldiers working all night to make them tenable by the troops of Stewart's corps."

This fort, designated only as "X" in Grant's map of April 12 (showing it unconnected by rifle pits to the other forts) came to be called Fort Hood sometime during the campaign. A Confederate artilleryman, Sgt. Edmund T. Eggleston of Cowan's Mississippi Battery, recorded in his diary for August 28, "We left Fort Hood today and moved to the left to join our division." Kurtz concludes that it was named for Hood (and not Johnston) because Southern troops completed work on it under Hood's tenure as army commander. Eventually the Confederates had six artillery forts from the tip of the salient (Fort X/Hood, now on the campus of Georgia Tech) to Peachtree Street, where "Battery K" was sited (now the Fox Theatre).

Hood ordered Lieutenant Colonel Presstman, his chief engineer, to inspect the line and report on its fitness. On the morning of July 21 Presstman's report came in: portions of Grant's works to the northeast—much like those to the northwest—were too close to the city and on ground which was too low. Hood therefore instructed him to erect a northeast salient, which added hundreds of feet of entrenchments and two artillery forts to Captain Grant's original perimeter, bringing it from an estimated ten and a half miles initially to now more than twelve. Authorities had been in the practice of impressing slaves to build the primary line of works encircling the city; for this eleventh-hour work they did the same, even more feverishly. The Yankees could see this activity from afar, or at least heard of it. Years later a Northern cavalry officer recalled, "Negroes were brought in from all directions and made to work night and day until the defenses assumed a form that would last for years."

Not only Negroes, but Confederate soldiers. When Gen. Daniel Reynolds's Arkansas brigade marched into the city perimeter during the night of July 21–22, it took position west of the Western & Atlantic. "My men were placed in 1½ ranks and we commenced the work of fortifying after midnight," Reynolds entered in his diary on the 21st. The next morning he expressed satisfaction that he had "a good skirmish line and good abatis in front of my line and traverses dug to protect against any enfilading fire." The Arkansans did not stop there. "Still at work on our fortifications to make them strong," Reynolds wrote on July 23. "Strengthening our lines by abatis and also our picket lines," he recorded the next day. "In response to a call from Gen. Hood," he added, "the whole army is busy at work."

The fortifications around Atlanta were so important as to command the attention of the higher-ups in Richmond. Stephen Presstman had served capably, but evidently the administration wanted someone better to assist Hood in defending Atlanta: Maj. Gen. Martin L. Smith, then serving as chief engineer for no less than Lee's army in Virginia. According to one source, the president himself asked Lee to release General Smith for duty with Hood; Lee regretfully acceded. The transfer took place quickly. On July 20, Col. Walter Taylor, Lee's A. A. G., issued orders relieving Smith of duty in Virginia so he could go to Georgia "in accordance with instructions from the honorable Secretary of War." It would take more than a week for Smith to arrive in Atlanta and assume his new position (on August 1); in the meantime, Lieutenant Colonel Presstman continued as chief engineer. Evidence of President Davis's interest in the officer transfer is seen in his telegram to Hood, July 23, informing him that Smith was being sent. Davis commended him to Hood, stating that with Lee, Smith "has won the highest reputation." The next day Hood answered, "I shall be happy to have the services of Major General M. L. Smith." More help came a week later, when the War Department ordered another engineer, Lt. Col. Bushrod W. Frobel, from Wilmington to join Hood's army. Frobel arrived in Atlanta, and reported for duty to Maj. Gen. Smith, becoming assistant engineer-in-chief for Hood's army.[41]

---

[41] Grant map, "Sketch of the City of Atlanta and Line of Defenses, Apr. 12. 64," in *OR Atlas*, plate 51, fig. 2; Thomas J. Wood to J. S. Fullerton, July 14, *OR*, vol. 38, pt. 5, 138; Hood report, February 15, 1865, *OR*, vol. 38, pt. 3, 631; Hood, *A & R*, 173–74; S. Davis, *What the Yankees Did to Us*, 112–13; Kurtz, "Fort Hood and the Ponders House," *Atlanta Constitution Sunday Magazine*, July 10, 1932, 5, 14; E. T. Eggleston, "Excerpts from the Civil War Diary of E. T. Eggleston," 349 (since published in Eggleston, *To Succeed or Perish*, 58); "Map Showing Battle Fields of July 20, 22 & 28, 1864" (showing six battery sites); Venet, *A Changing Wind*, 145 (slave impressment); Brackett, "With Sherman at Atlanta," 6:435 ("Negroes brought in"); Reynolds, *Worthy of the Cause*, 138–39; Dyer, *Gallant Hood*, 258 ("All during this day Hood was laboring to build new inner defenses against Thomas and Schofield north and east of the city"); A. L. Hull, "Maj. Gen. Martin L. Smith," 530 (President Davis...desired Gen. Lee, if possible, to spare Gen. Smith"); Bergeron, "Martin Luther Smith," 5:180 ("Davis, perhaps upon the recommendation of Robert E. Lee, sent Smith to Atlanta"); W. H. Taylor, Special Orders No. 169, July 20, 1864, *OR*, vol. 40, pt. 3, 787; Saml. W. Melton Special Orders No. 170, Richmond, July 20, *OR*, vol. 38, pt. 5, 898; Falconer Special Field Order No. 64, August 1, *OR*, vol. 38, pt. 5, 940 (Smith assigned to duty in Atlanta); Davis to Hood, July 23, *OR*, vol. 38, pt. 5, 898; Hood to "The President," July 24, Telegram Book I, Hood Personal Papers; Shoup to S. D. Lee, September 4, *OR*, vol. 38, pt. 5, 1018 (Colonel Presstman staying with the army, commanding a regiment of engineers); Special Orders No. 183 (Richmond, August 4), *OR*, vol. 38, pt. 5, 945 (Frobel transfer); [Frobel], "The Georgia Campaign," (May 1868): 241 (assistant engineer-in-chief).

Sherman was pleased with McPherson's progress on July 21. "Still more good results will flow from your pressing hard," he had instructed his friend on the night of the 20th; Mac had pressed hard enough to take the bald hill. "From this hill," Sherman wrote Thomas, "he has an easy range of town"; Sherman looked forward to a bombardment of the city. At the same time McPherson on the afternoon of the 21st reminded the commanding general of the significant enemy force in his front. ("General Leggett reports having seen at least ten regiments of infantry passing in that direction" [which would have been Maney's troops].) In a sarcastic vein that would have been impudent if Cump and Mac had not been such good friends, McPherson added, "I will simply remark in closing…that the whole rebel army, except Georgia militia, is not in front of the Army of the Cumberland."

Sherman issued orders on July 22 for his three armies to continue pushing forward in their concentric arc north and east of the city. "Each commander will endeavor to advance his line if it can be done without a direct assault on the enemy's parapets if held in force." All the while, Federal cannoneers were instructed to "open a careful artillery fire on the town of Atlanta, directing their shots so as to produce the best effect."

Sherman's intent is not clearly expressed in ordering a shelling of a city still filled with at least several thousand civilians, including old men, women and children. It did not need to be. By this time, the third year of the war, bombardment of cities had become commonplace. Grant had shelled Vicksburg during his siege the year before; as Sherman neared Atlanta, both Charleston and Petersburg had come under bombardment by Union artillery.

A generation after the war Confederate veterans in Atlanta were still referring to Fort Hood, as when Capt. Robert L. Barry wrote that during the enemy's semi-siege his "Lookout Battery" was ordered "to fall back into the trenches around Atlanta and command Fort Hood, on Marietta Street" (Barry, "The Lookout Battery," 385). Since the passing of the war generation the name *Fort Hood* has taken hold. In their key to Civil War-related sites marked in I. U. Kauffman's map of Atlanta, Franklin Garrett and Wilbur Kurtz described point No. 99 as "Site of Fort Hood. Northwestern salient of the city fortifications. Location of Capt. R. L. Barry's Confederate Battery during the defense of the city, July, August, 1864" ("Key to Map," 19). See also "Embattled Atlanta Street & Highway Map as of 1930," *Atlanta Constitution*, July 20, 1930, in which site 135 is "Fort Hood adjacent to the Ponders mansion."

Similarly, the fort at the opposite (southeast) side of the Grant-Presstman perimeter, Battery R, sometime afterward came to be called "Fort Walker" (*Battles of Atlanta*, 11). These two batteries are the only two, among the more than two dozen in the perimeter, which bear Confederate generals' names ([Kurtz], "Forts to Defend Atlanta," *Atlanta Journal Sunday Magazine*, May 25, 1941, 6; Marguerite Steedman, "Atlanta Fort Restored," *Atlanta Constitution*, May 2, 1937, 7). It is worth noting that none of the forts, so far as we know, came to carry the name of Joseph E. Johnston.

Sherman did not send into Atlanta any notice of his intent to shell it. Technically, he did not have to: accepted practice provided that if a place were not besieged (not completely surrounded) and only partly invested (leaving escape routes open), a commander did not need to warn a citizenry of impending bombardment.[42]

Sherman's plans for July 22 included another important element: sending Brig. Gen. Kenner Garrard's cavalry division off to further wreck the railroad toward Augusta. Garrard's cavalry had previously wrecked track west of Stone Mountain; Sherman wanted him to tear up more in the vicinity of Covington and Lithonia, farther to the east. Specifically, he charged Garrard with burning the bridges over the Yellow and Ulcofauchee [Alcovy] Rivers, torching the train depot at Covington, and destroying locomotives and cars if he could. "I want you to put your whole strength on this," he ordered, "and to do it quick and well." Sherman issued his orders at midnight on the 20th–21st; Garrard got them at 1:30 A.M. Twelve hours later, Garrard started for Decatur and within hours was riding away with his brigades toward Covington.

Garrard's departure meant that McPherson's army had no cavalry to guard its left flank, a fact which Mac pointedly informed Sherman at 3 P.M. This vulnerability led him to worry that the Rebels might attack him or raid into Decatur to capture his trains. McPherson therefore ordered Dodge to send a brigade (Brig. Gen. John W. Sprague's) to guard the wagons. To Blair, whose XVII Corps held the left, McPherson issued warning at 5:30 P.M. "Impress upon the command the importance of being on the alert at all times to repel an attack, especially about daybreak," he ordered, adding that the troops during the night should entrench and clear away brush in front of them. The Federals did so; Blair recorded that "unusual precautions were deemed essential from the absence of our cavalry force from this flank."[43]

---

[42] Sherman to McPherson, July 20 (8 P.M.), *OR*, vol. 38, pt. 5, 208; Sherman to Thomas, July 21, *OR*, vol. 38, pt. 5, 212; McPherson to Sherman, July 21 (3 P.M.), *OR*, vol. 38, pt. 5, 219; L. M. Dayton, SFO 40, July 21, *OR*, vol. 38, pt. 5, 222; S. Davis, *What the Yankees Did to Us*, 96–105.

[43] Sherman to Garrard, July 20–21 (midnight), *OR*, vol. 38, pt. 5, 209; Garrard to Sherman, July 21, *OR*, vol. 38, pt. 5, 221; Garrard report, July 24, *OR*, vol. 38, pt. 2, 809; D. Evans, *Sherman's Horsemen*, 91 (leaving Decatur 4–5 P.M.); McPherson to Sherman, July 21, 3 P.M., *OR*, vol. 38, pt. 5, 219; McPherson to Dodge, July 21, *OR*, vol. 38, pt. 5, 220; McPherson to Blair, July 21 (5:30 P.M.), *OR*, vol. 38, pt. 5, 220; Blair report, [undated] 1864, *OR*, vol. 38, pt. 3, 544; S. Davis, "Gen. Sherman's Blunder Led to McPherson's Death," 18–19; Whaley, *Forgotten Hero*, 155–56 ("It is difficult to understand Sherman's ordering all of the cavalry so far away at such a critical time").

McPherson's vulnerable predicament gave Hood an even better opportunity for attack than he had had at Peachtree Creek. Sometime on the afternoon of July 21—probably shortly after 2:30—Wheeler's scouts brought word to headquarters that the enemy cavalry were gone and that McPherson's left was in the air. Contemporary documentation of this intelligence is not to be seen in the *Official Records*, in which General Wheeler only claims to have notified Hardee of Garrard's departure on the morning of July 22. In his campaign report Hood makes no mention of Wheeler's intelligence or of the discovery that McPherson's flank was open to attack. On the other hand, Hood does make the statement in his memoir, which is historians' accepted source.

> Through the vigilance of General Wheeler, I received information, during the night of the 20th [*sic*: afternoon of the 21st], of the exposed position of McPherson's left flank; it was standing out in the air, near the Georgia Railroad between Decatur and Atlanta, and a large number of the enemy's wagons had been parked in and around Decatur. The roads were in good condition, and ran in the direction to enable a large body of an Army to march, under cover of darkness, around this exposed flank, and attack in rear. I determined to make all necessary preparation for a renewed assault, to attack the extreme left of the Federals in rear and flank....

Given the slips of memory that appear elsewhere in *Advance and Retreat*, some scholars have frankly not accepted Hood's account of Wheeler's intelligence. A number of writers imply that Hood more or less pulled his flank attack battle plan out of thin air. "Hood had, by the evening of July 21, prepared a bold plan," is Connelly's terse explanation. "During the day the Texan developed a plan to meet this threat," is McMurry's, referring to McPherson's advance. Ecelbarger makes no reference at all to Wheeler and the air-flank throughout his recent narrative of the battle. Other scholars from Horn to Castel, however, cite *Advance and Retreat* in accepting Hood's statement as to the origins of his battle plan. Stewart Bennett of Blue Mountain College, Mississippi, who is at work writing on the engagement of July 22, astutely observes, "Without this information, why would Hood plan a movement around the Union left if he was not sure that it was 'in the air'?"[44]

[44] Hood, *A & R*, 173; Wheeler to Hardee, July 22, *OR*, vol. 38, pt. 5, 900; T. L. Connelly, *Autumn of Glory*, 445; McMurry, *John Bell Hood*, 130; Ecelbarger, *Day Dixie Died*, 48–50 (no mention of Wheeler's intel on Garrard's departure); Ecelbarger, "An Ambitious Goal Indeed," 24 (Hood's plan "modeled after Jackson at Chancellorsville," but, again, no mention of Garrard's departure having left McPherson's left flank in the air); Horn, *Army of Tennessee*, 354; Castel, *Decision in the West*, 385; Stewart Bennett email to author, April 24, 2014.

In his memoir, Hood describes his battle plan:

• After sundown July 21, Stewart's and Cheatham's corps were to withdraw from the army's outer line south of Peachtree Creek into the main perimeter of works around the city;

• Hardee with his corps would also retire from the outer line and, during the night of July 21–22, march through the city, then south on the McDonough road, crossing Intrenchment Creek at Cobb's mill; he would then turn northeast in the direction of Decatur;

• his objective was to completely turn the left of McPherson's army so as to attack it from the rear, "even should it be necessary to go to or beyond Decatur," if need be, to gain the advantageous position;

• Wheeler with his cavalry to accompany the march, but riding on to Decatur was to attack the enemy's wagon trains there;

• when Hardee had forced back the enemy, Cheatham would take up the attack; and if all were successful, Stewart would join in the general advance against the retreating enemy.[45]

The closest we have to a written Confederate attack order for July 22 is a document recently found at the Atlanta History Center. The famed Civil War relic collector Beverly DuBose did not record its provenance, and it may be inauthentic. But he apparently acquired it in the belief that it was sent by Hood to Hardee. The handwritten note reads:

July 21/64

Gen Hardee

I have decided to attack the enemy tomorrow. You will hold your Corps in readiness to move by the McDonough road to-night under cover of darkness; be cautious, and for further order and plan Report to these headquarters en route

Hood

Gen Commanding

C. S. forces

("Hood to Hardee," DuBose Family Collection)

[45] Hood report, February 15, 1865, *OR*, vol. 38, pt. 3, 631.

Castel believes that Hood conceived his plan on the *morning* of July 21 (*Decision in the West*, 386). Such an assumption ignores the fact that Garrard did not begin riding toward Decatur until at least 1:30 P.M. that day; only after that could Confederate cavalry begin reporting McPherson's vulnerability. Moreover, Hood as late as 2:30 P.M. was instructing Wheeler on the need to occupy ground being held in the outer line by Cleburne's and Maney's divisions, and that he intended to fill the gap that night with infantry—a move he would not have planned had he known at that time that his army would withdraw into the city's main works (Smith to Wheeler, July 21, *OR*, vol. 38, pt. 5, 899).

The withdrawal of Confederate troops from their outer line during the night of July 21–22 involved stealth and silence. Col. James Nisbet of the 66th Georgia (Walker's division) himself crept along his picket post, quietly passing the order to fall back. He issued no general withdrawal order, however, lest some deserter carry the news to the enemy.

The shrinking of the Confederate lines meant that Hood had to move his headquarters. The Alexander Luckie home lay outside the fortified perimeter, and clearly the commanding general had to relocate. Hood moved to the Austin Leyden house on Peachtree between Ellis and Cain Streets. There, at a certain point—late in the afternoon or early evening of the 21st—Hood called a council of his three corps commanders, plus generals Wheeler and Smith. According to *Advance and Retreat*, Hood reviewed the battle plan in detail "again and again, until each officer present gave assurance that he fully comprehended his duties." Then at nightfall, Stewart's troops began marching back toward the Grant-Presstman line, which they found still so incomplete that they worked throughout the night to strengthen the works. As Bate's and Walker's divisions marched into the city via Peachtree Street, General Walker made a point, when passing Hood's headquarters, of stopping to call upon the commanding general. In the coming battle Walker pledged his full support; Hood must have given encouraging words because when Walker came outside, Maj. Joseph B. Cumming, his adjutant, noticed that Walker "was aglow with martial fire."[46]

Hardee's four divisions marched in the order of Bate, Walker, Cleburne, and Maney. "At dark that evening, in obedience to orders," Bate later reported, "though much fatigued, moved through Atlanta." The only battery Bate brought with him was Capt. Cuthbert H. Slocumb's battery of the Washington Artillery. Pvt. Philip Stephenson was one of Slocumb's loaders joining in the march that night. He remembered that the moon was full and that the battery boys were in good spirits. They sang "Annie Laurie" and other tunes; the infantrymen joined in. Around them, however, were somber scenes. The darkened city streets seemed abandoned. "Those tall dwellings remained as close and silent as convents!" Stephenson recalled; "not a head visible, nor shutter, nor door open." Many Atlantans had fled; those remaining huddled in their cellars when Sherman's shells had started shrieking. They did that

[46] Cone, *Last to Join the Fight*, 135–36; R. K. Brown, *To the Manner Born*, 263 (placing Hood's HQ at the Austin Leyden house); S. Davis, *What the Yankees Did to Us*, 174, 176–77; "Here Are Some Last Farewell Views of the Old Leyden House," *Atlanta Journal*, February 23, 1913, 8 ("General John B. Hood…occupied it"); Hood, *A & R*, 176–78; Ecelbarger, *Day Dixie Died*, 57–58; Castel, *Decision in the West*, 388.

night; Stephenson thought that the soldiers' singing had brought on the Yankees' cannonfire.[47]

The sight of so many soldiers marching through downtown unnerved the citizens, who suspected that the army was evacuating the city. "Everybody thought the army was running," recorded Lt. W. L. Trask of Hardee's staff. "The citizens became alarmed at the prospect and hurriedly packed up such things as they could carry and began to leave the city in every direction." In the confusion, "many houses were left wide open with furniture, carpets, bedding, tableware of every description," Trask noted. The opportunity for looting was too great, and some of Hardee's infantrymen broke ranks to help themselves to roadside plunder. The worst looting was committed by Wheeler's cavalrymen, who were riding with Hardee's column through the city. Samuel Richards was a downtown bookseller with a store on Whitehall Street that was looted. "A lot of cavalry robbers broke into the stores and stole everything that they took a fancy to," Richards ruefully recorded.[48]

After Bate's and Walker's divisions came Cleburne's and Maney's. Colonel Roy had written Cleburne at 7:30 P.M. that he was to get his division moving "at dark," but the troops did not leave their lines outside the city until midnight. Once inside the main lines the men were allowed a little rest, but then Hardee sent orders to Cleburne that he should set out at 1 A.M. "on the road which will be indicated by the guide."

Hood and his aides had staked out a march route that would take Hardee's corps through the city and well to the southeast of it, then to the northeast toward the flank and rear of the enemy line. Wheeler's objective lay beyond that, at Decatur. Most Southern infantrymen would not have known the length of their itinerary, although a soldier in Bate's division a few weeks later wrote his father, "during the night of the 21st we marched 12 miles." Nevertheless, all understood their route to be long. Hood had called for the assault to begin at dawn of July 22, but it soon became clear to everyone that such a timetable was impossible. Before leaving the city, therefore, sometime between 10 P.M. and midnight, General Hardee with Cheatham and maybe Wheeler paid a call upon Hood at his new headquarters. Hardee asked for a modification of plan. A surprise attack upon the enemy *rear* was unfeasible the next morning, but there was at least a possibility that, after a hard night of marching, Hardee could get his infantry in position for an assault on the enemy *flank*. "My troops had been marching, fighting and working the night

[47] Castel, *Decision in the West*, 389 (order of march); Waters, "Bate Reports," 216; Hughes, *Pride of the Artillery*, 200; Stephenson, *Memoir*, 216–17.

[48] Trask war journal, entry of July 24; Venet, *Sam Richards*, 228 (entry of July 22).

and day previous, and had little rest for thirty-six hours," Hardee later wrote, "and it was deemed impracticable to make so long a march in time to attack on the following day." Hood later contended, in another of those ungracious gestures which do not reflect well upon him, that he made no such adjustment to his plan, and that "Hardee failed to entirely turn the enemy's left as directed." In his defense of Hardee Colonel Roy corresponded after the war with Cheatham, and secured from him a statement that "on account of the lateness of the hour, the distance to be travelled, and the condition of the troops, he [Hood] finally consented that Genl Hardee should have discretion to make the assault on the flank and rear of McPherson." All this is enough to persuade McMurry, who flatly concludes, "Hood modified Hardee's orders."[49]

At the same time Hood had to deal with a tactical situation that offered as much threat as opportunity. McPherson's capture of the bald hill meant that enemy forces east of Atlanta were only seven miles northeast of the key railroad junction at East Point. Hood could not be sure that Sherman was not planning a rapid march by his left flank to cut the Macon & Western. "The demonstrations of the enemy upon our right, and which threatened to destroy the Macon Railroad—our main line for receiving supplies—rendered it imperative that I should check, immediately, his operations in that direction," he wrote in *Advance and Retreat*, "otherwise Atlanta was doomed to fall at a very early day." The threat was understood to be so dire that Hood's headquarters issued instruction on the 21st to Brig. Gen. Marcus J. Wright, post commandant of Atlanta, to prepare "for an evacuation of Atlanta, should it become necessary."

John Bell Hood had not been present at Chancellorsville, but like all Confederates, he knew of Stonewall Jackson's legendary flank march of May 2, 1863, which led to Lee's great victory. Hood himself was most sanguine that his vision of a flanking attack "would result not only in a general battle, but a signal victory to our arms." Numerous writers have commented on the Jacksonian elements of Hood's plan. "General Hood had conceived a move worthy of Stonewall Jackson, attempting to strike and crush Sherman's left wing in a flank movement," writes Captain Buck of Cleburne's staff. British historian Alfred H. Burne calls the engagement of July 22 "Hood's 'battle of

[49] Symonds, *Cleburne*, 223–24; T. B. Roy to Cleburne, July 21 (7:30 and 11 P.M.), *OR*, vol. 38, pt. 5, 899–900; Ives, *Civil War Journal and Letters*, 69 ("marched 12 miles"); Ecelbarger, *Day Dixie Died*, 59; Hardee report, April 5, 1865, *OR*, vol. 38, pt. 3 699; Hood report, February 15, 1865, *OR*, vol. 38, pt. 3, 631; Roy to Cheatham, October 15, 1881, quoted in Hughes, *Hardee*, 226; McMurry, *John Bell Hood*, 131.

Chancellorsville.'" Horn terms the plan "brilliant in its conception." John P. Dyer, Wheeler's biographer, affirms that Hood's "was a maneuver similar to Jackson's flank march and attack at Chancellorsville." More recently, Stephen M. Hood credits the Confederate commander with having designed "a broad, sweeping maneuver into Sherman's left and rear."[50]

Hood's flank march and attack at Atlanta, however similar in premise and concept to Jackson's at Chancellorsville, failed to bring about the same resounding Confederate victory, as we shall see now.

Problems began with Wheeler's cavalry. To be sure, one cannot completely blame General Wheeler. Throughout July 21, he and his troopers had been hotly engaged against McPherson; there had been little time for recon. The Southern cavalry's discovery of the enemy flank in the air was a providential stroke indeed—a gift from the "God of Battles" to whom Hood supplicated. But once the march commenced, Wheeler's performance sagged. Riding through the darkened streets of Atlanta, a number of Wheeler's horsemen were distracted by the vacant houses and silent storefronts and broke from column to indulge in looting. According to one disgusted Confederate, the cavalrymen so burdened themselves with stuff, the next morning their horses were "piled up and almost concealed by the plunder they had on them." Riding farther along with the infantry, Wheeler's horsemen got in the way and slowed down the march. "The long column of cavalry pursuing the same road," complained one staff officer, "much impeded the march of the infan-

---

[50] Ecelbarger, *Day Dixie Died*, 49; Hood, *A & R*, 175–76; Maj. A. P. Mason to Wright, July 21, *OR*, vol. 38, pt. 5, 900; Hay, *Cleburne and His Command*, 234; Burne, *Lee, Grant, and Sherman*, 105–106; Horn, *Army of Tennessee*, 304; Dyer, *"Fightin' Joe" Wheeler*, 178; Hood, *Rise, Fall, and Resurrection*, 238; Neul, "Battle Most Desperate and Bloody," 33 (Hood saw "a golden opportunity to create a Chancellorsville of his own").

Given the agreement in the literature about Hood's plan for a "Jacksonian" flank attack on July 22, William C. Davis's generalization that Hood launched "desperate and ill-conceived attacks around Atlanta" seems surprisingly careless and certainly outdated (W. C. Davis, *The Generals of the Civil War*, 13).

Castel notes that Federals began calling the engagement fought on July 22 "the battle of Atlanta" (as in Dodge, *The Battle of Atlanta and Other Campaigns*, and in Fuller, "A Terrible Day: The Battle of Atlanta, July 22, 1864," 546) and that historians have as well. Castel disagrees in that usage and suggests instead that it be called the "Battle of Bald Hill" (*Decision in the West*, 410). But the bald hill in east Atlanta is gone, bulldozed in the 1950s for an interstate highway. And as for the "battle of Atlanta," some writers, including Atlantans who should know better, use the phrase "battle of Atlanta" to include all operations around the city leading to its fall (e.g., Bonds, *War Like the Thunderbolt*). Given all the blurriness, I like the conclusion of Arkansas artillery captain Thomas J. Key: "As I know no name I call it the battle of the 22nd of July" (Cate, *Two Soldiers*, 99). I will henceforth use this appellation.

try." Making matters worse was Wheeler's apparent uncertainty as to his primary mission. During the morning of the 22d, when he sent back to Hardee confirmation of Garrard's departure, Wheeler asked if he should chase the Yankees. "I cannot spare you or any force to pursue Garrard now," was Hardee's irritated reply. "We must attack, as we arranged, with all our force."

As another factor, after Peachtree Creek, Hood entered battle east of Atlanta with diminished confidence in Hardee. To begin with, one must wonder at Hood's decision for Hardee and his corps to deliver the flanking blow, given Hood's criticism of Hardee's performance just the day before. The process of elimination would have been the key factor. Stewart's corps was still battered from the previous fight; Cheatham was too new to corps command to be given so important an assignment. Moreover, Hood believed that Hardee's troops had only "skirmished" with the enemy at Peachtree Creek, and would thus have been fit for another combat. Yet Hardee, as we have seen, believed Hood's initial idea for a morning attack into the enemy rear ("even should it be necessary to go to or beyond Decatur") to be unrealistic, and had appealed for it to be modified. After his nighttime meeting with Hood, Hardee, having laid out the difficulties and obstacles before him in his night march, did not come away with confidence. After the war, Hood and Hardee were still arguing over what the battle plan of July 22 entailed.

Hardee's troops marched all night with little time for rest and rations. From the very beginning, the whole march lagged behind schedule. Cleburne's division, after a morning's hard fighting and facing an aggressive enemy that afternoon, could not disengage until well after darkness; it began its march in the middle of the night. Maney's, the last division in line, did not get on the McDonough road leading out of the city until three in the morning. Hardee's march-pace was a mile and a half an hour. But the demands of a morning attack allowed no time for rest; his troops had to keep moving. In the darkness, the march column was anything but orderly. "Halts came," remembered Private Stephenson, "the men jamming up against each other each time (for they were sleepy) just like cars on a freight train." Almost as bad, Hardee's marched on a hot night. Temperatures in Georgia mid-July can stay into the nineties into the evening. The men, moreover, trod on hard roads of red clay, whose ankle-deep dust rose amid marching boots and clopping hooves to bedevil men and animals alike.

Hood, as we have seen, was insistent upon a dawn assault, but that clearly would not happen. Indeed, the commanding general may not have had a clear idea of the route he laid out for Hardee, for he grossly underestimated its length. In his memoir, Hood claimed that Decatur was "only six miles from

Atlanta," and that Wheeler's horsemen were so familiar with the routes to be taken by the infantry "that I considered Hardee's move one merely within the lines of our cavalry." Such a statement must be judged as absurd. In 1880, Colonel Roy invoked no less than Atlanta mayor William L. Calhoun to demonstrate that the distance from Atlanta via Cobb's mill southeast of the city and thence to Decatur measured fourteen to fifteen miles. In 1930, the indefatigable Wilbur G. Kurtz drove the distance from the Confederates' outer defense line, from which Bate's and Walker's divisions began their march, to the starting points for their attack; he metered fifteen miles.[51]

By dawn Bate's division, at the head of Hardee's column, had marched over twelve miles to Widow Parker's, three miles southwest of Decatur. Walker was a mile behind at Grandma Akers's. Hardee, riding with Cleburne, reached Intrenchment Creek; nearby were the mill and residence of William Cobb. There Hardee called a halt so his men could rest for an hour or so as he sought further direction for the advance toward the enemy. Hardee, Cleburne, Wheeler, and Walker called upon Cobb and asked his help. "Thus far the route had been sufficiently clear," writes Kurtz, "but the road northward from the mill threaded a tangled wilderness of forest and undergrowth that was anything but propitious." Cobb offered his services as guide and called one of his mill hands, Case Turner, to help as well. From them Hardee got a lay of the land: thick woods, ground brush, and, farther to the east, Sugar Creek and a wide mill pond. Cobb was assigned to accompany Cleburne and Maney. Turner would guide Walker and Bate, for Hardee directed that a mile or so ahead the column would split; Cleburne would take the Flat Shoals Road, and Walker's and Bate's divisions would stay on the Fayetteville, then turn to the northwest to approach the positions from which they would launch their attack.[52]

Hardee's deployment was difficult and time-consuming. It did not help that Wheeler rode off toward Decatur, leaving the infantry behind. As Kurtz

---

[51] S. Davis, *Atlanta Will Fall*, 134 ("God of Battles"), 138–46; Patrick, *Reluctant Rebel*, 200 ("horses piled up"); Evans, "Atlanta Campaign," 44 ("long column of cavalry"); Hughes, *Hardee*, 227, 230–31; Ecelbarger, *Day Dixie Died*, 55, 58, 61–62; Hood, *A & R*, 176–178; Roy, "General Hardee," 356; Stephenson, *Memoir*, 217; Castel, *Decision in the West*, 391 ("the night is hot and humid"; "dust on the road is ankle-deep"); Kurtz, "Major-General W. H. T. Walker," *Atlanta Constitution Sunday Magazine*, July 27, 1930, 6; Kurtz, "Atlanta in summer 1864," 10 ("that fifteen miles of night marching").

[52] Castel, *Decision in the West*, 389–91; Kurtz, "Walker," 6 (10.3 miles to Cobb's mill), 16; Kurtz, "The Death of Major General W. H. T. Walker," 176; Hinde, "The great night march of Hardee's Corps," 5 (Kurtz drawing, "General Hardee and escort at the William Cobb house, inquiring about routes to the battlefield, July 22, 1864").

put it, Hardee's attack "might have yielded the same results [as Jackson's] if Hardee had known precisely where the Federal lines were placed and how to get in rear of them." To be sure, Wheeler was following orders—to attack the enemy wagon train at Decatur. To the extent that sending him off was a blunder, the fault was Hood's, product of an inexperienced army commander who hoped for too much and who underestimated the difficulties Hardee would encounter. "Perhaps the big mistake was in sending Wheeler's cavalry off to burn McPherson's wagon trains," writes Bruce Catton. "The cavalry might have provided the guidance that would have eliminated a lot of guesswork; to destroy a few Yankee wagons, that dawn of July 22, was not nearly as important as putting the infantry where it could wreck McPherson's whole army," Catton concludes, making a good point.

Hardee was forced to take some three hours to get his troops in position through the difficult woods that William Cobb had predicted. The piney Georgia woods were virtually impenetrable. "I marched in line for two miles through a dense forest," Hardee later wrote, "where I could not see ten paces. Of course it was impossible to keep up an alignment." As his troops struggled through the underbrush, Walker ("Fighting Billy," according to Private Stephenson of the Washington Artillery) found Hardee and suggested a change in direction. "No, sir!" Hardee exclaimed. "This movement has been delayed too long already! Go and obey my orders!" An angry Walker rode off, muttering to Major Cumming that Hardee "must answer me for this…I shall make him remember this insult." A belated apology from Hardee, delivered by an aide, failed to mollify Walker.

Swampy terrain further delayed Bate's and Walker's deployments. East of Sugar Creek, Castel writes in his inimitable present tense, "Bate's troops are floundering through underbrush, muck and knee-deep water." West of the stream, about eleven o'clock, Walker "faces a far worse obstacle—a huge, deep, utterly impassable lagoon filled with dead trees," Terry's mill pond. Fighting Billy was so angry that he threatened to shoot his guide, the astonished Case Turner, before Major Cumming cooled him down. Things were hardly better to the left, where Cleburne's men "claw and grope their way through a jungle of wild blackberry bushes." With good reason, Stewart Bennett writes that the first battle Bate's and Walker's divisions had to fight was "the battle against topography."[53]

---

[53] S. Davis, *Atlanta Will Fall*, 143–45; Catton, "The Battle of Atlanta," 261; Roy, "General Hardee," 360 ("could not see ten paces"); Castel, *Decision in the West*, 393–94; R. K. Brown, *To the Manner Born*, 264–65 (including Stephenson on "Fighting Billy" Walker); Mosser, "I

Conditions of the night march, the terrible terrain, and the difficult deployment all whittled down Hardee's strength. In the woods southwest of Decatur by 11:30 when it was getting into final position, Hardee's corps had lost from straggling close to a quarter of its effective strength. Understandably hundreds fell out of the line of march, exhausted. "I never saw as much straggling from our Corps," recorded Martin Van Buren Oldham, a Tennessean in Maney's division. "Hardee's column, which should have been able to put 14,000 officers and men into battle formation, was down closer to 11,000 troops readily available," according to Ecelbarger. The Federals of the XVI and XVII Corps, which Hardee's troops engaged that day, numbered at least that many.[54]

More important still, at Atlanta on July 22 McPherson expected to be attacked. The day before, Lt. Samuel Edge, acting signal officer in the XV Corps, had set up an observation platform in a tall pine tree near the Federal lines. The position was a hundred feet above the ground, high enough even to be seen by the Rebels ("they ascend by means of a rope ladder," noted General Shoup in his journal). From his station of observation on the morning of the 22d Edge saw enemy columns moving and reported them at 10 A.M. to McPherson and Logan. An hour later he sent another message concerning "additional movements of an alarming nature." About that time McPherson directed Brig. Gen. Thomas W. Sweeny's troops toward his threatened flank so that by noon, when Bate and Walker opened their attack, the 2d Division, XVI Corps (Sweeny's) was directly across their path. How this occurred deserves consideration, as it speaks well of McPherson's generalship but very poorly of Sherman's.[55]

Shall Make Him Remember This Insult," 24; Hughes, *Hardee*, 229; Ecelbarger, *Day Dixie Died*, 73–74; S. Bennett, "A Ghost on Horseback," 4:205.

[54] Ecelbarger, *Day Dixie Died*, 63, 65 (Morrill's brigade + Sweeny's division, 5,000 men), 122 (Smith's division 2,000), 134 (Wangelin's brigade, filling gap between XVI and XVII Corps, 1,000), 137 (Leggett's division 3,000).

[55] Edge report, September 12, *OR*, vol. 38, pt. 3, 122; Shoup journal, *OR*, vol. 38, pt. 3, 691.

Lt. Edge's observation station is an interesting footnote to the campaign in that we know its site. Theodore Davis, the *Harper's Weekly* illustrator, apparently liked to ascend to Edge's viewing platforms and sketch scenes from them. (One, appearing in *Harper's* on August 6, shows generals Hooker, Logan, and Peter Osterhaus with Lt. Edge on July 7 gazing at a distant Atlanta from north of the river.) It is likely that Davis climbed atop Edge's platform east of the city as well. Twenty years later, as one of the artists working on the Cyclorama painting of the battle of July 22, Davis returned to Atlanta. From the observation tower which the artists had had constructed, Davis (Kurtz believes) drew a large view of the area, and pinpointed the "signal tree" behind the Troup Hurt house. In the 1950s Kurtz sited the position as at 1166 DeKalb

As we have seen, Sherman intended no decisive move on July 22, only to keep pressure on the Rebels and to intensify his bombardment of the city. He hoped that maybe Hood, having been repulsed at Peachtree Creek, might give up the city. "If Hood was as roughly handled by Thomas this afternoon as reported," he wrote McPherson at 1 A.M. on the 21st, "I would not be astonished to find him off in the morning." Then, twenty-four hours later, early on the 22nd, the Federal commander actually believed Hood was giving up the city. The nighttime withdrawal of the Confederate infantry from their outer line to the Grant-Presstman perimeter allowed Union pickets probing in the early morning to report that the Rebels had abandoned their defenses. As early as 3:15 A.M. General Howard was writing army group headquarters, "the enemy has evacuated his works." Schofield sent in a similar report. Sherman was soon ordering Howard, Palmer, Hooker, and Logan to pursue the enemy, whom he assumed were probably retreating toward East Point. Within a couple of hours, of course, the Federals were able to see that Hood very much still held onto Atlanta. As for Sherman's jump to conclusion, which has generally escaped criticism from his admiring biographers, Castel trenchantly observes that the Union commanding general "as always is eager to believe that his opponents are doing what he wants them to do." Evidently the commanding general's unfounded optimism seeped down the ranks. Along the XV Corps line, near the Georgia Railroad, some officers suggested that precautions be made against a possible enemy attack. Division commander Brig. Gen. Morgan L. Smith, who had made his headquarters at the unfinished brick house of Troup Hurt, dismissed the idea, predicting that they "would take dinner in Atlanta" that evening.[56]

General Sherman was able to see the Rebels himself by 7 or 7:30 A.M. By that time, he had established headquarters at a prominent residence a mile

---

Avenue; "General Sherman's Advance" and Davis's illustration, "General Sherman's Advance," 501–502; Kurtz, "How the Cyclorama was Painted," 45; "Atlanta on Canvas," *Atlanta Constitution*, October 20, 1885; copy of the "Saucer Drawing" of the Cyclorama by Theodore Davis (with Kurtz's handwritten note, "it is at once apparent that Davis made this drawing" [thanks go to Josh Hogan, archivist at the History Center for locating this illustration]); Kurtz, historical marker text, "Federal Signal Station," box 36, folder 1 (Scruggs, *Georgia Historical Markers*, does not include this marker text).

[56] L. M. Dayton, SFO 40, July 21, *OR*, vol. 38, pt. 5, 222; Sherman to McPherson, July 21 (1 A.M.), *OR*, vol. 38, pt. 5, 219; Castel, *Decision in the West*, 389; Howard to Brig. Gen. Whipple, July 22 (3:15 A.M.), *OR*, vol. 38, pt. 5, 224; D. Evans, *Sherman's Horsemen*, 94–95 (Schofield's pickets); Sherman to Howard, July 22 (4:40 A.M.), Whipple to Palmer and to Hooker, July 22, *OR*, vol. 38, pt. 5, 225, 227–28; McPherson to Logan, July 22 (6 A.M.), *OR*, vol. 38, pt. 5, 231 ("retreating in the direction of East Point"); Woodworth, *Nothing but Victory*, 542.

or so from the main enemy lines. It had been built before the war by plantation owner Augustus F. Hurt but was now empty. It had for a while been occupied by a Thomas Howard, and hence was called by numerous Federals the "Howard house." Today, the house site—fixed on maps from the 1930s by Wilbur Kurtz—is the northeast corner of the Carter Center and Presidential Library grounds. (The president is proud to be able to view Atlanta as once did General Sherman.)

From the Hurt/Howard house, Sherman sent McPherson a note that further showed the commanding general's failure to grasp tactical reality. Mac was worried about a surprise attack on his left, even if Sherman was not. On the evening of the 21st he had ordered a brigade of Dodge's XVI Corps (Col. John Morrill's) to encamp "half a mile in the rear of the extreme left and at right angles to Blair's lines," as Dodge put it—in other words, facing south, the very direction from which the Rebel attack was feared. Then, early on the 22nd, McPherson visited Dodge and gave him verbal orders to send Tom Sweeny's division (which had been "crowded out of line by the contraction of our lines as we neared Atlanta") to the threatened flank. Sweeny did so, and aware that his two brigades were the army's extreme left, arrayed them facing south. McPherson also told Brig. Gen. John Fuller, commanding another division in Dodge's corps, to put a brigade on Sweeny's right. But a half-mile gap still existed between Sweeny's/Fuller's position and the left of the XVII Corps' main line.

Sherman got word of Mac's plan and around 7 or 7:30 sent him this note, written himself:

> Headquarters Military Div. of the Miss.
>
> In the Field at Howard House,
>
> Near Atlanta, July 22, 1864
>
> General McPherson, Army of the Tennessee:
>
> General, —Instead of sending Dodge to your left, I wish you would put his whole corps at work destroying absolutely the railroad back to and including Decatur. I want that road absolutely and completely destroyed; every tie burned and every rail twisted; and as soon as Garrard returns, if the enemy still holds Atlanta, I will again shift you round to the extreme right, with Turner's Ferry as a depot. Explore roads, etc., with that view.
>
> Yours,
>
> W. T. Sherman,
>
> Major-General.

A quarter-century later, Lt. Col. William E. Strong, staff officer for McPherson, published this letter, which was still in his possession and for this reason does not appear in the *Official Records.* In the 1950s Sherman's handwritten note was donated to the Georgia Department of Archives. At the time it was hailed as an epistolary demonstration of Sherman's "iron will and his readiness to order destruction" of railroads. What it should be regarded as, however, is a demonstration of Sherman's tactical myopia: bull-headed obsession with destroying railroads instead of protecting McPherson's army from flank attack and possible disaster. If John Bell Hood had committed such a blunder, he would certainly not have been hailed as a general of commendable "iron will," but of censurable wooden-headedness.[57]

Upon receiving Sherman's order to send Dodge's infantry railroad-wrecking, McPherson rode to meet the commanding general to try to talk him out of it. Sherman's recollection was that "of course I assented at once," and that McPherson then rode off when they began to hear the sound of musketry coming (improbably) from the southeast. But Sherman was not quite won over, according to Union Capt. Rowland Cox. After Mac left Sherman at the Hurt house, he received another message timed at 12 noon, this one also in Sherman's own hand, agreeing that Sweeny's division could stay where it was, but that he wanted Fuller's division "back on the line of the railroad between here and Decatur to destroy it." (Cox commented, "the importance of the destruction of the railroad more than the safety of his flank was still in General Sherman's mind.") General McPherson had taken no steps to see Sherman's order was carried out when the enemy attack he had feared fell upon his left flank.[58]

Bate's attack began a little after noon. Exchange of skirmishers' fire quickly convinced the Confederates that they had not taken the Yankees in their flank-rear, but that the enemy lay ahead of them and in force. "I was ignorant of what was in my front but believed the enemy was without defenses," related General Bate. "In this we were mistaken…. The enemy had a

---

[57] Sherman, *Memoirs*, 2:75 ("in front of the Howard House…from which we could plainly see the whole rebel line of parapets"); Kurtz, "The Augustus F. Hurt House," 5–6; [Kurtz,] "Map of Atlanta as of 1938" (site #38, Hurt house/Sherman's HQ, July 22); Ecelbarger, *Day Dixie Died*, 87 ("yawning half-mile gap"); Strong, "The Death of General James B. McPherson," 515; Hendrick, "Expert gives Carter Civil War site lesson," *Atlanta Constitution*, July 10, 2013, B5; Dodge, "The Battle of Atlanta," 489–90; McKee, "New Sherman Letter Shows General's Iron Will," *Atlanta Journal*, January 24, 1956, 14.

[58] Sherman, *Memoirs*, 2:75–76; S. Davis, *What the Yankees Did to Us*, 122 (McPherson telling Sherman on the 22nd that he planned for his artillery to shell the Atlanta Rolling Mill, whose "high smoke-stack" he could see); R. Cox, "Snake Creek Gap," 343–44.

strong force with breast-works and heavy batteries crowning the eminence." "We now discover that the yanks have two lines of battle in front of us, both strongly entrenched," lamented Johnny Green, one of Bate's Kentucky orphans. "They pour their lead into us [and] their three batterries open on us slaying our men right & left." Exhausted, weakened in numbers by straggling, perplexed by the fight shown by the supposedly outflanked enemy, Bate's troops gave way.

General Walker was shot dead even before his men launched their attack after Bate's repulse. Disordered by their advance through the swampy thickets and without their commander, Walker's troops were repulsed and out of the fight by 1:15. In addition to their numerical parity—5,000 Yankees against as many Rebs—Castel lists other reasons for the Southerners' repulse: "the Federals were rested, skillfully commanded, and enjoyed the enormous advantage of artillery support, whereas Bate's and Walker's troops were weak from fatigue, their attacks were uncoordinated, and of necessity they had to rely solely on small arms" (the woods were too dense for Bate to call in Slocumb's battery). The Federals' leadership was another factor. Generals Dodge, Fuller, and Sweeny were all on the field directing troops. "Though fired on constantly by skirmishers," Sweeny later boasted, "[I] knew where to put every gun and regiment" and so positioned them.

In the end, for all of these reasons, the Confederates in the initial phase of their assault were soundly beaten, with casualties of 800 to 1,000 men in seventy minutes. Ecelbarger observes that with the two sides in this phase of the battle about equal in numbers, Northern artillery "made all the difference." In his report General Dodge specifically commended Battery H, 1st Missouri, and the 14th Ohio Battery for their role in repelling Bate's assault. The Southerners "stood so little chance of success," concludes Castel, "that it would have been remarkable had they achieved it." So much for the opening of Hood's Chancellorsville attack.[59]

Wheeler approached Decatur about noon, hoping to scoop up the Yankee wagons. He was surprised to find enemy infantry, dug in, supported by artillery—Colonel Sprague's brigade plus the Chicago Board of Trade Battery and a section of Battery C, 1st Michigan Artillery. The Confederates dismounted, attacked, and were initially repulsed. "We received a most galling fire from both infantry and artillery," Wheeler later wrote. But the Southerners attacked again, and the outnumbered Yankees fell gradually back to the

---

[59] Ecelbarger, *Day Dixie Died*, 74, 78, 86, 97, 103; Castel, *Decision in the West*, 395, 397–98; Waters, "Bate Reports," 218; Green, *Orphan Brigade*, 148; Dodge report, August 1, *OR*, vol. 38, pt. 3, 372.

courthouse. "The rebs came up on both sides of us, through lots, behind fences, houses, etc., and shot our men like dogs, dropping them all around the guns," recalled an artillerist in the Chicago battery. But Sprague's fight allowed Northern teamsters to rush their train northward to escape capture. Wheeler's men were able to push Sprague's troops through and out of town. Most residents of Decatur had left, but some stayed. One of them was young Mary Gay; when she saw the Yankees fleeing from the town square ("the flying blue-coats"), she was thrilled. Wheeler was chasing the enemy when, a mile north of Decatur, staff officers caught up with him and announced that General Hardee asked his return and support. Wheeler did so, and this phase of Hood's flank attack ended with the same lack of hoped-for result as the first. Colonel Sprague reported 242 killed, wounded, and missing. Wheeler claimed to have taken several hundred prisoners, a 12-pounder and six wagons with teams; his loss probably totaled 300–400.[60]

General McPherson had watched the repulse of Bate and Walker. Around 12:45 he sent Lieutenant Colonel Strong to Frank Blair for more information as to his position. From Brig. Gen. Giles Smith, Strong learned that the Rebels were feeling their way toward the half-mile gap between the XVI and XVII Corps. Strong returned with the news and accompanied McPherson as he rode westward. Mac then ordered Strong off to bring up a brigade and continued riding. It was about 1:30 P.M.

Just then Cleburne's division started its attack. Brigadier Daniel Govan's Arkansas brigade advanced against Col. William Hall's Iowa brigade. Govan's left-most regiments charged enemy troops, entrenched with *abatis* and a cleared fire-field ahead of them. Here the Confederates were repulsed. On their right, Govan's troops overlapped the enemy left and advanced through woods. They caught an enemy battery—F, 2nd U.S. artillery, six 3-inch rifles—en route and sent the captured Yankees, their cannon, caissons, and teams to the rear. Then the Southerners met and drove back Union infantry

---

[60] Ecelbarger, *Day Dixie Died*, 98–102; Wheeler report, October 9, 1864, *OR*, vol. 38, pt. 3, 952–53; D. Evans, "The Fight for the Wagons," 16–22; D. Evans, *Sherman's Horsemen*, 94; Gay, *Life in Dixie during the War*, 140; Sprague report, July 25, *OR*, vol. 38, pt. 3, 507; Poole, *Cracker Cavaliers*, 132.

Phil Secrist implies that Wheeler, after having sent Sprague's infantry retreating, might have run down the escaping Union wagon trains had it not been for Hardee's summons. "Success crowned his efforts at Decatur," he writes, "but before it could be properly exploited he was recalled to East Atlanta to assist Hardee's stalled attack" (Secrist, "Cavalry in Atlanta Campaign," 521). Just how Wheeler, with his 2,500 cavalrymen, could have possibly helped Hardee's assault at 1 or 1:30 P.M. has not been explained (Ecelbarger, *Day Dixie Died*, 98 [3,000 cavalrymen, reduced by casualties]).

of Col. Benjamin Potts's brigade and pressed Hall's so hard that the Iowans were ordered to fall back. Some were too slow in complying and were gobbled up by Govan's hard-pressing assault: 240 men of Lt. Col. Addison Sanders's 16th Iowa plus two Napoleons of the 2nd Illinois Light Artillery. Brig. Gen. James A. Smith's Texas brigade came in on Govan's right, effectively moving into the gap between the two Union corps which had so concerned General McPherson.

McPherson with a few staff officers rode right into the advancing Rebel troops. Called to halt and surrender, Mac turned about and was shot in the back; the bullet exited his chest. He fell to the ground and died within minutes. The Confederates captured an officer and orderly; the others got away. After taking a few of the general's effects—hat, field glasses, sword belt—the Southerners pushed on. An officer with McPherson, Lt. William H. Sherfy, later claimed that the general had been shot at 2:02 P.M.—the time on his watch when Sherfy, escaping on his frightened horse, ran into a tree. Lieutenant Colonel Strong thought it occurred earlier; Gary Ecelbarger concludes it happened about 1:45.

James B. McPherson has the distinction of being the only Union army commander killed in combat (unless one includes Brig. Gen. Nathaniel Lyon, killed at Springfield, whose "Army of the West" in August 1861 was only about the size of a corps by 1864 standards). McPherson was one of ten Federal major generals to die in combat; the most notable include Kearny (Chantilly), Mansfield (Antietam), Reynolds (Gettysburg), and Sedgwick (Spotsylvania). While the Confederates responsible for high-ranking enemy officers' deaths can be hard to identify, McPherson's assailant is established: Cpl. Robert R. Coleman of the 5th Confederate Regiment (Tennesseans), part of Smith's brigade. Coleman was given the order to fire by Capt. Richard Beard, commander of the 5th.[61]

---

[61] Ecelbarger, *Day Dixie Died*, 87, 103–104, 107–108, 110–118 (including good map, 112), 260 n.23 (detailed analysis of the time of McPherson's mortal wound, concluding it as having occurred at "or *before* 1:45 P.M."); Scaife, *Campaign for Atlanta*, 163 (six 3-inch rifles); Castel, *Decision in the West*, 398; Strong, "Death of McPherson," 522–34; Woodworth, *Nothing but Victory*, 550–52; Kurtz, "McPherson's Last Ride," *Atlanta Constitution Sunday Magazine*, June 29, 1930, 8–9, 14; Whaley, *Forgotten Hero*, 159–61; Hess et al., *Wilson's Creek, Pea Ridge, and Prairie Grove*, 4, 71; Symonds, *Cleburne*, 228 ("the only Union army commander killed in battle during the war,"); R. K. Brown, *Fallen in Battle*, 174–76; "Incidents of Gen. McPherson's Death," 118–19; Porter, *Tennessee*, 137–38; "Capt. Richard Beard,"425; Frazer, "Fifth Confederate," 1:151 ("Coleman singled out Gen. McPherson and sent a ball through him").

While General McPherson's body lay on the ground in what is today east Atlanta (Monument and McPherson avenues), Smith's Texas brigade advanced into the gap between the two Union corps. From the east, Brigadier General Fuller directed his artillery to open up on the Rebels and ordered a charge by two regiments (which was quickly repulsed). The Texans kept moving warily northwestward, actually parallel to the XVII Corps line, which was a thousand yards to the west. They chanced upon a single Illinois regiment that had been ordered into the gap and quickly overwhelmed it. General Smith reported that the enemy fought only a few minutes before they "fled in great confusion." Among the prisoners taken was Col. Robert Scott, commanding 2nd Brigade, XVII Corps, who had ridden out to the detached regiment. The Illinoisans—called "Egyptians" as they hailed from southern Illinois, nicknamed "Little Egypt"—nearly lost their regimental flag; one Pvt. Thomas Yates won a medal for saving it. The Confederates chased the fleeing Yankees with "ungovernable enthusiasm," as General Smith put it. But then they ran into trouble: Col. Charles C. Walcutt's brigade (Brig. Gen. William Harrow's division, XV Corps) directly confronted them northeast of the bald hill, a thousand infantrymen with a dozen pieces of artillery. The Confederates quickly took fire and stalled; Smith called for help from Cleburne (it did not come). Another Union brigade, Col. John Oliver's, reinforced Walcutt and stepped up the pressure on the Rebels. Smith had ordered a withdrawal when he received a wound that took him off the field. Then a Federal charge swept the 17th/18th Texas and 5th Confederate regiments, capturing 182 officers and men and the colors of both regiments. Smith's brigade, whose initial advance had promised much, was knocked out of action by "Harrow's coup," as Ecelbarger calls it, which "removed a quarter of Cleburne's punch and did so with few recorded casualties."[62]

It was about this time that General Hood and his chief of staff rode out of the city to see what he could of the battle. In *Advance and Retreat* he claims to have ridden to Cheatham's lines east of the city "at daybreak." Authorities

---

[62] McCarley, *The Atlanta Campaign*, 56–57; Pollock, "The Battle of Atlanta: History and Remembrance"; Ecelbarger, *Day Dixie Died*, 117–18, 126–31 (including map, "James A. Smith's Brigade Penetrates Gap, Repulsed by Harrow's Division, 2:00–2:15 P.M."); Tuthill, "An Artilleryman's Recollections of the Battle of Atlanta," 440–41 (identifying the 31st Illinois); Smith report, August 5, *OR*, vol. 38, pt. 3, 747; M. D. Leggett report, July 25, *OR*, vol. 38, pt. 3, 565 (Col. Scott's capture); Woodworth, *Nothing but Victory*, 23 ("Little Egypt"); General Orders No. 13, August 28, 1864, *OR*, vol. 38, pt. 3, 557 (Yates); "Return of a Confederate Flag," 302 (including statement by Sgt. Maj. Andrew LaForge, 15th Michigan, on capture of the 17th/18th Texas colors); Maryniak, "Private Charles F. Sanscrainte," 54–56 (on capture of the 5th Confederate's flag).

such as McMurry accept this, but Lieutenant Mackall recorded in his journal that Hood and his uncle did not ride out until early afternoon: "About 2 P.M. Genls Hood & M ride to rear Cheatham's line." The commanding general and his chief of staff may have gone first to the home of Confederate Brig. Gen. Lucius J. Gartrell on Decatur Street in east Atlanta. The house was on high ground ("Gartrell Hill") and from the place's upper story Hood may have tried to scan the action far off. But as Atlantan Dr. Robert E. Zaworski observes, the fighting was almost two miles away "and this was July, with the trees in full leaf." Hood and General Mackall rode a little way to a higher spot, the house of James E. Williams, south of the railroad near the City Cemetery, from whose second story window they saw Cheatham's entrenched line before them and heard distant musketry.[63]

While Smith's Texans were being pushed back by Harrow's two brigades, roughly 2 to 2:15, Maney's division attacked Giles Smith's division, entrenched south and southeast of the bald hill. The Tennessee brigade of Otho F. Strahl initially drove back the 15th Iowa, but then came upon Col. Benjamin Potts's brigade and was halted by a fire "too destructive," according to Lt. Edwin H. Rennolds of the 5th Tennessee. General Strahl was seriously wounded; his successor called off the attack and the Southerners retired to their starting point, woods to the south. Maney was not on the scene to countermand the order.

Additional pressure on the Federals was applied by the advance of the brigade of Alfred J. Vaughan (who had been wounded at Smyrna, July 4); Col. Michael Magevney now commanded. Magevney's line of attack was essentially the same as Govan's left the hour before, against Hall's brigade. This time the Southerners were able to drive the Iowans back. "The enemy Commenced retreating up thier works as soon as we Charged them," recorded Capt. Alfred Fielder of the 12th Tennessee. Because the Southerners' front crossed the Federal trench line as they pushed northwestward, Hall's troops took fire from both sides of their works. "They would Jump first on one side of the works and then on the other," Captain Fielder observed, "but we being

---

[63] Hood, *A & R*, 179; McMurry, *John Bell Hood*, 131; McMurry, "Mackall Journal," entry of July 22; Garrett, *Atlanta and Environs*, 1:621 (Gartrell estate); Zaworski, *The General and the House on a Hill*, 39, 42, 46; Kurtz, historical marker texts, "Where Hood Watched the Battle of Atlanta," box 36, folder 1 (Scruggs, *Georgia Historical Markers*, 233).

The Purdues note that Kurtz had interviewed two aged Atlantans who had seen Hood at the Williams house on the 22nd (*Pat Cleburne*, 365–66 n.51). One of them was Richard Fickett, as Kurtz recorded in his notebook #7, 43. The other was Maj. Charles Hubner, Hood's telegraph chief (Kurtz notebook #3, 101).

on both sides and pouring upon them such a galling fire they continued steadily to give way firing back at us as they went." After pushing the enemy back toward the bald hill, Magevney's attack finally sputtered out. Federals still held the hill. Hardee's attack, in Ecelbarger's words, was now at a stalemate.[64]

Cleburne and Hardee were south of the XVII Corps left flank and could see that Govan's and Magevney's assaults had failed. To break the enemy position Cleburne ordered Brig. Gen. Mark P. Lowrey with his brigade "to move rapidly to the front and charge the works." On his left Lowrey was supported by several Tennessee regiments of Col. John C. Carter's brigade (Maney's division); on his right Lowrey picked up some Texans of Smith's brigade, which he observed "had been repulsed and was reforming." Altogether, Lowrey had some 2,000 troops for his assault, which began about 2:15, just after the repulse of Magevney.

The Confederates were advancing toward the northwest. Remnants of Hall's brigade, Potts's, and a small brigade from Leggett's division faced them south of the bald hill. The Federals were in their trenches, but this time facing to the east, with their backs toward Atlanta. The 45th Alabama of Lowrey's brigade and the 38th Tennessee of Carter's led the assault on the Confederate left. Col. Harris Lampley charged to the very enemy works—so close that he was actually hauled over the parapet by his coat collar and captured. Lt. Col. Andrew Gwynne of the 38th Tennessee was also taken prisoner. The Confederates could not sustain their position and had to fall back.

On the bald hill were two brigades of Leggett's division: that of Brig. Gen. Manning Force and of Col. Robert Scott. Scott had been captured when Smith's charge overran the Illinois "Egyptians"; Lt. Col. Greenberry Wiles had succeeded him. Now Force's and Wiles's troops opened up as Lowrey's attacking troops approached the hill around 2:15. General Force stood among his men exhorting them, clutching a Union flag, when he was struck in the face by a minie ball. The bullet entered his left cheekbone below the eye and exited the lower jaw under his right ear, smashing bone and teeth. Force never lost consciousness as he was carried to the rear.

Despite the Federals' heavy musketry, some Confederates got to the enemy works, where hand-to-hand fighting ensued. "We fixed bayonets," recalled one Northerner, "and then and there we had it with clubbed muskets, fisticuffs, and wrestling." Col. John Wilkinson of the 8th Mississippi, killed near the enemy works, was only one of the Southern officers to fall in the heavy fighting. After perhaps 45 minutes, Lowrey had to send in orders for

---

[64] Ecelbarger, *Day Dixie Died*, 120–26; Castel, *Decision in the West*, 402; Peterson, *General Vaughan*, 248–49.

retreat. "The thing attempted was impossible for a thin line of exhausted men to accomplish," he concluded. Lack of sleep, the demanding march, and hot sun all took their toll on his brigade. "It was a direct attack by exhausted men against double their number behind strong breast-works," Lowrey wrote (correctly, except in regard to numbers, which were roughly equal on this part of the field, 2,000:2,000). General Blair observed that he had been saved by the Rebels' lack of coordination in their assaults. "If the enemy had concerted his attacks from front, flank, and rear, so as to strike my line at the same moment with his different lines of battle," he later wrote, "it would have been extremely difficult, if not impossible, to hold our ground."[65]

During the early afternoon Hood sent a staff officer to track down Hardee and learn how the battle was going. The report came back: some gains, but stalled progress due to unexpected enemy resistance and strength. At this point, around 3 P.M., Hood ordered Cheatham's corps into the fight. Hood's decision to commit 14,000 relatively fresh troops three hours after Hardee had begun his attack has been criticized by a number of writers, who assume that Cheatham should have been ordered to advance simultaneously with Hardee. "For some never explained reason," Scaife accuses, "Hood did not order Benjamin F. Cheatham's corps out of the trenches to support Hardee's attack, for almost three hours." "Instead of coordinating his planned attacks, he let Hardee's troops advance and spend themselves against the Federal position," writes John Cannan. Wilbur Kurtz errs further in making the statement, "Had Hood, at this critical moment, been able to launch a force from the Atlanta fortifications on the 15th and 17th Corps, *as planned....*" (emphasis added; Hood's battle plan called for no such simultaneity). Castel is convincing in his explanation that Hood all along had planned to send in Cheatham after Hardee had presumably begun to roll up the enemy left flank. This, as we have seen, Hardee was unable to do, so Ecelbarger reasons that Hood came up with "a new plan": pinning down the Union XV Corps as diversion, to prevent it from aiding the XVII as Hardee continued to hammer the enemy position at the bald hill. (Hood himself, in his report written months later, termed Cheatham's attack "a diversion").

Whether as diversion or as a grand attack *en echelon*—either explanation works better than Scaife's "never explained reason"—Cheatham's three divisions soon marched out eastward from their works. Cheatham ordered an

[65] Ecelbarger, *Day Dixie Died*, 133–47; Lafantasie, "The Gentleman General," 35 (on Force's wound); J. Welsh, *Medical Histories of Union Generals*, 117 (postwar neurologic changes due to the facial wound); Lowrey report, July 29, 1864, *OR*, vol. 38, pt. 3, 732; Blair report, [undated] 1864, *OR*, vol. 38, pt. 3, 546.

echelon assault of his own: Maj. Gen. Carter Stevenson's division on the right advancing first, then Brig. Gen. John C. Brown's in the center, and finally Maj. Gen. Henry Clayton's to the left. Stevenson advanced at 3:30, Brig. Gens. Alexander W. Reynolds's and Alfred Cumming's brigades in front, with two others behind them. Reynolds's and Cumming's troops attacked without *elan* and after taking a relatively small number of casualties began to fall back. The other half of the division never got into the fight before Stevenson called the whole thing off, not knowing that the enemy (Harrow's division on and to the north of the bald hill) were outnumbered and had no artillery. "Stevenson proved true to form as a mediocre offensive tactician," is Ecelbarger's lamentable but justified conclusion. To Stevenson's right, Maj. Gen. Gustavus Smith's Georgia militia also advanced; they too retired with slight loss.

To the north, Confederates of Brown's division found greater success. The XV Corps held the lines north and south of the Georgia Railroad, but its top command was shifting around. After he learned of McPherson's death, Sherman ordered XV Corps commander Maj. Gen. John A. "Black Jack" Logan to take charge of the Army of the Tennessee. That meant that Maj. Gen. Morgan L. Smith, leader of the 2nd Division, succeeded Logan, and Brig. Gen. Joseph A. J. Lightburn, leading the 2nd Brigade, succeeded Smith. All of these changes were taking place just as Cheatham prepared his assault. More important, the Federals were weaker in number. In the half-mile sector south of the railroad, the XV Corps line had been reduced by some 1,500 troops and ten pieces of artillery, sent southward to buck up the Union positions under attack by Cleburne and Maney. As Ecelbarger has noted, Hood's echelon attack was actually succeeding in textbook manner: striking the Yankees' left, causing them to send reinforcements from their center, then attacking the weakened center.

The mismatch allowed Brigadier Arthur M. Manigault's South Carolina-Alabama and Col. Jacob Sharp's Mississippi brigades to break the enemy line at the railroad. "At the last rush, most of the enemy broke and fled," recalled General Manigault. The Southerners branched out in both directions, widening the breach and sending Lightburn's infantry packing at least a quarter-mile to the east. In the process they captured the six Napoleon guns of Battery A, 1st Illinois Artillery. North of the railroad, near the unfinished brick house of George M. Troup Hurt (Augustus's brother), Southerners drove off the infantry supports of Francis DeGress's Battery H, 1st Illinois Light. DeGress had his four 20-pounder Parrotts spiked, but before withdrawing had his men try to drag them to the rear. "But they were to heavy and my men were to much worn out," he later reported, so the cannon had to be abandoned to the onrushing Rebels. Lt. Robert M. Gill, of the 41st Mississippi (Sharp's brigade), the day after the battle wrote his wife, "I never enjoyed a thing better in my life. We had the pleasure of shooting at Yankees as they ran without being shot at much." Soon came the order to halt the pursuit and fall back on the enemy works. Lieutenant Gill commented, "I never saw the like of Knapsacks, Blankets, oil cloths and Canteens in my life. Our men supplied themselves." (A soldier gave Gill a rubber rain cloth, but he was already so laden that he had to throw it away.) Col. Abda Johnson's brigade reinforced Manigault's and Sharp's breach. Farther south, Col. John Coltart's Alabama brigade also occupied Yankee works. By 4:30 Confederates occupied half to three-quarters of a mile of Union entrenchments.[66]

General Logan, now commander of the Army of the Tennessee, responded to news of his skedaddling troops with typical alacrity. He ordered Dodge to send a brigade from his XVI Corps (Col. August Mersy's), then galloped to rally Lightburn's skedaddlers. By that time, Clayton's division, the leftmost in Cheatham's alignment, had entered the battle. Its infantrymen supported Manigault's and Sharp's, but the Confederates pushed no farther. Artillery fire from the north (1st Division, XV Corps) came in, causing General Manigault to observe bitterly, "the artillery practice of the enemy was splendid."

Then the Federals rallied. Black Jack rode among his dispirited men, according to one, "galloping down the ranks like a man of iron." Logan arrived about a half mile east of the Troup Hurt House in time to meet the advance of Mersy's brigade and sent them forth. Generals Logan and Morgan Smith also organized a counterattack, as did Harrow north of the bald hill. All along the line, between 4:30 and 5, the four Confederate brigades that had occupied the enemy works were pushed back and gave way, from Manigault's near the Troup Hurt house to Coltart's area north of the bald hill. With enemy troops (Brig. Gen. Charles Woods' 1st Division, XV Corps) massing on their left, ready to join in the Union counterattack, generals Brown and Clayton sent orders for their divisions to retire back to their original position. General

---

[66] Hood, *A & R*, 180–81; Castel, *Decision in the West*, 404–408; Ecelbarger, *Day Dixie Died*, 148–72 (including 150, "Fourteen thousand officers and men...not fatigued by an overnight march"); Scaife, *Campaign for Atlanta*, 97–99; Cannan, *The Atlanta Campaign May–November 1864*, 125–26; Kurtz, *The Battle of Atlanta July 22, 1864*, 10; Hood report, February 15, 1865, *OR*, vol. 38, pt. 3, 631; Scaife and Bragg, *Joe Brown's Pets*, 34–35; Kurtz, "The Broken Line and the DeGress Battery," 4–5; Manigault, *Narrative*, 227; DeGress, "Report of the Battle of Atlanta," 29; Wiley, "3 Officers" (including Lt. Gill), 33; Wiley, *Embattled Confederates*, 73 (Gill's regret, "During the fight...I done some heavy swearing I am told").

Manigault later commented, "There was nothing left for us to do but obey, and I never saw men obey an order so unwillingly."

In the process the Southerners had to give up the four Parrotts (DeGress's) they had earlier overrun, along with two guns of the Illinois battery at the railroad. (By the end of the day, DeGress had three guns unspiked and back in use, shelling the Rebels; his fourth gun had burst.) Still, the Southerners carried away four guns of the battery, bringing the Confederates' haul for the day to a dozen pieces of artillery—an impressive achievement on any battlefield (Lee's haul at Chancellorsville amounted to fourteen guns). Cheatham's attack had led to 500 Northern soldiers killed or wounded, and as many more taken prisoner. But the Southerners had sustained casualties of almost 2,000. For all this, as the recent authority on the battle has written, by 5 P.M. the Confederates "were back in the same appearance and position they held at the midmorning."[67]

The chief problem in considering Cheatham's assault of 3–5 P.M. as a "diversion" lies in its timing. If intended to divert the enemy's attention from Hardee's last effort to take the bald hill, it was way too early. Cleburne did not begin arranging his troops for the final assault until five o'clock, and the attack itself did not begin until 6. Cleburne managed to put together a battle line from his own division (Govan's tired but still combative brigade), as well as a brigade from Maney's (Col. Francis M. Walker's) and from W. H. T. Walker's division (Mercer's brigade, now under Lt. Col. Morgan Rawls). The Southerners, some 3,000 strong, outnumbered the Federals holding the bald hill, but the Northerners were strongly entrenched ("Leggett's fort"), and had artillery (Battery H, 1st Michigan Light, the "Black Horse Battery"). And they were resolute. "The hill must be retained at all hazards and at whatever cost," General Leggett ordered. The Confederates were equally resolved. Colonel Walker led his men up the slope to its very crest. More hand-to-hand combat led to a Tennessean capturing the colors of the 30th Illinois. Colonel Walker was killed exhorting his troops, but the Confederates could not hold

---

[67] Ecelbarger, *Day Dixie Died*, 131, 173–90; Manigault, *Narrative*, 228–29; Evans, "DeGress Report," 30; Polhemus, *The Francis DeGress Story*, 28; Brig. Gen. Henry J. Hunt report of Chancellorsville, August 1, 1863, *OR*, vol. 25, pt. 1, 253 ("Fourteen guns were lost, but the most of these losses [eight] occurred in the rout of the Eleventh Corps").

Among the Confederates captured by the Federals in their counterattack was an African American, as recalled by Col. Robert Adams of the 81st Ohio: "After the firing ceased, we found fifty Rebels crouching behind the works, who preferred Northern prisons to the risk of an attempted retreat. Among the prisoners was a darkey, and the only darkey I saw during the war firing the wrong way" (Adams, "The Battle and Capture of Atlanta," 480).

their position. Their retreat ended the battle around eight o'clock, as the summer sun was setting. Southerners who could not retire safely kept up a musket fire well into the darkness, when they slipped back to their lines.

At the close of the eight-hour battle of July 22, Hood's aborted flank attack had taken a division-length of enemy works south of the bald hill (which the Federals would have abandoned anyway, Castel uncharitably judges). Despite its failure to achieve Hood's lofty hopes, the attack of Hardee and Cheatham had accomplished some notable results: blunting the Army of the Tennessee's approach to the city from the east, which had for several days been the swiftest and most threatening of Sherman's three armies; effectively preventing a fast march from McPherson's left to break the East Point railroad (which had been such a worry that Hood had warned General Wright to prepare for a sudden evacuation of the city); inflicting casualties that Logan eventually put at 3,722; the killing of McPherson, perhaps Sherman's favorite army commander, and the consequent unsettling of his army's command structure; and capturing three batteries of Northern artillery. This latter was not inconsequential to Hood, who a few days after the battle reminded his troops in a general field order that "guns and colors are the only unerring indications of victory."

To be sure, the cost in casualties for these trophies was high. As with Peachtree Creek, Sherman's count of Rebel losses (8,499) was accepted for many years by a number of writers. Some have gone even higher, such as Bruce Catton ("close to 10,000 men, in killed, wounded and missing"). Even McMurry in his biography gives Hood's casualty figures as "about 8,000." Fortunately, in the last few decades, scholars (including McMurry) have reexamined the sources and come to agree with Castel that Confederate losses totaled around 5,500. (Chief source for this number is Colonel Roy's statement of 3,299 men lost in Hardee's corps on July 22.) The proportion of 5,500 Confederates to 3,722 Federals *hors de combat* (about 1.5:1) was actually pretty reasonable by standards of Civil War attacking battles. Yet the toll among Southern officers was not inconsiderable. Major General Walker was killed, and the list of dead included Col. Francis Walker and Col. Samuel Benton; both were about to be promoted to brigadier general. Five more colonels or lieutenant colonels, regimental leaders, were killed or mortally wounded. The number of Southern officers wounded but surviving attests to the fierce fighting that day: Ecelbarger lists twenty-seven Confederates of rank from brigadier general (States R. Gist, James A. Smith, and Otho F. Strahl) to majors commanding regiments. The number of Union brigade or regimental commanders wounded/surviving was but eight.

Then there were the soldiers. Kate Cumming was matron at an army hospital that took in a number of Hood's men wounded in the battle east of the city. She recorded in her journal on July 27:

> A day or two ago we received a lot of badly wounded; some of them are shot near the spine, which paralyzes them so that they can neither use hands or feet. There is one very large man, named Brown, who is as helpless as an infant. Another, Captain Curran, is almost as bad. A fine-looking young man from Kentucky has lost a leg and arm; there is but little hope of his recovery. Mr. Pullet, a Georgian, is wounded through the lungs; the least movement causes the blood to run in streams from his wound; the doctors have little hope of saving him. Mr. Thomas is wounded through the head; his brain is oozing out, and at times he is delirious. Mr. Orr is injured in the spine, and is perfectly helpless. Mr. Summers of Mississippi is badly wounded in the right hand, and can not feed himself. Mr. Harper is badly wounded, and can scarcely eat any thing. Mr. Latta, his friend, has had his leg amputated.... We have so many poor, helpless fellows, that it is heart-breaking to look at them.[68]

Despite his heavy losses, Hood's battle of July 22 gave the Southern nation something it had not yet seen in the Georgia campaign: a major Confederate attacking battle that actually produced positive results. Johnston had never done this; the government and people hungered for it. General Hood fueled the public's exhilaration with his initial report on the engagement in which he greatly magnified his assault's measurable but modest achievements. "General Hardee, with his corps, made a night march and attacked the enemy's extreme left at 1 o'clock to-day," began his telegram to Richmond sent around 10 or 10:30 that night; Hardee's troops "drove him from his works, capturing 16 pieces of artillery and 5 stand of colors." Of course, Hardee had driven only *part* of the enemy's force from *part* of his works, and the numbers of guns and flags captured were not so numerous as Hood claimed. But as

---

[68] Ecelbarger, *Day Dixie Died*, 191–207, 233–45; Castel, *Decision in the West*, 409–12; Sherman, "Grand Strategy," 253; Logan report, September 10, *OR*, vol. 38, pt. 3, 38 (3,722 killed, wounded, or missing); Hood's General Field Orders No. 7, July 25, *OR*, vol. 38, pt. 3, 909; Catton, "The Famous Cyclorama of the Great Battle of Atlanta," 43; McMurry, *John Bell Hood*, 132; Roy, "General Hardee," 367; McMurry, *Atlanta 1864*, 155 ("about fifty-five hundred casualties" on July 22); Warner, *Generals in Gray*, 26 (on General Benton); Allardice, *More Generals in Gray*, 231–32 (on Francis Walker); Cumming, *Kate: The Journal of a Confederate Nurse*, 212.

Occasionally, writers backslide to Sherman's estimate of Rebel casualties on July 22. Recent example is Hess, "On the War Path," 26 ("The bloody battle of Atlanta, also known as the battle of July 22, resulted in 3,722 Union casualties and 8,500 Confederate losses").

students of military history know, generals like to inflate their accomplishments in their after-action reports, and Hood was certainly no exception, particularly after the months-long drought of good news from the Army of Tennessee. "Major-General Cheatham attacked the enemy at 4 P.M. with a portion of his command," Hood continued; he "drove the enemy, capturing six pieces of artillery." "During the engagements we captured about 2,000 prisoners," he claimed, another high number. Hood could not give a count of his own losses, but among the dead he named was Major General Walker, and among the wounded he stated there were brigadier generals Smith, Gist, and Mercer. "Our troops fought with great gallantry," he concluded.

It did not take much for a report of this kind to be turned into a heated, hyperbolic headline, distributed by the Confederate Press Association to Southern newspapers:

FROM ATLANTA.

GLORIOUS SUCCESS OF OUR ARMS.

CAPTURE OF ARTILLERY, PRISONERS,

&c.

The accompanying column repeated Hood's account of driving the enemy, capturing 22 pieces of artillery, and the taking of 2,000 prisoners, adding, "Prisoners report McPherson killed."

When posted on the bulletin board of a public building in Richmond, the Press Association column "diffused general joy through the community." "Everybody is applauding Gen. Hood and his noble army," read one report, "for the skill and gallantry displayed by them in this engagement." The news of Hood's "victory" prompted four of the five daily newspapers in the capital to print extras; the *Evening Whig* "made journalistic history," Andrews writes, by printing two editions. One of Lee's soldiers wrote home on the 24th, "No news—but of Hood's victory—which you have doubtless heard." Amidst all of this activity, even General Lee took notice. "If the news of the glorious victory at Atlanta, reported this morning, prove true," Lee wrote to President Davis on July 23, he (always thinking of his army) hoped that this would lead to improved food supplies to his troops from Alabama and east Mississippi. Lee was too sage a soldier to believe the first rumors of victory from the battlefield, and, sure enough, even Hood had to issue a restatement the day after the big fight: General Mercer had not been wounded; his troops had taken thirteen enemy cannon, not twenty-two, as it turned out. But the stands of colors taken were not just five as first reported; the impressive tally stood at thirteen. Such details tended not to matter as news of Hood's "victory" spread through the Southern press. Typical editorial comment was voiced by the *Richmond Dispatch*: "The news from Atlanta *is* (or *are*, if you choose to have the word plural) glorious, indeed. Two thousand prisoners, twenty-five guns, seven standards! Why, this is a great victory.... It is most auspicious...the skies begin to brighten in the direction of Atlanta." Even Wheeler's unsuccessful attempt to capture the wagons of McPherson's army came in for hot-breathed accolade. "LATER FROM ATLANTA. THE VICTORY OF GEN. WHEELER. GLORIOUS RESULTS OF FRIDAY'S FIGHT," ran the C.P.A.'s telegraphic column issued on the 23rd. "Gen. Wheeler last evening attacked the enemy's line in the neighborhood of Decatur, and drove them back," the statement read, "capturing five hundred wagons with supplies and a large number of prisoners. He is still pursuing."[69]

In such an atmosphere, it was to be expected that Hood's officers and men shared their commander's exultation over the results of the battle of July 22. "We gained a great victory yesterday," wrote Lt. Andrew Neal, adding that he had heard that Hardee's corps had "captured 3500 prisoners and 22 pieces of artillery & the enemies killed & wounded amounted to twice our own." A week later another Confederate wrote his wife, "general hardea whiped the yankeys and taken four thousand prisoners." One soldier reported to his family that Sherman's "'blue coats' have the mortification of being driven back in disorder to their ditches." General Bragg, traveling back from Alabama to Atlanta on the 25th, reported the local mood to the president: "the moral effect of our brilliant affair of the 22d has been admirable on our troops." But not all of Hood's men were so upbeat. Cavalryman William Nugent wrote his wife on July 26, "We have now a new General—Hood from Texas and he is for fighting all the time. He has so far done very well and gives evidence of some ability—but at what fearful cost. We have lost about *8,000* killed and wounded in the last two or three days. This sort of fighting,

---

[69] Hood to Seddon, July 22 (10:30 P.M.), *OR*, vol. 38, pt. 3, 900; "Telegraphic" and "The News from Atlanta. How It Was Received in Richmond," *Augusta Chronicle & Sentinel*, July 24, 1864; Andrews, *South Reports*, 457; Chamberlayne, *Ham Chamberlayne—Virginian*, 244 ("Hood's victory"); Lee to Davis and Hood to Seddon, July 23, *OR*, vol. 38, pt. 3, 900; "The Situation," *Richmond Dispatch*, July 27; telegraphic column in "Southern Newspaper Accounts of These Battles," 213.

Hood's telegraphic reports of the battle did not specify casualty numbers, but Confederate officers knew that attacking battles carried a heavy cost. Mrs. Chesnut recorded having heard Maj. Gen. Mansfield Lovell remark in the streets of Richmond, "No country can stand many such victories as Hood's on the 22nd" (Woodward, *Mary Chesnut's Civil War*, 626 [entry of July 26]).

unless we meet with some more decided success will dissipate our army very soon."[70]

Despite the public euphoria over his army's "glorious success," Hood had to come to terms with the failure of the battle of July 22 to achieve all that he had hoped for. To be sure, the Federal advance on the city from the east had been blunted and McPherson himself killed. "While the grand results desired were not accomplished," he wrote in February 1865, "the movements of McPherson upon my communications were entirely defeated," referring to the Macon railroad, but not to the Augusta one, which had already been broken. Besides, he claimed, "this engagement greatly inspired the troops and revived their confidence." Even Hardee seemed satisfied with the battle's outcome on July 22; he later termed it "the only decided success achieved by the army at Atlanta." Hood never acknowledged McPherson's providential placement of Sweeny's division, which had blunted the flank attack, and instead blamed Hardee for the overall failure of the battle. "Hardee failed to entirely turn the enemy's left as directed, took position and attacked his flank," he wrote in his campaign report. He expounded upon this charge in *Advance and Retreat*: "It had rested in his [Hardee's] power to rout McPherson's Army by simply moving a little further to the right, and attacking in rear and flank instead of assaulting an entrenched flank."

This is a manifestly unfair accusation, as Hood's perspective and argument are not sustained by the basic facts of what happened on July 22. Fellow Confederates realized this. Captain Buck, Cleburne's A. A. G., called Hood's contention "unfair." Colonel Roy, as to be expected, came to Hardee's lengthy defense, blaming Hood for not throwing Cheatham in against the XV Corps "as soon as Hardee became hotly engaged." Even Federals defended Hardee. "Hardee did not know of the presence of Dodge's corps till he struck it," wrote General Jacob Cox after the war. "The attack was to the full as vigorous and persistent as Hood's own in front of Marietta on the 22d of June" (a comparison Cox knew would have impact). Castel minces no words in judging Hood's accusation against Hardee, calling it "as absurd as it is unfair."[71]

---

[70] McMurry, "Confederate Morale," 236; McDonough and Jones, *War So Terrible*, 263; Bragg to Davis, July 25, *OR*, vol. 38, pt. 3, 908; Nugent, *My Dear Nellie*, 190.

[71] Hood report, February 15, 1865, *OR*, vol. 38, pt. 3, 631; Hardee report, April 5, 1865, *OR*, vol. 38, pt. 3, 699; Hood, *A & R*, 180; Hay, *Cleburne and His Command*, 244–45; Roy, "General Hardee," 361; J. D. Cox, *Atlanta*, 176; Castel, *Decision in the West*, 412.

After the war, a number of Southerners raised objection to Hood's efforts to blame Hardee for the lack of hoped-for success on July 22. "Not only did Hood endeavor to shoulder

Hood's biographers offer varied views of the battle. O'Connor quotes Union General Blair, who after the war complimented Hood's flank march as "a very bold and a very brilliant one, and was very near being successful" (Hood proudly printed Blair's long letter in *Advance and Retreat*). Dyer credits the Confederates' attack as "admirably if not brilliantly planned," and also quotes Blair's flattering statement from Hood's memoir. "The most critical historian," he concludes, "can find no real grounds for serious criticism of his strategy thus far [meaning, July 22], unless the position is taken that he should not have fought at all." McMurry, however, is more negative, terming Hood's "failure" to coordinate Cheatham's and Hardee's two assaults "the most serious flaw of the day," and places the blame squarely on Hood's shoulders.[72]

Hood's battle plan for July 22 was, in the words of Atlantan Paul Hinde, "one of masterly strategy and daring," but it was also overly ambitious. In Connelly's judgment, "Hood's planning, fashioned after a well-oiled command structure in Virginia, simply was too sophisticated for the new western command." Or as Castel puts it, "the fundamental reason for Hood's failure on July 22 is that he tried to do too much with too little in too short a time." In describing it after the war, there is no hint that Hood thought that he was expecting too much of Hardee and his men, yet clearly he was. To this extent

---

upon Hardee and others the blame of his failures," W. J. Milner told an audience of Confederate veterans after the war, "but he even pronounced a slander upon that gallant body of men, the Army of Tennessee" ("Lieut. Gen. William Joseph Hardee," 363).

[72] O'Connor, *Cavalier General*, 211; Dyer, *Gallant Hood*, 258; McMurry, *John Bell Hood*, 131.

Earl J. Hess adds to the questions raised of Hood's generalship on July 22 by asking why he failed to order the Georgia militia into the battle; General Smith acted on his own ("Civilians at War," 336–37).

Albert Castel raises a better point: while McPherson was fighting for his life (literally), why did Sherman not order Thomas forward against the Rebel lines north of Atlanta? "The Army of the Cumberland, with its 40,000 fine infantry, spent the day skirmishing with and entrenching in front of Stewart's Corps, which at most numbered 10,000 men," he writes. "Had they been explicitly ordered to storm Atlanta," he conjectures, "almost surely they would have advanced against the city's fortifications in such force as either to carry them or at least to compel Hood to abandon his attempt to crush the Union left" (Castel, *Decision in the West*, 414).

Castel is hardly the first author to make this observation; Lt. Col. Henry Stone of General Thomas's staff made it a century earlier. Speaking of Sherman's "careless confidence," Stone charges, "the most serious failure was that no attempt was made to use the rest of the army to secure—what was, all day, within its power—Hood's overwhelming defeat" on July 22. Indeed, "from 11 o'clock in the morning till darkness closed the day, 60,000 men stood idle spectators and auditors waiting for the word which never was given" (Stone, "Strategy of the Campaign," 158).

Hood—young, ambitious and as-yet relatively untested in army command—was committing a fundamental error: placing a naïve hopefulness in his plan without a veteran commander's appreciation that in battle, things will almost certainly go wrong. The Prussian military theorist Carl von Clausewitz had a phrase for this phenomenon: the friction of war. "Everything is very simple in War," he writes, "but the simplest thing is difficult." Orders get lost; a rainstorm blows in; roads turn bad; the enemy does something unpredictable. "The knowledge of this friction is a chief part of that...which is required of a good General."[73]

Another way of looking at it is to remember Castel's nifty analysis of why so many hoped-for flank attacks (every infantry commander's offensive dream) turned into bloody, failed frontal assaults. For a flank attack to succeed, as Jackson's had done at Chancellorsville, four elements had to be present: 1) the enemy had to be caught off-guard; 2) attackers had to know the enemy's position precisely; 3) the commander of the flanking column had to be extremely skillful and experienced; 4) and, finally, the attacker had to have a lot of luck on his side.

As we have seen, Hood lacked the three most important of these elements (possessing only Hardee's experience). Most of all, the Confederates on July 22 had been unlucky; the Federals had had the good luck that day, at least in their troop deployment at the time of the battle's opening. Everyone recognized that the providential placement of Sweeny's division, XVI Corps (and Morrill's brigade of the XVII) was a gift to the Union from the God of Battles. Union General Blair credited the fortunate placement of Sweeny's division with blunting the initial Rebel attack. "It was a most fortunate circumstance for the whole army," he wrote, "that the Sixteenth Corps occupied the position" it did. Maj. W. H. Chamberlin of the 81st Ohio, writing years later, was more dramatic. "Had Dodge's men not been where they were," Chamberlin wrote, "there would have been absolutely nothing but the hospital tents and the wagon trains to stop Hardee's command from falling unheralded directly upon the rear of the XV and XVII corps in line. Upon what a slight chance, then, hung the fate of Sherman's army that day." Hardee's biographer terms the event "one of the most fortunate moves made by the Union armies in the war." Steven Woodworth explains why:

> Had the Sixteenth Corps been busy tearing up the railroad as Sherman had intended, Bate and Walker would have gobbled up the portion of the

---

[73] Hinde, "Govan's heroic charge; Battle of Atlanta," 5; T. L. Connelly, *Autumn of Glory*, 447; Castel, *Decision in the West*, 413; von Clausewitz, *On War*, 164, 166.

> wagon train present with the army and then taken the Seventeenth squarely in rear, while, as per Hood's plan, the troops of Maj. Gen. Benjamin F. Cheatham's corps attacked it and the Fifteenth head-on. If all went according to plan, the Army of the Tennessee, assailed front, flank, and rear, would collapse, with vast numbers of its personnel killed and wounded, many times more captured, and the rest scattered in headlong flight not likely to stop this side of the Chattahoochee.

Sagest perspective of all comes from Union Capt. David Conyngham, who wrote columns for the *New York Herald*. "Men are judged, not so much by the wisdom of their acts or counsels, as by the result," Conyngham wrote several months after the war, complimenting Hood's plan for July 22. "It was so with Hood; he failed and is execrated; had he succeeded, he would have been idolized." One thing was certain, and it was affirmed by Capt. John C. Van Duzer, Sherman's chief telegraph officer in a message sent to Washington on the night of the battle: "Hood fights his graybacks desperately."[74]

A more extended conclusion would compare Hood's/Hardee's flank attack at Atlanta, July 21–22, 1864, and Lee's/Jackson's flank attack at Chancellorsville, May 1–2, 1863.

The main commonality between the Confederate sides in the two battles involves, of course, their identity as that rare Civil War phenomenon, a planned flank attack. The enemy army's sending off of a sizable cavalry force figured in both. Yet even here there lay a difference in Lee's and Hood's responses. Maj. Gen. Joseph Hooker, commander of the Army of the Potomac, had sent off Maj. Gen. George Stoneman, his cavalry commander, with most of the army's mounted troops (six of seven brigades), in a raid against the Rebels' railroads south of Lee's army. On April 29, the day after they set out, Jeb Stuart, Lee's cavalry chief, saw the enemy column and, after leaving a small force to follow it, rode back to join the army. Lee took no action in response to the news. In Hood's case, Wheeler's report on the afternoon of July 21 that Garrard had departed on a raid immediately gave Hood the idea of launching a flank attack on McPherson's exposed left flank.

Thus Hood planned for a flank attack from the beginning. But at the end of his first day of battle at Chancellorsville, May 1, Lee was still discussing with General Jackson for the 2d an assault against the enemy's front near the Chancellor house. He sent two staff officers, Maj. T. M. R. Talcott and Capt.

---

[74] Castel, "Mars and Longstreet," 112–14; Davis, "Reappraisal," 76–77; Blair report, [no date] 1864, *OR*, vol. 38, pt. 3, 545; Chamberlin, "Hood's Second Sortie at Atlanta," 4:326; Hughes, *Hardee*, 228; Woodworth, *Nothing but Victory*, 549; Conyngham, *Sherman's March*, 168; Van Duzer to Eckert, July 22, *OR*, vol. 38, pt. 5, 232.

J. K. Boswell, to recon the enemy's position there. Before they returned, Stuart rode up with news from Fitz Lee: "the enemy's flank," as Freeman writes, "was 'in the air,' resting on no natural barrier, and therefore could be turned." The two officers' report, soon delivered, that Hooker's front was too strong, "settled the question in Lee's mind." Lee and Jackson then began discussing the next morning's flank march.[75]

After that, we may discuss why Lee's flank attack at Chancellorsville succeeded and Hood's at Atlanta did not. There are more than a dozen contrasts.

1. The first is *Lee's experience as tactical planner, and Hood's inexperience.* Planning and executing brilliant flank attacks is something Lee had already done at Second Manassas. Hood had no such experience; indeed, this would be only his second battle as an army commander.

2. *Understanding of and enthusiasm for the plan by the flanking column commander.* After they resolved to attack the right flank of Hooker's army in the Wilderness—Maj. Gen. Oliver O. Howard's XI Corps—"Lee and Jackson understood that an important opportunity lay before them, and they were eager—even excited—to exploit it," as Mackowski and White put it. On the other hand, even before he left Atlanta, Hardee worried that a morning attack in the Federal rear was impossible, so he sought a change in orders to allow him to launch a flanking assault.

3. *Working relationship between army commander and flanking column leader.* In their year of working together, Lee and Jackson in Virginia had forged a genuine bond that enhanced their individual soldierly strengths. As Ernest Furgurson extravagantly describes, "American history offers no other pair of generals with such perfect rapport, such sublime confidence in each other." On the other hand, in Georgia there is evidence of Hardee's jealousy of Hood earlier in the campaign, when General Johnston had spent so much time with him, seeking his counsel. And just the day before, after Peachtree Creek, Hood had begun to form negative judgments of Hardee's conduct of his attack and may have expressed them. But for his flank attack planned for July 22, Hood had no other experienced corps commander to turn to.

4. *Tactical talent of the column commander.* Jackson, in the Shenandoah Valley and at Second Manassas in1862, had shown himself to be a master of daring maneuver. Hardee thus far in the western theater had had no opportunity to demonstrate any flair for independent command, and it may not be

---

[75] Sears, *Chancellorsville*, 166, 173; Freeman, *Lee*, vol. 3, 519–20; Talcott, "General Lee's Strategy at the Battle of Chancellorsville," 18 ("the last hope of a successful attack in front was abandoned, on the information obtained by Captain Boswell and myself as to the strength of the enemy's position").

unfair to reckon that he did not possess it. Jackson was the genius, Hardee the "old reliable." Historian Phil Gottschalk touches on this when he observes, "Hardee, although a capable general, was not the inspirational leader and military genius that was Jackson." Hood himself was crueler in his memoir when he wrote that because he considered his battle plan so reasonable and its goals so attainable, "no special quality, such as Jackson possessed, was required" in his assault leader, General Hardee. Hood's biographer Richard O'Connor carries the contrast further: "The bitter truth was that Hardee was not Jackson—and Hood was no Lee."[76]

5. *Condition of the men starting out.* Jackson announced on the night of May 1 that he would get his troops marching at 4 A.M. But he gave them several more hours of rest, so that they set out around 7, after having enjoyed a cooked breakfast to boot. Hardee's troops, though, had to march all night, after a hard day's work. Cleburne's division had even marched the night before, July 20–21, hustling to reinforce Wheeler at the bald hill, and then fighting hard (unsuccessfully) to hold it on the morning of the 21st.

6. *Distance of the march-route.* Jackson's march route through the Wilderness was twelve miles long—very reasonable for Stonewall's "foot cavalry" (recall that Jackson's troops had marched fifty-four miles in thirty-six hours to reach Manassas Junction). Hardee's route was longer, fifteen miles from outer line to point of attack. Hood totally failed to grasp this point. In his memoirs he goes so far as to state that he associated Hardee's march route with the distance between Atlanta and Decatur—six miles. "I considered Hardee's move one merely within the lines of our cavalry," he adds, absurdly.[77]

7. *The commanding general's expectations of attack start-time; time allowed for the march.* In their planning meeting on the night of May 1, Lee did not tell Jackson when to begin his assault, giving him discretion to map out his route and pace his march. Jackson chose the standard infantry march-pace, which called for two miles per hour, with ten minutes for rest. Col. Porter Alexander, with Jackson's artillery, remembered that the marchers enjoyed three breaks of twenty minutes each. Hood, on the other hand, called for a morning attack, "at daylight or as soon thereafter as possible," he wrote in his

[76] Mackowski and White, *That Furious Struggle*, 47; Furgurson, *Chancellorsville 1863*, 146; Gottschalk, *In Deadly Earnest*, 393 ("not the inspirational leader"); Hood, *A & R*, 178; O'Connor, *Cavalier General*, 213.

[77] Furgurson, *Chancellorsville 1863*, 142 ("nearing seven o'clock, not at four as Jackson had promised"); Freeman, *Lee's Lieutenants*, 2:546 ("estimated at 11 or 12 miles"); Stackpole, *Chancellorsville*, 207 ("twelve miles through the Wilderness"); Hennessy, *Return to Bull Run*, 13; Hood, *A & R*, 178.

report. This was a demanding schedule which forced Hardee to keep his tired men moving throughout the night.

8. *Weather.* Early May mornings in central Virginia are generally cool. Stonewall's mapmaker, Maj. Jedediah Hotchkiss, recorded in his diary that May 1 was "quite pleasant, a genuine May day." The day after that would have been, too. However, in Georgia, on a night in mid-July, the weather can be anything but pleasant. Humid evenings can stay hot, with temperatures sometimes still reaching 90 degrees.[78]

9. *Road conditions.* At their unrushed pace, Jackson's men marched over soft roads, "just wet enough to be easy to the feet and free from dust," according to South Carolinian J. F. J. Caldwell. Hardee's troops marched on roads of hard red clay, from which men and horses kicked up clouds of choking dust.

10. *Cavalry service.* Stuart and Fitz Lee provided invaluable help in scouting the march route, then guiding Jackson's infantry along it into position. As they neared their objective, Fitz rode up to Stonewall: "General, if you will ride with me, halting your columns here, out of sight, I will show you the enemy's position." Wheeler's performance was less sterling. Hood credited his cavalry as being "thoroughly familiar with the different routes" in the area of Hardee's march, and in this regard they rendered at least some service. But as they accompanied the infantrymen, Wheeler's troopers rode in and amongst them, interrupting their march and impeding it. Then, when Hardee needed cavalry most—his deployment on unknown ground ahead—Wheeler had to leave the column under Hood's order to ride on to Decatur and attack the enemy wagon trains.[79]

11. *Luck.* Sears observes that the absence of dust on the roads used by Jackson's column was a "double blessing": the Confederates would kick up no dust that might reveal their march to observant Federals. Hood had no such luck. When Federal Captain Edge saw dust clouds in the distance, he correctly figured they were enemy infantry, and so reported to headquarters; McPherson, already wary, promptly acted on the intel. Officers in Hooker's army did not. Starting at 8 A.M., Union observers in tall trees spotted Rebels

---

[78] Freeman, *Lee's Lieutenants*, 2:548; Alexander, *Memoirs*, 330; Hood report, February 15, 1865, *OR*, vol. 38, pt. 3, 631; Hotchkiss, *Make Me a Map of the Valley*, 137; Furgurson, *Chancellorsville 1863*, 155 (80 degrees); Sears, *Chancellorsville*, 233–35, 242 ("breezy and pleasant").

[79] Mackowski and White, *That Furious Struggle*, 67 ("recent rain kept the dust down"); J. Caldwell, *The History of a Brigade of South Carolinians*, 76; Freeman, *Lee's Lieutenants*, 2:552; Furgurson, *Chancellorsville 1863*, 162 ("Fitz Lee's brigade formed the screen for Jackson"); Hood, *A & R*, 178.

on the march, and so reported; their officers failed to respond. Maj. Gen. Daniel Sickles, III Corps commander, even shelled the Rebels his troops spotted moving through the Wilderness woods. Hooker sent word to General Howard that morning, "We have reason to suppose that the enemy is moving to our right. Please advance your pickets." Howard did not and later denied he had gotten any warning from Hooker. Hooker himself "displayed a curious indifference to the matter." For all of these reasons, Stephen Sears concludes that at Chancellorsville on May 2, "Dame Fortune was smiling on the South this day."

The "curious indifference" displayed by Hooker and other Union higher-ups on May 2 leads A. Wilson Greene to speculate that even if more of the army's cavalry—especially George Stoneman's six brigades off raiding toward Richmond—had been in the Wilderness that day, events would not probably have changed. "Would Hooker have placed his cavalry beyond the right flank of the Eleventh Corps? How could more cavalry have penetrated the Wilderness tangle any better than did Pleasonton's troopers [the brigade left with the army]? ...Did not Union infantry spot Jackson's flank march even in the absence of a larger number of mounted troops?"[80]

12. *Maps and local guides' knowledge of the land.* After Lee and Jackson had decided on their flank attack against the enemy right in the Wilderness, Stuart consulted the Rev. Beverly Tucker Lacy, who knew the area well—his family owned land there, and Lacy had once ministered to a local church. Lacy prescribed a march route that would be long and wide, but sure to keep Jackson's march from being detected. Jed Hotchkiss at some point joined the two generals in their several-hour meeting and gave them a map of the area. The next morning, General Jackson got up early and called for Reverend Lacy to point out his suggested route on Hotchkiss's map. But when Lacy pointed it out, Jackson saw that it would take his column too close to the enemy pickets seen by Confederate cavalry the day before. Was there no alternative? Lacy called on Charles C. Wellford, local landowner, for help. On Hotchkiss's map, Wellford pointed out a road recently opened by him that would avoid the enemy pickets. Hotchkiss and Lacy returned to HQ and explained to Lee and Jackson what they had learned. This was what the two generals were

[80] Sears, *Chancellorsville*, 232–34, 244–45 ("observers perched in tall trees"), 247, 254, 271 ("Dame Fortune"); Greene, "Stoneman's Raid," 99.

looking for, "the hidden route to the enemy's rear," in Sears's words. Lee began issuing orders for the movement; Jackson told him he would make it with his entire command.[81]

There is no similar story associated with Hardee's march east of Atlanta, because Hardee had no map. (This was indeed the key reason that General Hood at headquarters could be so misinformed on the length of Hardee's march route.)

13. *Difficulty of deployment and terrain.* Jackson's deployment went smoothly. Having covered a dozen miles in eight hours, Jackson was on the enemy flank by 3 P.M. He spent more than two hours getting his troops in just the alignment he wanted. The Wilderness is thickly forested (indeed, General Howard blamed dense woods for Jackson's ability to mass his forces unobserved), but its terrain and underbrush offered no vexing obstacles as the Confederates arrayed themselves in optimum order, perpendicular to Howard's unsuspecting flank.

Hardee's deployment was anything but smooth. The Confederates east of Atlanta thrashed through woods, briars, and swamps, not knowing where the enemy was ahead of them. Just one example: "because the country around Sugar Creek is an enormous briar patch," writes Castel, General Walker asked if he could find another approach. Hardee refused. "No, sir!" With his attack grossly behind schedule, Hardee told Walker, "This movement has been delayed too long already. Go and obey my orders!" Like Jackson, Hardee may have also spent two or more hours getting his men into position, but it was not with Jackson's calm sense of precision—it was out of anxiety that time was running out. This same sense of urgency meant that Hardee could not spend time, as Castel phrases it, "to conduct a reconnaissance for the purpose of ascertaining the location and strength of the enemy lines and the best way of approaching them." The Southerners were going in blind.[82]

14. *Numbers engaged.* As in all battles, opposing strengths are paramount. When he gave the order to advance about 5:15 P.M., Jackson had deployed at least 28,000 infantry in the Wilderness. The effective infantry of Howard's

---

[81] Freeman, *Lee*, 3:585 (meeting "lasted for hours"); Freeman, *Lee's Lieutenants*, 2:545 ("Hotchkiss had made a sketch"; "he spread the map"); Roper, *Jedediah Hotchkiss*, 55.

[82] Sears, *Chancellorsville*, 258–62; Howard report, May 13, 1863, *OR*, vol. 25, pt. 1, 630; Furgurson, *Chancellorsville 1863*, 171 (5:15 P.M.); Robertson, *Stonewall Jackson*, 721 ("The attack was to begin at 5:15"); Stackpole, *Chancellorsville*, 236–37 (map, situation 5 P.M. [reproduced in Cullen, "The Battle of Chancellorsville," 23]); Castel, *Decision in the West*, 391, 394; Ecelbarger, *Day Dixie Died*, 67 ("During the final hour before noon, General Hardee finished aligning his corps for the assault").

XI Corps numbered about 11,000 soldiers, so Jackson held an important numerical advantage. Jackson's strength in numbers was magnified because Howard's three divisions were strung out in line *perpendicular* to the Confederates' mile-and-a-half front, arrayed in such a way as to be routed in detail, one upon another.

In his attack, Hardee did not enjoy this kind of 2+:1 advantage. Indeed, he possessed no advantage at all. At the time of his assault, reduced by straggling, Hardee's corps took into battle maybe 11,000 troops—against 14,800 in the Union XVI and XVII Corps, according to Ecelbarger's calculations.[83]

15. *Condition of the enemy troops.* The XI Corps attacked by Jackson's force on the afternoon of May 2 was anything but battle-ready. With its large number of German-Americans, many of the men were still despondent that their popular commander, Maj. Gen. Franz Sigel, had left them a few months before. With the battle raging off to the east on the afternoon of May 2, Maj. Gen. Carl Schurz, division commander in the XI, had his German regimental bands play martial and patriotic airs, hoping to lift the men's spirits. More important than spirit was experience: nearly a third of the troops in Howard's corps had never been in battle.

Not so in McPherson's army, which was made up of veterans who shared a broad optimism and enjoyed high morale. The officers and men would have approved of Sherman's statement to his wife, Ellen, written a week before the battle: "My army is as strong as the day we left Chattanooga and full of confidence."[84]

16. The most important difference between Jackson's and Hardee's attacks involves *the enemy's awareness and preparation.* At Chancellorsville, after he received word that Rebel troops were moving to his right, Hooker eventually came to the conclusion that Lee was *retreating.* "The enemy is fleeing," he wrote as late as 4:10 that afternoon. "I think it is a retreat," Dan Sickles echoed. And even though Howard had earlier in the day sent word back to Hooker, "I am taking measures to resist an attack from the west," Howard took no such defensive measures. He could only later acknowledge, "I was

[83] Furgurson, *Chancellorsville 1863* ("28,000 Southern troops"); Sears, *Chancellorsville*, 239 (29,400 infantry, plus cavalry and artillery, totaled 33,000); 269 ("fewer than 11,000 men"); abstract for Army of the Potomac, April 30, 1863, *OR*, vol. 25, pt. 2, 320 (12,170 "present for duty equipped"); Ecelbarger, *Day Dixie Died*, 63 (Hardee's 11,000); 209–211 (XVI 6,000 + XVII 8,800).

[84] Sears, *Chancellorsville*, 263, 269–70; Simpson and Berlin, *Sherman's Civil War*, 669.

deceived." The Union commanders did not expect to be attacked, and as consequence, the troops of the XI Corps did not, either.[85]

In Atlanta, as we have seen, McPherson and his officers were ready. They had deployed in the very position designed to meet the Rebel attack squarely in its front. This preparation, more than any other factor, doomed Hood's/Hardee's "flank" attack on July 22.

All of these factors conspired to deny John Bell Hood the crushing victory he had so hoped for. The genesis of Hood's flanking attack had been the same as the genesis of Lee's: receipt of news about the departure of Garrard's cavalry division from McPherson's left flank led to Hood's quick decision to attack it. Fitz Lee's news on the afternoon of May 1 that Hooker's right flank was "in the air" got Lee thinking about a flank attack against it. For the aforesaid reasons, the two Southern assaults ended very differently. At the same time, at least a few of the reasons for the failure of the Confederate attack at Atlanta further suggest faults in Hood's generalship. It is important for our purposes to sort these out.

Hood cannot be blamed on July 21–22 for having no broad experience such as Lee had; on July 21, 1864, he had been in command of the Army of Tennessee all of three days. But he can be held responsible for making decisions that reflect that lack of experience.

Chief among those poor decisions is that Hood's plan was too ambitious, approaching the impossible in what he asked Hardee and his corps to do. An attack at dawn ("daylight or as soon thereafter as possible") after an all-night march? And against the enemy *rear*? Hardee was so fretful of his prospects for failure that early on he asked Hood to modify his orders and "settle" for a flank attack.

For instance, Hood seems to have had no concept of the length of Hardee's march-route as his corps worked to get into position for its assault. Hood's statement in *Advance and Retreat*—"I considered Hardee's move one merely within the lines of our cavalry"—shows that years after the war he still didn't get it.

Hood wanted it *all* in his first, great opportunity for a great victory, and his greed got the best of him. His order for Wheeler to leave Hardee, just when his troops were about to deploy, and ride on to Decatur to bag the enemy's trains left Hardee without the kind of reconnaissance and guide services that Fitz Lee gave Jackson during his deployment.

---

[85] Mackowski and White, *That Furious Struggle*, 61, (Sickles), 74 ("fleeing"), 76 ("deceived"); Stackpole, *Chancellorsville*, 224 ("taking measures").

All of this armchair parsing, of course, was not on Hood's mind as he departed from the battlefield following Hardee's final repulse. At some point in the late afternoon or evening, Hood left his perch in the Williams house and rode into the city; Lieutenant Mackall noted that he tarried near the Car Shed "until sunset." (Hood may have spent time at the Gate City Hotel across the street from the depot. There the army's telegraph office was located; Hood may have dictated the wire that was sent to Richmond that night about the day's battle.) Then the general and his staff rode to the new headquarters they had chosen: the house on Whitehall Street of John S. Thrasher, the Confederate Press Association superintendent, in which he had spent a week the previous autumn before traveling to Richmond. Location of the J. S. Thrasher home on Whitehall cannot be fixed precisely, but a clue comes from the Union spy Milton Glass, who reported on August 3 that "General Hood's headquarters is in White Hall street, near Rodger's tannery." That night Lieutenant Mackall recorded in his journal that the general was now at "Thrasher's House—suburbs of Atlanta."[86]

In the first days after the battle of July 22, both Hood and Sherman dealt with the loss of major generals in their respective armies. Hood's task was simpler; instead of replacing Walker, he, Hardee, and Bragg agreed to break up the division, transferring its three brigades to the other three divisions in Hardee's corps. To replace Clement Stevens, Brig. Gen. Henry R. Jackson was called from Savannah. A few other changes were made. Possibly to secure a division commander more experienced than John C. Brown, General Bragg ordered Patton Anderson back to the army from Florida. Meanwhile Lt. Gen. Stephen D. Lee was on his way from Mississippi to take command of Hood's old corps.

On the other side, Sherman obviously could not break up the Army of the Tennessee (which he had himself once commanded). General Logan, as senior corps commander in the army (over Dodge and Blair), was entitled at least to the spot promotion that Sherman made in the midst of the battle on the 22nd. Logan had performed well so far in the campaign, and Sherman may have preferred to keep him permanently as commander of the army. But Logan was a "political general" (Democrat from Illinois), not a West Pointer, and Sherman believed top posts in his army should be held by professional

---

[86] McMurry, "Mackall journal," entry of July 22; S. Davis, *What the Yankees Did to Us*, 89 (Gate City Hotel as the army's telegraph office); Ed. C. Denig to W. D. Whipple, August 3, *OR*, vol. 38, pt. 5, 348).

Rodgers's Tannery was near Whitehall and Garnett Street, five blocks south of the Car Shed, the center of the city ("Vincents Subdivision Map of the City of Atlanta").

soldiers. Sherman took his time in making a decision. Joe Hooker, commanding the XX Corps, felt entitled to be McPherson's successor; after all, he had commanded a far larger army in Virginia. But Fighting Joe was quarrelsome and vain. During the battle of July 22, after Sherman ordered Thomas to push forward upon the enemy's abandoned outer works north of the city, John Newton, commanding a division in the IV Corps, complained that his advance had stalled because Hooker had "cut in front of me." There was another factor. Sherman was already formulating his next tactical operation. Instead of pushing the Army of the Tennessee from its position east of Atlanta farther to the south to cut the Macon railroad, he planned to swing the army around to the rear of Schofield and Thomas west of Atlanta, then push south to cut the rail line between the city and East Point. He doubted Hooker's ability to command such a decisive movement. To cap it all, it was said Sherman did not like Hooker; and if he did not, he could not expect Schofield and Thomas to do so (Sherman wanted his three army commanders to be able to work efficiently together).

That brought the issue back to Jack Logan, but then Thomas stepped in. Thomas and Logan had argued over use of the railroads at Chattanooga even before the campaign had begun. Moreover, Thomas, a solid West Pointer, knew that Logan had secured promotion because of his political power (as a Democrat). Pap instead supported O. O. Howard as McPherson's successor and, according to Lloyd Lewis, even told Cump that he would ask to be relieved if Logan were given the army command. Sherman would have had his own reasons to support Howard's elevation. He had seen him handle his corps capably thus far in the campaign, and appreciated his "ability to act harmoniously," as a biographer puts it, with other officers.

Without mentioning any of this, Sherman wired Halleck on the 24th:

> The sudden loss of McPherson was a heavy blow to me. I can hardly replace him, but must have a successor. After thinking over the whole matter, I prefer that Maj. Gen. O. O. Howard be ordered to command the Army and Department of the Tennessee. If this meets the President's approval, notify me by telegraph, when I will put him in command, and name others to fill the vacancies created.

Tellingly, in a perfect reflection of his supreme confidence in his ability to win the campaign, Sherman added, "After we have taken Atlanta I will name officers who merit promotion."

Howard's appointment to army command was announced on July 26; Maj. Gen. David Stanley was named to take over the IV Corps. General Logan was disappointed by the news—he later called Sherman "an infernal brute"—but he shored himself up and resolved to stay on with the XV Corps. Hooker, on the other hand, was incensed. The next day Hooker wrote headquarters, "I have just learned that Major-General Howard, my junior, has been assigned to the command of the Army of the Tennessee. If this is the case I request that I may be relieved from duty with this army." He added, "Justice and self-respect alike require my removal from an army in which rank and service are ignored." Hooker explained more in a letter to General Logan the same day: "I asked to be relieved from duty with the army, it being an insult to my rank and service. Had you retained the command I could have remained on duty without the sacrifice of honor or principle." "Honor or principle" really meant seniority: the year before, O. O. Howard had been Fighting Joe's subordinate (commanding the XI Corps at Chancellorsville). Sherman knew all this and went ahead anyway. "Make Hooker resign his post as commander of the Twentieth Corps," Sherman wrote Thomas on the 27th. When Hooker left—"I am ordered to report to the Adj't Gen-l—the hopper into which all of us have to go now-a-days," he wrote—Sherman passed over Alpheus Williams, senior division commander in the corps, and instead brought Maj. Gen. Henry W. Slocum from Vicksburg to lead the XX. (Williams commanded the corps until Slocum arrived, then returned to his division, having muttered a year earlier, "I have long since given up all hope of promotion from the present powers.")[87]

Sherman would have to deal with other hassles among his general officers. Brigadier General Sweeny did not get along with his superior, Major General Dodge, whom he called a "damned political general." Dodge in turn thought Sweeny was "cocky and overly self-satisfied, and whenever he exposed himself to deadly peril in the performance of duty he made sure that everyone knew just how deadly the peril had been."A few days after the battle of July 22nd, Sweeny, Dodge, and Brigadier General Fuller were discussing

---

[87] Bragg to Col. John B. Sale, July 24, *OR*, vol. 38, pt. 5, 907; R. K. Brown, *To the Manner Born*, 283–84; Falconer Special Orders No. 61, July 29, *OR*, vol. 38, pt. 5, 930 (H. R. Jackson); Elliott, *John C. Brown*, 111; Castel, *Decision in the West*, 418, 423; J. P. Jones, "The Battle of Atlanta and McPherson's Successor," 399–401; Castel, "'Black Jack' Logan," 44; J. P. Jones, *"Black Jack,"* 218–24; Sherman to Slocum, July 24, *OR*, vol. 38, pt. 5, 246; Lewis, *Sherman Fighting Prophet*, 388–89; Carpenter, *Sword and Olive Branch*, 69–70 ("ability to act harmoniously"); Sherman to Halleck, July 24, *OR*, vol. 38, pt. 5, 240–41; Special Field Orders, No. 44, July 26 (appointing Howard), *OR*, vol. 38, vol. 5, 266; Hooker to Brig. Gen. Whipple, July 27, *OR*, vol. 38, pt. 5, 271; Hess, *The Battle of Ezra Church and the Struggle for Atlanta*, 34 (George Balloch, Hooker's chief commissary officer: to "take a junior officer it was more than the old Genl could stand"); Sherman to Thomas, July 27, *OR*, vol. 38, pt. 5, 273; Daniel, *Days of Glory*, 418; McMurry, *Atlanta 1864*, 155; Hebert, *Fighting Joe Hooker*, 285 ("hopper"); Castel, "Alpheus S. 'Ol' Pap' Williams," 76.

the fight when Sweeny (an Irishman) told Fuller (an Englishman) that his brigade had not fought hard—in fact, that it had acted cowardly. Dodge took Fuller's side, and Sweeny turned on him.

"You are a God-damned liar, sir!" Sweeny swore at Dodge, then added, "You are a cowardly son of a bitch, sir!"

Dodge puffed up angrily. Sweeny then taunted further, "You are a God-damned *inefficient* son of a bitch, sir!"

That did it. Dodge swung and hit Sweeny in the face. The Irishman hit back with his one arm (he had lost his right arm in the Mexican war), and Fuller joined in. Mortified staff officers watched the three generals scuffling and finally succeeded in breaking them up. General Sweeny was arrested and sent to Nashville for court martial (which eventually acquitted him, though he was never restored to command). Brig. Gen. John M. Corse, inspector general for the Army of the Tennessee, was appointed to replace Sweeny.[88]

This is not to say that Hood did not have some "general" issues of his own. Though there is no record of conversation after the battle of July 22 between Hood and Hardee, the latter was doubtless smarting from the former's criticism about Peachtree Creek, and it is likely that Hood had also begun to express to Hardee his disappointment at the failure of his attack. Besides, Hardee had not consoled himself at having been passed over as Johnston's successor the week before. In late August he confided to General Mackall, "the President is endeavoring to create the impression that in declining the command at Dalton, I declined it for all future time"—in other words, he had turned down the offer to assume command of the army in December 1863, but he might not have done so in July 1864. The prospect of continuing to serve under Hood (his junior in age and rank) irritated Hardee, especially when Hardee knew that in the first months of the campaign, "Hood was, of

---

[88] Maryniak, "Fighting Tom Sweeny, the Irish Bulldog," 55; Davis, "Hood Fights Desperately," 26; Anders, "Fisticuffs at Headquarters," 13–14; Morgan, *Through American and Irish Wars*, 103–104.

If one accepts the above text of the heated exchange between Sweeny and Dodge—there are slight variants—the inference to be drawn is that one could call Grenville Dodge a "son of a bitch," but you should not call him *inefficient.*

General Sweeny was known among his soldiers for flying off the handle. Sweeny—"the boys called him 'the Corporal,'" wrote one officer—

> had a manner of giving his orders in staccato, reminding one of electricity or dynamite, and if any hesitancy was shown in obeying, his comments were rendered in three languages, English, Irish-American and profane, thickly punctuated with exclamation points—making it decidedly unpleasant for the person addressed." (Compton, "The Second Division of the 16th Army Corps," 112)

all others, in favor of retreating," as he wrote to Mackall. "If General Johnston had followed his advice he would have crossed the Chattahoochee two or three weeks before he did." When General Bragg arrived back in Atlanta on July 24, Hardee requested transfer to another command. Hood agreed that Hardee's departure would be beneficial, so Bragg on the 27th recommended to President Davis that the lieutenant general be transferred. "There does not exist [between Hood and Hardee] that cordiality and mutual confidence and support" needed for effective leadership, Bragg explained. He suggested that perhaps Lt. Gen. Richard Taylor could be brought from Louisiana and Hardee sent there, an idea that appealed to Hardee. "It would suit us exactly to get command of the Dept. of Miss.," he wrote his wife, Mary; "we could be together, and I should have a suitable command." For the time being, the administration took no action on Hardee's request. Hood's senior corps commander would have to deal with his disgruntlement.

Unhappiness in the army's officer corps surfaced elsewhere. On the 24th, General Bragg arrived at the Thrasher house to meet with Hood. Brigadier General Mackall was no friend of Bragg, and he resented especially Bragg's obvious role in the removal of Johnston. When Hood called on Mackall to join him at headquarters that afternoon, Mackall approached Bragg and Hood on the Thrasher front porch but refused to shake hands with Bragg. This gave Hood the opportunity to relieve Mackall. As his nephew Tom recorded, "in 30 mins. is handed order relieving him." The next day General Mackall left for Macon. (Lieutenant Mackall moved to the staff of General Clayton.) As his new chief of staff, Hood appointed Brigadier General Shoup, his chief of artillery, then moved Col. Robert Beckham to Shoup's post. Amid the administrative matters being tended to, Polk's Army of Mississippi was designated Stewart's corps. About that time, July 26, S. D. Lee also arrived in Atlanta to take command of Hood's old corps, temporarily being led by Frank Cheatham (who returned to his division). A lot was going on. "Within the space of one day," writes Hess, "Hood had one brand new corps commander [Lee], another corps commander who had lost his independent status [Stewart], a new chief of staff [Shoup], and a new chief of artillery [Beckham]. And Hood himself had been in command of the army for only eight days."[89]

---

[89] Davis, "Hood Fights Desperately," 126; Hardee to W. W. Mackall, August 24, 1864, *OR*, vol. 38, pt. 5, 987; "Lieut. Gen. Hardee and His Position," *Atlanta Intelligencer*, August 12 ("It was strange then to the minds of many, not only in the army but out of it, that Gen. Hardee was not the successor of General Johnston"); Bragg to Davis, July 27, *OR*, vol. 52, pt. 2, 713; Hughes, *Hardee*, 243–44 (including Hardee to wife, n.d.); Castel, *Decision in the West*, 422–23; McMurry, "Mackall Journal," entries of July 24–25; Kinloch Falconer, Special Field

After having left Atlanta on July 15, General Bragg had traveled to Alabama. He spent a week in Montgomery, reporting back news of the enemy cavalry raid (Rousseau's), which was wrecking the railroad leading to West Point, Georgia. On the 23rd, he wired Richmond that he was in Columbus, having sent a brigade of reinforcements to General Hood's army. Meanwhile the strength of the Army of Tennessee became the subject of official speculation. Hood had never reported anything to Richmond about his casualties for July 22 beyond "loss not fully ascertained" on the night of the battle. This may have led someone in the C. S. War Department, perhaps Secretary Seddon himself, to worry about the strength of the Army of Tennessee, given the two big engagements recently fought. On July 23 the Adjutant and Inspector General's Office issued an odd special order: "The superintendent of the army intelligence office is authorized to send an agent to the Army of Tennessee to prepare a list of casualties occurring in said army." The inference to be drawn is that Seddon, who authorized the inquiry, had become suspicious that Hood was concealing—at least carelessly failing to report—the army's recent losses.

Even if Hood was not falsifying his loss numbers, Bragg was. He was at least misleading in his wire to the president of July 25 concerning the recent battle: "I am happy to say our loss was small in comparison to the enemy's."

Orders No. 56, July 24, *OR*, vol. 38, pt. 5, 907 (Mackall relieved; Shoup assigned); Hood to Cooper, July 26, *OR*, vol. 38, pt. 5, 910 (Beckham as chief of artillery); Hess, *Ezra Church*, 24.

Why General Shoup wanted to give up his post as the army's artillery commander is explained in a letter he wrote General Hood a year after the war. "I believe now, as I have from the first, that the final death blow to our glorious struggle was given upon the Dalton to Atlanta retreat," Shoup explained in his missive of April 5, 1866.

> After we crossed the Chattahoochie there was no reasonable hope. Had Gen Johnston broken Sherman, as I firmly believe he could have done anywhere between Dalton and the Etowah, the result of the presidential election would probably have been different and our cause saved.
>
> In any event a mainly defensive retreat was certain destruction, as the result proved, while the chances of successful battle were at least even.
>
> The artillery was in capital condition. It would have worked wonders in a fair open fight. My design and hope was to fight it in mass, and I believe it was in condition and discipline to warrant my confidence in it.
>
> As day after day passed till all the country favorable to its armies was left behind my disgust became extreme. It was battered and torn to pieces to advance into the works till it was sadly broken in spirit and material. My position as comm'dg the art'y became humiliating, as a general thing the whole of it was [illegible] and I had nothing to do but look after my baggage wagons. It was for this reason that I determined to construct the works at the Chattahoochee, and that finally I was glad to get rid of the comm'd. (Shoup to Hood, April 5, 1866, S. Hood, *Lost Papers*, 57)

Two days later he wrote the president a confidential letter, stating that since taking command Hood had lost "about 3,000." He added that some 5,000 "extra duty men and convalescents" had arrived to more than make good these losses; "more are coming in daily." In point of fact, Hood's headquarters issued general field orders on July 24 calling for "detailed men employed as clerks in the staff department of the army able to perform a soldier's duty in the field" to step forth and take a musket. Bragg mentioned to Davis that the Georgia militia also continued to come in for field duty. Governor Brown wrote Hood on July 23, expressing satisfaction with the recent service of G. W. Smith's militia. "Thousands of others have now been assembled in response to my call," he added, "and will be armed and sent forward as rapidly as possible." Hood wired back that he needed all the aid Georgia could furnish. (If Bragg was undercounting Hood's casualties, Brown was probably overcounting the state troops he could send to reinforce Hood's army.) Of final note, report of the army intelligence office superintendent on Hood's casualties July 18–22, called for by Seddon, is not to be seen in the *Official Records.* (Maybe Bragg squelched the investigation.)

"Tone of army fine and strength increasing daily," Bragg wired from Atlanta on the 26th. He added details in a long letter to President Davis, written on the 27th before setting out back to Montgomery to meet with Dabney Maury. He noted that S. D. Lee had arrived yesterday "and is most cordially welcomed by army and commander." "General Hood has found much to do" in shaping up the army, which Bragg stated was in "sad condition" when he took over "for want of administration" by Brigadier General Mackall (now gone). Shoup, his successor, "is decidedly fond of this kind of work and is very efficient at it." With extra duty men and convalescents returning to the ranks, Hood's biggest task was finding arms for these soldiers, whose return demonstrated a rising level of morale in the army. "Stragglers who left despondent are returning with revived confidence." With statements such as this, Bragg's report on the state of Hood's army was unusually sunny. Castel is justified in commenting, "Bragg, more than any one else except Johnston, is responsible for Hood's being commander of the Army of Tennessee. Now he finds himself in the position of having to justify his handiwork." Castel's conclusion is rapier-sharp: "His method is to deceive Davis as to what Hood has done and is capable of doing at Atlanta."[90]

[90] Bragg to Col. John B. Sale, July 17 and 23, *OR*, vol. 38, pt. 5, 887, 904; Samuel W. Melton, Special Orders No. 173, July 24, *OR*, vol. 38, pt. 5, 906; Castel, *Decision in the West*, 423–24 (on Mackall); Bragg to Davis, July 25, *OR*, vol. 38, pt. 5, 908; Bragg to Davis, July 27, *OR*, vol. 52, pt. 2, 712–13; A. P. Mason, General Field Orders No. 5, July 24, *OR*, vol. 38, pt.

The advance of McPherson's, now Howard's, army had been blunted by the Rebel assaults of July 22, but Sherman still held the initiative. He predicted to Logan on the 24th that the enemy, bloodied and weakened by attack, "will not again attempt it, but will await our action." Sherman already had that action in mind: to swing the Army of the Tennessee around the city to its west, then marching south with aim to strike and break the Macon & Western Railroad, which, as he wrote Logan, "you know to be the only road by which Atlanta can be supplied." Rousseau had cut the line to Montgomery; Garrard returned on the afternoon of July 24 reporting destruction of the railroad bridges across the Yellow and Alcovy Rivers, and six miles of track between them, further wrecking the road to Augusta. The W. & A. was supplying Sherman's army, even though the Confederates had burned the railroad bridge across the Chattahoochee. Col. George Buell, in charge of a brigade of pioneers, was overseeing its rebuilding. In the meantime, two pontoon bridges below Peachtree Creek allowed wagons to cross over rations and supplies brought down by train from Chattanooga.[91]

That left the Macon & Western for Hood and his army. "None remains to him but the Macon Road," Sherman wrote Grant, "and I think I will have that soon." He wanted to let his cavalry rest a few days, then get his troops in motion on the 27th. Orders for the movement went out on July 25, prefaced by what the several armies and bodies of cavalry should do if the Rebels suddenly abandoned Atlanta (Sherman still failed to grasp the tenacity of his opponent). Early in the morning of the 27th, the Army of the Tennessee would withdraw from its position on the left and, corps by corps, pass behind and beyond Thomas, extending his right. Schofield would draw back and take position as the army group's new left flank. As part of the plan, Sherman ordered Maj. Gen. George Stoneman (former commander of the Army of the Potomac's cavalry, now leading one of Sherman's four mounted divisions) and Brig. Gen. Edward M. McCook (another division commander) to lead two

---

5, 908; Hood to Brown, July 23; Brown to Hood, July 23; and Hood to Brown, July 24, *OR*, vol. 38, pt. 5, 906; Bragg to Sale, July 26, *OR*, vol. 38, pt. 5, 911.

Bragg's "tour of inspection," which began with his departure from Richmond around July 9, lasted more than a month. He spent a week in Montgomery, c. July 16–22, and was in Atlanta for several days after July 24 before returning to Montgomery by the 29th. He was in Columbus August 1 and in Macon August 3, heading back on trains to the capital, which he reached by August 16 (Seitz, *Bragg*, 450–57; Bragg's telegrams, July 15–August 6, *OR*, vol. 38, pt. 5, 881–948 *infra*.).

[91] Sherman to Logan, July 24, *OR*, vol. 38, pt. 5, 242–43; Sherman to Halleck, July 24, *OR*, vol. 38, pt. 5, 240; D. Evans, *Sherman's Horsemen*, 192; Buell to Whipple, July 24, *OR*, vol. 38, pt. 5, 242.

strong cavalry columns south of Atlanta to cut the railroad. McCook would ride down the Chattahoochee, cross, and by way of Palmetto hit the M. & W. near Lovejoy's Station, about twenty-two miles south of Atlanta. Stoneman's was the eastern column, to ride from Decatur toward Covington before turning south to rendezvous with McCook at Lovejoy's. When they struck the railroad, Sherman instructed that it "must be substantially destroyed for a space of from two to five miles"; the troopers were also to tear down telegraph lines. If the cavalrymen set out at daylight on July 27, they could be on the road on the 28th and tear it up before riding back to their respective flanks of the army group. Before heading out, Stoneman asked permission, after he wrecked the rails, to continue riding on to Macon. There he hoped to overwhelm the small Rebel prison and free its 1,500 captive Union officers. Then, if that went well, Stoneman proposed riding even farther, fifty-five miles southwest to the huge prison camp at Andersonville. With luck he might be able to drive off the guard garrison and free the 30,000 Northern soldiers wasting away in confinement there. Sherman assented, but emphasized that Stoneman's primary mission was wrecking the Macon & Western. Freeing the POWs could come later.

Unknown to Sherman, Stoneman had no intention of following orders. After Chancellorsville he had been removed from command of the Army of the Potomac's cavalry. Now he hoped to restore his reputation with a glorious exploit, the freeing of thousands of Union soldiers. As it turned out, George Stoneman would have done well to heed Cump's orders.[92]

Howard's infantry began their march from the Union left to the right before dawn on July 27. The XVI Corps marched out first, then the XVII and XV. Wagon and cannon wheels were wrapped with sacks for noise control; Sherman wanted to keep the enemy unaware of his new tactical move as long as possible. It turned out to be not long at all. On the basis of cavalry picket activity, Lawrence S. ("Sul") Ross, one of Wheeler's brigadiers, had reported the day before that the Federals "seem desirous of extending their line down the river." At 4:15 A.M. of the 27th, with Howard's march barely in progress, Hood's headquarters notified Wheeler, "indications are that the enemy will attack our left."

[92] Sherman to Grant, July 25, *OR*, vol. 38, pt. 5, 247; Sherman to Halleck, July 24, *OR*, vol. 38, pt. 5, 240 ("as soon as my cavalry rests"); L. M. Dayton, Special Field Orders No. 42, July 25, *OR*, vol. 38, pt. 5, 255; Sherman to Garrard, July 25, *OR*, vol. 38, pt. 5, 251; Davis, "Hood Fights Desperately," 28–29; D. Evans, *Sherman's Horsemen*, 202, 205–206 (Stoneman and Andersonville); Castel, *Decision in the West*, 437.

As the Army of the Tennessee marched behind Schofield's and Thomas's lines, pickets and batteries kept up a lively fire to divert the Rebels' attention. "Our skirmishing and artillery were just enough to make things interesting," Sherman recorded that night. Dodge's infantry marched past Thomas's right, extending the Union line southward. Blair's and Logan's fell into position so that by nightfall Sherman's right flank had been pushed almost two miles closer to the railroad.

During their semi-siege of Hood's army in its fortifications, the Federals occasionally sortied against the Rebel lines. One such incident occurred on the afternoon of July 27. General Hooker and staff were riding along their lines north of the city when they came into view of Confederate sharpshooters posted in houses across no-man's-land. The Southerners opened fire. Hooker was untouched, but several in his escort were wounded, one officer severely. Angrily he ordered a sally against the Rebels. Union generals Alpheus Williams and Thomas Ruger were "crooching under the works and taking quints" when the 13th New Jersey, wielding turpentine and torches, charged forth. They did so with such impetuosity, while Federal artillery pounded the Rebel position, that "the Jersey boys pounced right onto the Johnnies," as one Northerner recorded. The Jerseymen set fire to two houses and brought back thirty-three prisoners. "The planters' homes was in ashes in a short time, negro huts included," one Federal wrote. "This was the most spectacular exhibition we had witnessed during this campaign," recalled Sgt. Rice C. Bull of the 123rd New York, with some overstatement. "While it was a small affair it was exciting while it lasted."[93]

July 27 was a busy day for Hood and his new chief of staff, General Shoup. Lieutenant General Lee, who had arrived in Atlanta the day before, was now formally announced as taking over Hood's/Cheatham's corps. Wheeler, who days before had led a belated pursuit of Garrard toward Covington, again had his eyes on the Yankee cavalry raiding east (Stoneman's column). At 11 A.M. Hood gave his permission to pursue the enemy, but a couple of hours later Shoup added, "You must not go in person, unless you think it important." That night, even as Shoup urged Wheeler to "take the smallest number of troops possible," he instructed the cavalry chief to personally lead the column of pursuit after all.

---

[93] Castel, *Decision in the West*, 424; Davis, "Hood Fights Desperately," 30; Ross to Jackson, July 26 (10:45 A.M.), *OR*, vol. 38, pt. 5, 911; Smith to Wheeler, July 27 (4:15 A.M.), *OR*, vol. 38, pt. 5, 913 ("indications are that the enemy will attack our left"); Sherman to Halleck, July 27 (8:30 P.M.), *OR*, vol. 38, pt. 5, 272; S. Davis, *What the Yankees Did to Us*, 156; Bull, *Soldiering*, 155–56.

Hood quickly realized that the enemy's extension of his right had to be stopped. Twelve hours after the Federals began their stealthy early morning march, Hood was already planning to launch an attack. Unfortunately, he would be using the infantry corps whose commander had barely just taken over, Stephen D. Lee. At 1:30 P.M. on July 27, General Shoup instructed Lee to prepare Brown's and Clayton's divisions for action. Lee was told to move his men westward quietly and carefully, using the fortifications to keep the enemy from observing their withdrawal. They would halt behind Stewart's line between Peachtree and Marietta roads, "and be in readiness to move at any moment." Troops from Hardee's corps and the Georgia militia would replace them. Each man was to receive sixty rounds of ammunition. Later that afternoon Shoup wrote Stewart, "from present appearances the enemy is preparing to attack our left. Please have everybody on the alert and in readiness." Stewart was also told to issue sixty cartridges to each soldier.

Hood did not want merely to receive the enemy's attack; he wanted to launch one of his own. "The enemy...now seems to be working as if to turn our left," Bragg wrote the president before leaving Atlanta. "Hood is preparing a move for him which I hope will soon relieve that danger." During the afternoon Hood formed his plan; a little before 7 P.M. Shoup called both General Lee and General Stewart to headquarters to discuss it.[94]

Of Hood's instruction to his two corps commanders on the evening of July 27 we have little documentation. In his campaign report, Hood merely states that Lee's corps was to "hold the enemy in check" along Lick Skillet road, with Stewart's corps "ordered to support." General Lee adds little more. Stewart's recollection is key. After Lee gained the Lick Skillet road on July 28, "at an early hour next morning we were to move out on that road, turn to the right, pass in rear of the enemy, and attack." Hood in his memoirs does not explain the battle plan in this detail. Ironically, it is General Howard, Hood's enemy, who states in his autobiography that Stewart on the morning of the 29th was to be on the Lick Skillet road, "pass beyond Lee, gain ground,

---

[94] Bragg to Sale, July 26, *OR*, vol. 38, pt. 5, 911; Lee, General Orders No. 58, July 27, *OR*, vol. 38, pt. 5, 917; D. Evans, *Sherman's Horsemen*, 192; Smith to Wheeler, July 27 (11 A.M.), *OR*, vol. 38, pt. 5, 913; Hess, *Ezra Church*, 38–39; Shoup to Wheeler, July 27 (9 and 9:30 P.M.), *OR*, vol. 38, pt. 5, 914; Shoup to Lee, July 27 (1:30 P.M., sixty rounds), *OR*, vol. 38, pt. 5, 915–16; Shoup to Hardee, July 28 (9 A.M.), *OR*, vol. 38, pt. 5, 918 ("The militia did not quite reach the Peach Tree road. If it be possible for you to fill it and relieve Lee's troops, the general will be greatly pleased"); Shoup to Stewart, July 27 (4 P.M.), *OR*, vol. 38, pt. 5, 916; Shoup to Stewart, July 27 (1:30 P.M.), *OR*, vol. 38, pt. 5, 915 (sixty rounds); Bragg to Davis, July 27, *OR*, vol. 52, pt. 2, 714; Shoup to Stewart (6:40 P.M.), and to Lee (6:45 P.M.), July 27, *OR*, vol. 38, pt. 5, 916–17.

and attack, as far as possible, beyond my right flank." After the war, in a "sketch" of the Army of Tennessee's campaigns, General Stewart affirmed, "On the 28th General Hood attempted another flank movement."

Hood's biographers can be helpful. With characteristic generality, O'Connor states simply that "Hood sent two corps…to smash the Federals at Ezra Church." Dyer understood the intricacy, if not artfulness, of Hood's plan, especially the flanking nature of Stewart's march:

> Lieutenant General S. D. Lee, commanding Hood's old corps, was to march directly on Howard on the twenty-eighth. The next morning Stewart was to swing south of Ezra Church and come up in Howard's rear. Thus the Federals would be ground to dust between two millstones. It was another ill-fated plan, perfect in concept but poorly executed.

Herman Hattaway, Stephen D. Lee's biographer (1976), summarized Hood's plan accurately: with Lee's corps "holding the Federals" on the 28th, "the next morning Stewart's corps would swing around Lee's left and strike the enemy flank."

McMurry completes the details of what, in concept at least, Hood devised as another flank attack upon the enemy.

> The Lick Skillet Road ran west from Atlanta to the Chattahoochee. At Ezra Church, three miles from the city, it crossed a north-south road. Confederate control of this intersection would probably block Sherman's extension to the southwest. Hood ordered Lieutenant General Stephen D. Lee, newly assigned to replace Cheatham at the head of Hood's old corps, to move west on the Lick Skillet Road on the twenty-eighth. At Ezra Church, Lee was to halt and entrench a line facing to the north, establishing Confederate control of the crossroads. Then, on the twenty-ninth, Stewart's corps, following Lee, would pass out the Lick Skillet Road beyond Lee's left and swing around to attack the right of the line that Sherman would have formed facing Lee.

Castel adds a few more details: Hardee was to occupy the works from Peachtree to the Decatur road, the entire northeast quadrant of the Confederate position; Stewart was to withdraw Walthall's and Loring's divisions from their lines northwest of the city for the march to Lick Skillet. "Then, early on the morning of July 29 and after being reinforced by French's and Bate's divisions, he [Stewart] is to strike the Federal right wing from the rear and crush it." "In short," Castel adds, "Hood seeks to do to the Army of the Tennessee on July 29 what he was unable to do a week earlier. To make sure that it is done this time, he has assigned the task to Stewart, whose troops performed so magnificently on the twentieth." Earl Hess, the most recent and thorough chronicler of the battle, reinforces the current understanding of Hood's plan: "Lee...on the morning of July 28...was to merely hold Sherman at bay, not attack him that day. Stewart meanwhile would move two of his divisions (Walthall's and Maj. Gen. William W. Loring's) out along the Lick Skillet Road and set up to attack the Union flank and rear on the morning of July 29."[95]

Such is the current scholarship on Hood's plan at Ezra Church, but it has not always had this perspective. The old view of the battle was expressed by Joseph B. Mitchell a few generations ago: "In this instance there had been no plan of battle. It was just a head-on meeting engagement." Fortunately, in the last quarter-century, the literature has solidified on Hood's plan for a flank attack at Ezra Church on July 29, including my own writing ("Stewart's flank attack could rout a whole wing of the Yankee army much as Jackson had done at Chancellorsville").[96] Walthall's and Loring's divisions, states David Evans, "would pitch into the enemy's right flank." According to Hood's plan, drawn from textbooks of the "Lee and Jackson School," David Coffey explains, "Stewart moved around the Federal flank to strike Howard on the morning of the 29th." "Lee had been specifically instructed not to attack unless attacked first and then only the Union right flank," adds Thomas J. Brown. Even Prof. Brian Craig Miller, not known as a military historian, gets it: "Hood requested that Stephen D. Lee entrench his soldiers and wait for Alexander Stewart's corps to flank and rout a group of the Union flanking force."[97]

Hood's plan was set and promised much if effectively carried out by his subordinates. The problem, as on so many other battlefields, was that the Dag Gone Yanks got in the way. Even though General Sherman had dismissed the prospect of the Rebels launching yet another attacking battle, Howard

---

[95] Hood report, February 15, 1865, *OR*, vol. 38, pt. 3, 632; Lee report, Jan. 30, 1865, *OR*, vol. 38, pt. 3, 762; Stewart report, Jan. 12, 1865, *OR*, vol. 38, pt. 3, 872; Hood, *A & R*, 194; Howard, *Autobiography*, 2:22; Stewart, "The Army of Tennessee," 1:101; O'Connor, *Cavalier General*, 213; Dyer, *Gallant Hood*, 265; Hattaway, *Stephen D. Lee*, 128; McMurry, *John Bell Hood*, 133; Castel, *Decision in the West*, 426; Hess, *Ezra Church*, 240–41.

[96] J. B. Mitchell, *Military Leaders in the Civil War*, 212; Davis, "Hood Fights Desperately," 30.

[97] Evans, "Atlanta Campaign," 53; Coffey, *Struggle for Atlanta*, 91; T. J. Brown, *John Bell Hood*, 50; Miller, *John Bell Hood*, 130.

Herman Hattaway fell into the old-school trap about Lee's eventual frontal attack at Ezra Church when he wrote a decade before his biography of the general: "Probably he should have stopped his attack sooner than he did, but he was obeying orders to the letter" ("Stephen Dill Lee—A Profile," 23).

disagreed, saying that he knew Hood at West Point, "and that he was indomitable." All during the 27th, he oversaw the careful extension of his line, making certain that each division took its place so as to protect the right of the preceding one. Early on the 28th Howard continued his cautious extension, so that by late morning his three corps were aligned on a north-south ridge: XVI Corps on the left, near the extreme right of Thomas's army; then Blair's XVII along the road leading to a small Methodist country chapel named for the prophet Ezra. There Woods's division began the XV deployment. When Harrow's and Morgan Smith's divisions came up, they were arrayed on a ridge running west from the chapel. Howard's position thus developed into a refused line, a capital L in reverse. Smith and Harrow were prepared to repulse the enemy assault just like Tom Sweeny had done on the 22nd, if the Rebels attacked again. When Federal skirmishers approaching the Lick Skillet road encountered enemy cavalry, General Howard's caution proved justified. By 11:30 he became convinced that he was about to be attacked and sent word to General Sherman. His troops began to entrench in their east-west line near the Lick Skillet road—the very position which S. D. Lee hoped to have taken first.

Lee clearly had not. When his units began to come up toward Ezra Church, Red Jackson, the cavalry commander in charge, told him he thought the Yankee force in their front was small. Lee deduced that "the enemy's works were slight, and besides they had scarcely gotten into position." Despite orders to form and hold purely on the defensive, he launched a frontal assault on the entrenched enemy. ("I believed that he would yield before a vigorous attack," he later explained.) Herman Hattaway, Lee's biographer, blames his subject for not informing Hood of the enemy's unexpected position ("as he should have") and for failing to coordinate the assaults of his troops, who went in piecemeal. When Stewart's troops began to arrive, Lee threw them in too, believing that the enemy needed "only to be attacked with spirit to be put to utter rout." Sam Elliott, Stewart's biographer, similarly faults his subject for committing his troops before seeing the tactical situation for himself. "In Stewart's defense, however," Elliott adds, "the young corps commander on the field asked for help. Stewart gave it, but in a different context than Hood had originally anticipated."

The day was as hot as the enemy fire. "I think that the intensity of the heat did more to whip our brigade than the enemy," one Reb later wrote; "the men were beginning to faint from the excessive heat before we arrived in 500 yards of the enemy." The predictable repulse occurred in a few hours. The Southerners lost nearly 3,000 men; afterward even the troops heard of their casualties. "I suppose that our loss was quite heavy," wrote one Confederate on July 30, "but do not credit near all that I hear in regard to it." Howard's army suffered a fifth (632) of Hood's loss, which was the usual bloody disproportion in such Civil War combats when the defenders were fortified. Howard credited his small casualties to the "rough barricades" his men had constructed from dirt, logs, and even the wooden benches dragged out of Ezra Church. "Our men were partially covered, while the enemy were exposed," Sherman telegraphed Halleck that night.[98]

Lieutenant General Stewart and Major General Loring were both wounded in the battle. Stewart was struck in the forehead by a spent bullet; blood flow temporarily blinded him, and he had to be taken from the field (he returned to duty in a couple of weeks). Loring's wound to the chest was more serious, knocking him out of the campaign. Brig. Gen. George D. Johnston was severely wounded, shot in the leg. After Johnston, within minutes Col. John G. Coltart, then Col. B. R. Hart also went down; command of the brigade devolved upon Lt. Col. Harry T. Toulmin. As for Brig. Gen. John C.

---

[98] Green, *Orphan Brigade*, 150 (Jim Bemis on the "Dag Gone Yanks"); Hess, *Ezra Church*, 31–32, 42–44, 56 (Lee's "snap decision"); Howard, "The Struggle for Atlanta," 4:319; Castel, *Decision in the West*, 426, 428–34; S. Davis, *Atlanta Will Fall*, 151, 153, McMurry, *Atlanta 1864*, 156–57; Howard report, September 17, *OR*, vol. 38, pt. 3, 41; John C. Brown report, July 31, *OR*, vol. 38, pt. 3, 767 ("General Jackson…indicated the enemy's infantry to be small"); Lee report, January 30, 1865, *OR*, vol. 38, pt. 3, 763; Hattaway, *Stephen D. Lee*, 128; Elliott, *Soldier of Tennessee*, 213–14; Wiley, "Billy Yank and Johnny Reb in the Campaign for Atlanta," 21 ("intensity of the heat"); "Letter Written in War Times," 151 (from unnamed soldier, 41st Tennessee); Howard to Sherman, July 28, *OR*, vol. 38, pt. 5, 283; Sherman to Halleck, July 28, *OR*, vol. 38, pt. 5, 279.

As at Peachtree Creek, Sherman at Ezra Church initially won the battle of casualty numbers. He put Confederate losses at 4,632 (note the specificity) in his article for *Battles and Leaders.* Even Livermore calculated 4,300 killed, wounded, and missing. More recently, a grosser figure has jumped around: "Another five thousand men to Sherman's six hundred" (Key, *Battle of Atlanta*, 68); "At least five thousand men" (T. L. Connelly, *Autumn of Glory*, 455; Carter *The Siege of Atlanta, 1864*, 254); "Estimated to be 5,000 killed, wounded or captured" (Bailey, *Battles for Atlanta*, 136). Albert Castel and I went through the *OR* and came up with 2,800–3,000 (Davis, "Hood Fights Desperately," 33; Castel, *Decision in the West*, 434). Since publication of our work, logic has settled in, *viz.*, "Hood's army suffered 3,000 losses to a mere 632 Union casualties" (Ecelbarger, *The Day Dixie Died*, 219). See also Sherman, "Grand Strategy," 4:254; Livermore, *Numbers and Losses*, 124.

Sherman's figure for Confederate casualties, much higher than the 3,000 or so currently accepted, was nevertheless accepted fifty years ago by some Atlantans. As he had done in the text he wrote about Peachtree Creek for the Georgia Civil War Commission plaques placed in 1964 at Tanyard Branch Park, Allen P. Julian echoed Sherman's 4,632 number in his text for metal markers placed at Mozley Park, the area of the battlefield of Ezra Church. They can be seen there today.

Brown, slightly injured in the battle, on the 29th Hood asked President Davis to promote him to major general so he could take over Loring's division.

The commander of the whole operation, Lieutenant General Lee, emerged from Ezra Church unscathed save in reputation. His decision to launch an unauthorized attack has rightly been attributed to his youth (thirty years) and inexperience at large command. A West Pointer (class of 1854, same as O. O. Howard) and veteran of the old army, he had steadily risen through the C. S. ranks from his initial captaincy. When promoted to lieutenant general in June 1864, he was the youngest Confederate to attain that rank. At the time he commanded the Department of Alabama, Mississippi, and East Louisiana. On July 14, just before he was summoned to Georgia, Lee led some 5,000 troops (and an equal number of Forrest's) in an attack on the stronger entrenched force of A. J. Smith at Tupelo. The Southerners were bloodily repulsed, but Smith retreated the next day back to Memphis. Lee claimed that the battle was "a drawn one," yet because it led to the Federals' retreat, Tupelo had been called a Confederate victory (Bragg termed it a "very important" one). Historians know otherwise. Blaming Lee for the "slaughter" at Ezra Church, Castel writes that the battle of July 28 is "the exact duplicate of his conduct two weeks ago at Tupelo, from which experience he obviously has learned nothing."[99]

We cannot pinpoint Hood's whereabouts on July 28, when Lee strayed from orders and launched his attack at Ezra Church. Lieutenant T. B. Mackall, who had carefully recorded in his journal Hood's shift of headquarters on the evening of July 22 to Thrasher's house on Whitehall Street, ended his diary after he offered his services to General Clayton on the 25th. General Shoup, W. W. Mackall's replacement as chief of staff, kept a journal, July 25–September 7, but its contents lack such details as to where the commanding

---

[99] J. Welsh, *Medical Histories of Confederate Generals*, 29, 144, 206; Hood to Seddon, August 13, *OR*, vol. 38, pt. 5, 961 ("Lieutenant-General Stewart returned to duty to-day"); Raab, *Loring*, 169; "Journal of Brig. Gen. Francis A. Shoup, C. S. Army, Chief of Staff, of operations July 25-September 7," entry of July 28, *OR*, vol. 38, pt. 3, 688 (Stewart, Loring, Johnston, and Brown wounded); Brown report, July 31, *OR*, vol. 38, pt. 3, 767–68 (Johnston shot); Hood to Davis, July 29, *OR*, vol. 38, pt. 5, 924–25; Castel, *Decision in the West*, 416–17, 428, 435; Hattaway, "Stephen Dill Lee," 4:59, 63; Hattaway, *Stephen D. Lee*, 123–24; Lee report, July 14, *OR*, vol. 39, pt. 1, 320.

Woodworth is only one scholar to criticize Sherman's combat performance, in this case "his failure to anticipate Hood's attacks at Atlanta and Ezra Church." "Throughout the war, Sherman had difficulty in calculating what the enemy was doing outside of the range of his direct observation," Woodworth observes, adding that in these two engagements he was fortunate that "his younger lieutenants McPherson and Howard" had second-guessed him (and had been proven correct) (*Sherman*, 138).

general spent the 28th. We can surmise, given that he expected no general engagement that day, that Hood received messages and issued orders by courier at Thrasher's in the suburbs. There, into the early afternoon Hood and Shoup remained unaware of Lee's break with their plan. At noon Shoup sent note to Lee, advising that "if the enemy should make an assault upon our left the general directs you to strike him in flank." Thirty minutes later Shoup reminded Hardee, "General Lee is directed to prevent the enemy from gaining the Lick Skillet road, and not to attack unless the enemy exposes himself in attacking us." General Lee, aware that he had not acted in accordance with Hood's instructions, seems to have sent misleading word back to headquarters as to what was happening on the battlefield. That night in his journal Shoup wrote, "about 1 P.M. the enemy, who had massed a heavy force near Ezra Church on our left, advanced for the purpose of driving us from Lick Skillet road. After a hot contest, which lasted until nearly dark, the enemy were repulsed and position retained."

Only in this context—with Hood and Shoup believing that the enemy had attacked Lee at Ezra Church, rather than the other way around—can the following messages from Shoup be read:

> General Hood directs that you hold the enemy in check. The object is to prevent him from gaining the Lick Skillet road.
> —Shoup to Lee, July 28 (2:20 P.M.)

> General Hood directs me to say that if you are pressed you can draw a brigade or so from Bate....He desires you to hold the enemy....
> —Shoup to Stewart, July 28 (3:25 P.M.)

> General Hood directs me to inform you that he desires you not to allow the enemy to gain upon you anymore than possible, and that General Stewart has directions to support you fully.
> —Shoup to Lee, July 28 (4 P.M.)

By five o'clock, headquarters knew that General Stewart was wounded and that the situation on the battlefield was dire. Bate's and French's divisions were considered as reinforcements; "the object is to hold the enemy in check," Shoup wrote at 5 P.M. Cheatham was called to headquarters, but to manage the crisis at Ezra Church, Hood summoned Old Reliable: General Hardee. "I received a note in the afternoon saying Hood wanted to see me without delay," Hardee wrote his wife afterward. "I hurried on, and before reaching his quarters received another note, asking me to come as soon as possible." Hardee arrived at Thrasher's, learned of the situation on the Lick Skillet road, and was instructed to go there "and look after matters." Hardee set out, but by the time he arrived the battle was ended.[100]

As a rule, generals try to obscure bad news coming in from the battlefield, and Hood exercised this unbecoming tendency in his report on the night of July 28 to Secretary Seddon.

> The enemy commenced extending his right about 8 this morning, driving in our cavalry. Lieutenant Generals Stewart and Lee were directed to hold the Lick Skillet road for the day with a portion of their commands. About 1:30 o'clock a sharp engagement ensued with no decided advantage to either side. We still occupy the Lick Skillet road.

Such a gross misrepresentation of the fight at Ezra Church reflects poorly on Hood's character. (A more magnanimous officer would have admitted failure; think of Lee on July 3, 1863, and the note Eisenhower drafted in case of disaster on D-Day.) Hood should have explained the botched flank attack to Richmond, given estimates of his casualties, and assumed responsibility for the outcome. Instead, in his report of February 1865, he merely stated that Lee engaged the enemy on July 28 in order to seize the position which they had taken and that the battle ended with "the road remaining in our possession." Hood's evasiveness was at least a disservice to the 600 Southern soldiers killed at Ezra Church, and the 2,200 more men who had been wounded or captured. Even worse was his statement in his memoirs that General Lee told him later that the troops had not fought with complete *elan*; "whilst one brigade fought gallantly, another failed to do its duty." (To be sure, Brigadier General Brown, reporting for his division, commended Jacob Sharp's and Col. William Brantley's brigades and noted that "the major portion of Johnston's brigade behaved well," but added, "the greater portion [of] Manigault's brigade behaved badly.") Hood acknowledged such "lack of spirit" as a reflection of the state of morale in the army at the time.

Complicating the situation is a reference in General Shoup's journal, dated July 28, 11:50 P.M., to "the proposed night attack." Lieutenant Mackall had entered a similar entry into his journal on the night of July 20. Was Hood, despite the painful repulse suffered on the 28th by Lee and Stewart, thinking

---

[100] McMurry, "Mackall Journal," entry of July 25; S. Davis, *Atlanta Will Fall*, 153; Shoup to Lee, July 28 (12 A.M.) and to Hardee, July 28 (12:30 P.M.), *OR*, vol. 38, pt. 5, 919; Shoup journal, entry of July 28, *OR*, vol. 38, pt. 3, 688; Shoup to Lee and Stewart, July 28, *OR*, vol. 38, pt. 5, 919–20; Shoup to Cheatham, July 28 (5 P.M.), *OR*, vol. 38, pt. 5, 920; ("General Stewart is wounded"); Shoup to Walthall, July 28 (5 P.M.), *OR*, vol. 38, pt. 5, 921; Hardee to wife, July 30, quoted in Roy, "General Hardee," 370; Hughes, *Hardee*, 233; Hardee report, April 5, 1865, *OR*, vol. 38, pt. 3, 699.

of ordering a nocturnal assault? Historians have read the *OR* for over a century, including Shoup's journal, but have not discussed this nebulous passage and its possible implications about Hood's generalship. I offer this passing mention only.[101]

Hood's callous disregard for his soldiers' courage and sacrifices at Ezra Church was coupled with his devious attempt to hide from the administration the unhappy truth about the battle. "This fight of the 28th is mentioned by General Hood in terms to leave an impression of its success," Hardee later observed, "but it was well known throughout the army that so great was the loss in men, organization, and morale in that engagement that no action of the campaign probably did so much to demoralize and dishearten the troops engaged in it." The diary entry of Alabaman Hiram S. Williams supports Hardee's statement. "A few more such affairs as this and that of the 22nd and we will have no army left," Williams wrote on the 28th. "This day's work has done more to demoralize our army than 3 months under Gen Johnston." "We lost a great many men for mity little gain," wrote Thomas Warrick of the 34th Alabama the day after the battle. "This Battle Discouraged our men Badly," observed Arkansas Capt. John W. Lavender. On the other hand, word among Hood's troops about the fight at Ezra Church was not completely negative. "It is said that our losses in the fight on the left of the army on the 28th will amount to 2,500," wrote Capt. Tom Key in his diary. "Our forces found the enemy in advance of their fortifications and attacked them, driving them back and even taking a portion of their works." Andy Neal was even more upbeat. "Gen. Hood watches his flanks closely and has twice whipped the flanking columns out," he wrote his sister Ella on August 4. Tennessee surgeon James M. Brannock was more upbeat still in a letter to his wife the night of the battle: "since Genl' Hood has fought the enemy so successfully he has gained the entire confidence almost to as great an extent as Johnston had done." "Gen. Hood is fast gaining the confidence of the army & all are satisfied that we are done falling back," Alabaman Newton Davis wrote Bettie on August 1. "My private opinion is—you must not mention it to any one," confided William Elliott of South Carolina to his wife on July 29, "that Gen. Hood was put in command too late."

---

[101] Hood to Seddon, July 28, *OR*, vol. 38, pt. 5, 917; Brown report, July 31, *OR*, vol. 38, pt. 3, 768 (111 killed, 583 wounded, and 113 missing = 807 in his division); Clayton report, September 16, *OR*, vol. 38, pt. 3, 821 ("one half of their original numbers" lost in his division); Walthall's report, January 14, 1865, *OR*, vol. 38, pt. 3, 927 ("152 officers and nearly 1,000 men" as casualties in his division); Hood report, February 15, 1865, *OR*, vol. 38, pt. 3, 632; Hood, *A & R*, 194–95; Shoup journal, entry of July 28 (11:50 P.M.), *OR*, vol. 38, pt. 3, 919.

At this point criticism of Hood's generalship ends—or should. "The true culprit is Stephen D. Lee," Castel observes. "Historians have uniformly criticized Lee for starting the battle in the first place and for conducting it in piecemeal fashion," adds Earl Hess; "Lee should have stopped the battle long before he did rather than press forward every available unit as it arrived on the field, duplicating failure the whole afternoon." To Lee's inexperience and committal to piecemeal attacks, Connelly adds criticism of his failure to confer with headquarters once he learned that the enemy already commanded the Lick Skillet road. (Hattaway agrees.) McDonough and Jones add another: Lee's failure to recon toward the enemy's right flank, seeking some alternative to frontal assault. In his recent book on Ezra Church, Gary Ecelbarger makes the surprising claim that in launching his frontal attack Lee was actually carrying out Hood's order that the high ground north of Lick Skillet road be carried. Such an interpretation, however, is not sustained by either Hood's or Stewart's reports.

Connelly observes that Peter Stewart was not without culpability. Like Lee, he "had also disobeyed orders, since his instructions were to sweep wide around Lee's left flank and strike the enemy in flank and rear." In the end, Hood's worst mistake west of the city in late July was to have sent Lee, his least experienced and tested corps commander, to go out and conduct a bold but complex maneuver. Indeed, generals Lee and Stewart *combined* had spent less than five weeks as corps commanders. For this and other reasons, Lieutenant Colonel Burne's observation on Hood's three attacking battles of July 1864 is worth recalling: "Hood's weapon was becoming steadily blunter, whereas Sherman's was becoming steadily sharper." For this reason alone, Sherman was justified in writing his wife the day after the battle that Hood "must be watched closer [than Joe Johnston], as he is reckless of the lives of his Men."[102]

After the battle, high-level command posts had to be filled. Peter Stewart, recovering from that spent bullet to the forehead, left his corps in need of

[102] Hardee report, April 5, 1865, *OR*, vol. 38, pt. 3, 699; Hess, *Ezra Church*, 173–75 (including Warrick, Lavender, Neal, and Brannock); H. S. Williams, *This War So Horrible*, 107; Cate, *Two Soldiers*, 105–106 (Key diary entry of July 30); Ecelbarger, *Slaughter at the Chapel*, 195 (Davis and Elliott); Castel, *Decision in the West*, 435; T. L. Connelly, *Autumn of Glory*, 454; Hattaway, *Stephen D. Lee*, 128 ("without exchanging messages with Hood, as he should have"); McDonough and Jones, *War So Terrible*, 259; Ecelbarger, *Slaughter at the Chapel*, 55–56; *OR*, vol. 38, pt. 3, 632 (Hood: "hold the enemy in check"), 762 (Lee: "check the enemy"); McMurry, *John Bell Hood*, 134 ("Lee, his least experienced lieutenant general"); Davis, "Hood Fights Desperately," 34 (less than five weeks); Burne, *Lee, Grant, and Sherman*, 111; Simpson and Berlin, *Sherman's Civil War*, 676 (to Ellen, July 29).

a leader. Hood called back Frank Cheatham to reassume the post he had recently relinquished to Lee. Major General French believed that the exit of both Stewart and Loring entitled him to corps command, and, upon learning of Hood's decision, asked to be relieved from service. His application was returned disapproved on August 1, whereupon the next day he wrote General Cooper in Richmond repeating his request. Two weeks later Hood wrote Bragg, stating that both he and Stewart favored French's departure. Bragg forwarded Hood's message to the president, who endorsed, "I doubt the propriety of relieving Genl. French." French thus stayed on with the army, for at least awhile. Stewart returned to resume command of his corps on August 15.

Loring's more serious wound prevented him from returning until September 10. In the interim Brigadier General Featherston led the division. Lieutenant Colonel Toulmin retained command of George Johnston's brigade. Maj. Gen. Patton Anderson, who had been summoned from Florida, arrived on the 30th to take over Brown's division. Not all officers' vacancies came from Ezra Church. A day before the battle, Brig. Gen. Matthew Ector, commanding a brigade in French's division in the trenches west of the city, was struck in the leg by a shell. Amputation was performed and Col. (soon to be brigadier) William H. Young took his place.[103]

It must be said that although bloody and botched, Lee's attacking battle at Ezra Church was at least effective in stopping Howard's march on the Macon & Western, at least for the time being. This caused Sherman to hope for good results from the cavalry raid he had planned days before. "It is time we heard something of our cavalry," Sherman wrote on July 29.

Stoneman's and McCook's cavalry raid indeed started out auspiciously. Stoneman, with 2,150 men, rode out from his camps on the south fork of Peachtree Creek at 3 A.M. on July 27. A few hours later McCook's column, numbering 2,700 to 3,000, also set out from Turner's Ferry on the Chattahoochee. The next day McCook fell behind schedule when his troopers wrecked two and a half miles of the Atlanta & West Point Railroad near Pal-

---

[103] Losson, *Tennessee's Forgotten Warriors*, 184; T. L. Connelly, *Autumn of Glory*, 456; S. G. French, *Two Wars*, 219–20; Hood to Bragg, August 16 and 22, *OR*, vol. 38, pt. 5, 966, 983; Davis endorsement, Crist, Williams, and Dillard, eds., *Papers of Jefferson Davis*, 10:618; Elliott, *Soldier of Tennessee*, 214; Wingfield, *A. P. Stewart*, 94; Shoup to Stewart, August 16, *OR*, vol. 38, pt. 5, 967; Raab, *W. W. Loring*, 171; Scaife, *Campaign for Atlanta*, 178 n.515 (Toulmin); Anderson report, February 9, 1865, *OR*, vol. 38, pt. 3, 769; Shoup to Brown, July 29, *OR*, vol. 38, pt. 5, 926 (Anderson assigned to command); McMurry, "Patton Anderson," 16; French report, December 6, 1864, *OR*, vol. 38, pt. 3, 904 (on General Ector); [Hood] to Cooper, July 28 (9:20 A.M.), Telegram Book No. III, 44 (recommending promotion of Colonel Young).

metto and captured 500 Confederate wagons parked on the road to Fayetteville. The wagons were burned, several hundred teamsters captured, and 800 mules sabered. Meanwhile Stoneman spent the 28th riding toward Macon (so intent was he on freeing Union prisoners that he did not even give a nod to Lovejoy's Station, the raid's presumed objective). On July 29, McCook's men reached Lovejoy's and obediently wrecked two and a half miles of track, ripped down five miles of wire, and burned two trains filled with food, tobacco, and ordnance. Then, their objective accomplished, McCook and his men started riding westward, back toward the army.

The Confederates—who measured the damage as a mile and a half—had the railroad working again in two days. The only sign that the railroad was temporarily out of operation came on the afternoon of the 29th when headquarters urged the three corps commanders to advise troops to use as little ammunition as possible. Nevertheless, Hood kept an anxious eye on the repair of both rails and wire. On the morning of the 29th, aware that the enemy had cut the telegraph, Shoup sent a message to "Telegraph Operator": "Please use every exertion to have line to Macon repaired. It cannot be seriously interrupted." At 10:50 that morning Hood called for the post commander at Jonesboro to "establish relays of couriers in the break." Hood had wired Macon, Griffin and Andersonville warnings of enemy raiders; telegrams were sent also as far as Columbus, Augusta, and even Charleston. To Maj. Gen. Howell Cobb, commanding at Macon, he telegraphed, "The destination of the raiding party is still unknown. Prepare for it."[104]

On the 29th, Stoneman's column reached Clinton, twelve miles north of Macon. The next day as he sent a detachment to damage the Central Railroad running toward Savannah (a mile of track, three locomotives, and two dozen cars destroyed), Stoneman approached Macon from the east. In front of him were the Ocmulgee River, some home guards, a thousand Georgia militia,

[104] Sherman to Schofield, July 29, *OR*, vol. 38, pt. 5, 293; D. Evans, *Sherman's Horsemen*, 207, 208, 219, 228–31, 289 ("two and a half miles of the Macon road and five miles of telegraph wire"), 296; Castel, *Decision in the West*, 437–38; Bragg, "The Union General Lost in Georgia," 16, 18; Nye, "Cavalry Operations Around Atlanta," 48 (two days); Secrist, "Cavalry in Atlanta Campaign," 524 ("The Macon Railroad was back in operation within two days"); S. W. Presstman to Captain Shoup, July 29 (10:30 P.M., "send me at once spikes for one and a half miles [1½] Rail Road"), [Hood] to Captain Irby, July 29 (10:50 A.M.), [Hood] to Cobb, Winder and Jones (at Charleston), July 27 (1:40 P.M.), all in Telegram Book No. III; Hood to Cobb, July 27 (7:20 P.M.), *OR*, vol. 38, pt. 5, 913, 916–17; Shoup to Hardee, July 29 (3:25 P.M.), *OR*, vol. 38, pt. 5, 925 (same to Lee and Cheatham); Shoup to Telegraph Operator, July 29 (9:20 A.M.), *OR*, vol. 38, pt. 5, 930; Hood to Bragg, July 27 (5:30 P.M.); Hood to Winder, July 27 (5:30 P.M.), OR, vol. 38, pt. 5, 913, 916.

and even a brigade of Tennessee regulars en route to Atlanta. Altogether General Cobb had some 2,000–2,500 men, plus artillery in a hilltop wooden blockhouse, guarding the river. This proved enough to stop the Federals. An ineffective Union cannonade and half-hearted charge on the blockhouse yielded no results, and after the "Battle of East Macon" (seventy casualties on each side), Stoneman ordered a withdrawal. Unable to cross the Ocmulgee, he gave up on his plan to free the Federal POWs in the city and ordered his column to head back the way it had come, toward Clinton. Stoneman may also have learned that Confederate cavalry from Atlanta were pursuing him.[105]

They were indeed. Wheeler had sent three brigades under Brig. Gen. Alfred Iverson to hound Stoneman. The day after the battle of Macon, the Federals on July 31 found Iverson's force dug in north of Clinton and ready to fight. Stoneman ordered an assault; again, it was fainthearted and failed. Then he heard that Rebels were pursuing him from Macon, approaching his rear. Envisioning encirclement, Stoneman panicked and ordered two of his brigades to cut and run. The Rebels then charged, and the two brigades trying to escape had their flight considerably hastened. Stoneman stayed with his remaining force. After some of his men managed to break out, he sent forth a rider with a white flag to arrange the surrender of what was left of his command, about 700 men. Confederate casualties at this battle of Sunshine Church added up to "less than eighty," Iverson reported. George Stoneman thus became the highest-ranking Union officer captured during the war. He was soon a guest in the very officers' prison at Macon he had hoped to liberate. Meanwhile the two brigades that had escaped capitulation had a hard time getting back to Sherman's army. One, that of Col. Horace Capron, on August 3, west of Athens, was attacked by Confederate cavalry sent by Iverson in

---

[105] D. Evans, *Sherman's Horsemen*, 309, 312 (map, "Battle of Macon July 30, 1864"); Castel, *Decision in the West*, 438–40; Iobst, *Civil War Macon*, 310–14; Bragg, "Lost in Georgia," 19–21.

A detachment of Stoneman's force succeeded in raiding to Gordon, twenty miles east of Macon, and burning the 700 foot-long railroad bridge across the Oconee River. "In that instant," writes Dr. Evans, "Atlanta became an isolated city. All its rail connections, east, south, and west, had been severed by Sherman's horsemen." General Hood in Atlanta ordered speedy rebuilding of the trestle. "Push the work forward as rapidly as possible," he wired Capt. J. C. Shoup on August 1; "Let there be no delay" (D. Evans, *Sherman's Horsemen*, 303; [Hood] to J. C. Shoup, August 1 [10:45 A.M.], Telegram Book No. III; [Hood] to Shoup, August 1 [8 P.M.], *OR*, vol. 38, pt. 5, 939; Bragg to Sale, August 3, *OR*, vol. 38, pt. 5, 942 ["It will take several days yet for the track and four weeks for the Oconee bridge"]).

General Shoup on August 2 suggested to General Cobb that a wagon/pontoon bridge be built over the river while the railroad bridge was being repaired. Cobb wrote to his wife on August 10 that repairs would take another two to three weeks (Iobst, *Civil War Macon*, 325).

pursuit and was considerably cut up. Over the next few days, the survivors of Stoneman's failed raid limped back to Union lines, "stiff, sore and worn out."[106]

After leaving Lovejoy's Station, McCook and his column fared little better. Wheeler and two of his brigades had arrived at Lovejoy's after dark July 29, seeing the damage done by the Yankees who had ridden away just hours earlier. Red Jackson and more troopers joined Wheeler, who then set out after McCook. On the 30th, at a place called Shakerag, the Confederates caught up with the enemy cavalry and attacked, driving them from the field. Then, later the same day, at Brown's Mill south of Newnan, Wheeler again attacked. At one point, McCook thought he was surrounded and considered surrender. Castel captures the action: "McCook, realizing that he is trapped, thereupon orders his men to escape any way they can. Most of them, including McCook, manage to slip through the enemy lines, then in small groups cross the Chattahoochee at various points. Even so, McCook's loss totals about 600, virtually all of them captives, and his division has been wrecked." It was "the most demoralized and the worst used up raiding party perhaps that operated during the campaign," judged Lt. S. B. Barron of the 3d Texas Cavalry, who had helped make them so.[107]

McCook and Stoneman accepted the fact that their cavalry raid had been an utter failure, and they so reported to General Sherman. "My loss very heavy. No co-operation from Stoneman," McCook wrote on August 3. Stoneman found a way to write his report while he sat in prison at Macon. In

---

[106] Longacre, *Cavalry of the Heartland*, 288–90; Bragg, "Lost in Georgia," 21–22; D. Evans, *Sherman's Horsemen*, 324–42; Castel, "In Quest of Glory," 124; Starr, *The Union Cavalry in the Civil War*, 3:470; Castel, *Decision in the West*, 440; Fordney, *George Stoneman*, 100 ("highest ranking officer ever captured by the Confederates"); S. Davis, *What the Yankees Did to Us*, 300 n.34 (not counting Union Maj. Gen. William B. Franklin, senior to Stoneman by eight months, but who escaped hours after his capture in Maryland, July 11, 1864 [Cooling, *Jubal Early's Raid*, 167, 170]).

[107] Longacre, *Cavalry of the Heartland*, 290–92; D. Evans, *Sherman's Horsemen*, 251, 257–72; Castel, *Decision in the West*, 438; Barron, "Wheeler's Cavalry in Georgia Campaign," 70; Guthrey, "Wheeler's Cavalry around Atlanta," 267 (200 Federals killed and wounded, 300 captured along with 600 horses); Haynie, "'I'll Be Damned If I Ever Surrender,'" 59.

Evans counts 587 Federals captured at Newnan, but adds that Wheeler also took "two cannon, eleven ambulances, several hundred horses and mules, and enough saddles, bridles, halters, pistols, carbines, and accoutrements to outfit the whole Texas brigade in true regulation style" (*Sherman's Horsemen*, 280–82). One chronicler of the fight goes so far as to say that "almost thirteen hundred Yankee prisoners were rounded up at Newnan and transferred to the prison camp at Andersonville" (Hale, *The Third Texas Cavalry in the Civil War*, 238). In between is John W. Rowell's observation that McCook's division numbered 2,100 before the raid; on August 8, 1,139 officers and men were counted (Rowell, "McCook's Raid," 5, 47).

those gloomy surroundings he confessed the calamity that had befallen him. "We were whipped," he admitted; "our loss in killed and wounded was quite large." To punctuate this dolorous admission, he signed it, "George Stoneman, Maj. Gen., U. S. Army, Prisoner of War." David Evans, the historian of Sherman's cavalry, counts the cost of McCook's and Stoneman's disaster: "When company clerks prepared the muster rolls for July and August, 1,230 of the 3,000 officers and men who had started south with McCook were listed as dead, wounded, or missing." Stoneman's losses were even worse, he adds: "No fewer than 1,329 of the 2,144 cavalrymen in his column had been killed, captured, or wounded."

Sixty-two percent casualties for Stoneman, forty-one percent for McCook; fifty percent overall for the Union raiders, all for a couple of miles of railroad damage that was repaired in two days. General Shoup followed the equestrian events very closely, judging from his lengthy journal entries. In fact, on July 28, Shoup wrote more about the enemy raid than he did about the battle of Ezra Church. His entry of the next day concerned nothing but the raid and runs a full half-page in the *Official Records.* On the 30th, word came that the Macon & Western was under repair and that resumption of train traffic was expected on the morrow. For July 31, Shoup recorded news of Wheeler's victory near Newnan, how he and Jackson "put them to rout, killing, wounding, and capturing about 1,000, also 2 pieces of artillery, 10 ambulances, releasing all the prisoners they had captured on the road, and completely breaking up the entire force." Shoup observed on August 1, "good news has flowed in from all distant points." In addition to the good news from Wheeler came that of Iverson's rout of Stoneman. "To-day deserves to be marked with a white stone," he exulted.[108]

Wheeler's defeat of Sherman's two-pronged cavalry raid on Hood's railroad line was "the greatest single feat of his military career," according to one scholar. Connelly agrees, calling the exploit Wheeler's "finest hour as head of the army's cavalry." Atlantan Bill Scaife terms the event "the most spectacular Confederate victory of the campaign," noting that enough horses were captured with Stoneman's surrender that a full brigade of Confederate infantry would be mounted with them. For his cavalry's running down of McCook and Stoneman, which Dyer terms "the one bright spot of the Atlanta Campaign," John B. Hood could take no credit. Still, Hood was delighted. He had already recommended that Red Jackson be promoted to major general. On

---

108 McCook to Sherman, August 3, *OR*, vol. 38, pt. 2, 762; Stoneman to Sherman, August 6, *OR*, vol. 38, pt. 2, 914; D. Evans, *Sherman's Horsemen*, 376; Shoup journal, entries of July 28–August 1, *OR*, vol. 38, pt. 3, 688–89.

August 1 he wired Secretary Seddon that Wheeler reported the enemy expedition to Macon had been "entirely broken up." He also forwarded Iverson's dispatch that Stoneman and 500 of his men had surrendered and that "the rest of his command are scattered and fleeing toward Eatonton." The cavalry's victories brightened an already satisfactory situation, which he summarized to Seddon: "the two recent engagements [of July 22 and 28] have checked his extension on both flanks." Hood could even think of taking the initiative. As soon as reinforcements and militia sent by Bragg and Brown arrived, Hood wrote, "I hope to strike the enemy with my main force." He also had plans for his cavalry. "I hope in a few days," he wrote Seddon on July 30, "to send Wheeler, with his cavalry, to break Sherman's communications." Indeed, Hood had developed the idea of sending Wheeler raiding on the afternoon of the 29th. "General Hood desires your return very much as soon as you can get through with those fellows in rear," Shoup wrote him. "He has most important service for you."[109]

At the same time Hood remained focused on the tactical situation facing him. Before the fight at Ezra Church he knew that an enemy army had been swung west. "I have a prisoner who belongs to Logan's corps," reported Sul Ross on the afternoon of July 28. "He says that three corps marched from their left last night, and are now flanking around to strike the railroad this side of Atlanta." Thus Hood was confirmed in the awareness from the Lick Skillet fight that Sherman's effort to cut the Macon & Western would be aimed at the track leading southwest to East Point. After fighting on the 28th ended, Lee's troops fell back from the battlefield but still covered the railroad. Early on the 29th they slipped back into the main city works, but not for long. Hood planned the construction of a new trench line extending southwest from the main works out of the city to protect the railroad running to East Point and Macon. That night the Southerners began to entrench from the western edge of the Grant-Presstman perimeter (now the Morehouse College campus) in a line of works running southwest, a mile and a half or so in front of the railroad. Hood instructed Lee to extend his entrenchments as far as the "Baugh house," the home of a Robert Baugh, which was fully three and a half to four miles southwest of the Grant-Presstman perimeter. Cheatham was to

[109] Longacre, *Cavalry of the Heartland*, 292; T. L. Connelly, *Autumn of Glory*, 452; Scaife, "Sherman's Great Cavalry Raid," 53; Dyer, *Gallant Hood*, 263; Hood to Seddon, July 30 and August 1, *OR*, vol. 38, pt. 5, 930, 936–37; [Hood] to Commanding Officer West Point, July 28 (7:40 P.M.), Telegram Book No. III, Hood Papers ("Have troops for this place left yet? If not hasten them forward"); Hood to Seddon, July 23, Telegrams Book I, Hood Personal Papers (General Jackson); Shoup to Wheeler, July 29 (3:45 P.M.), *OR*, vol. 38, pt. 5, 927.

cooperate by having the men of his corps work at night "to extend the line to the left." Hood placed much importance on these new works (which Wilbur Kurtz has called the "railway defense line"). On July 31 he specified to Cheatham and Lee the need to "fell the timber, large and small, in your front as far as practicable." The object, wrote Shoup, was "to form an impassable abatis," against a sudden infantry attack.[110]

For Sherman, the news of his cavalry's failure was slow to sink in. On July 29, with McCook and Stoneman still in the field, he telegraphed Halleck, "Our cavalry has not been out three days, and must have done its work about Griffin." Indeed, a hopeful Sherman even accepted a rumor that Wheeler had been killed. Until he heard the outcome of the raid, he ordered Thomas to throw two divisions beyond Howard's right to conduct a reconnaissance-in-force threatening the East Point railroad, but not to extend Howard's fortified line. "Until the result of our cavalry is known," he wrote, "I want the utmost activity on our flanks." For a few days Sherman actually believed his cavalry were succeeding in their mission. "I think General Stoneman has gone to Macon, east of Yellow River and that it is well," he wrote Thomas and Howard on the 31st. Then the news came in of McCook's disaster at Newnan. "Brownlow says all were killed or captured, except such as cut their way out," Sherman reported to Halleck (referring to Col. James P. Brownlow, one of McCook's brigade leaders). "I can hardly believe it," Sherman confessed. News of Stoneman's disaster trickled in much more slowly. On August 3, when General Grant in Virginia was reading in the Richmond papers of Iverson's capture of Stoneman, Sherman was telling Halleck he thought that the hope of rescuing the Union prisoners still had a chance of success. "No news from Stoneman," Sherman's telegrapher Van Duzer wired Washington on the night of August 3, "except from rebels, who say Wheeler caught and thrashed him fifteen miles from Macon." The next day Grant sent Sherman the statement in the *Richmond Dispatch* about Stoneman's capture and the rout of his command. Sherman answered that 900 of Stoneman's men had returned from their raid; "I fear the balance are captured as related in your dispatch." On August 5 Sherman still expressed hope that Stoneman and the balance of his command "will fetch up at Pensacola or somewhere." The next

[110] Ross to Jackson, July 28 (3 P.M.), *OR*, vol. 38, pt. 5, 923; S. Davis, *Atlanta Will Fall*, 156; Lee report, Jan. 30, 1865, *OR*, vol. 38, pt. 3, 763; Shoup to Cheatham, July 29 (3 P.M.) and to Lee, July 29 (9:20 P.M.), *OR*, vol. 38, pt. 5, 925; Kurtz, "Dugout Home in Atlanta," *Atlanta Journal Sunday Magazine*, July 10, 1932, 3; Hess, *Ezra Church*, 140, 182; *OR Atlas*, plate 60, pt. 2 (showing Baugh house); Scaife, *Campaign for Atlanta*, f.p. 126 (map, "Battle of Utoy Creek," also showing Robert Baugh house); Shoup to Cheatham, July 31 (5:35 P.M.), *OR*, vol. 38, pt. 5, 935 (same to Lee).

day Sherman finally had to admit to Halleck that Stoneman had surrendered with 700 of his men. The great cavalry raid he had hoped would break Hood's railroad had finally been acknowledged as a failure.[111]

[111] Sherman to Halleck, July 29 (8:30 A.M.), *OR*, vol. 38, pt. 5, 289; Sherman to Thomas, July 29, *OR*, vol. 38, pt. 5, 290; Sherman to Schofield, July 29, *OR*, vol. 38, pt. 5, 296; Sherman to Thomas and Howard, July 31, *OR*, vol. 38, pt. 5, 310; Sherman to Halleck, August 1 (8 P.M. and 10 P.M.), *OR*, vol. 38, pt. 5, 320; Grant to Halleck, August 3, *OR*, vol. 38, pt. 5, 339; Sherman to Halleck, August 3 (9 P.M.), *OR*, vol. 38, pt. 5, 340; Van Duzer to Eckert, August 3 (8:30 P.M.), *OR*, vol. 38, pt. 5, 349 ("rebels…say Wheeler"); Grant to Sherman, August 4, *OR*, vol. 38, pt. 5, 349; Sherman to Grant, August 4, *OR*, vol. 38, pt. 5, 350; Sherman to Granger, August 5, *OR*, vol. 38, pt. 5, 377; Sherman to Halleck, August 6, *OR*, vol. 38, pt. 5, 392.

Chapter 7

August

If cavalry could not break Hood's road, Sherman would have to turn to his infantry. Howard's army had extended its line as far to the right as it safely could. To get closer to the Macon railroad, Sherman planned to swing Schofield's army around from east of the city, as he had Howard's. The movement began on the night of August 1. The next day Schofield extended the Union right a mile and a half, almost reaching Utoy Creek. On the 3rd, Brig. Gen. Milo Hascall's division crossed the north fork of the creek at Herring's Mill. Confederate cavalry under Brig. Gen. Frank C. Armstrong resisted as long as they could. Sherman kept pressing for results; he wanted his forces to get within cannon-range of the railroad. He expected that the threat to the line would force Hood "to attack our right or be in a tight place." On the 4th he wired Grant that Hood "shows a bold front wherever I get at him."[1]

Hood did not feel himself to be in a "tight place," but he certainly showed a "bold front" as he kept his troops building the railway defense line. "The work of intrenching was pushed with vigor night and day," wrote Patton Anderson, "till a feeling of security and even defiance pervaded the whole line." Anderson's division after Ezra Church had its right on the Lick Skillet road and its left on Utoy Creek. On July 31, Bate's division of Hardee's corps was ordered out to extend the left of the line to Baugh's house. Anderson described how the line was dug. At first a string of shallow rifle-pits was dug hurriedly at night, from twenty to fifty feet apart. "A few nights' work, however, added much to their strength," he wrote, "and in the course of ten days

---

[1] Sherman to Howard, July 29, *OR*, vol. 38, pt. 5, 297 ("You may extend…as much as possible"); Sherman to Schofield, July 30, *OR*, vol. 38, pt. 5, 303; Army of the Ohio Special Field Orders No. 67 and 68, July 31, August 1, *OR*, vol. 38, pt. 5, 315, 327; Castel, *Decision in the West*, 454; Shoup to Armstrong, July 30 (12:30 P.M.), *OR*, vol. 38, pt. 5, 932 ("hold the crossing at Herring's Mill, if possible"); Sherman to Howard, August 2, *OR*, vol. 38, pt. 5, 336; Sherman to Thomas, August 3, *OR*, vol. 38, pt. 5, 342; Sherman to Grant, August 4, *OR*, vol. 38, pt. 5, 350.

or a fortnight the pits were gradually connected and the whole became almost one continuous line of intrenchments, with head-logs and loop-holes to protect our sharpshooters." Capt. James L. Cooper, on the HQ staff of a brigade in Bate's division, remarked that the troops in their works of late July were "well fed and in pretty good spirits, although we had hardly recovered from the loss of 'Old Joe.'"

Brigadier General Manigault, whose brigade held the left of Anderson's division, filled in details of how the railway defense line was constructed. "During the first two or three days after the 28th, much of our work was done in daylight, for although the enemy were in our front, yet their first lines of intrenchments, as also their picket lines, were at a considerable distance from us." But when the Yankees got closer, Manigault's troops had to work at night.

> Our works consisted of a ditch for the infantry, with the earth thrown outwards, forming an embankment of sufficient thickness to resist even the three-inch rifle shot, or solid shot from the 12-pounder Napoleon guns of the enemy, provided it struck below the exterior crest of the work. Head logs protected the men whilst firing. A double row of palisades covered our entire front, and beyond these an almost impenetrable "abattis," varying in width from fifty to one hundred feet.

In addition, behind the front lines were "traverses, zig-zags and the like," trenches allowing men to come up to the main works protected from enemy fire. "The ground in rear of our works was completely honey-combed," Manigault recalled.

Such was the new line extending toward East Point. By August 13, according to Milton Glass, the Northern spy, the Southerners' line extended five or six miles beyond East Point; he saw "large squads of negroes along the railroad from East Point to Fairburn, felling timber and throwing up breastworks." Confederates also strengthened their works around Atlanta. Fortifying went on day and night, according to General French, "the whole front obstructed by a double line of sharpened stakes firmly set in the ground at an inclination of about sixty degrees." *Abatis* was set out in front; even the trench used by advanced skirmishers had a stockade before it.

To assist his soldiers with all this work, Hood called for up to 2,000 slaves, appealing to General Cobb and Governor Brown for help in securing these laborers. He even asked ordnance officers at Macon to send blacks working there to Atlanta. Calls also went out to Savannah, Columbus, Charleston, even Andersonville, for the temporary employ of black workers.

Hood's energetic defense of his railroad drew commendation in the Georgia press. The *LaGrange Reporter and Bulletin* on August 3 declared that "the policy of Sherman, will now be, beyond all doubt, to send out large raiding parties to cut our communications, hoping to starve Hood out of Atlanta. This is well understood by our commander."[2]

The enemy's advance against his left did not deter Hood from pursuing his idea of sending cavalry against Sherman's railroad supply line in north Georgia. He advanced his plan to President Davis on August 2. "Since our late success over the enemy's cavalry," he wrote, "I hope now to be able, by interrupting Sherman's communications, either to force him to fight me in position or to retreat." Hood also asked Maj. Gen. Dabney Maury at Montgomery if he had any cavalry which could strike Sherman's railroad; "small parties could do great service." On the 5th, the president replied to Hood, agreeing with his plan for cavalry raids. He singled out as targets the railroad bridges of the Memphis & Charleston line running across north Alabama. The enemy could always forage for supplies, Davis acknowledged, but he agreed that a severe disruption of Sherman's rail service could force him to attack or withdraw northward. Furthermore, such a tactic would allow the Southerners to retain the initiative without having to launch an infantry assault on the enemy's lines. The president used his telegram to offer Hood a reminder that "the loss consequent upon attacking him in his entrenchments requires you to avoid that if practicable." (Hood wired back an acknowledgement that did not address this point.)

In the first week of August, President Davis was asked to weigh in on another issue: General Hardee's continued unhappiness serving under Hood. Hardee had reached the point of thinking about resigning from the army altogether. "It is a grave step to take in time of war and in the face of the enemy," he confided to Mary on August 2, but "I am prepared to do it." Even General Sherman heard about the brouhaha. When Southern deserters brought in the

---

[2] Anderson report, February 9, 1865, *OR*, vol. 38, pt. 3, 769–70; Shoup to Lee, July 31 (4:40 P.M.), *OR*, vol. 38, pt. 5, 934 (Bate to Baugh's); Kurtz, historical marker text, "Confederate Intrenchments: 1864," box 38, folder 1 (Scruggs, *Georgia Historical Markers*, 246); J. L. Cooper, "The Civil War Diary of Captain James Litton Cooper," 164; Manigault, *Narrative*, 238; "Statement of J. Milton Glass," August 18, *OR*, vol. 38, pt. 5, 579; French report, Dec. 6, 1864, *OR*, vol. 38, pt. 3, 905; Hood to Cobb (9:25 A.M.) and to Brown (9:30 P.M.), August 7, Hood to Maj. Gen. Samuel Jones, to "Commanding Officer Savannah" (7:40 A.M.) and to "Commanding Officer Columbus" (9:45 A.M.), August 7, all in Telegram Book No. III, Hood Papers; Hood to Col. Richard M. Cuyler and Col. John W. Mallet, August 1 (9 P.M.), *OR*, vol. 38, pt. 5, 939; "The Situation at Atlanta," *Mobile Advertiser & Register*, August 9 (quoting *LaGrange Reporter and Bulletin* of August 3).

rumor that Hardee had actually resigned, Sherman wired it to Halleck on the 3rd.

That day Hardee telegraphed the president and asked for an exchange of command with Lieutenant General Taylor; "I rely upon your kindness to relieve me from an unpleasant position." Davis wired back on the 4th expressing regret that Hardee felt that way. The president reminded him that the previous year he had been offered command of the Army of Tennessee, but that he had acquiesced to General Johnston's taking it instead: "Your letters, when commanding-in-chief, created the belief that the course adopted would be satisfactory to you." Davis closed by reminding Hardee, "the country needs every effort of all her sons." His words must have registered. On the 5th Hardee checked himself. "I am not entirely satisfied that I would be doing right just now to resign," he wrote Mary. "There is always a reflection on an officer who resigns in the immediate presence of the enemy, and more particularly when that enemy is invading his own country and his own State."

The day after that, August 6, the general again wired the president:

> No letter of mine while commanding-in-chief was intended to convey the impression that the appointment of a junior to command me would be satisfactory. The justice or propriety of that appointment I do not propose to question. I only ask to be relieved from a position which is personally humiliating. This I think due to an old soldier who has faithfully endeavored to perform his duty, and who is still willing to make any sacrifice for the cause except his self respect. I respectfully renew my request to be relieved. General Hood approves my application.

Telling Jefferson Davis that he was wrong concerning earlier correspondence, and what Hardee had meant in it, was just the kind of diversion the president savored. On the 7th, after expressing "regret and disappointment'" over Hardee's telegram, Davis explained, "I will look for your letters to send you by mail extracts showing what you said, and from which your intention was inferred." With that Davis had won the argument, but he still asked, "Is this a time to weigh professional or personal pride against the needs of the country?" He considered the matter closed, at least for the time being. On August 7, Davis wired Bragg that he had rejected the idea of swapping Hardee and Taylor. "In both cases it would be at the sacrifice of country and troops." Meanwhile Davis asked Hood to get along with his senior corps commander. "General Hardee's minute knowledge of the country, and his extensive acquaintance with the officers and men of the command, must render his large professional knowledge and experience peculiarly valuable in such a campaign as I hope is before you." (Hood did not acknowledge this advice.)[3]

Even before President Davis approved the plan for a cavalry raid, Hood was taking steps in its preparation. To Col. Hypolite Oladowski, supply officer in Macon, on August 3 he sent requisition for three hundred pounds of blasting powder, up to twelve kegs of it. To Col. George W. Rains, head of the government gunpowder works in Augusta, he sent a similar request on the 4th, specifying that it was for General Wheeler's use. Rains was to send the twelve kegs to Social Circle, Georgia, which was the most westward point to which the Georgia Railroad ran from Augusta after the Union cavalry raiders' damage to the line. The next day Hood repeated his request, adding two thousand feet of "quick match." In ordering this powder and fuse for Wheeler, Hood was enabling his cavalry, once behind Sherman's lines, to blow up bridges and maybe tunnels. Such destruction, if carried out, would surely "compel the enemy to attack you in position or to retreat," as Davis had advised Hood."[4]

Meanwhile Frank Armstrong kept headquarters informed of the enemy's progress along Utoy Creek. Hood asked Wheeler and Maj. Gen. Will Martin, one of his division leaders, to provide specific information as to what forces were there. As the enemy advanced farther, Hood asked Hardee to have ready more of his troops (Bate was already at Utoy). G. W. Smith and his militiamen would replace Hardee's troops in their sector of the line, east of Peachtree Road.[5]

Sherman hoped his forces could soon push forward and cut Hood's lifeline. On August 4 he ordered Schofield with his army, supported by Palmer's XIV Corps of Thomas's, to "move directly on the railroad" and not stop until he had "absolute control" of it. "If necessary to secure this end," the orders concluded, "ordinary parapets must be charged and carried, and every hour's delay enables the enemy to strengthen. Therefore let it be done to-day."

---

[3] Hood to Davis and to Maury, August 2, *OR*, vol. 38, pt. 5, 940–41; Davis to Hood and Hood to Davis, August 5, *OR*, vol. 38, pt. 5, 946; Hughes, *Hardee*, 244–45; Sherman to Halleck, August 3, *OR*, vol. 38, pt. 5, 340; Hardee to Davis, August 3 and 6; Davis to Hardee, August 4 and 7, *OR*, vol. 38, pt. 5, 987–88; Davis to Bragg, August 7, *OR*, vol. 39, pt. 2, 760.

[4] [Hood] to Oladowski, August 3 (1:15 and 9:15 P.M.) and to Rains, August 4 (1:10 P.M.) and August 5 (9:45 A.M.), all in Telegram Book No. III, Hood Papers.

[5] Shoup to Hardee, August 2 (10 P.M.), *OR*, vol. 38, pt. 5, 940 ("Armstrong says he [the enemy] has large camps on Utoy Creek"); Shoup to Wheeler and Martin, August 3 (9:45 A.M.), *OR*, vol. 38, pt. 5, 942; Shoup to Hardee, August 2 (10 P.M.) and to Smith, August 2 (11:30 A.M.), *OR*, vol. 38, pt. 5, 940–41.

It was not done that day or the next. Schofield issued orders for the move by the XXIII and XIV Corps, but General Palmer refused to obey them; Schofield was junior to him in rank (at least in date of commission). Sherman got involved, asking Palmer to cooperate. Palmer refused and offered his resignation, which was quickly accepted. Sherman eventually proposed to Halleck that Jeff. C. Davis, division leader in the XIV Corps, be promoted to major general and given the corps command.[6]

Amid the high-level squabbling, all that Sherman got from his order to break the railroad was a reconnaissance on the afternoon of August 4 by a brigade in Palmer's corps. The Federals advanced, captured two dozen Rebel pickets, then retired to their starting point. The next morning Brig. Gen. Absalom Baird's division of the XIV Corps moved forward and overran the Confederates' outer line, taking 140 prisoners with a loss of 83 killed and wounded. This mediocre achievement caused Schofield that evening to report to Sherman, "I am compelled to acknowledge that I have totally failed to make any aggressive movement with the Fourteenth Corps." He stated his resolve on the 6th to attack with his own XXIII corps and "try to recover what has been lost by two days' delay."

In those two days the Confederates of Major General Bate's division had been digging in on a ridge south of the Sandtown road that was out in front of the Confederates' main defensive line. The Southerners formed *abatis* and entanglements of underbrush and logs. Federals had ominously heard the "constant chopping and falling of trees," so, when they finally attacked on the morning of August 6, they suspected they would run into trouble. And they did. Col. James W. Reilly's brigade (Cox's division) charged Bate's line but got caught in the Rebels' obstructions—"an entanglement of the undergrowth, half cut off, bent down, and interlaced," as General Cox described it. Some Federals actually reached the Rebels' parapet before they were shot, General Schofield reported, "but the abatis and entanglements were such as to render success impossible." Reilly's troops fell back, then tried again, and again were repulsed. His brigade lost 76 men killed, 199 wounded, and 31 captured in the fight at Utoy Creek. Bate's casualties numbered all of 15 or 20.

Later that day Schofield's other division, Hascall's, advanced on Cox's right, found the flank of Bate's line and drove off Sul Ross's horsemen and a section of horse artillery guarding it. That night Bate pulled his troops back

[6] Dayton Special Field Orders No. 51, August 4, *OR*, vol. 38, pt. 5, 364; Castel, *Decision in the West*, 454–58; Sherman to Halleck, August 6 (3 P.M.), *OR*, vol. 38, pt. 5, 364; McMurry, *Atlanta 1864*, 161–62.

into the railway defense line. On the 7th elements of the XIV Corps tested the Rebels' picket line in a sally that bagged two hundred prisoners but cost the Federals double that number in casualties. Thus ended the fighting at Utoy Creek, which Castel terms a "Union fizzle."

It is worth noting that Utoy Creek was the fifth time in the Georgia Campaign that Sherman had ordered an infantry frontal assault against entrenched enemy lines. He had done so at Resaca, May 14–15; New Hope Church, May 25; Pickett's Mill, May 27; and Kennesaw Mountain, June 27. Each onslaught had been predictably repulsed with loss. Looking farther, we can see that each time Sherman had sent his troops to attack (Chickasaw Bluffs, December 1862; Vicksburg, May 1863; and Missionary Ridge, November 1863), he also had been repulsed. In fact, it is not too much to say that since the beginning of the war Sherman had not won a single victory as a result of his troops' combat. After Kennesaw Mountain, George Thomas had pointedly told Sherman, "one or two more such assaults would use up this army." Sherman seemed not to care. To Ellen he had written, "I begin to regard the death & mangling of a couple of thousand men as a small affair, a kind of morning dash." In other words, Sherman was fighting the way John Bell Hood has been criticized for doing (wasteful frontal assaults against enemy trenches) but seems to have gotten away with it in the campaign literature. The same literature has over the years had difficulty recognizing that Hood in July had fought three battles in which his infantry actually flanked the enemy line (July 20 and 22) or was intended to (July 28). Sherman, the bloody frontal attacker, versus Hood, the artful flanker? The ironies evident are too rich for further comment.[7]

Sherman took in stride his troops' failure to gain the railroad at Utoy Creek. He did not call for another attack on the Rebel works; Reilly's bloody repulse had shown them to be too strong. "The parapets of Atlanta present a

---

[7] Castel, "Union Fizzle at Atlanta," 29–30; Castel, *Decision in the West*, 456–60; Cox report and Schofield report, both September 10, *OR*, vol. 38, pt. 2, 518, 689; Scaife, *Campaign for Atlanta*, 124–25; Davis, "Hood Fights Desperately," 39.

W. T. Sherman benefits from a legend that during the entire Atlanta Campaign he ordered a frontal attack by his troops only once, at Kennesaw Mountain. The misconception goes back at least as far as Sherman's admiring biographer, Basil H. Liddell Hart. "A direct assault on an unshaken enemy in position has no justification in history," he writes, "although Sherman made it only once" (*Sherman*, 267). Years later he made the same claim: "only once in all these weeks of manoeuvre did he attempt a frontal attack, at Kenesaw Mountain" (*Strategy*, 150). The legend thrives among some to this day. In a recent telecast on the campaign, Todd Groce, president of the Georgia Historical Society, repeated the "only once" claim ("When Georgia Howled").

well-filled line wherever we approach them," he had written Grant. Nor did he order Schofield to continue pushing toward the Rebel railroad. "I do not deem it prudent to extend more to the right," he wired Halleck on the night of August 7, "but will push forward daily by parallels, and make the inside of Atlanta too hot to be endured."

This latter remark reflected Sherman's personal interest in the artillery bombardment of the city, which had begun on July 20. It had continued, with the Northern cannon firing more heavily on some days than others. Buildings were struck and people were hit, but because many civilians learned to find refuge in cellars or dug "bombproof" shelters in their yards, casualties were relatively low. An enterprising reporter for the *Atlanta/Macon Intelligencer*, Isaac Pilgrim, stayed in town throughout the bombardment and wrote daily reports on what he saw and heard. On August 4 he recorded that a shell perforated a house at Hunter and Loyd streets downtown and exploded in the cellar where a family was huddled; a Mrs. Gates and her child were wounded. Pilgrim heard that a lady had been killed one evening on Peachtree Street, but he was leery of hearsay. "I hear of a great many persons being killed," he wrote, "though I cannot trace them to any person who actually knows it to be a fact. On my arrival I learned that a lady had been killed by a shell. I called on the lady's sister, who is an acquaintance of mine, and she informed me that it was news to her as her sister was in the city of Macon." Others were also skeptical of rumored shell-deaths and injuries. The *Appeal* stated on the 8th that it had received word of three recent casualties, "but on investigation conclude it is without foundation." On the other hand, repeated reference to individuals in city-talk lead to the identification of actual shell-victims. Among them are the wife (wounded on July 23) of John M. Weaver, a railroad engineer living on Walton Street, and her child (killed). Both Southern and Northern newspapers carried the gruesome details of the deaths on the night of August 3 of Joseph F. Warner, superintendent of the city's gas works, and his young daughter Elizabeth. In the end, the judgment of Atlanta resident Samuel P. Richards has been accepted that "about twenty lives have been destroyed by these terrible missiles." Perhaps another hundred people were wounded.[8]

General Sherman paid little heed to the possibility that his artillery barrage was causing civilian casualties. He seemed to be much more interested in

---

[8] Sherman to Grant, July 29, and to Halleck, August 7, *OR*, vol. 38, pt. 5, 289, 408; S. Davis, "'A Very Barbarous Mode of Carrying on War,'" 78–79; Gibbons, "Life at the Crossroads of the Confederacy," 46; S. Davis, "How Many Civilians Died in Sherman's Bombardment of Atlanta?," 15; Venet, *Sam Richards*, 232; Venet, "From Gate City to Gotham," 168.

the damage and destruction of the buildings of Atlanta. Interdicting wagon traffic into the city from the south and preventing the running of trains in from Macon and the unloading of food, ammunition, and supplies at the downtown train depot (Car Shed) should have been Sherman's priority for his cannoneers, but very seldom did he mention these targets in his instructions for the army group's artillery. Only once in the length of the bombardment is there record of a Federal shell striking a Southern locomotive parked at the car shed downtown, when the locomotive *Swiftsure* was struck through the front (but the shell did not explode and did not perforate the boiler). Instead of targeting the central train depot, Sherman was targeting the entire city. "I would like to get a good battery as near it as possible that will reach the heart of Atlanta and reduce it to ruins, and to keep up a fire that will prevent wagon supply trains from coming into town," he wrote on August 8. But more often Sherman expressed keener interest in a general shelling that would wreck the city. As he put it to General Howard, "Let us destroy Atlanta and make it a desolation." Given Sherman's complex notions of *casus belli*, one may infer that he considered the buildings if not the people of Atlanta as legitimate targets of his cannonfire.

Stymied for a while in his attempts to cut the railroad south of Atlanta, Sherman determined to intensify his bombardment as the only tactic he could use to punish the enemy. "One thing is certain," he telegraphed Halleck, "whether we get inside of Atlanta, or not, it will be a used-up community by the time we are done with it." Not content with the heavy shells being fired by his 3-inch rifles, 10- and 20-pounder Parrotts, Sherman ordered a heavier caliber of cannon. "I have sent to Chattanooga for two 30-pounder Parrotts, with which we can pick out almost any house in town," he wrote on August 8. Sent to him instead were four 4.5-inch siege rifles. Sherman expected them on the 9th, and he looked forward to seeing them at work. "Get your guns well into position," he instructed Thomas, "and the moment the ammunition comes, let them open slowly, with great precision, making all parts of the town unsafe." The arrival of the guns and ammo was delayed, however. With typical impatience, Sherman ordered a heavy bombardment for August 9 with the rifled artillery he already had in place. "All the batteries that can reach the buildings of Atlanta will fire steadily on the town," he instructed; each gun was to fire fifty rounds. The bombardment, Sherman claimed, was to divert the Rebels' attention from Schofield's effort that day to "develop well the enemy's flank."[9]

Evidence, admittedly circumstantial, suggests that Thomas on August 9 wanted his cannon to fire on the city area where Hood's headquarters were thought to be. Already the Thrasher house on Whitehall had come under fire. On July 29, when Confederate Capt. Tom Key visited the army's chief of artillery at Hood's, he recorded that "shells were exploding near his headquarters." On August 3, the Northern spy, J. Milton Glass, informed Sherman's staff that Hood's headquarters were on Whitehall Street, near Rodgers Tannery (between Garnett Street and Grenville's alley). On the 8th, as he anticipated the arrival of the big guns, Thomas reported to Sherman that he had sited a position for them which "enfilades White Hall street, upon which is General Hood's headquarters."

On the day of the big bombardment, August 9, Thomas's 4½-inch siege rifles had not arrived, but as Sherman ordered, all other rifled guns fired into Atlanta. Whitehall Street, and Hood's headquarters, came under fire. "More shells were thrown into the city to-day than any day previously," noted General Shoup, "a great many falling immediately around General Hood's headquarters." "Hood's quarters came in for a large share of the shelling," noted Lieutenant Trask of Hardee's staff. The Right Reverend Henry C. Lay, Episcopal bishop of Arkansas, visited Hood that day, at invitation of General Shoup. "Shells are exploding all around Head Qrs (at Thrasher's House)," Lay entered into his diary. Chaplain Charles T. Quintard, who had established St. Luke's church downtown that spring, was with the bishop. "The city was being shelled by the Federals," he wrote in his journal, "and some of the shells fell very thickly about the General's headquarters." Quintard felt the setting "unhealthy," but observed that "the General and his staff did not seem in the least disturbed," so he and the bishop "stood it with the best of them." Lay observed that the enemy shelling induced Hood and his staff to move their headquarters a bit farther out from downtown. This they did, to "where White Hall street intersects Faith's alley," as Glass reported on August 12 to his superiors. The house chosen belonged to L. Windsor Smith, an early Atlanta landowner. The impressive two-story residence was two tall wooden structures, painted white and connected by portico and a towering roof. The

---

[9] S. Davis, *What the Yankees Did to Us*, 168, 170–71, 179, 235; "The Swiftsure," 1 ("had it done so [the shell exploded], the Swiftsure's career would have been brought to an end as sudden as it would have been disastrous").

place had plenty of room, yet according to longtime Atlantan James Bell, interviewed in the mid-1930s by Wilbur Kurtz, Hood stayed in a tent pitched in the back yard—perhaps to keep cool.

In Hood's room (probably at the Windsor Smith house), on the night of August 10, General Hood and three of his staff officers were confirmed as members of the Episcopal Church by Bishop Lay. Dr. Quintard spoke about Confirmation; the bishop recorded in his diary that "it was a singularly solemn service." General Hood was unable to kneel and was confirmed standing with his crutches. Capt. William W. Gordon, whom Hood invited to participate in the service, seeing "several high Generals & their brilliantly appareled staffs," was embarrassed by his "jacket, flannel shirt, weather stained pants, muddy shoes and shaven head," so he stood outside in the hall until called in to kneel before the bishop and Dr. Quintard. "The rite," wrote Gordon, "was most impressive." And it was important to Hood. Quintard wrote that the night before—having undergone the "unhealthy" shelling of that day—the general had said to him, "Doctor, I have two objects in life that engage my supreme regard. One is to do all I can for my country. The other is to be ready and prepared for death whenever God shall call me." Hood was not known to be an outwardly religious man, but according to James Bell, once when S. D. Lee visited Hood, he saw the commanding general reading his Bible.[10]

---

[10] Cate, *Two Soldiers*, 105; S. Davis, *What the Yankees Did to Us*, 176; Thomas to Sherman, August 8, *OR*, vol. 38, pt. 5, 419; Shoup journal (entry of August 9), *OR*, vol. 38, pt. 3, 690; Trask diary, entry of August 10; Bishop Lay diary (entries of August 9–10), Lay Papers (my thanks go to Elizabeth Shulman of the Southern Historical Collection's Research and Instructional Services Staff for copy of the diary entries); Quintard, *Doctor Quintard*, 1905 ed., 100; Denkler, ed., *Voices: Atlanta*, 140–41 (Barnard photograph of Smith house); Castel, *Decision in the West*, 466; Quintard, *Doctor Quintard*, 2003 ed., 87–88; Hoehling, *Last Train from Atlanta*, 285–86 (quoting Lay letter "to a friend in Charleston"); Lay to Gordon, August 9, 1864, and Gordon to wife, August 12, Gordon Family Papers; Kurtz notes of interview with James Bell, June 24, 1935, Kurtz notebook #10, 250, Kurtz Collection.

Some years ago an erroneous legend developed that during the bombardment Hood sent protest of the bombardment to Sherman and asked that he cease shelling the city. Example is found in Franklin M. Garrett, "Civilian Life in Atlanta," fifty years ago: "General Hood protested to Sherman the shelling of Atlanta, pointing out that…the destruction of Atlanta ought not to involve the taking of innocent lives" (31). Another is in Peggy Robbins's article in the 1970s: "In early August…Hood started sending enraged, indignant letters about the increased bombardment" ("Hood vs. Sherman: A Duel of Words," 25). Hood eventually criticized Sherman for his bombardment, but it was *after* the fall of the city, when the two generals were exchanging sharply worded letters over Sherman's expulsion of Atlanta's civilians. Garrett and other writers have thus misdated this epistolary exchange. See S. Davis, *What the Yankees Did to Us*, 306–308. "Fighting Words," 30–37, correctly dates five letters of Hood and Sherman during September 7–14, when they discussed the bombardment.

Hood thus endured Sherman's bombardment as stalwartly as the several thousand civilians still in the city. Sherman had clearly underestimated his opponent when at the start of the cannonade, July 21, he wrote Halleck, "I doubt if General Hood will stand a bombardment." Indeed, in his telegraphic updates to the War Department, Hood made scant reference to the enemy's shelling: "All quiet to-day except skirmishing, and the enemy occasionally throwing shell into the city" (July 23); "All has been quiet to-day except a little picket-firing and occasional shells thrown into the city" (July 24); "No important change in the past two days. Some skirmishing and a little shelling from the enemy" (August 4). The citizens of Atlanta expressed among family and friends their anxieties during the Yankees' bombardment, but they made no appeal to General Hood that he act to end it. After the war, he could not recall "one word from their lips, expressive of dissatisfaction or willingness to surrender." When, after the fall of Atlanta, he exchanged letters with Sherman about the fate of the city's residents, Hood asserted that during the barrage, "I made no complaint of your firing into Atlanta." When Sherman on August 7 bragged to Halleck that he intended to "make the inside of Atlanta too hot to be endured," he was then—as he had done several times before—misjudging the grit of the enemy commander opposing him.[11]

Sherman may have misjudged John Bell Hood's grit, but some of Sherman's own soldiers did not. As the two opposing armies settled into a kind of sitzkrieg in the second week of August, a member of the 20th Connecticut opined that "there was so little progress made in the siege that the idea became prevalent among the men of the Union army that 'Uncle Billy had got his match this time' and was going to back out of the attempt to take Atlanta." Sherman would do no such thing, yet he could not help feeling distraught that despite his superior numbers, his forces' progress had been halted by

---

A few years after the war, an officer with Hood's staff took note of Sherman's written reply to Hood's complaint about the bombardment. Lt. Col. Bushrod W. Frobel, assistant chief engineer for the Army of Tennessee ("my tent was in rear of headquarters, near the upper end of Whitehall street"), stated that "Gen. Hood had remonstrated earnestly against the wanton barbarity of such a proceeding, but was answered by General Sherman that the firing and shelling was directed against his works, and if the missiles reached the town it was accidental." Of course, Atlantans, notes the author, were incredulous of Sherman's alleged contention ([Froebel], "The Georgia Campaign," 756–57).

[11] Sherman to Halleck, July 21, *OR*, vol. 38, pt. 5, 211; Hood to Seddon, July 23, 24, and August 4, *OR*, vol. 38, pt. 5, 903, 906, 943; Hood, *A & R*, 203; Hood to Sherman, September 12, *OR*, vol. 39, pt. 2, 420; Sherman to Halleck, August 7, *OR*, vol. 38, pt. 5, 408.

Hood. "I feel mortified that he holds us in check by the aid of his militia," Cump wrote Schofield on August 11.[12]

Sherman exaggerated the defensive prowess of the Georgia militia, but he was correct in gauging that Hood had been numerically strengthened by their accession. Governor Brown worked to send militiamen to the front, and his proclamations eventually brought the number in the trenches to 5,000. Georgia newspapers commended this activity. "Two trains loaded with them passed up the road yesterday morning," announced the *Macon Telegraph* on July 21, "and the movement has only just fairly begun." "The State Militia are still pouring into Macon on every train," reported the *Augusta Constitutionalist* on July 26. Glass, the Northern spy, took note, seeing what he thought to be several thousand militia arriving on the night of August 2 and marching through the city. "The enemy seems to be putting every man into the ranks they can get hold of," he concluded. To be sure, Hood was too practiced an officer to place much confidence in militiamen if they were caught in an open-field battle. But in the strong works of the Grant-Presstman perimeter, given the multiplicative power of trenches for defending forces (see Bate's casualties, August 6), the militia were important adjuncts to Hood's veteran infantry.

"Rover," a correspondent for the *Augusta Chronicle & Sentinel*, wrote on August 10, "They occupy an important point on the lines, and with the aid of a judicious admixture of veterans, are doing good service." "The Militia are doing splendid service," wrote "Outline," a soldier-correspondent for the *Columbus Times*, "and the term 'Melish' has long since ceased to be a by-word of reproach in this army." Earl Hess seems to be a little dismissive of their value when he judges that "their true significance was a matter of morale"—meaning, the sight of train cars bringing old men and boys to the front lifted the spirits of both soldiers and civilians, and emboldened them to be more optimistic as to the defense of Atlanta. General Hood held the militiamen in somewhat higher esteem, or else he would not have called for more of them. To be sure, on July 30 he had wired Brig. Gen. John H. Winder at Andersonville that he needed every man he could spare from his guards and garrison, but that Winder could retain the militia if necessary. Yet two weeks later he was asking Governor Brown if he could send forward those very same militiamen from Andersonville; he added, "and the others, you have armed, as soon as possible." It is clear that Hood was not neglecting any resource in his

---

[12] Storrs, *The "Twentieth Connecticut,"* 142–43; Sherman to Schofield, August 11, *OR*, vol. 38, pt. 5, 461.

effort to sustain the strength of his army. "Can you send me any more Militia?" he wired Brown on August 14. "Answer."[13]

The Georgia Militia were not even included in the Army of Tennessee's returns for July 31. Not surprisingly, the numbers of "Present for duty" were reduced from the previous return of July 10—the date Johnston had retreated across the Chattahoochee. Since then Hood had fought three attacking battles. In them he had tried to roll up Thomas's army before it lodged securely south of Peachtree Creek; had struck the flank of McPherson's army as Jackson had done fourteen months before; and had ordered a flank attack on Howard's army west of Atlanta, which had gone awry. He had been repulsed each time, with the result that the army returns in the *Official Records* show the three infantry corps predictably reduced in strength ("present for duty") from 45,413 on July 10 to 38,722 at the end of the month. This loss hardly conformed to James B. McPherson's statement that in the three battles of July 20, 22 and 28 Hood suffered "15,000 casualties." It certainly does not sustain the math done by one writer, Mark Coburn, that the three battles in July resulted in some 17,800 Confederate casualties, or his consequent judgement, "While Hood had not yet destroyed the Army of Tennessee, he had made a significant start."

Moreover, Wheeler's cavalry corps maintained respectable strength, despite the arduous running down of McCook and Stoneman. "Present for duty" among the mounted troops figured as 13,226 the day Johnston crossed the river and 12,803 a month later. Counting the staff, escort, and engineer troops, the Army of Tennessee numbered 51,793 on July 31, and increased its strength slightly so as to number 51,946 on August 10.[14]

General Hood deserves credit for this feat, which he accomplished in several ways. To begin with, he directed medical officers to examine the thousands of soldiers languishing in army hospitals and to return to the ranks those

[13] Parks, *Joseph E. Brown*, 290, 292–93; T. L. Connelly, *Autumn of Glory*, 373; G. W. Smith, "The Georgia Militia about Atlanta," 334; McMurry, *John Bell Hood*, 140; Hood to Seddon, August 11, *OR*, vol. 38, pt. 5, 955 ("They furnish now about five thousand muskets in the trenches here"); "To the Front," *Macon Telegraph*, July 21, quoted in *Columbus Sun*, July 24; "The Militia," *Augusta Constitutionalist*, July 26; "Off for the Front," *Macon Telegraph*, July 27; Denig to Whipple, August 3, *OR*, vol. 38, pt. 5, 347 (Glass's report); "Rover," "From the Georgia Front" (Atlanta, August 10), *Augusta Chronicle & Sentinel*, August 14; "Outline," "Special Correspondence of the Times" (Atlanta, August 31), *Columbus Times*, September 2; Hess, "Civilians at War," 339; [Hood] to Winder, July 30 (5 P.M.), [Hood] to Brown, August 15 (4 P.M.), [Hood] to Brown, August 14 (6:35 P.M.), Telegram Book No. III, Hood Papers.

[14] "Abstract from returns of the Army of Tennessee, July 10, July 31, August 10," *OR*, vol. 38, pt. 3, 679–81; McPherson, *Battle Cry of Freedom*, 755; Coburn, *Terrible Innocence*, 100, 102, 105.

sufficiently well for service. It was even rumored that the general himself strolled through hospitals in Atlanta and identified fully fifteen hundred men he could send to the front. This article, appearing in the *Augusta Constitutionalist*, encouraged the editors to urge that Augusta's military hospitals be similarly purged of skulkers. The next day, however, the *Constitutionalist* had to retract its story. The paper had been told that Hood had not visited the Atlanta hospitals, as most of the hospitalized soldiers had been sent out of the city by General Johnston's orders. Besides, an Augusta surgeon had told the newspaper, "it is not according to usage for commanders to inspect the hospitals to turn out the well and able bodied." (The editors opined that "the thing should be changed"—meaning the custom that commanders did not check the hospitals for "skulkers.") Anyway, talk of General Hood scouring the wards for shirkers made for good press. The story of the fifteen hundred malingerers appeared as far away as the *Montgomery Advertiser.* And it gave rise to others. The *Macon Confederate* reported a decline in the number of sick and wounded soldiers in the city, due at least partly to "the stringent orders of Gen. Hood to forward every man able for duty." The *Macon Telegraph* ran a column about a Confederate surgeon in the city who went through a line of sick-leave applicants, dismissing their protestations of ill health. To one who asked for continued hospitalization, the surgeon rejoined, "Going to the country a little while, and perchance killing a Yankee will do you more good than anything else. Fall in."

To be sure, medical officers always strived to send able-bodied, hospitalized soldiers back to the ranks. Medical historians have run the numbers. In July, the Army of Tennessee's hospitals admitted 23,736 cases and returned 9,832 of them—a return rate of 41.4 percent. For August total admissions numbered 21,248 but of them 6,568 were returned to duty (30.9 percent). Hood's publicized effort in August to strengthen his ranks from the army hospitals thus met with only moderate success, but his alleged efforts to do so were reported far and wide. The *Buffalo Commercial Advertiser* carried a story brought to it by a Union officer recently released from the prison at Macon: Hood had conscripted "even hospital stewards." Women had been sent in to take their places, and to accommodate other patients still in need of care, sick and wounded soldiers had been dispatched to private homes.[15]

---

[15] "We have been informed…," *Augusta Constitutionalist*, August 2; "In conversation with the chief surgeon…," *Augusta Constitutionalist*, August 3; "A few days after Gen. Hood assumed…," *Montgomery Advertiser*, August 16; "Good Signs," *Richmond Enquirer*, August 27 (quoting *Macon Confederate*); S. Davis, *Atlanta Will Fall*, 161; Breeden, "Medical History," 55;

Hundreds, maybe several thousand additional troops were brought out from his own army: able-bodied men who had secured noncombat positions, such as clerks, cooks, and couriers. Newspapers hailed these steps, as publicized in the *Chattanooga/Griffin Rebel*:

> As an evidence of Gen. Hood's determination to increase the strength and efficiency of the army, an order has just been issued from headquarters, requiring all able bodied men in the different departments to report at once at the front; and requiring, whenever it can possibly be done, the heads of departments to perform their own clerical labor, and, when that is impracticable, the employment of men unfitted for the field. This order is directed, not only to Quartermasters, Commissaries, Commandants of Post, Provost Marshals, and Surgeons, but also to the Generals. Enforced it will bring several thousand additional men to the field.

This kind of pronouncement drew approbation from the veterans. "An order was issued allowing but one clerk to each commissary and the same number for other departments," recorded Captain Tom Key in his diary on July 26; "this returned many men to the ranks."

It takes a lot of cooks to feed an army, and Hood turned to them, too. "All of the cooks have been armed and returned to their respective commands," announced the Atlanta correspondent of the *Columbus Enquirer*, in a column reprinted in the *Intelligencer* and *Chronicle & Sentinel.* Negro laborers were taking the place of men heading to the trenches. "In Clayton's division alone the number of new men returned from the cooking detachments reached 300," the paper declared, "which at this time is considered a good regiment." The column very optimistically predicted "the increase from the cooks alone will reach 4,000 effective men."

The army needed couriers, as well, but even these were restricted in number by Hood's orders. A memorandum of August 3 limited the number of message-runners that could be retained by the various headquarters of corps, divisions, and brigades. Those couriers in excess of this limit were to hand over their horses and take up a musket.

Wagon-drivers who could serve as soldiers were replaced by African Americans. Newspapers published the army's intent to hire slaves for the work. "The Army of Tennessee is advertising for fifteen hundred negroes to act as teamsters," announced the *Augusta Constitutionalist* on August 11. "Gen. Hood has placed a gun in the hands of all his wagoners, and wants to

---

Freemon, "Medical Support System," 52; "The War," *Buffalo Commercial Advertiser*, September 6.

place the wagon whip in the hands of darkies." Lt. Col. A. H. Cole, inspector general of transportation, offered to pay slaveowners $25 a month to feed and clothe the hirees and to reimburse at appraised value if the slave were lost "by the casualties of war." Agents, assisted by local sheriffs, were sent to Macon, LaGrange, and other places in the state near the several railroads to obtain the laborers. "Rover," writing for the *Augusta Chronicle & Sentinel* on August 5, mentioned that "an effort is now being made to secure negro drivers, which, if it succeeds, will add three thousand more to the ranks."

Hood's soldiers appreciated all his efforts to procure manpower. "Genl. Hood is stirring up the detailed men ordering them all back to the army," wrote one approving Confederate; "it will add a good brigade, at least to our force." "B.," a member of the 29th Georgia, wrote for the *Savannah Republican* that "among the many steps Gen. Hood has made towards the reformation and the increase of the army, none deserves the commendation of his friends more than his efforts to replace the white teamsters with negroes." "The uninitiated," "B." concluded, "can form no idea of the large force it requires to supply an army with drivers."

Some enslaved African Americans volunteered for army service without coercion from their owners. "Gen. Hood having determined to place all his able-bodied teamsters, cooks, &c., in the army, and substitute blacks," ran a column in the *Chronicle & Sentinel*,

> has sent officers to different sections of Georgia to hire negroes for this purpose. We learn that an officer lately visited Putnam and some of the adjoining counties of Georgia for the purpose. In Putnam several of the planters proposed that the officer should try the system of volunteering. It was adopted, and in the course of a few days upwards of three hundred of the primest slaves had volunteered for the purpose. A gentleman who saw them says they were in the highest spirits and anxious to go to the front.

"General Hood is putting guns into the hands of every available man connected with the army," read another news report. Not every soldier, however, was enthusiastic about all this. When he called for one man from every artillery crew to step forth and join the infantry in the trenches, there was "considerable fluttering among the artillery boys."

General Bragg also did his part to increase the army's strength. Having sent "a large brigade of good troops" from Columbus by July 23, he continued to scour for more men, especially among those previously exempted by special detail. "Confederate" commended Bragg on the 25th for "stirring the skulkers from their hiding places and sending them to the front." "Confederate" stated that the soldiers in Hood's army were much pleased "to see these chaps brought out…and made to face the music." The *Appeal* also congratulated Bragg for his "good work…of sending to the front the thousands of detailed men, post officials, contractors, purchasing agents, and other parties who have long been filling easy, safe and remunerative positions." "Whitmarsh" sent a letter to the *Constitutionalist* urging Bragg to conduct his scouring for skulkers in Augusta and other cities. The writer suggested that government agencies should only employ wounded and disabled soldiers and added his own voice in complaint against able-bodied men employed in "petty clerkships."

Bragg earned his share of editorial compliments, but so did Hood. In a column titled "Hood Vacates the Bomb-Proofs," the *Augusta Constitutionalist* estimated that fully 8,500 men had been added to the army by the general's calling forth of the ex-cooks (4,000), supernumerary artillerists (3,000), and teamsters (1,500). These numbers, of course, were grossly inflated. The Army of Tennessee about this time had some 150 pieces of artillery, and bringing a crew member from each of the guns would provide only as many men. (Besides, the number of officers and men present for all the artillery of the army's three infantry corps totaled at the time no more than 2,900.) Nevertheless, Hood's call for supernumerary artillerists boosted morale, both in the army and among the people.

Hood suspected there were large numbers of men absent without leave who could be exhorted (or shamed) to return to duty. On August 2, he asked his friend John Thrasher, superintendent of the Confederate States Press Association, then in Macon, "Cannot the papers assist us by appealing to absent soldiers to return to their colors? If you think so, please give them the cue." Thrasher apparently gave the cue, for newspapers across the state quickly sounded the call. "Absentees to the Front!" cried the *Macon Telegraph.* The *Augusta Constitutionalist* urged friends and relative of soldiers on furlough to push them forward. The *Augusta Chronicle & Sentinel* appealed to all men "dallying at home, or straggling in out of the way places, or delayed by interrupted railway communication'" to return to the army. "Fill the trenches at Atlanta with the natural defenders of the State," intoned the *Appeal*, "and Gen. Hood can have at his command a movable army of veterans, with whom to inaugurate an offensive war that will soon rid us of the presence of the invaders." "Return to Your Colors," exclaimed the *Columbus Times.* "Loiterers in the rear!" shouted the *Columbus Enquirer*, as it summoned every absentee back to the army. "We must have more men at the front," asserted the Atlanta/Macon *Southern Confederacy.* "So far, all has gone well. Richmond is safe, and around Atlanta the balance trembles in our favor. A few more blows and the tide of war will turn."

General Hood directed other efforts. To bolster Lieutenant Colonel Presstman's battalion of engineers, on July 29 General Shoup called on conscript authorities at Macon to round up 200 men for Presstman's use. This was apparently done. The *Macon Telegraph* on August 5 noticed armed officers patrolling the city streets for several days, examining male civilians' papers, arresting those without verification of status and sending them to camp, "where they will be armed and dispatched to Atlanta."

Provost guards roamed the streets of Atlanta, also looking for men and checking their papers. Charles H. Smith was one who got nabbed. Smith had been an officer in Lee's army, sent home in spring 1863 on medical discharge. Returning to his hometown of Rome, Georgia, he and his family fled when the Yankees approached in May 1864. Smith worked for a while in Atlanta, writing funny columns in semi-literate dialect as "Bill Arp" for the *Southern Confederacy.* He also served as assistant editor of the *Knoxville Register.* When the paper left Atlanta, Smith did too. After lodging his family with friends, Smith worked in Macon as lawyer for the Confederate government. At least once he seems to have traveled to Atlanta, perhaps in August 1864, and there he got pinched.

He told about it as Bill Arp would have:

> Gittin on to Atlanty, a fule Irishman stoped me right at the edge of town and demanded my papers. I didn't have no papers. Nobody had axed me for my papers, but he wouldn't heer any argument.... I was sent off to Dekater with sum fifty konscripts, who wer all in mourning, xceptin their close.... Well, there warn't none of us discharged, for there was bran new orders calling for everybody for thirty days to go to the ditches. As I couldn't walk that fur, I were ordered to Andersonville to gard prisoners. At Macon I met General Cobb who had me transferred to his department....

Smith/"Arp" was lucky. Doubtless many men were sent under guard to "the ditches" during summer 1864 to aid in the defense of Atlanta.

Hood also looked beyond his own department for help. In mid-August he asked Bragg for help in bringing back "a large number of men in Genl Maury's Department, belonging to this Army." The same message went to Maury, closing with "I am in need of every man I can get." He even wrote Bragg about attempting to bring absentees back from as far away as the Trans-Mississippi. The idea was not so far-fetched as to prevent John Beauchamp Jones, the Rebel war clerk, from recording in his diary in mid-August the (false) rumor floating in Richmond that "Gen. Taylor will cross the Mississippi with 4000 men on the 18th of this month." To General Cooper, Hood suggested that western soldiers be granted pardon "if they will rejoin their commands on this side of the river." The press reported that Hood would use his influence so that authorities would not punish AWOL soldiers voluntarily returning.

All of this activity by the commanding general cheered the men. Indeed, despite the army's heavy casualties in the recent battles, General Hood seems to have been held in considerable regard by his soldiers. Captain Key recorded on August 11 having seen Hood and Cleburne riding together that morning. "I was surprised to find the former so sociable and affable to all persons. He had none of that disgusting stiffness which so frequently attaches to men when they reach high positions—He saluted the humblest private in the road, conversed freely upon the condition of the army," Key wrote, "and appeared in hopeful anticipation of ultimate victory."[16]

---

[16] "From the Front," *Augusta Chronicle & Sentinel*, July 29 (quoting the *Griffin Rebel*); Cate, *Two Soldiers*, 102, 113 (Key diary entry of August 11); "J.T.G.," "From our Correspondence with General Hood's Army" (reprinted as "Gen. Hood Increases his Army," *Atlanta Intelligencer*, August 10, 1864; same in *Augusta Chronicle & Sentinel*, August 12); Falconer General Field Orders No. 10, August 3, *OR*, vol. 38, pt. 5, 943; "Negro Teamsters," *Augusta Constitutionalist*, August 3; "Telegraphic," *Augusta Constitutionalist*, August 11; "1,000 Negro Teamsters Wanted" (advertisement), *Macon Telegraph*, August 16; A. H. Cole, "To the People of Georgia," and Capt. John G. Clark circular, *Macon Telegraph*, August 11; Kennett, *Marching through Georgia*, 249–50; "Rover," "Temporary change of Commanders…Reinforcements, etc.," *Augusta Chronicle & Sentinel*, August 9; McMurry, *John Bell Hood*, 140 ("stirring up the detailed men"); letter of "B.," 29th Georgia, near Atlanta, August 19, *Savannah Republican*, August 26 (reprinted in *Atlanta/Macon Intelligencer*, August 27 as "From Atlanta. Correspondence of the Sav. Republican," August 27); "Gen. Hood…," *Augusta Chronicle & Sentinel*, August 31; "Good for Cuffee," *Augusta Chronicle & Sentinel*, September 4; Bragg to Sale, July 23, *OR*, vol. 38, pt. 5, 904; "Confederate," "Bragg 'Stirs the Skulkers'" (July 25), *Columbus Sun*, July 30; "A Good Work," *Memphis Appeal*, August 1 (reprinted in *Macon Telegraph*, August 4; and in *Augusta Chronicle & Sentinel*, August 7); "Whitmarsh" and "Detailed Men," *Augusta Constitutionalist*, August 10; "Hood Vacates the Bomb-Proofs," *Augusta Constitutionalist*, August 20; abstract from returns, July 10, *OR*, vol. 38, pt. 3, 679 (number of artillerymen); Hood to Thrasher, August 2 (8:30 P.M.), Telegram Book No. III, Hood Papers; "Absentees to the Front!," *Macon Telegraph*, August 4; "To the Front," *Augusta Constitutionalist*, August 5; "Appeal to Soldiers Absent from their Commands," *Augusta Chronicle & Sentinel*, August 5; "Rallying," *Memphis Appeal* (n.d.), reprinted in *Augusta Chronicle & Sentinel*, August 6; "Return to Your Colors," *Columbus Times*, August 6; "To Your Posts," *Atlanta Southern Confederacy*, reprinted in *Augusta Constitutionalist*, August 11; Shoup to "Conscript Authorities at Macon, Ga.," July 29, *OR*, vol. 38, pt. 5, 930; "Hunting the Militia," *Macon Telegraph*, August 5; Parker, *Alias Bill Arp*, 14–17; "Editorial Accession," *Atlanta Intelligencer*, June 25 (Smith and the *Southern Confederacy*); [C. H. Smith], "Bill Arp in the Piny Woods," *Atlanta Southern Confederacy*, June 21, 1864; [C. H. Smith], *Bill Arp's Peace Papers*, 94–95; Hood to Bragg, August 15 (4:15 P.M.; same message to Maury), and Hood to Bragg, August 13 ("To whom and where shall the details to go to Trans Miss Dept for absentees…report?"), Telegram Book No. III,

Hood and Bragg, however, could be too zealous in their efforts to bring in more troops, as when they ordered skilled artisans and factory workers to take up muskets, despite their industrial importance and military innocence. Columbus, Georgia, was the site of a large naval iron works, government arsenal, and the largest shoe manufacturing plant in the Confederacy. It was evidently Bragg who sent Maj. E. H. Youngblood and a battalion of Columbus government mechanics (probably arsenal workers) to Atlanta; they were ordered to report to Walthall's division on July 26. Youngblood's battalion participated in the battle of Ezra Church two days later. In the attack of Daniel Reynolds's brigade, "Youngblood's battalion got in some confusion," General Reynolds reported, "and many of them left the field, though a number of them acted very well." Someone must have told Hood that the country's best interests lay in not taking its skilled laborers from the factories. The very morning after the battle he telegraphed the commanding officer at Columbus, "How many shoemakers have been sent to the front from Columbus? How can they be gathered together to return?" Some black workmen were evidently taken from Columbus as well; on the 30th Hood wired no less than Secretary Seddon that they were being sent back with overseers. There was apparently impatience in Columbus to get the workmen back. Headquarters assured Maj. George O. Dawson at Columbus on July 31, "Shoemakers to be returned immediately." Colonel Oladowski in Macon may have also expressed concern for the Columbus armory workers; the Macon armory had been out of operation since July 29 due to Stoneman's raid. On August 1 headquarters telegraphed Oladowski, "The men will be returned at once to Columbus Armory." Even Bragg seemed to want to know. Hood on August 4 wired him word that "the workmen from Columbus have already been returned." That day the *Columbus Times* reported the return of the detailed men, gone for all of ten days, appearing "in rather a dingy condition, but quite buoyant in spirits": factory companies, the shoemakers' battalion, a company from Louis and Elias Haiman's pistol factory, and even one from an oil cloth factory. According to the *Times* of the 4th, three companies of workmen were still to return, two of them assigned to the armory. Hood's and Bragg's experiment in enlisting Columbus government workers was at an end.[17]

Hood Papers; J. B. Jones, *Rebel War Clerk's Diary* (1958 ed.), 410 (entry of August 15); Hood to Cooper, August 2, *OR*, vol. 38, pt. 5, 940; "Gen. Hood...," *Augusta Constitutionalist*, August 23 ("promises to use his influence to procure forgiveness").

[17] Standard, *Columbus, Georgia, in the Confederacy*, 31, 38, 41–42; Walthall report, January 14, 1865, *OR*, vol. 38, pt. 3, 926; Reynolds report, August 2, *OR*, vol. 38, pt. 3, 939; [Hood] to commanding officer, Columbus, July 29 (9 A.M.), Hood to Seddon, July 30 (5:20 P.M.), [Hood] to "Commanding Officer Columbus Georgia," July 30 (9 P.M.) ("How can I ascertain

Not counting this latter failure allows one to list at least seven different ways General Hood sought to strengthen his forces at Atlanta in late July and early August 1864:

1. With the help of Governor Brown, he had brought in some 3,000 militiamen to help garrison the city's fortifications.

2. He urged medical officers to examine hospitalized soldiers more stringently and return them to the ranks. According to one account, he replaced able-bodied male hospital stewards with female attendants.

3. He ordered clerks, cooks, and couriers who could shoulder a musket to enter the trenches. Bragg helped look for detailed men in other districts. Hood's call for the seventh crew member serving each piece of artillery brought forth 150 or so men, but the call served notice that the commanding general was looking everywhere for additions to his infantry force.

4. He arranged for 1,000 to 2,000 slaves to be hired as teamsters, replacing enlisted men as wagon drivers so that they too could take their place in the lines.

5. He got newspapers throughout Georgia to urge soldiers straggling or loitering in the rear to return to the army. He ordered conscription officers to round up those absent without leave and send them back.

6. He sought soldiers in Maury's department of Alabama and Mississippi who should be returned to the Army of Tennessee and proposed to Bragg the possibility of doing the same in the Trans-Mississippi.

7. He asked General Cooper in Richmond to extend amnesty to those soldiers AWOL who returned to the army.

---

the names of shoemakers sent here from your post?"), [Hood] to Dawson, July 31 (9:35 P.M.), [Hood] to Oladowski, August 1 (11:45 A.M.), all in Telegram Book No. III, Hood Papers; Vandiver, *Ploughshares into Swords*, 245 (Macon Armory operations suspended July 29–August 2); Davis, *Atlanta Will Fall*, 163; Hood to Bragg, August 4, *OR*, vol. 52, pt. 2, 718; "Returned," *Columbus Times*, August 4; Albaugh, Benet, and Simmons, *Confederate Handguns*, 9–16 ("The Columbus Fire Arms Manufacturing Company, also known as L. Haiman and Brother").

Frank Vandiver criticizes Hood for requisitioning black workers from the Macon Armory, noting that "this maneuver crippled operations at Macon." "But Hood had not yet learned his lesson," Vandiver adds; he ordered the skilled and unskilled workmen at the Augusta ordnance works to field service with the army during the crisis at Jonesboro, August 31–September 1 ("Hood as Logistician," 4).

All of these measures tended to lift spirits of Southerners both in and out of the army. William E. Sykes of the 43rd Mississippi wrote his wife, Augusta, on July 26, "Gens. Bragg and Hood have enfused a very vigorous policy on the conduct of the Army—All hospitals have been cleared of convalescents—detailed men ordered in—clerks and couriers from Hd. Qrs.—quarter-master and commissary sergeants—All are to return to their commands." These reinforcements seemed to give Sykes hope: "it has been determined to hold Atlanta at all hazards."

"Our Army has been strengthened by the Militia," wrote Andrew J. Neal, lieutenant in Perry's Florida Battery (Hardee's corps), to his sister on August 4. Besides rumored reinforcements from Forrest (not true), Neal declared, "all the bands cook details extra artillery men drivers and dead heads have been brought to the front and given muskets" (true). Optimistically he added, "Thus we will be enabled to hold our long lines and have force enough to operate on the flanks." Word of Hood's activity reached Lee's army in Virginia. The same day that Neal wrote Ella, Maj. Edgeworth Bird (on the staff of Brig. Gen. Henry L. Benning) wrote his wife: "Hood's army is strengthening every week." "General Hood is so strong at Atlanta," recorded J. B. Jones in his war clerk's diary on August 23rd, "that he has promised to send, in an emergency, a brigade to Mobile." (Jones was correct; Hood wired General Maury, "send you a brigade, though I cannot spare it.") About this time Mary Chesnut talked with friends about Hood's situation. Louise Bartow (widow of Col. Francis S., killed at Manassas) offered an upbeat opinion: "After all, Hood is expected to hold Atlanta until reinforcements can be got there," she said, prompting someone to ask from where they would come. Others were as optimistic as Mrs. Bartow. "When we asked Brewster what Sam meant to do at Atlanta," Mary recorded on August 1, Hood's friend replied, "Oh, oh! Like the man who went, he says he means to stay there."

Still, there were worriers that Hood's army lacked the necessary manpower to drive the Yankees out of Georgia. "It follows, then, that if a decisive engagement is to be fought near Atlanta, that Gen. Hood must be the assailant," reasoned the *Appeal* on August 14; "But is he strong enough? Has he the men? We do not know. While he may be quite strong enough to hold the city forever against the mercenary minions of the Northern despotism, and to repel, with terrible loss to them, every attempt to take Atlanta, he may not be able to attack them in their trenches and defeat them."

Elsewhere in the press one could see expressions of confidence in Hood and his management of the situation. "The Confederate private has come to the conclusion that Atlanta can be held, and the citizen is making up his mind it will be," wrote "Rover" (who was possibly one of those very privates in the army) from Atlanta on August 17. "A change has come over both soldier and citizen, and a sense of security is enjoyed far different from the uncertainty that pervaded the breasts of nearly all three weeks ago"—meaning, probably, after the setback at Ezra Church.

One sees that Hood's orders led to an infusion of at least some new troops. Bragg had exaggerated in writing the president on July 27, "The increase by arrival of extra duty men and convalescents, &c., is about 5,000; more are coming in daily." But he was on the mark in writing, "The return of the 1st of August will show a gratifying state of affairs." Actually, it was the Army of Tennessee's later returns that attested to Hood's industry regarding his army's manpower. Recording 51,793 officers and men present for duty on July 31, and 51,946 ten days later (note the slight increase), the abstract of returns in the *Official Records* for August 20 showed only a slight decrease in numerical strength, 51,157, despite the weeks of skirmishing and casualties. Months of hard campaigning had indeed taken their toll on the army, and some regiments had dramatically lost strength. On August 17, Col. Newton N. Davis of the 24th Alabama wrote his wife, Bettie, "I had over three hundred men when we left Dalton and now I have but one hundred and twenty left." Nevertheless, the overall returns for the army told the tale, and those for August 30 counted 51,141 officers and men present for duty. Thus, according to the Confederate army's statistical records (there is no reason to question them), Hood's army maintained roughly the same level of strength throughout August. The only way it could have done so was through Hood's troop-finding efforts.

Hood was understandably proud of his work and called attention to it in a wire to Secretary of War Seddon on August 18. "Exaggerated statements of the losses of this army since I assumed command [had] been made in reports of Federal commanders," Hood asserted, which compelled him to submit his own numbers. When he took charge of the army on July 18, its effective strength totaled 49,012; "by last report" (presumably August 10) that figure was 43,091. Hood then did the arithmetic to show a decrease of only 5,921 officers and men. He also wanted to remind the administration of what he had done: he had held onto Atlanta another month, fought three big battles to hold the Yankees at bay, and launched a strong mounted column against the enemy's railroad communications—all done with a loss of under 6,000 soldiers. To boot, Hood added at the end of his telegram, "Reinforcements proper since that time [his assumption of command] have been sent back."

Another approach is to look at the numbers for Sherman's army group. McMurry argues convincingly that "present for duty" should be the category to use for Confederate forces and "effective strength" for the Federals'. He matches these two categories, and from them calculates Hood's strength as a percentage of Sherman's.

	June 30	July 31	August 31
Confederate	62,747	51,793	51,141
Federal	106,070	91,675	81,758
Percentage	59.15	56.49	62.55

By these calculations, while one sees a steady weakening of Sherman's strength in the final two months of the campaign, one sees that Hood succeeded in holding steady his army's strength. Indeed, in terms of percentages, Hood's army *gained strength* against Sherman's. Even Cump seemed to be aware of Hood's accessions. "Our Lines are extended ten miles," he wrote Ellen on August 9, "but still the enemy is beyond." To his foster father, Thomas Ewing Sr., two days later Sherman even made the fairly outrageous statement "I rather think today Hoods army is larger than mine."[18]

---

[18] Bell, *The Camel Regiment*, 168–69; A. J. Neal to "Dear Ella," August 4, box 1, folder 6, Neal Papers; Rozier, *The Granite Farm Letters*, 180; J. B. Jones, *Rebel War Clerk's Diary*, 413; Hood to Maury, August 24, *OR*, vol. 38, pt. 5, 987; Woodward, *Mary Chesnut's Civil War*, 625, 628; "Sherman's Policy—Hood's Necessity," *Columbus Enquirer*, August 19 (quoting *Memphis Appeal*, August 14); "Rover," "Letter from the Georgia Front" (Atlanta, August 17), *Augusta Chronicle & Sentinel*, August 21; Army of Tennessee abstract of returns, July 31, August 10, 20, and 31, *OR*, vol. 38, pt. 3, 680–83; Towles, "Dalton and the Rebirth of the Army of Tennessee," 91–92 n.29 (Newton Davis to Elizabeth Davis); McMurry, *Atlanta 1864*, 194–95; "Abstract from returns showing the effective strength of the army in the field under Maj. Gen. William T. Sherman, during the campaign against Atlanta, Ga., 1864," *OR*, vol. 38, pt. 1, 115–17; Hood to Seddon, August 18, *OR*, vol. 38, pt. 5, 972; Simpson and Berlin, *Sherman's Civil War*, 685, 690.

To be sure, not all of Hood's troops shared a sunny outlook. "There is much demoralization in the army," wrote Lt. Robert M. Gill on August 11. "Desertions are numerous," he noted; "some of the best soldiers are leaving." Brig. Gen. Josiah Gorgas, ordnance chief in Richmond, recorded in his journal on August 12, "An Artillery officer just from Atlanta gives me a gloomy view of the situation." The officer believed that the city would have to be abandoned, "as the enemy is closing in his lines about it. They extend from the river to near the Macon R. R." (Wiley, "3 Southern Officers," 34; Gorgas, *Journal*, 127).

Federals, of course, did not always have a handle on the size of Hood's army and its strength. Brig. Gen. William B. Hazen wrote a friend from near Atlanta on August 2 about the Rebel forces.

Officers throughout Hood's army saw the results of his efforts. One was Capt. J. B. Jordan, commanding Co. G, 36th Alabama. Jordan had deserted and consented to be questioned by his captors. He told them of demoralization in Hood's army, scant rations, and low supplies of ammunition. But as to its numerical strength, Captain Jordan was very clear, as stated in his interrogation report:

> No re-enforcements have been received except militia and men from convalescent camps and hospitals; says a large number of men have been added to the army from these sources; thinks the aggregate number since Hood has been in command, including cooks, teamsters, and other detailed men, will reach 15,000. His own company, Jordan's, numbered 14 men at Kennesaw, and yesterday mustered 34 guns; has received 12 men since the fight on the 22d.[19]

One of Hood's chief concerns was the adequate supply of rations for his troops. The Federals' "scout," Milton Glass, reported after he had again slipped through the lines that the Rebels "appear to have plenty of rations and forage from day to day, but there is no supply on hand"; everything was being brought in by wagons and trains from Macon. From Lieutenant Edge's treetop perch east of the city, Federal observers could see the arrivals of trains and even report on what they were carrying. On August 12, three trains came in loaded with troops. On the 14th a locomotive and fifteen cars chugged in. "Train ran slow; engine appeared to be working hard"; the implication was that it was heavily loaded with supplies. The next day Edge reported a train of eighteen cars, "five of them loaded with white sacks." Edge concluded, "Green corn was hauled in and issued to rebel troops in their works this p.m." Captain Jordan told his captors that the supply of bacon (sometimes beef) and corn bread issued to the troops was skimpy but regular; he had "often seen his men eat a day's supply at a meal and then not be satisfied." "Rover," the soldier correspondent for the *Augusta Chronicle & Sentinel*, wrote to his paper on August 8, "So far as meal and meat are concerned the fighting men of the army

---

> Johnston's veteran army by his official report June 25th, contained 46,628 arms-bearing men including 6,631 of Wheeler's cavalry. They have lost since that time 5,000 prisoners, and in their three separate assaults upon our works since arriving in front of this place, at least 20,000 men. They have received from Mississippi 3,500, and are receiving from Governor Brown's proclamation, about 8,000 militia. This gives them today an army of about 25,000 veterans and 8,000 militia; 33,000 in all. (*Buffalo Commercial Advertiser*, August 19, 1864, quoting *Cleveland Ledger* and Hazen's letter)

[19] Statement of J. B. Jordan, August 14, 1864, *OR*, vol. 38, pt. 5, 494–95.

are well fed. Rations of beef and bacon are alternated. The cooking is done by the details in the rear, and distributed along the trenches at night." "Rover" further stated that someone had uncovered in the city huge stashes of tobacco, doubtless hoarded and hidden by speculators. A commissary officer told him that there was enough tobacco to supply the army for two full months; "this will be joyful news to the soldiers." On the other hand, "Rover" complained about the shortage of vegetables—always a problem for armies in the field—and could only implore the people of Georgia to send their foodstuffs to the front. As always, it depended on whom you talked to about the important question of rations. One Arkansan in Cleburne's division wrote his parents on August 14, "we get a few vegetables and occasionally some fruit." Capt. William L. Nugent, in Ferguson's cavalry, and a staff officer to boot (hence able to ride about and "forage"), wrote his wife, Nellie, on August 15 that "with tolerable fare [and] little or no work to do," he was actually "growing fat."

Despite these selective shortages, as they endured Sherman's semi-siege and bombardment of Atlanta, the officers and men of the Army of Tennessee could not blame their commanding general for not seeing to their supply of sustenance. The army's commissary officers deserved great credit during this time, but General Hood should also be credited for not buckling under the Confederates' logistical difficulties.[20]

Meanwhile Hood looked to improve his tactical situation.

> Now that the enemy's cavalry is in some sort broken up, I think I shall have no great difficulty, and that in addition I shall be able to operate upon his communications. At present everything here looks well. I beg to assure you that I have no intention of abandoning this place, and that if no other recourse be left I shall certainly give the enemy battle before I leave it.

So wrote General Hood to Bragg on August 4. The message is important in four ways.

1. *"I shall be able to operate upon his communications."* This is a statement General Johnston certainly never made, for he had insisted that Wheeler's

[20] Glass statement, August 18, *OR*, vol. 38, pt. 5, 580; Howard to Sherman, August 12, *OR*, vol. 38, pt. 45, 480; Edge to Logan, August 14 and 15, *OR*, vol. 38, pt. 5, 501, 513; Jordan statement, August 14, *OR*, vol. 38, pt. 5, 495; "Rover," "Letter from the Georgia Front" (Atlanta, August 8), *Augusta Chronicle & Sentinel*, August 12; S. Davis, *What the Yankees Did to Us*, 202–203; Christ, *Getting Used to Being Shot At*, 98 ("a few vegetables and occasionally some fruit").

cavalry stay with the army, reconnoitering the enemy's movement and guarding his army's flanks.

2. *"At present everything here looks well"*; this is another statement Johnston did not make because he was always worried about being outflanked and having to retreat.

3. *"I have no intention of abandoning this place"*; Joe Johnston had never said anything so resolute to the administration in Richmond.

4. *"I shall certainly give the enemy battle before I leave."* This statement is a polar opposite from Joe Johnston's last message to the administration while he commanded the Army of Tennessee. One may indeed argue that if Johnston had stated this to President Davis on July 16 (instead of his idea of holding Atlanta with the militia so that the movements of his army could be "freer and wider"), he would have retained his command.

Secretary Seddon on July 17 had urged Hood to send mounted forces against the enemy's communications. Now he was complying. Nonetheless, how Hood determined to send off part of his cavalry is an important decision that deserves to be examined.

In his report he lays forth his logic: "Believing the enemy's cavalry well broken, and feeling myself safe from any further serious operations of a like nature, I determined to dispatch a force of cavalry to the enemy's rear, with the hope of destroying his communications." Hood ordered Wheeler to launch his raid with 4,500 horsemen; "this still left sufficient cavalry to meet the necessities of the army." According to the returns of August 10, the day Wheeler started north, the army's cavalry with horse artillery numbered 12,803 officers and men present; Hood was sending away *less than half* of his mounted troops. Wheeler later reported that he set out on August 10 with a force of 4,000. This would have left Red Jackson with plenty of material with which to perform the cavalry's functions. Hood had such high hopes for Wheeler's success in burning bridges on the W. & A. that, as his cavalry rode off on August 10, he instructed his chief engineer, Major General Smith, to assemble bridge timbers for rebuilding bridges from Atlanta to Chattanooga. The implication is that if Sherman fell back, Hood would advance.

Another factor in Hood's logic was Wheeler's success in leading an earlier raid on the enemy's supply lines. The previous October, when Rosecrans's army was pinned down at Chattanooga, Wheeler led 4,000 horsemen on a foray through the Sequatchie Valley in which the Confederates captured and destroyed a thousand wagons filled with stores, burned depots and bridges, and wrecked railroad track. Rosecrans's troops were seriously short of supplies; only the arrival of Grant and reinforcements prevented the Federal abandonment of Chattanooga.[21]

The administration in Richmond, Georgia's political leaders, the Confederate press, and the people as a whole had been hoping for just such a raid as Hood planned. After Secretary Seddon in mid-July had mentioned raiding the Federal rear, Hood answered him positively but asked for help: if S. D. Lee's cavalry "can co-operate with me by breaking the enemy's communication, I ask that it may be done with the least possible delay." Since then, Lee had joined Hood's army and the enemy were keeping Lee's cavalry (especially Forrest) busy, so Hood had to call upon his own horsemen for the raid. General Wheeler certainly wanted to do it; he had written Bragg on July 1 making that point. Then Hood had floated the idea of a Wheeler raid to the president, who assented, noting the possibilities of rail-wrecking in north Alabama. The people also wanted it; indeed, in late June a citizen had written S. D. Lee, providing details on bridges that could be burned by cavalry raiders in Tennessee.

Davis and Seddon gave their approval to an operation long asked for by Governor Brown. Throughout the campaign the governor had been pestering Davis with pleas for a raid into the enemy rear. "Let me again beg you to send a cavalry force to cut off Sherman's supplies, which must compel his retreat," he wrote on July 19, the day after Hood had taken command of the army. Other prominent Georgians called for such an expedition. "If his communication was cut for ten days his army would be destroyed," Howell Cobb predicted, "and Georgia, as well as Alabama and Mississippi saved, and Tennessee recovered."

Georgia's newspapers echoed these hopes. Sherman's dependence on the vulnerable single-track Western & Atlantic running for Chattanooga prompted an analogy by the *Macon Telegraph*: "A spider swinging from the ceiling by a single fibre of his web is not an inept illustration of the situation." The *Augusta Chronicle & Sentinel* was at once more literal and hyperbolic. "Surely there is enterprise, there is daring, there is determination, there is fury enough in the Southern heart," the editors cried in a column on July 14, "to

---

[21] Hood to Bragg, August 4, *OR*, vol. 52, pt. 2, 718; Hood report, February 15, 1865, *OR*, vol. 52, pt. 3, 632; abstract of August 10 returns, *OR*, vol. 38, pt. 3, 681; Wheeler report, October 9, 1864, *OR*, vol. 38, pt. 3, 951; Snider and Brooksher, "A Ride Down the Sequatchie Valley," 39; Brooksher and Snider, *Glory at a Gallop*, 183; Longacre, *Mounted Raids of the Civil War*, 214–17.

destroy that line of railroad, to burn every bridge, to tear up every rail, to annihilate every tie, and to blow up every tunnel between Atlanta and Louisville." Two weeks later the *Knoxville/Atlanta Register* added its voice to the clamor: "Where are our guerrilla bands that should be constantly hovering along this outstretched line of communication, and making it unsafe for the trains to move?" For his failure to launch such a raid, Wheeler had come in for criticism. The *Chattanooga/Griffin Rebel* complained on August 2, "the inefficiency of the cavalry of this army has been too serious a matter, and too patent to the commonest understanding to require silence." The press was only calling for what the citizenry wanted, too. "People have got tired asking why have not Sherman's communications been disturbed," "Rover" asked on August 4. "They have not....Why is it so?"[22]

With its solid logic, governmental backing, and public demand, Hood's decision to send Wheeler raiding into Sherman's rear has, remarkably, drawn criticism. Early complaints came from political opponents of the Davis administration. "Hood made the fatal mistake. He sent off Wheeler and *his entire* cavalry [emphasis added] to raid on Sherman's line of communications," asserted Edward Pollard, one of the administration's harshest critics. In his memoirs Hood expressed sensitivity to criticism of his decision by Davis's enemies. "At this period, I was charged by the Johnston-[Louis] Wigfall party, through the press, with having committed a serious blunder by sending off my cavalry," he wrote in *Advance and Retreat.* But he defended himself: "we had still a sufficient number, with Jackson, to protect not only the flanks of the Army, but likewise our communications against similar raids." Unfortunately, those who should know better have repeated Pollard's erroneous (and preposterous) claim that Hood sent *all* of his cavalry northward. "Hood thrust into the hands of his adversary the opportunity he had waited for," Otto Eisenschiml and Ralph Newman wrote in 1947. "He sent off his *entire* cavalry [emphasis added] toward Chattanooga to raid the enemy's line of communication."

---

[22] Seddon to Hood, July 17, *OR*, vol. 38, pt. 5; 885; Hood to Seddon, July 19, *OR*, vol. 38, pt. 5, 892; T. L. Connelly, *Autumn of Glory*, 317 ("he had the power to cut Sherman's line of supply if Johnston would only order him to do so"); Wheeler to Bragg, July 1, Bragg Papers; James E. Saunders to S. D. Lee, June 28, *OR*, vol. 38, pt. 4, 802–803; Brown to Davis, July 19, *OR*, vol. 52, pt. 2, 710; Cobb to Seddon, July 1, *OR*, vol. 38, pt. 5, 858; D. Evans, *Sherman's Horsemen*, 4 (quoting *Telegraph* and *Chronicle & Sentinel*); "Gossip from Atlanta," *New York Times*, August 16 (quoting *Register* of July 26); "From the Army of Tennessee," *Columbus Times*, August 4 (reprinting from *Chattanooga/Griffin Rebel*, August 2); "Rover," "Letters from the Georgia Front" (Atlanta, August 4), *Augusta Chronicle & Sentinel*, August 9.

More recent scholars, even those aware that Hood sent off no more than half of his mounted troops, still have judged his decision a mistake. J. Cutler Andrews knew the mood of the Southern press, but nonetheless opines that "Hood made the mistake of sending Wheeler's four thousand cavalrymen north of Atlanta." Considering Sherman's missteps with his own horse soldiers, Dyer, Wheeler's biographer, states that with Wheeler's raid Hood was "going Sherman one better in the unwise use of cavalry." Others have challenged Hood's belief that he had retained with Jackson enough cavalry to meet demands. Hood's dispatch of Wheeler "left only General William Jackson's division with Hood's army, and some isolated units," asserts Connelly, who fails to mention that Jackson's command numbered at least 5,000 men. Stephen Starr, historian of the Union cavalry, charges that "Hood made a mistake" with the raid. A recent historian of the Army of Tennessee's cavalry says the same thing. "Hood was gambling that while Wheeler was gone he could get by without adequate reconnaissance and intelligence-gathering assets," Longacre writes, terming the decision "another manifestation of Hood's recklessness." "With half the cavalry gone, Hood lacked enough troops to penetrate the curtain of blueclad cavalry screening the Federal movement west of Atlanta," Bailey judges in his volume for Time-Life. Shelby Foote weighs in, noting that sending Wheeler off deprived Hood "of half his cavalry during the critical final stage of the contest for Atlanta." "All Hood had done was deprive himself of 'his eyes and ears,' an action that Sherman would play to his benefit," chimes in John Cannan.

Most surprising of all is an early viewpoint by Castel. In sending Wheeler off on his raid, "Hood was left as strategically blind as Robert E. Lee was prior to Gettysburg, and with the same fatal results for the Confederacy." Castel would temper this judgment later in *Decision in the West*, as we shall see.[23]

These writers' assumptions and accusations should not overshadow evidence that in its first days, at least, Wheeler's raid portended success. Six hundred pounds of blasting powder, fuses, and other supplies had been sent from the Augusta powder works to Covington for the Confederates' destruction of bridges and rail facilities. This equipment would be hauled in wagons, but Wheeler otherwise ordered his men to travel light. On the day before they set out, he issued a circular stipulating that no personal items save "one single

---

[23] Pollard, *Southern History of the War*, 2:384; Hood, *A & R*, 201; Eisenschiml and Newman, *The American Iliad*, 628; Andrews, *South Reports*, 459; Dyer, *"Fightin' Joe" Wheeler*, 188; T. L. Connelly, *Autumn of Glory*, 457; Starr, *Union Cavalry*, 3:457; Longacre, *Cavalry of the Heartland*, 293; Bailey, *Battles for Atlanta*, 143; Foote, *The Civil War*, 3:520; Cannan, *Atlanta Campaign*, 144; Castel, "Union Fizzle," 31.

blanket and one oil-cloth" would be allowed per man. On August 12, Wheeler's troopers struck the Western & Atlantic at Big Shanty and tore up a mile of track. The next day, near Calhoun, they captured well over a thousand cattle, which were sent back to Hood's army. Several more miles of track were wrecked between Resaca and Tunnel Hill. On the 14th Wheeler captured Dalton; the men plundered shops set up by "Yankee storekeepers," according to one cavalryman, who remembered enjoying "condensed milk (the first I ever saw) and crackers and other good things." Three trains loaded with supplies were destroyed, and more track damaged between Dalton and Tilton. But the next day Maj. Gen. James F. Steedman brought 4,000 Union infantry down from Chattanooga, and the Confederates rode off to the north. "The large force sent from Chattanooga prevented our working at the tunnel," Wheeler reported on the 16th, implying that the blasting powder brought with his column may have been intended for use at Tunnel Hill. That day the Southerners tore up another half-mile of track between Dalton and Chattanooga before they headed northeastward for Cleveland, Tennessee. That was the extent of their railroad-wrecking, a task that evidently did not suit Wheeler's cavalrymen. "Tearing up a railroad is no easy job for a fellow inexperienced in the business," complained a member of the 5th Georgia Cavalry. The Confederates' spotty and minimal railroad damage between Big Shanty and Chattanooga was all repaired by Northern work crews by August 18.[24]

Back at headquarters, General Hood and his chief of staff Shoup strained for information on what Wheeler and his 4,000 troopers were accomplishing. Enemy prisoners and random reports from scouts provided only hearsay, but any scrap of good news was welcome. "The *on dit* throughout the army and city to-day is that a body of our cavalry has reached the enemy's rear, and captured and burned the enemy's stores at Marietta," "Rover" reported for the *Augusta Chronicle & Sentinel*. "The same rumor prevails at headquarters, but

---

[24] Wheeler "Circular," August 9, *OR*, vol. 38, pt. 5, 953; Records Relating to Ordnance Stores Powder Factory, 1862–1865, p. 406 (August 9); Records Relating to Ordnance Stores Powder Factory, 1862–1865; Record of Government Stores Received and Issued at Powder Factory 1862–1865, August 9 ("Shipped to Gen. Wheeler Covington 12½ bbls of Blasting Powder 600 lbs"); E. J. Thomas, "A Raid with Joe Wheeler," 169; Wheeler to Shoup, August 16, *OR*, vol. 38, pt. 5, 967; Steedman report, September 11, *OR*, vol. 38, pt. 2, 495–96; Lawson, *Wheeler's Last Raid*, 130; "Raid of the Rebel Wheeler," *Buffalo (NY) Commercial Advertiser*, August 22 ("the first train from the front, since the rebel General Wheeler's raid, reached Chattanooga on the 18th inst."); Merrill, "Block-houses for Railroad Defense in the Department of the Cumberland," 3:401 ("During this raid he swept along large portion of the railroad from Chattanooga to Atlanta, and tore it up in some places, but destroyed no bridge and attacked no block-houses").

no official information has been received, if such a feat has been accomplished." Nevertheless, Dr. A. J. Foard, medical director for Hood's army, wrote Dr. Samuel H. Stout on the 14th, "affairs are progressing finely here and everybody seems in good spirits. We expect much from Wheeler." "Everything looks brighter and more hopeful than at any time since the siege began," chirped the Press Association's telegraph report on the 15th. General Shoup recorded in his journal on August 16, "Wheeler reported to have torn up ten miles of railroad at Acworth and captured large drove of cattle." The latter part was correct; the former was not. To be sure, Shoup could only record what he and others heard in Atlanta. "Railroad in enemy's rear reported to be badly torn up," the general observed in his journal. On the 18th, Shoup recorded his most upbeat notice: "prisoners report the tunnel blown up by Wheeler, captured Dalton and Resaca, burned Etowah bridge, and are going up the railroad." That day a spy brought word of an alleged statement made by the Yankee general Cox when a citizen petitioned for some food. Cox said no; "I have been living on short rations for seven days, and now your people have torn up our railroad and stolen our beef-cattle, we must live a damned sight shorter."

In the absence of confirmed fact amid the presence of swirled rumor, Southern newspapers as often as not printed the rumor. The *Columbus Sun* printed a letter sent back by one of Wheeler's officers. "The railroad destruction," he wrote, "exceeds everything heretofore done by our troops." A correspondent for the *Atlanta/Macon Intelligencer*, riding with Wheeler, sent back a column dated August 17, in which he described damage done to the Western & Atlantic: "We have torn up the road, burnt the ties, and cut and destroyed the rails so that they cannot be used again, and so that Sherman will not be able to repair it for some weeks."

"S. D. L," a correspondent for the *Sun*, wrote home on the 20th of the numerous "reports" circulating about Wheeler's raid. "The most reliable is that he has captured Dalton, and blown up the Tunnel at that place."

If the Western & Atlantic had been substantially broken, Sherman's troops would soon be running low on supplies. The Press Association "Telegraphic" release of August 16 attributed the remarkable inactivity of the enemy's artillery to "the scarcity of ammunition, caused by the interruption of his communication." "A lady in enemy's lines told one of our scouts that enemy complained of short rations," Shoup wrote on the 17th. "Deserters from the enemy are coming in in large numbers," ran the P. A. "Telegraphic" column for August 20. "They say that the Yankees are living on roasting ears." Confederates eagerly circulated rumors that "the Yankees are starving, that they are living on green corn, that they are deserting by squads of ten and twenty." The *Memphis/Atlanta Appeal* stated that Yankee foraging parties were ranging as far out as Stone Mountain gathering provisions, "giving out as an excuse therefore that their army is on very short rations, and they are obliged to help themselves this way." Optimistic Southerners like Col. Ellison Capers could even predict (as he did to his wife, August 19) that Sherman would retreat. Col. Taylor Beatty of Hood's staff entered into his diary the same day that Wheeler's success "will cause Sherman to fall back." This kind of excitement filtered down to the men in the ranks. "If we can bee successful in getting in there rear and cut off their supplies," wrote one, "it may bee the means of making them fall back from Atlanta."

Col. Moses W. Hannon and his brigade of Alabama cavalry, leading the cattle captured near Calhoun back to Hood's army, returned to it on August 26. They had come on a circuitous route via Athens, harried by some pursuing enemy cavalry, who managed to recapture some of the beeves. But headquarters was delighted to receive a thousand steers to be butchered for the troops ("the cattle are very large, and in good order," reported the *Athens Banner*). Moreover, Colonel Hannon brought the first authentic news from Wheeler, and it was very good. "He spoke in the most hopeful terms of Wheeler's work," reported a correspondent of the *Atlanta/Macon Confederacy*, "and thought his work of destruction to the road was so thorough that it could not be repaired under six weeks or two months." A few days after that, the C.P.A. telegraphic column cited "official dispatches from Gen. Wheeler," to the effect that his raiders had destroyed twenty-five miles of railroad.[25]

---

[25] "Rover," "From the Georgia Front," *Augusta Chronicle & Sentinel*, August 14; A. J. Foard to S. H. Stout, August 14, box 2, folder 16, Stout Papers; "Telegraphic Report of the Press Association" (Atlanta, August 15), *Augusta Chronicle & Sentinel*, August 16; Shoup journal, entries of August 16–18, *OR*, vol. 38, pt. 3, 691; "Wheeler's Campaign," *Columbus Sun*, September 4; "From Sherman's Rear," *Augusta Chronicle & Sentinel*, August 26 (reprinting from *Macon Intelligencer*); "S. D. L.," "Correspondence of the *Columbus Sun*," *Columbus Sun*, August 25; "Telegraphic" (Atlanta, August 16), *Augusta Chronicle & Sentinel*, August 18; "From Atlanta...Yankee Army on Short Rations," *Augusta Chronicle & Sentinel*, August 20; Castel, *Decision in the West*, 484 ("the Yankees are starving"; Colonels Capers and Beatty); "From Stone Mountain," *Mobile Advertiser & Register*, August 30 (quoting "Atlanta Appeal Extra 26"); Longacre, *Cavalry of the Heartland*, 293–94 ("If we can bee successful"); Hood to Seddon, August 26, *OR*, vol. 38, pt. 5, 990 (Hannon's arrival); "Important from Wheeler's Command," *Augusta Constitutionalist*, August 23; "Capture of Yankees, Horses and Cattle," *Augusta Constitutionalist*, August 26, and "Particulars of the Cattle Capture," *Augusta Chronicle & Sentinel*, August 26 (both quoting the *Athens Banner* of August 24); "Sherman's Rear," *Mobile Advertiser & Register*, September 3 (quoting "Macon Confederacy" of August 27 on Hannon); "Further from the Front," *Augusta Chronicle & Sentinel*, August 30.

To their credit, some journalists looked askance at these glowing "reports." The lack of confirming intelligence forced "Rover" to acknowledge, "there are those who begin to doubt the truth of the favorable reports we have had." The *Augusta Chronicle & Sentinel* on August 27 printed a column on what a Yankee prisoner had said; that "Wheeler had tapped [note, *tapped*] the road twice, which was repaired immediately and trains are running through." After earlier and decidedly encouraging columns, the *Columbus Sun* was forced to conclude on the 28th that it had been wrong. "If trains are running through from Chattanooga to Sherman's lines near Atlanta," the editors acknowledged, "it is clear that the Tunnel has not been 'blown up,' and 'our cavalry' has not accomplished much so far." Henry Watterson, writing as "Grape" for the *Augusta Constitutionalist*, also had to acknowledge that "the blowing of the Tunnel was only a partial success. The Yankees repaired it in a very little time after." Given the lack of his meaningful results, the *Sun* expressed resentment that Wheeler was evidently riding off to Cleveland, Tennessee. "Burning Cleveland...may be a very brilliant affair," the editors explained, sagely adding, "but it is not easy to comprehend how the destruction of that town can effect Sherman when his depot of supplies is at Chattanooga and not Cleveland." Sagest of all observers was "Shadow," Henry Watterson, writing for the *Mobile Advertiser & Register*: "Perhaps he regarded the destruction of Cleveland as better game, and did not pause upon the doubtful labor of plucking tracks or trying to cram the tunnel."[26]

The newspapermen were onto something, for in riding off toward Cleveland, Tennessee, Wheeler was neglecting the chief object of the raid—damaging the enemy railroad—in favor of detouring deep into Tennessee. Though Confederates tarried in the area of Tunnel Hill for several days, they made no effort to blow the long Chetoogeta tunnel. The *Chattanooga/Griffin Rebel* criticized this failure, fixing the blame on Maj. Gen. Will Martin, one of Wheeler's division commanders. Then, too, while Hood's orders called for the breaking of the two rail lines running south from Nashville (the Nashville & Chattanooga and the Tennessee & Alabama), Wheeler was actually riding

---

[26] "Rover," "From the Georgia Front" (Atlanta, August 25), *Augusta Chronicle & Sentinel*, August 30; "From Atlanta. New Movements of the Enemy. Statement of Deserters," *Augusta Chronicle & Sentinel*, August 27; "Grape," "Letter from Atlanta" (Atlanta, August 26), *Augusta Constitutionalist*, August 30; "If trains...," untitled editorial, *Columbus Sun*, August 28; "Shadow," "Our Army Correspondence. Letter from Atlanta" (August 26), *Mobile Advertiser & Register*, August 30.

J. Cutler Andrews guessed that Henry Watterson was the veteran Southern journalist writing as "Shadow" (*South Reports*, 574). Watterson's identity was confirmed in Ellis and Dick, "'Who Was "Shadow"?,'" 947–62.

away from those lines. Indeed, in the next week, Wheeler led his column all the way beyond Knoxville before turning west. When he finally struck the Nashville & Chattanooga, it was August 31—the fateful day on which Federal infantry fell on Hood's own railroad to Macon and compelled the Confederates' evacuation of Atlanta within twenty-four hours. Wheeler may have attempted to justify his long ride as a recruiting jaunt—he issued a proclamation on August 30, urging Tennesseans to "rise to duty" and "join us"—but it would have been a thin, reedy excuse. Longacre is correct in calling Wheeler's wide-ranging route "a major deviation of his orders."

Wheeler compounded his dereliction of duty by lying about it. General Shoup recorded on August 28 that dispatches dated August 19 had been received from Wheeler. "He reports having captured Dalton and a lot of supplies, 300 fine mules, and destroyed 35 miles of railroad with the loss of only 30 men since his departure from this place," Shoup wrote. Hinting his incredulity at such boastfulness (thirty-five miles of track?), the general added, "on the whole the reports of his operations are very encouraging."[27]

Sherman had predicted for a good while that the Rebels would attack his supply lines. "I suspect Hood...will...endeavor to operate on our road with cavalry," he had predicted to Thomas on July 22, just a few days after Hood had acceded to army command. "No doubt he expects to cut to my rear," Cump had written Grant on the 25th. On August 1, Sherman had sent as far away as Nashville his calls for additional cavalry, "as the enemy will surely be on our railroad very soon." Sherman's apprehensions were confirmed when Garrard on August 11 sent a message that "a large part of their cavalry is now in the neighborhood of Covington, and it is generally believed that they will soon make a grand raid to Tennessee or Kentucky." Sherman did not panic at the news, for he had made his preparations. First, he had accumulated tons of rations at Allatoona; wagons could bring forth supplies from there if needed. To guard the stretch of track south of Chattanooga, in mid-June Sherman had put General Steedman in charge of a "District of the Etowah," with

---

[27] Longacre, *Cavalry of the Heartland*, 296 ("Wheeler lingered near Tunnel Hill, resting, for the next four days"); Lawson, *Wheeler's Last Raid*, 150; Wheeler to "TENNESSEANS," August 30, *OR*, vol. 38, pt. 5, 1004; Shoup journal (entry of August 28), *OR*, vol. 38, pt. 3, 694.

Edward G. Longacre offers a sensible hypothesis on why Wheeler, deviating from Hood's orders, strayed into east Tennessee. "Perhaps his recent, against-all-odds victory had given him an overly optimistic opinion of his troopers' capabilities," he writes. "Perhaps, too, his successful pursuit of McCook, during which his decisions, not his superior's, had won the day made him believe he had earned the right to substitute his own strategy for Hood's, revising his timetable and itinerary based on his own instincts" (Longacre, *A Soldier to the Last*, 169).

enough troops to parry raiders and deal with guerrillas. If the enemy made any breaks in the line, William W. Wright's Construction Corps assured Sherman that timely repairs would be made. Farther to the rear were big stores of supplies at Chattanooga, Bridgeport, and Decatur, Alabama; enemy cuts of the track north of these places to Nashville would not matter.

More important tactically, Sherman had ordered the construction of strong log-and-dirt blockhouses at bridges and other critical points of the railroads leading back to Nashville; twenty-two of them had been built south of Chattanooga alone. These little forts, constructed during spring 1864, were designed to withstand artillery bombardment and Rebels' calls to surrender for at least seven days, by which time reinforcements would arrive and drive off the enemy raiders. Indeed, Sherman called for blockhouse commanders "to fight with desperation to the last."

This was what the Union commander at Dalton, Col. Bernard Laiboldt, had done when Wheeler's forces rushed in on August 14, calling for surrender. Laiboldt refused, withdrew his troops into an earthen fort on a hill overlooking the town, and telegraphed General Steedman for help. Steedman's relief column arrived and drove Wheeler off into east Tennessee (and irrelevance, at least in the rest of the Atlanta Campaign).[28]

Now is the time to consider the following: "Nothing but the natural and intense desire to accomplish an end so inviting to one's feelings would have drawn me to commit a military mistake at such a crisis, as that of dividing and risking my cavalry, so necessary to the success of my campaign."

This would have been a perfect *mea culpa* from John Bell Hood, acknowledging his "military mistake…of dividing and risking my cavalry" when he sent Wheeler raiding. The problem is that this was written by *William Tecumseh Sherman*, concerning his decision to send McCook and Stoneman off raiding (the latter to Andersonville). Given all the historiographical polemics against Hood's "military mistake" in handling his cavalry, one wishes future

---

[28] Sherman to Thomas, July 22 (11 A.M.), *OR*, vol. 38, pt. 5, 223; Sherman to Grant, July 25, *OR*, vol. 38, pt. 5, 247; Sherman to Webster, August 1, *OR*, vol. 38, pt. 5, 320–21; D. S. Stanley to Sherman, August 11, *OR*, vol. 38, pt. 5, 457 (quoting report from Garrard on "grand raid"); Major, *Supply of Sherman's Army during the Atlanta Campaign*, 13; Mruck, "The Role of Railroads in the Atlanta Campaign," 267 ("large supply center in Allatoona"); Clauss, "Sherman's Rail Support in the Atlanta Campaign," 414–15; Kime, "Sherman's Gordian Knot," 104 ("numerous blockhouses"); S. Davis, *Atlanta Will Fall*, 21–22; Dyer, *"Fightin' Joe" Wheeler*, 190.

students would pay as much attention to *Sherman's* errors, which he candidly admitted, in handling his mounted troops.[29]

Sherman was delighted that Wheeler had ridden off. "Wheeler...may hurt some of the minor points, but, on the whole, East Tennessee is a good place for him to break down his horses, and a poor place to steal new ones." The news caused him to ponder a tactical ploy he might not otherwise have: yet another cavalry raid on the Macon & Western Railroad. After Garrard on the left and Brig. Gen Judson Kilpatrick (another cavalry division leader) on the right reported on August 16 only weak resistance from enemy mounted forces, Sherman wired Thomas that Kilpatrick, "with his own and Garrard's cavalry, could ride right round Atlanta and mash the Macon road all to pieces." "Wheeler is out of the way," he added, "and when shall we use cavalry, if not now?" That very day Kilpatrick had ridden to Fairburn and torn up some track before returning without loss. On the 16th Sherman had drawn up orders for a major new maneuver—nothing less than "the movement of the army against the Macon railroad" to begin two days later—but he suspended the movement and ordered Kilpatrick out. Sherman gave him the assignment, as he thought Garrard was not daring enough ("if he can see a horseman in the distance with a spyglass he will turn back"). Kilpatrick eagerly accepted the task, promising to knock out the Macon road for two weeks' time. Sherman committed to him as much cavalry as he could, directing him to head for Jonesboro and destroy as much track as possible.

Kilpatrick set out late on August 18 from Sandtown, southwest of Atlanta, with 4,700 riders (more than Wheeler had taken with him). Sul Ross's Texas brigade, numbering merely 400, contested the Yankees' advance and kept General Jackson apprised of it. Hood sent more cavalry to help, as well as Reynolds's infantry brigade traveling by train to Jonesboro, reinforcing Armstrong's cavalry brigade already there. He also wired post commanders along the rail line to be alert.

Unfortunately, miscommunication sent Armstrong and Reynolds off toward Lovejoy's Station, seven miles south of Jonesboro. On the afternoon of August 19, Kilpatrick's troopers thus rode into Jonesboro without a fight and began to tear up track. A rainstorm prevented the fires they would have used to heat and bend the rails; the Northerners could only pry them up and throw them aside. After five hours' work (5 to 10 P.M.), Kilpatrick ordered his men to mount up and head back. Near Lovejoy's they encountered Sul Ross's 400

[29] Sherman to Halleck, August 7, *OR*, vol. 38, pt. 5, 409; Mathews, *The McCook-Stoneman Raid*, 141 ("military mistake").

troopers and with upraised sabers literally ran over and through them. Kilpatrick's force then made its way back to Sherman's lines, which it entered at Decatur on the 22nd. Kilpatrick reported 237 casualties; Confederates lost about the same.

The key lay in how much track the Federals had damaged. Kilpatrick told Sherman he had destroyed "enough to disable the road for ten days." A Federal cavalryman thought it was about a mile; David Evans, historian of the topic, concludes about two. In either case, on August 20 Hood called Lieutenant Colonel Presstman to his headquarters and instructed him to take a brigade to Jonesboro to oversee repair of the road. The Confederates had all of Kilpatrick's damage restored by August 21, two days after the Yankees inflicted it, and even before they returned to Federal lines. Sherman soon learned of the raid's failure. Lieutenant Edge sent in word from his buzzard's roost at the Decatur road on the 23d that at 7:30 A.M. "two engines, with some cars, passed into Atlanta."

Sherman has not been criticized for sending his cavalry out one last time to try to break the Macon railroad, just as Hood should not be criticized for sending his cavalry off to attempt to break the rail line to Chattanooga. Both generals belatedly realized that mounted troops alone could not inflict enough damage to knock a railway out more than a couple of days. "I now became satisfied," Sherman recalled in his memoirs, "that cavalry could not, or would not, make a sufficient lodgment on the railroad below Atlanta." Hood followed suit in his reminiscences: "I thereupon became convinced, and expressed the opinion in my official report, that no sufficiently effective number of cavalry could be assembled in the Confederacy to interrupt the enemy's line of supplies to an extent to compel him to retreat." Hood, however, was not correct in part of his statement, for his report of February 15, 1865, does not address the limited capacities of cavalry. Instead, Hood claims that in his raid Wheeler "succeeded in partially interrupting the enemy's communications by railroad." Perhaps he was addressing Pollard and other critics and chose to overstate the effect of the raid. Hood furthermore declined to criticize Wheeler for his divergence from orders even though he states in his report, "I intended General Wheeler should operate...south of Chattanooga." Given Hood's heavy criticism of General Hardee for his performance in the battles of July 20 and 22, one can only wonder why the commanding general absolved his cavalry commander from blame for his erratic performance, given the high hopes which the Confederates had placed on a mounted raid against Sherman's railroad communications, and the very poor results Wheeler achieved in what would be the army's last cavalry raid of the war.[30]

While Wheeler and Kilpatrick rode off in their opposite directions, infantry held the critical lines southwest of Atlanta. After repulsing the enemy assault at Utoy Creek, the Confederates on their "railway defense line" settled into trench warfare of a rather stagnant, sunbaked sort. General Shoup's journal entries reflect this torpor.

> August 6. Enemy still moving toward our left. They have evidently abandoned their intention of direct attack, and propose the old game of flanking.
> August 7. Enemy still extending their line and massing on our left.
> August 8. Enemy still continue to move still nearer the West Point railroad on their right. They are now within three miles of the road.
> August 9. The enemy still continue to move slowly to our left.
> August 10. Our lines remain the same as yesterday.
> August 11. Our lines remain the same as yesterday.
> August 12. No change in our lines, excepting Vaughan's brigade moved from our right to extreme left.
> [On the 12th Hood wired General Maury at Meridian: "No news here Enemy comparatively quiet—all confident"]
> August 13. No change in our lines to-day.

---

[30] Liddell Hart, *Sherman*, 296; Castel, *Decision in the West*, 469–74; Sherman to Thomas, August 16, *OR*, vol. 38, pt. 5, 526; L. M. Dayton Special Field Orders No. 57, August 16, *OR*, vol. 38, pt. 5, 546; Hibbard and Castel, "Kilpatrick's Jonesboro Raid, August 18–22, 1864," 32–33; Ross to Jackson, August 16 (9:30 P.M.), *OR*, vol. 38, pt. 5, 968 (Kilpatrick's ride to Fairburn); D. Evans, *Sherman's Horsemen*, 430, 449–50; Curry, "Raid of the Union Cavalry," 607; [Hood] to Presstman and to Jackson, August 20, Telegram Book No. III, Hood Papers; S. Martin, "Kill-Cavalry," 56 ("Rebels had undone Kilpatrick's work in just 24 hours"); Edge to Logan, August 23, *OR*, vol. 38, pt. 5, 642; Sherman, *Memoirs*, 2:98; Hood, *A & R*, 198, 200; Hood report, February 15, 1865, *OR*, vol. 38, pt. 3, 632; Longacre, *Soldier to the Last*, 170 (map, "Wheeler's Atlanta Campaign Raid August 10–September 17, 1864").

After the battle of Lovejoy's Station, General Ross reported to Hood, "General, I got my brigade run over yesterday." Hood is said to have responded, "General Ross, you have lost nothing by that, sir. If others, who should have been there, had been near enough to…be run over, your men would not have been run over" (Crabb, *All Afire to Fight*, 249).

Shelby Foote quotes both Hood and Sherman on their belated recognition that cavalry alone could not achieve strategic results—that is, by striking railroads, cavalry raids could force an enemy army to retreat: "no sufficiently effective number of cavalry could be assembled in the Confederacy to interrupt the enemy's line of supplies to an extent to compel him to retreat" (Hood); "cavalry could not or would not work hard enough to disable a railroad properly" (Sherman) (Foote, *Civil War*, 3:520–21).

August 14. No change in our lines since yesterday.
August 15. No change has occurred in position of our lines.
August 16. Everything has been remarkably quiet to-day.
[Hood to Seddon, August 16: "No change in affairs in the past few days."]
August 17. No change in our lines to-day.
August 18. No change in our lines to-day.
August 19. All quiet along our lines.
August 20. No change in our lines to-day; all quiet along our lines.
August 21. All quiet along our lines.
August 22. Along our lines everything is quiet.
August 23. All quiet along our lines.
August 24. All quiet along our lines.
August 25. No important change has taken place on our left.

Soldiers took advantage of the lull in hostile operations by actually fraternizing with the enemy while on picket duty. Along the line held by his regiment, Illinoisan John Ferguson recorded in his diary that on the night of August 11, "our boys got to hollowing over to them and the two parties for the first time at this place, quit shooting and entered into conversation. They agreed to quit shooting and be more friendly while neather party were advancing." News of the men's spirit of friendliness made its way to the top, with the predictable result. "It has been reported to these headquarters," read the circular from S. D. Lee, "that, contrary to orders, intercourse between our pickets and those of the enemy is still kept up, and in some instances it has been agreed that they shall not fire at each other with intent to kill, but to shoot over each others' heads." Such practice was ordered to stop. Within a few days Ferguson noticed the change in behavior. "The Rebs would not talk to us today," he wrote. "Their officers have strickly forbiden it."

Shoup's observation, "everything quiet along the lines," in his journal on August 10 did not mean that there was no exchange of musketry. That day General "Old Grits" Bate took a rifle ball in the knee; a "chance shot" Shoup called it. Bate would be out of service for two months. Hood assigned John C. Brown to command the division. Casualties in the ranks also accumulated. "It is a picket fight all the time," wrote Mississippian William B. Honnell; "we loose some men every day in our brigade eather killed or wounded." John Young, Co. E, 15th Mississippi, was one such casualty. "Your noble and brave brother John was killed yesterday morning," wrote a relative to Young's sister on August 15. "John was shot by a Minnie ball over the right temple. We were in the act of preparing breakfast when the fatal shot struck him." It was dangerous on the other side as well. "We have splendid works & while I write the musket balls are whisteling over our heads," wrote a soldier in Walcutt's brigade, XV Corps, on August 6; "I tell you we have too keep low for the sharp shooters have good range on our works." And "everything quiet" also did not mean there was no exchange of cannonfire. A Texan wrote his parents on August 9, "the Bum Shells and Minnie Balls are constantly flying over us but the Boys are getting sorty ust to them but once and a while one gets hit." On the 11th Shoup wrote S. D. Lee that General Hood advised that traverses be dug every ten paces in his trench lines to better protect the men from enemy artillery.

Illness as well as injuries among higher-ups caused Hood and Shoup to shuffle general officers around. When Brig. Gen. James Cantey fell sick, Hood recommended Col. Charles M. Shelley of the 30th Alabama to succeed him; Shelley was promoted to brigadier a month later. There were other vacancies. Hood asked Richmond to promote Col. Joseph B. Palmer of the 18th Tennessee so he could assume command of a brigade (this was not done until after the campaign). Hood even asked Brig. Gen. Marcus J. Wright, overseeing the Macon Arsenal, if he could take charge of a brigade in the field (Wright could not).

High-ranking casualties occurred on the other side as well. General Dodge suffered a serious head injury on August 19. He went back to Indiana; Brigadier Thomas E. G. Ransom took command of the XVI Corps. Dodge never fully recovered from cranial and neurologic injury and did not return to Sherman's army.

Despite the enemy's comparative inactivity, Hood remained vigilant. Kilpatrick's short sortie to Fairburn August 16 stirred at headquarters new concerns of a Union cavalry raid. By now General Stoneman would have told his captors of his plans to free the Northern prisoners; worries about another such effort were doubtless behind Hood's suggestion to Bragg that "a part of the prisoners of war now at Andersonville" be moved elsewhere. Hood expressed concern for their protection, and on August 17 inquired of General Winder at Andersonville how the fortifications (begun at the time of Rousseau's raid) were progressing. Nevertheless, no action was taken to move the POWs until after the fall of Atlanta.

Reflecting his satisfaction with the status quo, General Hood even agreed to sending some of his troops to another threatened theater. After the battle of Mobile Bay in early August, Federal land and sea forces hovered near Mobile, threatening to attack and capture the city. Maj. Gen. Dabney Maury, commanding the defenses, pleaded with General Bragg to send veteran reinforcements to augment his militia. On August 24 Maury wired Hood directly: "Expect early attack on city.... Can you spare me the infantry?" Hood answered promptly, "Send you a brigade, though I cannot spare it. It goes at daylight." Hood was as good as his word. On the 25th Alpheus Baker's brigade of four Alabama regiments boarded the train, some 700 men. They reached Mobile three days later and were sent into the works protecting the city.

Meanwhile the men of the Army of Tennessee expressed confidence in their commander. Thomas B. Hampton, officer in the 63rd Virginia, wrote home on August 10. "Gen Hood has checked the forward movements of the Enemy & if the Enemy does not look sharp," he predicted, Hood "will give him a complete Flogging as the Troops are confident & determined." "The Boys are all well and in fine spirits," wrote William M. Bowden of the 5th Tennessee to his sister on August 24 (before adding darkly, "what few of them is left"). Bowden explained, "We are in full confidence of Driveing the Yankees out of Gorgia or at lest I dont think Sherman will go any furty in Gorgia." A few days later Alabama Sergeant Joel Murphree put the situation in perspective. "The change of Commanders no doubt caused the death or capture of those near and dear to us but I do believe it was the best for the success of our cause. It required hard fighting to check the enemy here after having pursued us so far and I have no idea Johnston would have made a stand at Atlanta."

Those from afar could take a different view of things. One such was Maj. Thomas J. Goree, General Longstreet's A. D. C. "I do not feel the same confidence about Atlanta that I do about Richmond & Petersburg," he confided to his mother from Lynchburg on August 26. "Johnston was removed because his policy was to give it up rather than run too great a risk in holding it. Hood was appointed because his policy was to hold it at all hazards." Goree agreed with Johnston's policy of giving up ground while holding his army intact. "By this means he was wearing away Sherman's army without a corresponding loss on his own." For this he had been relieved. "Well, Hood takes command and attacks the enemy in his entrenchments," Goree added. "He loses 8 or 10,000 men and inflicts a loss upon them of about 5,000. He captures a few stands of colors & pieces of artillery. He whips them on one part of the field and is whipped on another. What is gained? Nothing, but ground is lost. So much for a change of policy & commanders."

Goree feared that if Hood failed at Atlanta, "the blow will come nearer prostrating the Confederacy than any she has yet recd."

To forestall that blow, Hood kept his troops busy improving their earthworks, which were already strong. "Working parties every night to work on forts and reserve lines," observed Brig. Gen. Daniel Reynolds in his diary, August 8. The men were digging traverses behind their lines to protect themselves against enfilade cannonading, Reynolds recorded. But they were also working in front of their parapets. "About 50 yards in front of our breastwork we have a line of stakes drove down in the ground, firmly slanting from us," recorded Texas captain Samuel T. Foster in his diary on August 13, "and made sharp at the top and so close together that a man can not get thru, and high enough to strike him [a] little above his waist." Such obstacles would make an enemy infantry attack "very difficult," the captain believed. The Confederates also continued to entrench. "All night long the sound of shovel and pick might have been heard in the trenches," Lieutenant Colonel Frobel, the engineer, entered in his diary on August 21. The next morning General Hood rode up to inspect the work. "After riding along the greater portion of the defenses, examining each point with a critical eye," Frobel recalled, "he finally bade us good-morning." The next day an entire brigade arrived to assist in the labor. "Orders have been sent down from headquarters to push the work night and day until completed," the engineer noted. Hood returned to the site on August 23, "and personally urged that all diligence should be used."

Digging in took place not just in the Grant-Presstman works around the city, but along the railway defense line to East Point as well. "We are now at East Pt where we have been building forts and fortifying generally," recorded Alabama engineer Hiram Williams in his journal on August 19. A few days later he was near Utoy Creek, "ordered to make a lot of cheveau-de-frize's for the protection of our line." As if explaining things for posterity (which he probably was in his journal), Williams explained, "they are made something like a horse rack, consequently the boys have christened them by that name. Worked hard at it all day." "Still at work on 'racks,'" he wrote on August 24.

On the 20th, engineer Frobel was conferring with General G. W. Smith about where to site a new fort in the Confederates' perimeter of works. A civilian came up and tried to talk the officers out of using the site under consideration, which was his land and residence. "It is all I have in the world," he pleaded. General Smith sat down with the man and calmly explained the military necessity, whereupon the property owner walked home. Several hours later, when work crews arrived to start cutting down the house, as Frobel recorded in his diary, "I saw the man, accompanied by his wife, carrying away an old straw bed, together with some few pieces of broken and nearly worn out furniture." To Frobel he explained that with his few possessions out, the soldiers "could now proceed to tear down his nice little cottage and that he sincerely hoped the fort we were building would enable us to whip the Yankees."

Federal signal officers watched the Rebels further fortifying their northwest salient, east of the W. & A. Railroad. "The enemy are busy strengthening their works and building new ones in front of the Twentieth Corps," Lieutenant Edge wrote on August 22. The Confederates were building new artillery forts on the other side of the railroad as well. Edge mentioned a "new casemate battery...south of big gun" in front of the XVI Corps' left. The Southerners' "big gun" was a nine-inch Dahlgren cannon placed near Fort Hood. Firing huge sixty-four pound shells, it was called the "Growler" by the Federals, who by their own artillery fire were able to dismount it sometime during August. After that, the Confederates' heaviest artillery consisted of thirty-two pounders, whose activity General Hood himself monitored. Four of them were on French's front in the works west of the city, and their activity drew much Northern counterbattery fire. "The Yankees must be angry," French recorded in his diary on August 18, estimating that some two thousand shells were fired at his lines that day. Some, like those of Ransom's division artillery (XVI Corps), facing French, hit their targets. "Genl. Ransom with his usual energy has planted a battery...right in the teeth of the enemy," wrote Illinois Maj. Charles Loop to his wife on August 12; "I saw one of the enemy's guns thrown high into air from his shots."[31]

---

[31] Shoup journal, entries of August 6–25, *OR*, vol. 38, pt. 3, 690–93; Ferguson, *On to Atlanta*, 74, 76; General Orders No. 63 (Lee's Corps), August 12, *OR*, vol. 38, pt. 5, 960–61; Kelly, "Back in the Saddle," 32 (Bate's wound); Hood to Maury, August 12, [Hood] to Wright, August 10 (8:30 P.M.), [Hood] to Governor Brown, August 16 (2 P.M.) ("the Enemy seems to be threatening a raid on our left"); [Hood] to Bragg, August 16, [Hood] to Col. George Rains, July 27 ("Can you send us three hundred [300] shells for rifle thirty two pounder [32 pr]?"), all in Telegram Book No. III, Hood Papers; Castel, *Decision in the West*, 482–83 (Honnell); Wynne, *A Hard Trip*, 126; Hood to Seddon, August 16, *OR*, vol. 38, pt. 5, 966; J. Welsh, *Medical Histories of Confederate Generals*, 15–16 (Bate returned October 10); Marshall, *A Life of William B. Bate*, 136 (borne to the army hospital in Barnesville GA); Hood to Cooper, August 11 and 12, *OR*, vol. 38, pt. 5, 956–57; Orendorff, *We Are Sherman's Men*, 98–99; Carlock and Owens, *Tenth Texas Cavalry*, 160 (Henry Watson to parents, August 9, "Bum Shells"); Shoup to Lee, August 11, *OR*, vol. 38, pt. 5, 956; Hood to Seddon, August 14 (5:30 P.M.); Ransom report, September 14, *OR*, vol. 38, pt. 3, 391; J. Welsh, *Medical Histories of Union Generals*, 100–101; Futch, *History of Andersonville Prison*, 83–86, 95–96; Bergeron, *Confederate Mobile*, 151–52, 158; Maury to Hood and Hood to Maury, August 24, *OR*, vol. 39, pt. 2, 796; Mathis, *In the Land of the Living*, 168–69; Frank Gardner to Samuel Cooper, August 29, *OR*, vol. 39, pt. 2, 803 ("about 700 men"); Phillips, *Diehard Rebels*, 108–109 (Hampton); Daniel, *Soldiering*, 146 (Bowden); McMurry, "Confederate Morale," 237; Cutrer, *Longstreet's Aide*, 134–35; Reynolds, *Worthy of the Cause*, 141–42 (diary entries, August 8, 12); Foster, *One of*

Meanwhile General Sherman planned his last resort. Even before he sent Kilpatrick out on one last equestrian effort to break the Macon & Western,

---

Cleburne's Command, 121 (Foster diary, August 13); Lane, *People's Principles*, 200–201 (Frobel's diary); H. S. Williams, *This War So Horrible*, 108–109; Lt. Ch. H. Fish to Edge, August 22, *OR*, vol. 38, pt. 5, 631; S. Davis, *What the Yankees Did to Us*, 197–98 and n.70 (including newspaper report, "near the state road we have planted...a 64 pounder"); S. G. French, *Two Wars*, 221; Loop, "'I Was There,'" 60.

More than one Yankee succumbed to sixty-four-pounderitis—the belief that the Rebels' big Dahlgren gun was in their front. In August 1864, the 7th Pennsylvania Cavalry held the far left of the Federal line, near the Augustus Hurt house. Its colonel wrote of his regiment "being almost constantly under fire from a Confederate battery of sixty-four pound siege guns" (Herberich, *Masters of the Field*, 195).

Lieutenant Colonel Frobel's construction of a new Confederate fort in late August raises the question: how many such artillery positions ringed Atlanta by the time of the city's fall? Southerners made no maps of the city's defenses after mid-April 1864. During the occupation, Captain Poe, Sherman's engineer, surveyed the Rebel works and mapped them to include the forts constructed in the northwest and northeast salient ("Map Illustrating the Siege of Atlanta," in Cowles, *OR Atlas*, plate 88, no. 1). In a recent paperback on Atlanta's fortifications, Lawrence Krumenaker counts thirty-six artillery forts on Poe's map (*Walking the Line*, 5, 39).

Another source needs to be cited: the map of Robert K. Sneden (1832–1898). Sneden joined the 40th New York Infantry, serving in Virginia as mapmaker for the Army of the Potomac's III Corps. He was captured in November 1863, and confined at Andersonville and Camp Lawton until exchanged in December 1864.

After the war, based on maps and sketches he had made, Sneden painted hundreds of vivid watercolors depicting war scenes. He never found a purchaser of them and died a pauper. In 1994, Sneden's works were found in a Connecticut bank vault and were acquired by the Virginia Historical Society. Much of Sneden's remarkable work has been published by in Bryan and Lankford, *Eye of the Storm*, and in Bryan, Kelly, and Lankford, *Images from the Storm*.

Sneden painted many maps based on those he had seen during the war. Afterward, he painted other maps of sites he had never visited, drawing on army maps or surveys he somehow got access to. Such is the case of "Atlanta. Georgia and Its Rebel Defences." This colorful map is in neither of the aforementioned books but is in *America's Civil War* 23/4 (September 2010): 31. What catches the eye are Sneden's notations, "Abatis" outside the northwest salient and "slashing" outside the Confederate perimeter's eastern side. How did Sneden know this if he never visited Atlanta? Answer lies in Sneden's subtitle for this map, "Copy from a Survey made by the Engineers of the 20th Corps U.S.A. after surrender—Septr 1st 1864." In other words, Federal engineers during the XX Corps' occupation of the city, September–November 1864, had the opportunity to study the Rebel fortifications and chart them. Just a few months after the fighting, "abatis" and "slashing" would still have been visible.

How Sneden came upon a XX Corps engineers' survey of Atlanta's works after the war, when he painted "Atlanta. Georgia and Its Rebel Defences," remains to be seen. But given the details in his watercolor map, he clearly had resort to a source which otherwise is lost today. For instance, he includes the number of Rebel cannon which had been positioned in the forts and along the parapets. This he could only have done by following Federal engineers' count of the individual earthen embrasures. Counting them leads to around 130 pieces of artillery—well within the number of guns possessed by the Army of Tennessee in mid-1864.

the Federal commander laid out his intentions: leading almost all of his infantry on a wide-swinging march from his right flank, heading southeast to strike the railroad below East Point. He explained to General Halleck that he would leave one corps at the Chattahoochee to guard the army group's wagon trains and with the rest, "60,000 men, reduced to fighting trim," he would "try the Macon road, without which Hood cannot feed his army." This bold move, he reckoned, "will bring things to a crisis." The day of Kilpatrick's return, August 22, Sherman wired Halleck that he would launch his infantry raid "to make the matter certain." Orders for "the grand movement by the right flank," as he called it, were issued the next day.[32]

The Federal maneuver began during the night of August 25–26. The XX Corps withdrew to the Chattahoochee while the IV, on Thomas's left, started marching westward. The XIV, XV, XVI, and XVII pulled back the next night from their positions west of the city and headed south. The XXIII Corps was last to set out, at noon on the 28th.

Sherman's "movement round Atlanta by the south"—another of his phrases for the maneuver—began without a hitch. Even though Howard's artillery west of the city fired during the night, some Confederates could hear rumbling of artillery and wagons behind the enemy lines. A picket of Brig. Gen. Joseph Palmer's Tennessee brigade, posted near Peachtree Street, reported unusual noises from the Federal lines during the night; Palmer ordered further scouting and thus learned of the disappearance of the enemy in his front. When pickets on the morning of the 26th confirmed that the Yankee trenches north of the city were empty, Hood and Shoup jumped to the conclusion that perhaps Wheeler's railroad damage may have had its hoped-for impact, forcing Sherman to fall back. "General Hood and his Chief of Staff 'Shoup' are in high glee," observed Lieutenant Trask; "they think that he is now preparing to retreat." According to Hardee, Hood even issued orders to Red Jackson to harass the retreating enemy's rear. The fact that Union artillery ceased to fire into the city on the 26th seemed to confirm the conclusion. Later that morning, however, headquarters had to conclude that their initial assumption had been too hasty. "We learn that Wheeler has gone into Middle Tennessee without doing Sherman's rear much harm," Trask wrote, "So his raid, too, has turned out a failure." As proof, Southern soldiers roaming about the Yankees' abandoned campgrounds found not only plenty of food left behind, but tasty delectables to which they greedily helped themselves. "To-day

---

[32] Sherman to Halleck, August 13; to Brig. Gen. William Vandever, August 15; and to Brig. Gen. John E. Smith, August 16, *OR*, vol. 38, pt. 5, 482, 517, 540; Sherman to Halleck, August 22, *OR*, vol. 38, pt. 5, 628; Sherman report, September 15, *OR*, vol. 38, pt. 1, 79.

our troops are feasting on sardines and lobsters, canned fruit of every kind, candies and raisins," Trask gleefully recorded, "besides many other good things their stomachs had long been strangers to."[33]

Hood and Shoup sifted through all the reports being sent in by infantry pickets, cavalry vedettes, or anyone observant and trustworthy enough to be termed a "scout." By 8 A.M. they knew that there were no enemy troops north of the city and east of the W. & A.; Stevenson's pickets occupied the abandoned trenches. But the Yankees were still in force south of the railroad, to the west of the city. Within an hour Hood directed Jackson to send out cavalry on both flanks of the army "to ascertain what is going on." Cleburne, on the extreme left, was instructed to "push out your scouts and ascertain what the enemy is doing." By the end of the day Hood could report to Secretary Seddon, "His left now rests on the Dalton railroad. He has not extended his right at all." Stevenson's and Maney's divisions, holding the works facing the now-abandoned enemy position, were alerted for possible movement as the situation developed. Smith's militia would fill in the positions they vacated.[34]

On the morning of August 27, Hood knew that the situation was shifting. Pickets brought word that the Federals during the night had disappeared from in front of Stewart's and Lee's Corps, but they were still in force southwest of it, toward East Point. French sent two brigades with artillery toward the Chattahoochee; they found a strong enemy bridgehead at Turner's Ferry, three miles downriver from the railroad. Prisoners said they belonged to the XX Corps; this information was also forwarded to HQ.

That afternoon Sul Ross's cavalry brought in the news that enemy infantry columns were marching south from Sandtown and Mt. Gilead. Frank Armstrong's cavalry was directed to oppose the enemy advance. Midafternoon Ross reported enemy infantry had reached Camp Creek, seven miles southwest of Atlanta, and were digging in. Jackson was told to learn all he could

---

[33] Castel, *Decision in the West*, 485–86; S. Davis, *Atlanta Will Fall*, 175; S. Davis, *What the Yankees Did to Us*, 225 (night of August 25 as last shelling of city); Anderson report, February 9, 1865, *OR*, vol. 38, pt. 3, 771; ("rumbling of artillery, wagons, &c."); Neff, *Tennessee's Battered Brigadier*, 102–103; Trask diary, entry of August 26; Hardee report, April 5, 1865, *OR*, vol. 38, pt. 3, 700 ("General Hood believed the enemy to be retreating for want of supplies").

[34] Shoup to Stewart, August 26 (8 A.M.), *OR*, vol. 38, pt. 5, 990; Shoup to Maney, August 26 (7:45 A.M.), *OR*, vol. 38, pt. 5, 992; [Shoup] to Ferguson, August 26 (8:15 A.M.), and to Jackson, August 26 (9 A.M.), *OR*, vol. 38, pt. 5, 992; [Hood] to Cleburne, August 26 (9 A.M.), Telegram Book No. III, Hood Papers; Evans, "Atlanta Campaign," 58 ("Hood ordered Jackson's cavalry to locate the Federal army"); Hood to Seddon, August 26, *OR*, vol. 38, pt. 5, 990; [Shoup] to Stewart, August 26 (3 P.M.), *OR*, vol. 38, pt. 5, 990; [Shoup] to Maney, August 26 (7:45 A.M.) and to Smith (3 P.M.), *OR*, vol. 38, pt. 5, 992.

about the enemy infantry forces and to ride to Hood's headquarters that evening to discuss the situation. On the night of August 27, Hood wired Richmond that the enemy was moving troops toward the right, but he could not offer details. "The exact intention of the enemy has not yet been ascertained," recorded General Shoup in his journal that night. He added, "the general commanding has disposed of his troops so as to be prepared for any emergency."

Even the press picked up on the situation. "The enemy disappeared from our entire front last night, except on the extreme left," the Press Association's "Telegraphic" column announced on the 27th, "and it is supposed they are moving upon our left." The *Appeal Extra* of that date (probably written by John Dumble, the editor who stayed in Atlanta after his big press had been moved to Montgomery) stated that only a few folk yet entertained the hope that the Yankees' movement meant a retreat: "The probability is that the troops thus withdrawn…will be found extending their lines to their right, in the direction of the West Point and Macon railroads."[35]

At this point, Hood knew that Sherman's forces had stretched themselves a dozen miles from their Chattahoochee bridgehead toward the southwest. Camp Creek, where Ross reported them entrenching, is well to the west of East Point. Trimble's mill, where Hood instructed Jackson to prevent the enemy's creek-crossing, is farther southwest still. It thus appeared that Sherman was eyeing a new target. East Point had seemed to have been his objective thus far. Cleburne commanded at that place, and at noon of the 28th Brown's (Bate's) division was also ordered there—but the Yankees were clearly marching beyond East Point. Armstrong and Ross were contesting their advance and reported when they reached the Atlanta-West Point railroad ("in quite a large force, consisting of cavalry, artillery, and infantry," Shoup noted). As precaution, most of Brown's division was sent to Rough and Ready (three and a half miles south of East Point). Daniel Reynolds's and Joseph Lewis's brigades were ordered to Jonesboro late on the 28th. That night Hannon's cavalry regiment was sent there as well.

---

[35] [Hood] to Cleburne, August 27 (dawn), *OR*, vol. 38, pt. 5, 994; Shoup to Lee, August 27 (7:30 A.M.), *OR*, vol. 38, pt. 5, 993 ("Hardee reports the enemy still in his front"); S. G. French, *Two Wars*, 221; Ross to Jackson, August 27 (12 m.), *OR*, vol. 38, pt. 5, 995; Ross to Armstrong, August 27 (3 P.M.), *OR*, vol. 38, pt. 5, 995; [Hood] to Jackson, August 27 (4:30 P.M.), *OR*, vol. 38, pt. 5, 994; [Hood] to Jackson, August 27 (5:45 and 6 P.M.), Telegram Book No. III, Hood Papers; Hood to Seddon, August 27, *OR*, vol. 38, pt. 5, 993; Shoup journal, entry of August 27, *OR*, vol. 38, pt. 3, 693; "From Atlanta. Enemy Massing on Our Left" (Atlanta, August 27), *Augusta Chronicle & Sentinel*, August 28; "The Immediate Front," *Appeal Extra*, August 27, quoted in *Columbus Enquirer*, August 31.

Even after Kilpatrick's cavalry raid, Hood worried about another one, which he believed might be headed for Jonesboro. That morning he asked Armstrong, "Do you think Cavalry starting on a raid?" At 12:15 he sent a wire to commanders of the posts at Jonesboro, Griffin, and Macon, "Look out for raid." The two brigades of infantry and the cavalry regiment sent to Jonesboro, as Shoup recorded, were "to co-operate with General Armstrong in repelling raids." By nightfall of the 28th, Hood had dismissed the notion of a quick enemy strike to the Macon road. "No immediate danger of raid," he telegraphed General Wright in Macon. "Yankees have not left West Point railroad."[36]

Indeed they had not. The head of Logan's XV Corps had struck the railroad at noon of the 28th at Shadnor Church two miles north of Fairburn (which via rail line was twelve miles southwest of East Point). The XVII Corps also reached the rail line north of Shadnor. General Blair reported that after digging a line of rifle pits, "a large fatigue party set to work destroying the railroad, which was thoroughly accomplished for a distance of about two miles." The railway to Montgomery, though recently repaired from Rousseau's damage, had not resumed activity as a major supply route for Hood's army, which continued to receive most of its supplies from Macon. Nonetheless General Sherman wanted it wrecked. In fact, he ordered his troops to continue working on the railroad all day on August 29, even though (as Castel writes) the Rebel cavalry's opposition to his advance was so slight that if he had marched swiftly, his infantry could have approached the Macon & Western, which was his true objective. "We will remain here tomorrow," Sherman that evening informed Thomas, whose IV and XIV Corps would reach the railway on the 29th. "I wish the railroad thoroughly destroyed as far forward as possible, and to the rear until you meet General Howard's troops. Let the destruction be so thorough that not a rail or tie can be used again." Then, as if he did not already know how Sherman wanted the work done, Thomas received another lecture from the commanding general.

---

[36] [Hood] to Jackson, August 27 (4:30 P.M.), *OR*, vol. 38, pt. 5, 994 ("oppose stoutly the enemy should he attempt to cross the creek at Trimble's"); [Shoup] to Lee, August 28 (12 m.), *OR*, vol. 38, pt. 5, 997; Shoup to Armstrong, August 28 (6 P.M.), *OR*, vol. 38, pt. 5, 998; [Hood] to Wright, August 28 (9:40 P.M.), *OR*, vol. 38, pt. 5, 999; Shoup journal, entry of August 28, *OR*, vol. 38, pt. 3, 693; [Hood] to Armstrong, August 28 (10:50 A.M.) and to "Commanding Officers Jonesboro, Griffin, Macon," August 28 (12:15 P.M.), Telegram Book No. III, Hood Papers; [Shoup] to Brown, August 28 (6 P.M.), *OR*, vol. 38, pt. 5, 998; [Hood] to Wright, August 28 (9:40 P.M.), *OR*, vol. 38, pt. 5, 999.

> My own experience demonstrates the proper method to be: to march a regiment to the road, stack arms, loosen two rails opposite the right and two opposite the left of the regiment, then to heave the whole track, rails and ties, over, breaking it all to pieces, then pile the ties in the nature of crib work and lay the rails over them, then by means of fence rails make a bonfire, and when the rails are red-hot in the middle let men give the rail a twist, which cannot be straightened without machinery. Also fill up some of the cuts with heavy logs and trunks of trees and branches and cover up and fill with dirt.

Sherman wanted this work to begin at dawn and ordered Howard's troops to join in it. "You can't do too much of it," he stressed.[37]

The Yankees did a number on the Montgomery line. Men of the XIV Corps wrecked three miles of track between Fairburn and Palmetto and another mile and a half north of Fairburn. General Stanley thought that his IV Corps' damage extended to within three and a half miles of East Point (it was not that close). General Davis reported that some of his XIV Corps soldiers "were kept vigorously at work during the day, destroying the railroad track." Blair reported no movement by his corps on August 29; "the men were kept busily engaged during the day completing the destruction of the railroad." Logan personally supervised the work of his XV Corps: "I directed the cuts in the road to be filled up with dirt and logs, and that percussion-shells be concealed in them in such a manner as to explode should the enemy attempt to repair the road." While all of this mayhem was being wreaked on the 29th—the Federals tore up twelve and a half miles of track between Red Oak and Fairburn—Confederate cavalry could only look on.[38]

---

[37] Logan report, September 13, *OR*, vol. 38, pt. 3, 107; Blair report, September 12, *OR*, vol. 38, pt. 3, 554; Castel, *Decision in the West*, 489; Sherman to Thomas, August 28 (6:45 P.M.), *OR*, vol. 38, pt. 5, 688–89; Sherman to Howard, August 28, *OR*, vol. 38, pt. 5, 695.

[38] Ransom report, September 14, *OR*, vol. 38, pt. 3, 391; Stanley report, September 15, *OR*, vol. 38, pt. 1, 213; Davis report, September __ [no date], 1864, *OR*, vol. 38, pt. 1, 512; Blair report, September 12, *OR*, vol. 38, pt. 3, 554; Logan report, September 13, *OR*, vol. 38, pt. 3, 107; Sherman to Halleck, August 31 (8 A.M.), *OR*, vol. 38, pt. 5, 717 ("We reached the West Point railroad, and broke up twelve miles of it thoroughly"); Castel, *Decision in the West*, 492 (12½ miles).

Castel considers Sherman's tarrying on the Atlanta & West Point Railroad August 28–29 to be one of the six blunders he committed in the last days of the campaign ("the first and worst muff"). Francis McKinney agrees, although he overstates the tactical mobility Hood exercised while the Federals were fixed in their rail-wrecking: "Sherman's obsession that his combat infantry must twist rails permitted Hood to march, countermarch and march again with impunity along the front of the Federal army" (Castel, *Decision in the West*, 539; McKinney, *Education in Violence*, 359).

Sherman's march to the West Point Railroad allowed Hood to begin forming an idea of his enemy's purpose. On the 27th Shoup had written, "the exact intention of the enemy has not yet been ascertained." On August 28, after the Federals' column reached the railroad at Fairburn, Shoup entered in his journal, "Every precaution has been taken by the commanding general to keep our line of communications from being cut by the enemy." At 2 P.M. on the 29th, Hood received a dispatch from Frank Armstrong apprising him of the situation. Armstrong was keeping his eyes on the enemy; Hood called him to headquarters at dark to talk it over. Red Jackson, too, was on the *qui vive*; headquarters told him how to deploy his troopers to meet the unfolding events. Meanwhile Lieutenant General Lee was asked to continue sending out scouting parties, indicating Hood's uncertainty as to whether all of the enemy had left his front. Hardee at East Point was instructed to "use his own discretion in the disposition of troops in that quarter" (Shoup's wording). Information was coming in from all sources. Shoup recorded that a correspondent of the *Chicago Tribune*, "named Reed," had been captured; he told headquarters that Henry Slocum had formally assumed command of the 20th Corps.

When Hardee was called to Hood's headquarters in Atlanta on the night of the 29th, Hood and Shoup had an accurate understanding of the situation. The next morning Hardee wrote his wife what he had learned in the briefing: "we know that five certainly, perhaps six Corps are on the West Point R. Road." This was indeed the case. At darkness of August 29 the IV, XIV, XV, XVI, and XVII were all tearing up track. Schofield's XXIII Corps, having left the lines last, spent the day catching up with the rest of the army; it bivouacked that night south of Camp Creek.[39]

With by far the bulk of the enemy infantry on the West Point railroad six miles southwest of East Point, Hood knew that Sherman aimed to break the Macon line well below that place. Even though General Hardee left the headquarters briefing feeling uncertain of Sherman's plans ("I suppose by tonight we shall know more of his designs," he confided to Mary), Hood had begun to figure them out. The enemy were moving, Shoup recorded, "in the direction of Jonesborough and Rough and Ready" (the latter a station ten

[39] Shoup journal, entries of August 27–29, *OR*, vol. 38, pt. 3, 693–94; [Shoup] to Hardee, August 29 (5 P.M.), *OR*, vol. 38, pt. 5, 999 ("dispatch received from General Armstrong at 2 P.M."); [Hood] to Armstrong, August 29 (4:10 P.M.), Telegram Book No. III, Hood Papers; [Hood] to Hardee, August 29 (2 P.M.), *OR*, vol. 38, pt. 5, 999; Slocum to Whipple, August 27, *OR*, vol. 38, pt. 5, 678 (formally assuming command of the XX Corps, which he in fact had done July 30); Hardee to "My own dear Mary," August 30, Hardee Papers; Schofield report, September 10, *OR*, vol. 38, pt. 2, 518.

miles by line of railroad north of Jonesboro). "The general commanding, in his opinion, has taken all necessary precautions, and made such disposition of his forces as to prevent either of the above-named places from falling into the enemy's hands." The "necessary precautions" involved shifting the army's strength to its left. In the evening of the 29th Patton Anderson received orders to march his division at 4 A.M. toward East Point (Hardee's headquarters). Hardee, holding the army's left, seemed satisfied, as he told Mary: "We have made some dispositions to meet the exigency," even though no one could know Sherman's precise objective.

Even the soldiers in the ranks understood the situation and its lack of clarity at the moment. On August 29, Robert L. Bliss wrote his father from the lines of the 16th Alabama, Lowrey's brigade. "The Yankees have evacuated all of their position in front of Atlanta," he explained. Some of his comrades still thought they were retreating, yet "others think they are concentrating on the left for the purpose of extending their lines across our railroads in rear, and thus cutting our communications." Sergeant Bliss came to the same conclusion as had General Hardee: "A few days will tell the tale."[40]

Six weeks into his new role as commander of a Confederate army, John Bell Hood was facing his most critical test. He had an accurate idea of the enemy's position: one corps at the Chattahoochee, the other six southwest of East Point, at the railroad between Fairburn and Red Oak, and evidently heading for the Macon & Western. He also had to realize that he did not have enough troops both to hold the city (and its fortifications against the enemy infantry still at the Chattahoochee) while stretching his outnumbered forces out beyond East Point (where Sherman's infantry was marching). It was his numerical weakness, not a failure to grasp the tactical situation, that hobbled Hood in the last week of August. To this extent Sherman was off the mark when he wrote Schofield, "I don't think the enemy yet understand our movement. They have made no effort to stop us, only cavalry holding the road."[41]

At the same time, if Hood could not be certain of Sherman's objective, it was at least partly because Sherman himself was a little vague about it. A few days before the grand maneuver was to begin, he wrote General Howard

---

[40] Hardee to wife, August 30, Hardee Papers; Shoup journal, entry of August 29, *OR*, vol. 38, pt. 3, 694; Anderson report, February 9, 1865, *OR*, vol. 38, pt. 3, 772; Bliss to "Dear Father," August 29, 1864, Bliss Letters.

[41] S. Davis, *Atlanta Will Fall*, 179; Sherman to Schofield, August 28, *OR*, vol. 38, pt. 5, 694–95; Sherman to Halleck, August 28 (4 P.M.), *OR*, vol. 38, pt. 5, 688 ("Enemy has made no serious opposition to our movement").

about his plan of "swinging our whole army by the right across by Fayetteville or Jonesborough" (Fayetteville is eight miles southwest of Jonesboro). On the evening of the 26th, he sent a telegram to Halleck that summarized his progress on the first day of the march and predicted that tomorrow his army would reach Fairburn, but he did not specify any objective beyond that. On the afternoon of August 26, he wrote Slocum of how he intended to wreck the Atlanta & West Point line: "We will break it good and move on to the other one at once," but he did not say where. It is apparent that Sherman wanted his forces advancing east and southeast on a wide front. "We will move to-morrow for our next objective," he wrote Schofield on August 29, finally specifying his targets: Howard would move toward Jonesboro; Thomas's line of march would be "via Shoal Creek Church, Mrs. Long's and Couch's, on the road from Morrow's Mill, toward Fayetteville." Morrow's Mill is five miles south of East Point; Couch's place is two miles below that. Sherman's front was thus a broad one. Hood had responded on August 28 by sending three infantry brigades to Rough and Ready and two more with some cavalry to Jonesboro. "The general commanding, in his opinion, has taken all necessary precautions," Shoup wrote on the 29th, although "in his opinion" can be read as Shoup's (or perhaps Hardee's) quiet expression of concern that perhaps *not* all needed measures had indeed been taken.[42]

The situation as understood by Confederate headquarters on the night of August 29 was fluid, to be sure. "The enemy is maneuvering to get us out of Atlanta, or to get on our communications," Hardee wrote to his wife on the morning of August 30. Hood, according to General Shoup, had concluded that Sherman aimed to hit the railroad somewhere toward Jonesboro or Rough and Ready. Given what we know now about what Hood knew then, criticisms by earlier writers seem not only unfair but also uninformed. "Hood seems to have been completely in the dark as to Sherman's plans and intentions," Thomas R. Hay intoned in a seminal essay published in 1923. Stanley Horn was harsher two decades later: "Hood had apparently been completely bewildered by Sherman after August 25." "It was August 30 before Hood...realized that Sherman was attacking, not retreating," McDonough asserts in his biography of Schofield. Dyer is even worse, claiming that it was not until the afternoon of September 1 that Hood "at last completely com-

---

[42] Sherman to Howard, August 23 (9:15 A.M.), *OR*, vol. 38, pt. 5, 642; Sherman to Halleck, August 26 (6:45 P.M.), *OR*, vol. 38, pt. 5, 669; Sherman to Slocum, August 28 (4 P.M.), *OR*, vol. 38, pt. 5, 691; Sherman to Schofield, August 29, *OR*, vol. 38, pt. 5, 705; Shoup journal, entry of August 29, *OR*, vol. 38, pt. 3, 694.

prehended Sherman's plans." "He had no idea that practically the entire Federal army was about to march on the Macon & Western," Bailey remarkably claims in his Time-Life volume. Even McMurry is off-base in his biography, stating, "For several days Confederate headquarters in Atlanta simply lost contact with the Northern army." Given what Hood and Shoup told General Hardee at headquarters on the night of August 29—five, perhaps six, infantry corps near Fairburn—McMurry elsewhere also strays: "When the Rebels learned of Sherman's presence on the West Point Railroad, they realized that his next objective would be the Macon & Western. Hood, however, without cavalry to ascertain the enemy's strength and intentions, seems to have believed that the force below Atlanta was only a raiding party." More recently, in their narrative on the war, Murray and Hsieh repeat the old saw that Hood was blind because he had sent Wheeler off raiding. "With minimum cavalry available, Hood had no idea of what was transpiring other than the fact that much of the Union host had pulled back from their positions surrounding Atlanta."

The proper perspective on how Hood responded to Sherman's "grand movement by the right flank" includes a recognition that General Jackson's cavalry division—three brigades, 4,856 officers and men present on August 20—was not only adequate but attentive to Hood's need for reconnaissance. Even General Cox, the enemy, gets this point.

> By the evening of the 27th, it will be seen, all of Sherman's army except the Twentieth Corps was between Atlanta and Sandtown, echelloned along that road. Hood had not interfered with the movement, and had only followed it with light cavalry reconnaissances. But his horsemen, skirmishing with Garrard on the north and with Kilpatrick on the south, were able to locate the National forces with sufficient accuracy.

Another perspective recognizes that Hood's range of possible responses was limited by manpower. "Somehow he hoped to hold Atlanta and keep the railroad open at the same time—an impossibility considering his inferior forces," O'Connor writes. "Hood, in Atlanta, realized at the time what was taking place," George Osborn argues, "but the lack of men prevented all save the most perfunctory of defense action."[43]

---

[43] Hardee to wife, August 30, Hardee Papers; Hay, "Atlanta Campaign" (Part Two), 113; Horn, *Army of Tennessee*, 367; McDonough, *Schofield*, 95; Dyer, *Gallant Hood*, 270; Bailey, *Battles for Atlanta*, 143; McMurry, *John Bell Hood*, 147; McMurry, "The Atlanta Campaign," 252; Murray, and Wei-Siang Hsieh, *A Savage War*, 443; Army of Tennessee returns, August 20, 1864, *OR*, vol. 38, pt. 3, 682; J. D. Cox, *Atlanta*, 197; O'Connor, *Cavalier General*, 215; Osborn, "The Atlanta Campaign, 1864," 285.

Given these points, Connelly and Castel come to the same conclusion: "Hood seemed to do all that he could possibly do by August 30," Connelly writes. Castel agrees.

> Far from being deluded by the Union withdrawals into thinking that Sherman is retreating, as contemporary critics and critical historians will subsequently charge, Hood realizes that Sherman most likely is launching, or is about to launch, another large-scale flanking movement to the south. What Hood does not know, and, under the circumstances, cannot know is exactly what Sherman intends to do, where, when and how. To strike the railroad at Rough and Ready? Or at Jonesboro? Or at some other point? Or, conceivably, does Sherman seek to draw the Confederates out of Atlanta, then have the XX Corps swoop down to seize it? Until Hood can, with reasonable confidence, answer these questions, all he will be able to do is what Shoup, in his journal entry for August 27 states he had done: dispose "his troops so as to be prepared for any emergency." Indeed, at this juncture, it would be foolish for Hood to do anything other than this.

Despite these difficulties, the mood at Hood's headquarters remained optimistic. A correspondent for the *Columbus Times*, writing from Atlanta, noted, "the enemy is still engaged in the movement upon our left, and is advancing slowly but steadily."

> It is hardly necessary to repeat, what I stated in my last letter, that Gen. Hood is keenly alive to all the movements of the enemy, and that the necessary arrangements have been made to foil him upon whichever point he may conclude to advance in force.
>
> Our left at present rests upon the Macon road in the vicinity of Jonesboro' twenty miles below the city, and active operations in the future, will scarcely alarm the citizens of Atlanta—it is out of the range of shot and shell.
>
> While these movements are going on, you must not imagine, that Atlanta has been left out in the cold. The line of fortifications around the city still presents a glorified array of frowning cannon and bristling bayonets, and any number of Yankees attempting to storm them at present, would have cause to regret the folly of the undertaking.[44]

---

[44] T. L. Connelly, *Autumn of Glory*, 461; Castel, *Decision in the West*, 487; "Outline," "Special Correspondence," *Columbus Times*, September 2.

The assumption persists in the literature that Hood misread Sherman's maneuver toward the Macon & Western. See Bell, *The Camel Regiment*: "five days late, Hood realized that Sherman was headed toward the Macon and Western Railroad at Jonesboro" (172).

Hood may have taken all tactical steps he thought necessary to prevent the enemy from cutting the Macon railroad, but logistical cautionary measures had to be taken as well, in case the M. & W. were indeed broken. On August 30 General Shoup began ordering chiefs of departments to start packing their stores and equipment for loading on southbound trains by daylight of August 31. Chief of ordnance Lt. Col. J. M. Kennard assembled his material—a dozen pieces of artillery, two score caissons, limbers and forges, and at least 12,000 artillery shells, plus another 1,500 in damaged state—the Army of Tennessee's entire reserve ordnance train (the army's regular ordnance train had already been sent to Hardee at Jonesboro). Kennard's work was completed in time for the arms and ammunition to be inventoried and loaded on twenty-eight cars by midday on the 31st.

On August 30, all eyes at Confederate headquarters were still on the enemy infantry headed for the Macon railroad. Hood instructed Jackson at eight that morning, "Should the enemy move to the railroad you must detain them as long as possible." He added, "Flint River gives us great advantages in protecting the Macon railroad." Armstrong, eight miles west of Jonesboro, reported that "a very heavy column of infantry" stood in his front. Yet he believed that Yankee cavalry seemed poised to launch a raid on Griffin (twenty miles south of Jonesboro). This caused Brigadier General Lewis, commanding the infantry at Jonesboro, to worry that Armstrong would ride off and leave him without mounted troops. (Hood cautioned Jackson accordingly.) Then, in the early afternoon, Hood concluded that the situation on his extreme left was not so grave. Shoup wired Hardee, "General Hood does not think the necessity will arise to send any more troops to Jonesborough today."

Hood would be proven wrong, but his generalship in these critical days of the campaign should be faulted only if generals were always expected to divine their enemies' plans and purposes. After Hardee reported enemy forces advancing against him, Hood ordered Lee to send Patton Anderson's division to his support. Still, Shoup wrote Jackson at 1:20 P.M., "General Hood does not think there can be a large force advancing upon Jonesborough." Then, twenty-five minutes later, Hood wired Hardee, "It may become necessary for you to send another brigade and battery to Jonesborough." Hardee wanted more than that, though. After shifting Cleburne's division three miles south of East Point, he asked that Lee send Anderson's division all the way to Mount Gilead Church, where it would take position on Cleburne's left and thus become the army's extreme left flank. Hood subsequently directed Anderson to form on the right of Maney's division as Hardee shifted his line to the left. As part of the southward extension, Hardee moved his headquarters to Rough and Ready; Lee was ordered to take his place at East Point.

This is the Confederate line depicted in the Dodge-Ruger map (Kurtz's term) for the final stage of the campaign in the *Official Records Atlas*: a continuous line of earthworks stretching for five miles from East Point to Mount Gilead. As a result of the army's pitch to the left, Featherston's and Walthall's divisions were withdrawn from the Grant-Presstman perimeter of fortifications. General French noted that on August 31 only his division and Smith's Georgia militia defended the city that day. Nevertheless, stretched as it was, Hood's railway defense line only covered the road as far as Quick Station. Between Hardee's left and the 2,000 troops at Jonesboro, only cavalry opposed the Yankee advance.[45]

Hardee remained nervous enough to send a scouting patrol out to feel the enemy. Col. William D. Pickett brought back word of enemy columns marching on Rough and Ready and Jonesboro. Hardee sent this information to Hood. Precisely what Hardee wrote in his dispatch is not known, but he probably asked for a meeting at headquarters to discuss the rapidly unfolding tactical situation. Hood replied, "Will send engine for you; had anticipated it before your dispatch was received." Generals S. D. Lee and W. H. Jackson

---

[45] Jno. Withers, A. A. G., "Findings of the Court of Inquiry upon the loss of Confederate stores at Atlanta," March 2, 1865, *OR*, vol. 38, pt. 3, 991–92; W. D. Humphries, "Approximate statement of ordnance stores destroyed in the evacuation of Atlanta, September 1, 1864," September 16, *OR*, vol. 38, pt. 3, 685–86; Vandiver, "Hood as Logistician" 5 n.23 ("Hood's regular ordnance train was under Hardee's care at Jonesboro"), 9 (Shoup ordering reserve ordnance to be packed on train August 30); Hardee report, Apr. 5, 1865, *OR*, vol. 38, pt. 3, 701 ("I was encumbered by the immense subsistence and ordnance trains of the army, which had been sent for safety from Atlanta to Jonesborough"); [Hood] to Jackson, August 30 (8 A.M.), *OR*, vol. 38, pt. 5, 1004; Armstrong to Lewis, August 30 (11 A.M.), and Lewis to Jackson, August 30, *OR*, vol. 52, pt. 2, 726–27; Hood to Jackson, August 30 (10 A.M.), *OR*, vol. 38, pt. 5, 1005; [Shoup] to Hardee, August 30 (1 P.M.), *OR*, vol. 38, pt. 5, 1000; [Shoup] to Lee, August 30, (1:10 P.M.), *OR*, vol. 38, pt. 5, 1002; [Shoup] to Jackson, August 30 (1:20 P.M.), *OR*, vol. 38, pt. 5, 1005; [Hood] to Hardee, August 30 (1:45 P.M.), *OR*, vol. 38, pt. 5, 1000; Hay, *Cleburne and His Command*, 250; Cate, *Two Soldiers*, 124 (Key diary entry of August 30, re: Cleburne's movement); [Shoup] to Lee, August 30 (2:40 P.M.), *OR*, vol. 38, pt. 5, 1002; [Hood] to Lee, August 30 (3:25 P.M.), *OR*, vol. 38, pt. 5, 1003; [Shoup] to Lee, August 30 (1:25 P.M.), *OR*, vol. 3, pt. 5, 1002; J. T. Dodge and H. H. Ruger, "Map Illustrating the Fifth Epoch of the Atlanta Campaign," *OR Atlas*, plate LX, pt. 2; S. G. French, *Two Wars*, 222; Green, *Orphan Brigade*, 154n. (2,000 Confederates at Jonesboro, August 30).

were also summoned to headquarters for the meeting. Already Hood was deciding how he would deal with the emergency. "Have your command under arms," he instructed Lee.[46]

Even before the generals' conference on August 30, events became more ominous still. Mid-afternoon Howard's army was at Renfroe's, five miles northwest of Jonesboro. The other three corps were to the north; their advance was being slowed by "very obstinate" Rebel cavalry. To maintain alignment among his forces Sherman ordered Howard to halt, but the men of the XV, XVI, and XVII Corps were thirsty on this hot summer day, and the Flint River ahead promised replenishment of their canteens. Howard told his troops they could keep marching. Blair's corps stayed at the Renfroe Place, but the XV and XVI pressed on. Against Armstrong's cavalry they covered ground quickly. Toward nightfall they approached the Flint. Hazen's division of the XV crossed the river, drove off the Rebels, and got within a quarter-mile of Jonesboro before dark. Some Yankees even attacked Lewis's line, "but were easily repulsed," according to General Shoup.[47]

These alarming developments are reflected in the stream of messages sent out by Hood's headquarters—a dozen from 5 to 10 P.M.—about the situation at Jonesboro. Lewis and Armstrong were directed to prevent the enemy from crossing the river. At 6:10 Hardee was told to move his corps to Jonesboro. Lee would follow with his. Lewis got this: "Hold your position at all hazards. Help is ordered to you." Then an hour later: "Has the Enemy Crossed Flint River at any point?" At 9:30 headquarters was still worried that the Yankees had gotten across the river: "Is there any of the Enemy remaining on this side of Flint River?"[48]

---

[46] Hughes, *Hardee*, 234–35; [Hood] to Hardee, August 30 (3:15 P.M.), *OR*, vol. 38, pt. 5, 1001; Hood to Lee, August 30 (3:45 P.M.), *OR*, vol. 38, pt. 5, 1003; [Hood] to Jackson, August 30 (4:15 P.M.), Telegram Book No. III, Hood Papers.

The indefatigable Wilbur G. Kurtz learned from old time Atlantan James Bell that the locomotive sent to pick up the generals at East Point was named the N. C. Monroe (Kurtz, "The Jonesboro Campaign," typescript of address delivered before the Atlanta Civil War Round Table, June 18, 1957, box 51, folder 10, Kurtz Collection).

[47] Castel, *Decision in the West*, 494; S. Davis, *Atlanta Will Fall*, 180 (map); Stanley report, September 15, *OR*, vol. 38, pt. 1, 213; Blair report, September 12, *OR*, vol. 38, pt. 3, 554; Woods report, September 8, *OR*, vol. 38, pt. 3, 567 ("encamped near the Renfroe Place"); Logan report, September 13, *OR*, vol. 38, pt. 3, 108; Ransom report, September 14, *OR*, vol. 38, pt. 3, 391; Shoup journal, entry of August 30, *OR*, vol. 38, pt. 3, 694.

[48] [Hood] to Lewis (5:15 P.M.); [Shoup] to Armstrong (5:15); [Hood] to Hardee (6:10 P.M.); [Hood] to Lee (6:35 P.M.); [Hood] to Lewis (6:35 P.M.), all August 30, *OR*, vol. 38, pt.5, 1001–1005; [Hood] to Lewis, August 30 (7:35 and 9:30 P.M.), Telegram Book No. III, Hood Papers.

Hood had been proven wrong. He had told Red Jackson earlier that afternoon that he did not think a large enemy force was marching on Jonesboro. Five hours later, he was telling Hardee and Lee they would have to march their two infantry corps to that point during the night. The Yankees would doubtless cross the Flint on the morrow; Hardee and Lee were "to attack and drive the enemy at that place across the Flint River." Hardee's troops headed out at 4 P.M. but did not begin to reach Jonesboro until sunrise on August 31. Lee's corps followed, arriving between 10 A.M. and 1 P.M. Their attack began at 3 P.M. It is worth noting that General Sherman initially credited Hood with having stymied his lunge at the railroad. "The enemy is too smart for us, and we may have to maneuver thus down to Macon," he wrote Howard at 12:30. "It may be that some accident will happen, of which we can take advantage."

August 31, 1864, was the decisive day of the Georgia Campaign, but not because of the battle fought at Jonesboro. As Confederate soldiers began their attack against Howard's entrenched troops about 3 P.M., those of Jacob Cox's division of the XXIII Corps reached the Macon & Western Railroad eight miles north of Jonesboro. "The advance was sharply resisted by the enemy's cavalry," Cox reported, "but no infantry force was found." Indeed, to get to that point on the railroad, a mile south of Rough and Ready, the Union infantry would have had to march over the very line of rifle pits between East Point and Mount Gilead that Hardee's soldiers had abandoned the previous evening. Cox's men drove off the Rebel cavalry and began tearing up the long-sought-after Macon line, "being the first of the army to reach that road," Cox proudly recorded.

Hood and Shoup had gotten warning of the bad news to come. The chief of staff received a report by 12:15 that the enemy "in strong force" was advancing against Southern troops at Mount Gilead. A few hours later, two trains southbound from Atlanta were stopped below Rough and Ready. One of them bore the army's reserve ordnance, trying to get out of Atlanta (among the two locomotives was the General, which had been stolen by the Andrews train thieves two years before). Southern horsemen told the engineers that Yankees were on the road ahead and that they could go no farther. The trains promptly steamed in reverse back on the single-track line toward Atlanta. When one returned by 5 P.M., Hood knew that the enemy had finally cut his last railway supply artery. "Rover," a reporter for the *Augusta Chronicle & Sentinel*, told the story:

> About that hour [noon]...[a] train left for below, made up from rolling stock of the West Point road; but about four o'clock it returned to the city,

> bringing intelligence that, having proceeded a short distance below Rough and Ready, it was stopped by some of our retreating cavalry, who reported the enemy occupying the road a short distance below. The train was at once stopped...in a very short time the engineer backed up the road, and soon reached the city.

Further confirmation that the Macon railroad had been cut came when the second southbound train, carrying Lieutenant Colonel Kennard's ordnance, chugged back to East Point. Brig. Gen. John T. Morgan, in command there, notified headquarters; Hood wired back, "Move the ordnance train to this place at once."

Hood knew that Atlanta was doomed even before he learned that Hardee's and Lee's attack at Jonesboro had failed. The Federals had cut the telegraph, so no word came in from Hardee late on the 31st. Shoup closed his journal entry for that day by writing, "The result is not yet known."[49]

What Hood and Shoup also did not know was that the enemy had made a second lodgment on the Macon & Western at Morrow, four miles down the line from Quick Station. Several regiments of Baird's division, XIV Corps, struck the road about 6 P.M., drove off the Rebel cavalry, and fortified across the tracks. Federals began tearing up the railroad early the next morning.

Historians have yet to catch up with these events. Castel fails to note the Confederates' extension of their lines August 31 to Mount Gilead. Like so many others, his focus is on Jonesboro, and he devotes ten pages to the Confederate attack against Howard at Jonesboro before noting (in seven lines) that it was Cox's troops who actually broke the Macon & Western well to the north. Like Castel, McMurry treats Jonesboro extensively before concluding that "the battle was a useless event."

In the future, chroniclers of John Bell Hood's Confederate record will do well to keep in mind that while Howard's three corps were marching on

---

[49] S. Davis, *Atlanta Will Fall*, 184–85; Hood report, February 15, 1865, *OR*, vol. 38, pt. 3, 633; Sherman to Howard, August 31 (12:30 P.M.), *OR*, vol. 38, pt. 5, 726; Cox report, September 10, *OR*, vol. 38, pt. 2, 692; S. Davis, *What the Yankees Did to Us*, 251–52; handwritten notes of Wilbur G. Kurtz interview with James Bell, Summer 1935, notebook #12, 37–38, Kurtz Collection (identifying the engines *General* and *Missouri*); "Rover," "A Day of Excitement and Rumors," *Augusta Chronicle & Sentinel*, September 6; [Shoup] to Hardee, August 31 (12:15 P.M.), *OR*, vol. 38, pt. 5, 1007; [Hood] to Morgan, August 31, *OR*, vol. 38, pt.5, 1008; Shoup journal, entry of August 31, *OR*, vol. 38, pt. 3, 694.

Given that Hood had sent the army's ordnance train to Hardee at Jonesboro, and that Shoup ordered on August 30 the reserve ordnance to be loaded on train, one may disagree with Connelly's judgment that "Hood had simply waited too late to get many of his stores out of Atlanta" (*Autumn of Glory*, 467).

Jonesboro, the rest of Sherman's flanking force—the IV, XIV, and XXIII Corps—were marching toward Morrow and Quick Station, points four and eight miles north. And they should keep in mind that those three corps were far larger numerically than Howard's: 35,293 to 20,777 officers and men present on August 31. It was thus Thomas's and Schofield's heavier columns, not Howard's, that decided the fate of Atlanta. To ignore these facts and to focus only upon Hood's defense of Jonesboro is to disregard the multiple challenges Hood faced in trying to hold the city in late August 1864. In this light, Wiley Sword's assertion that "Hood carelessly allowed Sherman's men to gain a lodgment on the vital Macon and Western Railroad near Jonesboro before responding" is most certainly unfair.[50]

---

[50] Davis report, September [undated], 1864, *OR*, vol. 38, pt. 1, 513; Baird report, September 7, *OR*, vol. 38, pt. 1, 749; Hunter report, September 8, *OR*, vol. 38, pt. 1, 770–71; Sherman, *Memoirs*, 4th ed. (1892), 507 (Fullerton diary, September 1, "4 A.M.—Our working parties commenced to destroy the railroads"); S. Davis, *Atlanta Will Fall*, 180 (map of "Situation—August 30," which I sketched for cartographer Gary Joiner, specifically to show the number of Federals in the several infantry corps, based on officers and men present in Sherman's return of August 31), 191; Castel, *Decision in the West*, 494–504; McMurry, *John Bell Hood*, 147; McMurry, *Atlanta 1864*, 173; "Effective strength of the army…August 31," *OR*, vol. 38, pt. 1, 116; Sword, *Southern Invincibility*, 266.

General Hardee at Jonesboro was unaware of the enemy's breach of the railroad near Rough and Ready, and thus could not figure out why Hood called for Lee's corps to return to Atlanta. "It is to be remarked," he later wrote to President Davis, "that this order was issued when the result of the fight at Jonesboro could not have been known to the Comdg. Gen." (Hardee to Davis, September 27, Crist, Williams, and Dillard, eds., *Papers of Jefferson Davis*, 11:71). Errol Clauss's statement that Hood ordered Lee back because he "suddenly deduced that a major Federal attack on Atlanta was imminent" seems equally unaware of the tactical situation ("The Battle of Jonesborough," 20).

Local writers chronicle the battle of Jonesboro, such as Kurtz, "Last Battle Near Atlanta," *Atlanta Journal Sunday Magazine*, February 26, 1933, 8, 19; and Hatfield, "Battle of Jonesboro," 65–70.

James L. Stokesbury is one of the few writers on the campaign to avoid Jonesboropia, as I call the persistent myth that the battle of August 31–September 1 sealed Atlanta's fate. "Schofield, the inner element of the wheeling movement," he writes of Sherman's "grand wheel" toward the Macon railroad, "broke it at Rough and Ready, below Eastpoint, then Thomas was across it," reflecting the fact that troops of the XXIII and IV Corps were the first to break the Macon line (Stokesbury, *A Short History of the Civil War*, 280). A recent instance of Jonesboropia is in D. A. Johnson, *Decided on the Battlefield*, 184–87; Johnson never mentions the Federal lodgment on the M. & W. as the reason Hood evacuated Atlanta.

A scholar who gets it right is J. Britt McCarley, chief historian for the U. S. Army Training and Doctrine Command. "On August 31, the blue columns reached their objective, the M&W south of East Point and north of Jonesboro, and began destroying that line," he writes in his essay on Sherman's logistics. "Atlanta was now isolated," rendering Hardee's attack at

Hood himself fell victim to Jonesboropia (as I call it). In his first telegram to Bragg in four days, sent on September 3, he reported that Hardee's and Lee's assault on the 31st failed to drive off the Federals from their lodgment on the Flint, and that "this made it necessary to abandon Atlanta." As he had done earlier, Hood here too manifested his unbecoming tendency to belittle the heroism of his soldiers by equating it with their casualty numbers: "our loss on the evening of the 31st was so small that it is evident that our effort was not a vigorous one." Two days later he compounded this regrettable statement: "To let you know what a disgraceful effort was made by our men in the engagement of August 31," he wrote Bragg, "I give you the wounded in the two corps: Hardee's corps, 539; Lee's 946; killed a very small number." Among the wounded was Maj. Gen. Patton Anderson, hurt so severely as to be knocked out of the service until almost the end of the war.[51]

At 6 P.M. of the 31st, without having heard that Hardee's and Lee's assault had been repulsed, Hood sent courier to Jonesboro ordering Lee's corps back toward Atlanta. The cut telegraph and returning trains had proven that Federal forces were interposed on the Macon railroad. Atlanta's railway supply line was finally broken; late that afternoon he issued orders for the city to be abandoned beginning the next day. Moreover, the Army of Tennessee was now split. Lee's approaching infantry would at some point link up with Stewart's corps and the militia marching south from the city. Hardee would have to fend for himself for a few days until the army was able to reunite. Fortunately, Sherman was too obsessed with wrecking the Macon & Western to exploit the opportunity he had to attack and destroy Hood's army in detail.[52]

Jonesboro later that afternoon "an empty gesture, as Sherman already had cut Atlanta's last railroad" ("'The Great Question of the Campaign Was One of Supplies,'" 44).

Evidence of the persistence of Jonesboropia, however, is seen in the text of the Georgia Historical Society's historical marker, "The Battles for Atlanta," which was placed on the grounds of the Carter Center in Atlanta in 2011: "U.S. forces swept west and then south and at Jonesboro (August 31–Sept. 1) cut the last railroad supplying Atlanta, forcing the Confederates to abandon the city."

[51] Hood to Bragg, September 3 and September 5, *OR*, vol. 38, pt. 5, 1016, 1021; J. P. Anderson "Autobiography of Gen. Patton Anderson, C.S.A.," 70; Hewitt, "James Patton Anderson," 1:24. To replace Anderson, the War Department sent Maj. Gen. Edward Johnson, division commander in Lee's army who had been captured on May 12, 1864, when the Angle at Spotsylvania was overrun (R. K. Krick, "Edward Johnson," 3:187). Johnson was exchanged by August 5, the day Mrs. Chesnut wrote that he would pay a call. "Old Edward Johnson sent to Atlanta against his will," she recorded in her diary (Woodward, *Mary Chesnut's Civil War*, 632, 634). He was assigned to command Anderson's division on September 1 (Falconer Special Field Orders No. 95, September 1, *OR*, vol. 38, pt. 5, 1013).

[52] Castel, *Decision in the West*, 505; McMurry, *Atlanta 1864*, 173–75;

Chapter 8

September

Early on September 1 Confederates in Atlanta began packing up what stores they could carry away, dispensing foodstuffs to civilians and burning the rest. Troops began marching out of town around 5 P.M. Cannon in the lines too heavy to be wheeled away were spiked. That night the last Confederates in Atlanta were Ferguson's cavalry and Smith's Georgia militia. The state troops, marching out of the city via the Decatur road, set fire to the rolling mill east of the city and to the locomotives and train cars parked nearby on the Georgia Railroad (according to "Rover," reporter for the *Augusta Chronicle & Sentinel*, the Georgia Railroad depot was also burned about 11 P.M.). Twenty-eight cars of ammunition went up in fiery blast, which alerted Sherman, twenty miles to the south, that Atlanta was being evacuated by the Rebels. "At 2 A.M. our brigade left Atlanta at a rapid gait," recalled Capt. R. N. Lea, "and just as we were leaving the suburbs the explosion of the magazines shook the city from center to circumference." With the troops abandoning and the shells outside exploding, civil order in the city naturally evaporated. Capt. William L. Nugent, Ferguson's A. A. G., lamented what he saw during the night of September 1–2:

> I never shall forget what I witnessed when our army evacuated the Gate City. Hundreds of women congregated on the streets, and recklessly broke open and robbed store houses and private dwellings. Barefooted in the main, poorly clad, with brazen faces they would saunter along jostling against and pushing the soldiers from the sidewalks. As soon as the coveted plunder was exposed to view there was a general rush and grab. It was really a pitiful sight to see the women tumbling over one another, quarreling, calling names, etc. I hope I may never witness such a sight again.

After Lee's infantry started marching northward, Hardee and his corps could only dig in at Jonesboro, where on September 1 Howard's infantry attacked. The Confederate line bent under the heavy assault—Brig. Gen. Daniel Govan and some 900 men were captured, as well as eight field pieces—but

Hardee ended the second day of the battle of Jonesboro holding his ground. At night he was able to withdraw and march to rejoin the rest of the army, which was headed for Lovejoy's Station. General Hood rode with his despondent troops, leaving behind much of his headquarters baggage that could not be carried out for want of wagons to haul it. The next morning, September 2, Union troops discovered the abandoned Rebel fortifications and marched into the city. Mayor James Calhoun ruefully surrendered it.[1]

A review of the last month and half of the Georgia Campaign will conclude that Sherman had used his forces smartly and in concerted determination to cut the Rebels' supply lines and compel their abandonment of Atlanta. In the end, he has been criticized for allowing Hood's defeated, beaten army to withdraw southward to reunite and fight again. "There could have been no difference of opinion as to the first duty of the armies of the Military Division of the Mississippi," General Grant wrote years later in *Battles and Leaders.* "Johnston's army was the first objective, and that important railroad center, Atlanta, the second." But Sherman had taken the prized city, and this was what President Lincoln and the authorities in Washington most wanted before the Northern presidential election in November. "Atlanta is ours," he telegraphed General Halleck, "and fairly won."

For his part, Hood had lost Atlanta, but an evaluation of his performance as commander of the Confederate army trying to hold it actually leads to a favorable assessment. True, Hood's three attacking battles in July led to some 11,000 casualties. But the government had asked him to seek opportunity to give battle on favorable terms. At Peachtree Creek, July 20, both because of the two-mile gap in the enemy line and because of his troops' rightward shift,

---

[1] Smith to Governor Brown, September 5, *Supplement to the Official Records*, vol. 7, 151–52; S. Davis, *What the Yankees Did to Us*, 253–59; "Rover," "The Evacuation of Atlanta," *Augusta Chronicle & Sentinel*, September 6 ("near Decatur, Friday morning, September 2"); Nugent, *My Dear Nellie*, 219.

Casualties for the battle of Jonesboro total approximately 2,200 Confederate and only 172 Federal for the first day, when the Southerners frontally attacked Howard's defenders, who had worked on their fortifications for hours before the assault. Hardee lost some 1,400 on September 1, most of whom were prisoners. Livermore puts casualties for the Federals, who did the attacking on the second day, at 223 killed, 946 wounded and 105 missing (Castel, *Decision in the West*, 503, 524; Livermore, *Numbers and Losses*, 126).

Among the prisoners taken on the second day was Brig. Gen. Daniel Govan. Three weeks later he was released as part of the prisoner exchange arranged by Generals Hood and Sherman, "being exchanged at Rough and Ready, Georgia, for Major General George Stoneman," as Edwin C. Bearss has written ("Daniel Chevilette Govan," 3:18). The Govan-Stoneman exchange would not have been a one-for-one swap. On the basis of the established cartel, a major general would equal a brigadier general plus ten privates (Speer, *Portals to Hell*, 102).

his orders for an echelon attack by division from the right (in a plan that Johnston later tried to take credit for) actually resulted in a flank attack (Bate assaulting the left of Newton's division). East of Atlanta and at Ezra Church six days later he had sought to deliver flank attacks. In both cases the enemy had gotten in the way: in the former, with the providential placing of Sweeny's division by a worried McPherson; in the latter with Howard's seizure of the key crossroads that would have been S. D. Lee's defensive position before Stewart's flanking march. Nevertheless, in the battle of July 22, Hood's forces delivered a Jacksonian assault that mauled the enemy army and came so close to victory that it was hailed as such by newspapers throughout the Confederacy. And despite his repulses, Hood had taken satisfaction that the battles had slowed or blunted Sherman's encroachment upon his railroad.

During August his cavalry had defeated three separate raids on the Macon line, with only brief interruption of supply service to the army. He then had done what the government had also suggested, and which President Davis approved: sending Wheeler off with half of the army's mounted troops to try to cut Sherman's railroad network. Wheeler failed to do this, but through his own shortcomings, not Hood's. Then Hood had found reinforcements for his ranks which did much to make good the losses of July and which boosted the morale of his army and of the Southern people at large. Finally, when Sherman began his decisive circuit south of Atlanta, Hood had used his cavalry to stay informed of the enemy's movement and had stretched his infantry as far as he could to block it. He ultimately failed because Sherman's numerical superiority allowed his marching columns to advance on a front so broad that their objective point(s) could not be predicted by Hood, much less defended by his inferior numbers. He committed the bulk of his infantry to parrying the threat posed by Howard's advance on Jonesboro but did not have enough troops to prevent Thomas and Schofield from breaking the road north of that point. As Errol Clauss has written in his survey of the campaign, "Given Hood's position, relative strength and strategic restrictions, he probably did all that was possible under the circumstances." More recent historians have weighed in on the near-impossibility of Hood's holding Atlanta. "By the time Davis replaced Johnston with Hood," writes McMurry, "it was too late for the Rebels." William C. Davis makes the same judgement: "Hood at least tried, but only when Johnston's retreat had already made Atlanta's fall practically a foregone conclusion."

A few weeks after the fall of Atlanta, President Davis wrote Sen. Herschel V. Johnson of Georgia. "No one was more anxious than myself to prevent the fall of Atlanta," he admitted. "I was not among those who deemed that result inevitable as soon as the enemy had crossed the Chattahoochee, and I was not willing that it should be yielded before manly blows had been struck for its preservation." A few weeks after that, Ben Hill wrote Senator Johnson expressing the same opinion: "Atlanta had virtually been lost when Hood took command."[2]

The story of Atlanta's loss was laid out several months later in a long report submitted by the secretary of war to the president, which is usually not seen in the campaign literature. Seddon made a number of points. At the start of the campaign, General Johnston had been reinforced "exceeding the relative disproportion which has usually existed with all our armies"—meaning he was outnumbered by Sherman but enjoyed better numerical odds than Confederate armies usually faced. "Prudence or strategy led the general intrusted with the conduct of the campaign to prefer a retreating and defensive line of action"—the secretary was here being most diplomatic regarding Johnston's eternal retrograde. His troops had gotten in some licks, "severely checking and punishing their adversaries, whenever venturing direct assault." Yet Johnston had been pressed back to Atlanta. When it "appeared about to be abandoned, a change was made in the commander"—Seddon's simplified narrative avoided mentioning the president's week of agonized decision-making. Placed in command, General Hood did what he could: "a strenuous effort was made to hold and defend, by force of arms, this place." Then came Sherman's wide maneuver around south of the city to cut the Macon & Western Railroad. The secretary's efforts to describe it without criticizing Hood led to a statement of unabashed abstruseness: "a hazardous movement of the enemy, which, with as much success to our arms as there was promptitude in seizing

---

[2] Castel, *Decision in the West*, 540 ("Sherman enabled Hardee to escape"); Clauss "Atlanta Campaign" 380, 384 ("In his failure to capture or destroy Hood's Army of Tennessee…Sherman failed to achieve Clausewitz's first objective of war: 'To conquer and destroy the armed power of the enemy'"); Grant, "Preparing for the Campaigns of '64," 4:99; Sherman to Halleck, September 3, *OR*, vol. 38, pt. 5, 777; McMurry, *Atlanta 1864*, 208; W. C. Davis, *The Cause Lost*, 131; Davis to Johnson, September 18, in Rowland, *Jefferson Davis Constitutionalist*, 6:338; Hill to Johnson, October 13, 1864, quoted in T. L. Connelly, *Autumn of Glory*, 408 n.12 (from Johnson Papers, Duke).

Sherman's failure to pursue Hardee's corps on September 2, when it was separated from the rest of Hood's army, qualifies as a topic for comment by David Frey in his recent study, *Failure to Pursue.* "By Monday, September 5, 1864, Sherman had pretty much disregarded Grant's orders to destroy the Confederate army, instead opting to forego any further pursuit efforts against Hood's army" (*Failure to Pursue*, 115).

Edward H. Bonekemper III counts Sherman's failure to pursue and attack Hood's army after the Confederate evacuation of Atlanta as among the ten most consequential blunders of the entire war. See *The 10 Biggest Civil War Blunders*, 193–204.

the occasion by our commander, must have caused his ruin, resulting in victory, compelled our evacuation of the city and its occupancy by the enemy."[3]

Perhaps Senator Hill was right: the fall of Atlanta was been inevitable after Joseph E. Johnston crossed the Chattahoochee River—but so was President Davis in crediting John Bell Hood for having struck manly blows, and more, in trying to save one of the Confederacy's last prize cities.

[3] Seddon to Davis, November 3, 1864, *OR*, ser. 4, vol. 3, 757.

Works Cited

Abel, E. Lawrence. *Singing the New Nation: How Music Shaped the Confederacy, 1861–1865.* Mechanicsburg, PA: Stackpole Books, 2000.

Adams, Robert N. "The Battle and Capture of Atlanta." In Kerkis, comp. *Atlanta Papers,* 465–86.

———. "Campaign for Atlanta." In Kerkis, comp. *Atlanta Papers,* 199–212.

Adolphson, Steven J. "An Incident of Valor in the Battle of Peach Tree Creek." *Georgia Historical Quarterly* 57/3 (Fall 1973): 406–20.

Albaugh, William A. III, Hugh Benet, Jr., and Edward N. Simmons. *Confederate Handguns: Concerning the Guns, the Men Who Made Them, and the Times of Their Use.* Philadelphia: Riling and Lentz, 1963.

Alexander, Bevin. *How the South Could Have Won the Civil War: The Fatal Errors that Led to Confederate Defeat.* New York: Crown Publishers, 2007.

Alexander, E. P. "The Battle of Fredericksburg." *Southern Historical Society Papers* 10/10–11 (October–November 1882): 445–64.

———. *Military Memoirs of a Confederate: A Critical Narrative.* Dayton, OH: Press of Morningside Bookshop, 1977.

Alfriend, Frank H. *The Life of Jefferson Davis.* Cincinnati: Caxton Publishing House, 1868.

Allan, William. *The Army of Northern Virginia in 1862.* Boston: Houghton Mifflin and Co., 1892.

Allardice, Bruce S. *More Generals in Gray.* Baton Rouge: Louisiana State University Press, 1995.

Allen, Felicity. *Jefferson Davis: Unconquerable Hero.* Columbia: University of Missouri Press, 1999.

Anders, Leslie. "Fisticuffs at headquarters: Sweeny vs. Dodge." *Civil War Times Illustrated* 15/10 (February 1977): 8–15.

Anderson, Frank. "A Courier at the Battle of Resaca." *Confederate Veteran* 5/6 (June 1879): 297–98.

Anderson, James P. "Autobiography of Gen. Patton Anderson, C. S. A." *Southern Historical Society Papers* 24 (January–December 1896): 57–70.

Andrew, Rod, Jr. *Wade Hampton: Confederate Warrior to Southern Redeemer.* Chapel Hill: University of North Carolina Press, 2008.

Andrews, J. Cutler. *The South Reports the Civil War.* Princeton: Princeton University Press, 1970.

Axelrod, Alan. *America's Wars.* New York: John Wiley & Sons, 2002.

———. *Generals South, Generals North: The Commanders of the Civil War Reconsidered.* Guilford, CT: Lyons Press, 2011.

Bailey, Ronald H. *Battles for Atlanta: Sherman Moves East.* Alexandria, VA: Time-Life Books, 1985.

Baldwin, James T. *The Struck Eagle: A Biography of Brigadier General Micah Jenkins.* Berryville, VA: Rockbridge Publishing Company, 1996.

Ballard, Michael B. *Pemberton: A Biography.* Jackson: University Press of Mississippi, 1991.

Banks, Robert W. "Civil War Letters of Robert W. Banks: Atlanta Campaign." Edited by George C. Osborn. *Georgia Historical Quarterly* 27/2 (June 1943): 208–16.

Barnes, W. T. "An Incident of Kennesaw Mountain." *Confederate Veteran* 30/2 (February 1922): 48–49.

Barnwell, Valentine T. *Barnwell's Atlanta City Directory.* Atlanta: Intelligencer and Job Office, 1867.

Barron. S. B. "Wheeler's Cavalry in Georgia Campaign." *Confederate Veteran* 14/2 (February 1906): 70.

Barry, Robert L. "The Lookout Battery." *Confederate Veteran* 30/10 (October 1922): 385.

"Battle of Cassville, Ga., May 24, 1864." *Kennesaw Gazette* 1/9 (September 1886): 3–4.

Battles of Atlanta. Short Sketch of the Battles Around, Siege, Evacuation and Destruction of Atlanta, Ga., in 1864. Prepared under the Direction of the Committee of the Atlanta Camp, United Confederate Veterans. Atlanta: n.p., 1895.

Bearss, Edwin C. "Daniel Chevilette Govan." In Davis, *The Confederate General*, 3:17–19.

———. "Sherman and Hood at Atlanta." Address to Chicago Civil War Round Table, October 13, 1990. Audio recording courtesy of Hal Ardell, Chicago.

Beatty, Taylor. Diary. #00054, Southern Historical Collection, The Wilson Library, University of North Carolina at Chapel Hill.

Bell, W. Scott. *The Camel Regiment: A History of the Bloody 43rd Mississippi Volunteer Infantry, CSA 1862–1865.* Gretna, LA: Pelican Publishing Company, 2017.

Bennett, Stewart. "A Ghost on Horseback: The Many Wounds and Curious Death of William H. T. Walker." In Hewitt, *Confederate Generals in the Western Theater*, 4:195–215.

Bennett, William W. *A Narrative of the Great Revival Which Prevailed in the Southern Armies During the Late Civil War between the States of the Federal Union.* Philadelphia: Claxton, Remsen & Haffelfinger, 1877.

Bergeron, Arthur W., Jr. *Confederate Mobile.* Jackson: University Press of Mississippi, 1991.

———. "Mansfield Lovell." In Davis, *The Confederate General*, 4:101–102.

———. "Martin Luther Smith." In Davis, *The Confederate General*, 5:179–80.

Bierce, Ambrose. "The Crime at Pickett's Mill." In *Ambrose Bierce: The Devil's Dictionary, Tales and Memoirs*, edited by S. T. Joshi, 684–92. New York: Library of America, 2011.

Biggs, Greg. "The 'Shoupade' Redoubts: Joseph E. Johnston's Chattahoochee River Line." *Civil War Regiments* 1/3 (1991): 82–93.

Bill, Alfred Hoyt. *The Beleaguered City: Richmond 1861–65.* New York: Alfred A. Knopf, 1946.

Black, Robert C., III. *The Railroads of the Confederacy.* Chapel Hill: University of North Carolina Press, 1952.

Bliss, Robert Lewis. Civil War Letters, 1861–1865. Alabama Department of Archives and History, Montgomery, AL. Cited as Bliss Letters.

Blomquist, Ann K., and Robert A. Taylor, eds. *This Cruel War: The Civil War Letters of Grant and Malinda Taylor 1862–1865.* Macon, GA: Mercer University Press, 2000.

Blount, Russell W., Jr. *The Battles of New Hope Church.* Gretna, LA: Pelican Publishing Company, 2010.

———. *Clash at Kennesaw: June & July 1864.* Gretna, LA: Pelican Publishing Company, 2012.

Blythe, Robert W., Maureen A. Carroll, and Steven H. Moffson. *Kennesaw Mountain National Battlefield Park Historic Resource Study.* N.p., 1995.

———. *Master of War: The Life of General George H. Thomas.* New York: Simon & Schuster, 2007.

Bogle, James G. "The Western & Atlantic Railroad and Its Boom Town." *Blue & Gray* 6/4 (April 1989): 14.

Bohannon, Keith S. "'A Bold Fighter' Promoted beyond His Abilities: Gen. John Bell Hood." In *Leaders of the Lost Cause: New Perspectives on the Confederate High Command*, edited by Gary W. Gallagher and Joseph T. Glatthaar, 249–87. Mechanicsburg, PA: Stackpole Books, 2004.

———. "'Dirty, ragged, and ill-provided for': Before Sharpsburg, Lee's Supply Officers Fought a Battle Every Day." *Civil War* 74 (June 1999): 18–25.

———. "Dirty, Ragged, and Ill-Provided For: Confederate Logistical Problems in the 1862 Maryland Campaign and Their Solutions." In *The Antietam Campaign*, edited by Gary W. Gallagher, 101–42. Chapel Hill: University of North Carolina Press, 1999.

———. "'Witness to the Redemption of an Army': Reenlistments in the Confederate Army of Tennessee, January–March 1864." In Gordon and Inscoe, *Inside the Confederate Nation*, 111–27.

Bohmfalk, Frederick H. "Cleburne's Victory at the Pickett Settlement, May 27, 1864." In Joslyn, *A Meteor Shining Brightly*, 211–32.

Bollet, Alfred J. *Civil War Medicine: Challenges and Triumphs.* Tucson, AZ: Galen Press, 2002.

Bomar Family Papers. MSS 86. Manuscripts, Archives and Rare Books Library, Emory University. Cited as Bomar Papers.

Bonds, Russell S. "Pawn Takes Bishop." *Civil War Times Illustrated* 45/3 (May 2006): 52–58.

———. "Sherman's 1st March through Georgia." *Civil War Times Illustrated* 46/6 (August 2007): 30–37.

———. "The W & A: Sherman's Lifeline." *Civil War Times Illustrated* 46/6 (August 2007): 35.

———. *War Like the Thunderbolt: The Battle and Burning of Atlanta.* Yardley, PA: Westholme Publishing, 2009.

Bonekemper, Edward H, III. *The 10 Biggest Civil War Blunders.* Washington: Regnery History, 2018.

Bonin, John Aubrey. "Lost Victories: Johnston and Sherman at Cassville." *Blue & Gray* 13/6 (August 1996): 28–37.

Borcke, Heros von. *Memoirs of the Confederate War for Independence.* London: W. Blackwood and Sons, 1866.

Brackett, Albert G. "With Sherman at Atlanta." In Peter Cozzens, ed., *Battles and Leaders of the Civil War* (Urbana: University of Illinois Press, 2004) 6:431–35.

Bragg, Braxton. Papers. Manuscript Collection No. 2000, Western Reserve Historical Society, Cleveland, OH. (Microfilm at John C. Hodges Library, University of Tennessee, Knoxville.) Cited as Bragg Papers.

Bragg, William Harris. *Joe Brown's Army: The Georgia State Line, 1862–1865.* Macon, GA: Mercer University Press, 1987.

———. "The Union General Lost in Georgia." *Civil War Times Illustrated* 24/4 (June 1985): 16–23.

Braley, Chad O. *The Battle of Gilgal Church: An Archeological and Historical Study of Mid-Nineteenth Century Warfare in Georgia.* Athens, GA: Southeastern Archeological Services, 1987.

Breckinridge, W. C. P. "The Opening of the Atlanta Campaign." In Johnson and Buel, *Battles and Leaders of the Civil War*, 4:277–81.

Breeden, James O. "Medical History of the Later Stages of the Atlanta Campaign." *Journal of Southern History* 35/1 (February 1969): 31–59.

Brew, Cowan. "Death Deep in the Woods." *Civil War Quarterly* 1/2 (Early Winter 2013): 14–23.

Bridges, Hal. *Lee's Maverick General: Daniel Harvey Hill.* New York: McGraw-Hill, 1961.

Bright, John M. "The States in the Confederate War." *Confederate Veteran* 17/8 (August 1909): 393–99.

Brooksher, William R., and David K. Snider. *Glory at a Gallop: Tales of the Confederate Cavalry.* Gretna, LA: Pelican Publishing Company, 2002.

Brown, Barry L., and Gordon R. Elwell. *Crossroads of Conflict: A Guide to Civil War Sites in Georgia.* Athens: University of Georgia Press, 2010.

Brown, Dee Alexander. *The Bold Cavaliers: Morgan's 2nd Kentucky Cavalry Raiders.* Philadelphia: Lippincott, 1959.

Brown, Kent Masterson. *Retreat from Gettysburg: Lee, Logistics and the Pennsylvania Campaign.* Chapel Hill: University of North Carolina Press, 2005.

Brown, Russell K. *Fallen in Battle: American General Officer Combat Fatalities from 1775.* Greenport, CT: Greenwood Press, 1988.

———. *To the Manner Born: The Life of General William H. T. Walker.* Athens: University of Georgia Press, 1994.

Brown, Thomas J. *John Bell Hood: Extracting Truth from History.* Bloomington, IN: Xlibris, 2012.

Bryan, Charles F., Jr., and Nelson D. Lankford, eds. *Eye of the Storm: A Civil War Odyssey Written by Private Robert Knox Sneden.* New York: Free Press, 2000.

Bryan, Charles F. Jr., James C. Kelly, and Nelson D. Lankford. *Images from the Storm: 300 Civil War Images by the Author of Eye of the Storm.* New York: Free Press, 2001.

Buck, William Pettis. "Headquarters Staff: Cleburne's Division, Hardee's Corps." In Joslyn, *A Meteor Shining Brightly*, 89–112.

Buell, Thomas B. *The Warrior Generals: Combat Leadership in the Civil War.* New York: Crown Publishers, 1997.

Bull, Rice C. *Soldiering: The Civil War Diary of Rice C. Bull, 123rd New York Volunteer Infantry.* Edited by K. Jack Bauer. San Rafael, CA: Presidio Press, 1977.

Burne, Alfred H. *Lee, Grant, and Sherman: A Study of Leadership in the 1864–65 Campaign.* New York: Charles Scribner's Sons, 1939.

Burt, Jesse C. "Sherman, Railroad General." *Civil War History* 2/1 (March 1956): 45–54.

Busey, John W., and David G. Martin. *Regimental Strengths and Losses at Gettysburg.* Hightstown, NJ: Longstreet House, 1994.

Butkovich, Brad. *The Battle of Pickett's Mill: Along the Dead-Line.* Charleston: History Press, 2013.

Bynum, Hartwell T. "Sherman's Expulsion of the Roswell Women in 1864." *Georgia Historical Quarterly* 54/2 (Summer 1970): 169–82.

Caldwell, J. F. J. *The History of a Brigade of South Carolinians, Known First as "Gregg's" and Subsequently as "McGowan's Brigade."* Dayton, OH: Press of Morningside Bookshop, 1974.

Candler, Allen D. "Watch on the Chattahoochee: A Civil War Letter." Edited by Elizabeth Hulsey Marshall. *Georgia Historical Quarterly* 43/4 (December 1959): 427–28.

———, ed. *The Confederate Records of the State of Georgia.* 6 vols. Atlanta: Charles P. Byrd, 1909–12.
Cannan, John. *The Atlanta Campaign May-November 1864.* Conshohocken, PA: Combined Books, 1991.
"Capt. Richard Beard." *Confederate Veteran* 39/11 (November 1931): 425.
Carlock, Chuck, and V. M. Owens. *History of the Tenth Texas Cavalry (Dismounted) Regiment 1861–1865.* N. Richland Hills, TX:Smithfield Press, 2001.
Carpenter, John A. *Sword and Olive Branch: Oliver Otis Howard.* Pittsburgh: University of Pittsburgh Press, 1964.
Carroll, Edward. "There Will Be Home Made Thunder Here." *Civil War Times Illustrated* 28/4 (Summer 1989): 29.
Carson, E. Scott. "Hampton's Legion and Hood's Brigade." *Confederate Veteran* 16/7 (July 1908): 342.
Carter, Samuel III. *The Siege of Atlanta, 1864.* New York: St. Martin's Press, 1973.
Casdorph, Paul D. *Prince John Magruder: His Life and Campaigns.* New York: John Wiley & Sons, 1996.
Castel, Albert. "Alpheus S. 'Ol Pap' Williams: The Fighting and Writing General from Michigan." In Castel, *Articles of War*, 68–78.
———. *Articles of War: Winners, Losers and Some Who Were Both during the Civil War*. Mechanicsburg, PA: Stackpole Books, 2001
———. "The Battle That Should Not Have Been" [Resaca]. *Civil War Times Illustrated* 31/5 (November/December 1992): 41–46, 68, 70, 72, 74–78, 82–83.
———. "'Black Jack' Logan." *Civil War Times Illustrated* 15/7 (November 1976): 4–6, 8–10, 41–45.
———. "Death Comes to the Bishop: When Luck Ran Out for Leonidas Polk." In Castel, *Articles of War,* 150–60.
———. *Decision in the West: The Atlanta Campaign of 1864.* Lawrence: University Press of Kansas, 1992.
———. "In Quest of Glory: George Stoneman's Attempt to Free the Andersonville Prisoners." In Castel, *Articles of War,* 114–24.
———. "Mars and the Reverend Longstreet: or, Attacking and Dying in the Civil War. *Civil War History* 33/2 (June 1987): 103–114. Reprinted in Castel, *Winning and Losing,* 119–32, and in Hubbell, *Conflict and Command*, 312–24.
———. "Prevaricating through Georgia: Sherman's *Memoirs* as a Source on the Atlanta Campaign." In Castel, *Winning and Losing*, 89–116.
———. *Tom Taylor's Civil War.* Lawrence: University Press of Kansas, 2000.
———. "Union Fizzle at Atlanta: The Battle of Utoy Creek." *Civil War Times Illustrated* 16/10 (February 1978): 26–32.
———. *Winning and Losing in the Civil War: Essays and Stories*. Columbia: University of South Carolina Press, 1996.
Cate, Wirt Armistead, ed. *Two Soldiers: The Campaign Diaries of Thomas J. Key, C.S.A. December 7, 1863–May 17, 1865 and Robert J. Campbell, U.S.A. January 1, 1864–July 21, 1864.* Chapel Hill: University of North Carolina Press, 1938.
Cathey, M. Todd, and Gary W. Waddey. *"Forward My Brave Boys!": A History of the 11th Tennessee Volunteer Infantry C.S.A. 1861–1865.* Macon: Mercer University Press, 2016.
Catton, Bruce. "The Battle of Atlanta." *Georgia Review* 10/3 (Fall 1956): 256–64.

———. "The Famous Cyclorama of the Great Battle of Atlanta." *American Heritage* 7/2 (February 1956): 33–44.

———. *Mr. Lincoln's Army.* Garden City, NY: Doubleday & Company, 1951.

———. *Never Call Retreat.* Garden City, NY: Doubleday & Company, 1965.

"Causes of the Defeat of Gen Lee's Army at the Battle of Gettysburg." *Southern Historical Society Papers* 4/2 (August 1877): 49–50.

Chamberlayne, John H. *Ham Chamberlayne—Virginian: Letters and Papers of an Artillery Officer in the War for Southern Independence 1861–1865.* Edited by C. G. Chamberlayne. Richmond: Press of the Dietz Printing Co., Publishers, 1932.

Chamberlin, W. H. "Hood's Second Sortie at Atlanta." In Johnson and Buel, *Battles and Leaders of the Civil War*, 4:326–31.

Chapman, W. S. "Comparing Two Great Armies." *Confederate Veteran* 7/3 (March 1899): 100–101.

Cheeks, Robert C. "Carnage in a Cornfield." *America's Civil War* 5/2 (July 1992): 30–37.

Chesnut, Mary Boykin. *A Diary from Dixie.* Edited by Ben Ames Williams. Boston: Houghton Mifflin Company, 1949.

Chiles, Paul. "Artillery Hell! The Guns of Antietam." *Blue & Gray* 16/2 (December 1998): 6–16, 24–25, 41–44, 49–59.

Chilton, F. B., ed. *Unveiling and Dedication of Monument to Hood's Texas Brigade.* Houston: self-published, 1911.

Chollet, David G. "Advance and Retreat: Rage or Reason?" *Atlanta Historical Bulletin* 20/1 (Spring 1976): 53–78.

Christ, Mark K., ed. *Getting Used to Being Shot At: The Spence Family Civil War Letters.* Fayetteville: University of Arkansas Press, 2002.

Cisco, Walter Brian. *States Rights Gist: A South Carolina General of the Civil War.* Shippensburg, PA: White Mane Publishing Company, 1991.

Clampitt, Bradley R. *The Confederate Heartland: Military and Civilian Morale in the Western Confederacy.* Baton Rouge: Louisiana State University Press, 2011.

Clark, Donald A. *The Notorious "Bull" Nelson: Murdered Civil War General.* Carbondale: Southern Illinois University Press, 2011.

Clark, Sam L., and H. D. Riley, Jr. "Outline of the Organization of the Medical Department of the Confederate Army and Department of Tennessee." *Tennessee Historical Quarterly* 16/1 (February 1969): 55–82.

Clark, Walter A. *Under the Stars and Bars or, Memories of Four Years Service with the Oglethorpes, of Augusta, Georgia.* Jonesboro, GA: Freedom Hill Press, 1987.

von Clausewitz, Carl. *On War.* Edited by Anatol Rapaport. London: Penguin Books, 1982.

Clauss, Errol MacGregor. "The Atlanta Campaign, 18 July–2 September 1864." Ph.D. dissertation, Emory University, 1965.

———. "The Battle of Jonesborough." *Civil War Times Illustrated* 7/7 (November 1968): 12–23.

———. "Sherman's Rail Support in the Atlanta Campaign." *Georgia Historical Quarterly* 50/4 (December 1966): 413–20.

Clayton, H. D. Maps of Alabama, Georgia, and Tennessee, 1862–1864. Photocopies of originals. Alabama Department of Archives and History. Location No. CD-5 (oversized map folio).

Cleaves, Freeman. *Rock of Chickamauga: The Life of General George H. Thomas.* Norman: University of Oklahoma Press, 1948.

Coburn, Mark. *Terrible Innocence: General Sherman at War.* New York: Hippocrene Books, 1993.

Coco, Gregory A. *A Strange and Blighted Land—Gettysburg: The Aftermath of a Battle.* Gettysburg: Thomas Publications, 1995.

———. *A Vast Sea of Misery: A History and Guide to the Union and Confederate Field Hospitals at Gettysburg, July 1–November 20, 1863.* Gettysburg: Thomas Publications, 1988.

Coddington, Edwin B. *The Gettysburg Campaign: A Study in Command.* New York: Charles Scribner's Sons, 1968.

Coffey, David. *John Bell Hood and the Struggle for Atlanta.* Abilene: McWhiney Foundation Press, 1998.

Cohen, Stan, and James G. Bogle. *The General & the Texas: A Pictorial History of the Andrews Raid, April 12, 1862.* Missoula, MT: Pictorial Histories Publishing Co., 1999.

Colston, F. M. "Gettysburg as I Saw It." *Confederate Veteran* 5/11 (November 1897): 551–53.

Commager, Henry Steele, ed. *The Blue and the Gray: The Story of the Civil War as Told by Participants.* Indianapolis: Bobbs-Merrill Company, 1950.

Compton, James. "The Second Division of the 16th Army Corps, in the Atlanta Campaign." In Kerksis, comp., *The Atlanta Papers*, 237–57.

Cone, Daniel. *Last to Join the Fight: The 66th Georgia Infantry.* Macon: Mercer University Press, 2014.

Connelly, Donald B. *John M. Schofield and the Politics of Generalship.* Chapel Hill: University of North Carolina Press, 2006.

Connelly, Thomas Lawrence. *Autumn of Glory: The Army of Tennessee, 1862–1865.* Baton Rouge: Louisiana State University Press, 1971.

———, and Archer Jones. *The Politics of Command: Factions and Ideas in Confederate Strategy.* Baton Rouge: Louisiana State University Press, 1973.

Conyngham, David P. *Sherman's March through the South.* New York: Sheldon and Co., 1865.

Cook, Ruth Beaumont. *North across the River: A Civil War Trail of Tears.* Birmingham, AL: Crane Hill Publishers, 1999.

Cooke, John Esten. *Wearing of the Gray; Being Personal Portraits, Scenes and Adventures of the War.* Bloomington: Indiana University Press, 1957.

Cooling, B. F. *Jubal Early's Raid on Washington: 1864.* Baltimore: Nautical & Aviation Publishing Company of America, 1989.

Cooper, Edward S. *William Babcock Hazen: The Best Hated Man.* Madison, NJ: Fairleigh Dickinson University Press, 2005.

Cooper, James L. "The Civil War Diary of Captain James Litton Cooper, September 30, 1861 to January 1865." Edited by William T. Alderson. *Tennessee Historical Quarterly* 15/2 (June 1956): 164.

Cooper, William J., Jr. *Jefferson Davis, American.* New York: Alfred A. Knopf, 2000.

Coppee, Henry. *General Thomas.* New York: D. Appleton and Company, 1893.

Cormier, Steven A. *The Siege of Suffolk: The Forgotten Campaign April 11–May 4, 1863.* Lynchburg, VA: H. E. Howard, 1989.

Cox, Jacob D. *Atlanta.* Dayton, OH: Morningside, 1987.

Cox, Rowland. "Snake Creek Gap, and Atlanta." In Kerksis, comp., *The Atlanta Papers*, 327–51.

Coxe, John. "In the Battle of Seven Pines." *Confederate Veteran* 30/1 (January 1922): 25.

Cozzens, Peter, ed. *Battles and Leaders of the Civil War, Volume 6.* Urbana: University of Illinois Press, 2004.

———. *General John Pope: A Life for the Union.* Urbana: University of Illinois Press 2000.
———. *No Better Pace to Die: The Battle of Stones River.* Urbana: University of Illinois Press, 1990.
———. *This Terrible Sound: The Battle of Chickamauga.* Urbana: University of Illinois Press, 1992.
———, and Robert I. Girardi, eds. *The New Annals of the Civil War.* Mechanicsburg, PA: Stackpole Books, 2004.
Crabb, Martha L. *All Afire to Fight: The Untold Tale of the Civil War's Ninth Texas Cavalry.* New York: William Morrow, 2000.
Crist, Lynda Lasswell, Kenneth H. Williams, and Peggy L. Dillard, eds. *The Papers of Jefferson Davis.* 14 vols. Baton Rouge: Louisiana State University Press, 1971–2018.
Crozier, Granville H. "A Private with General Hood." *Confederate Veteran* 25/12 (December 1917): 556–58.
Cullen, Joseph P. "The Battle of Chancellorsville." *Civil War Times Illustrated* 7/2 (May 1968): 4–24, 29–40.
———. "Gaines's Mill." *Civil War Times Illustrated* 3/1 (April 1964): 10–17, 24.
Cumming, J. B. Cumming War Recollections 1861–1865. #2650-Z. Southern Historical Collection, Wilson Library, University of North Carolina.
Cumming, Kate. *Kate: The Journal of a Confederate Nurse.* Edited by Richard Barksdale Harwell. Baton Rouge: Louisiana State University Press, 1959.
Curry, W L. "Raid of the Union Cavalry, Commanded by General Judson Kilpatrick Around the Confederate Army in Atlanta, August 1864." In Kerksis, comp., *The Atlanta Papers*, 599–621.
Cutrer, Thomas W., ed. *Longstreet's Aide: The Civil War Letters of Major Thomas J. Goree.* Charlottesville: University Press of Virginia, 1995.
Cutting, Elisabeth. *Jefferson Davis, Political Soldier.* New York: Dodd, Mead and Company, 1930.
Dabney, T. G. "Gen. A. P. Stewart on Strong Topics." *Confederate Veteran* 17/1 (January 1909): 31–32.
———. "When Hood Superseded Johnston." *Confederate Veteran* 22/9 (September 1914): 406–407.
Daniel, Larry J. *Days of Glory: The Army of the Cumberland, 1861–1865.* Baton Rouge: Louisiana State University Press, 2004.
———. *Soldiering in the Army of Tennessee: A Portrait of Life in a Confederate Army.* Chapel Hill: University of North Carolina Press, 1991.
Davis, Archie K. *Boy Colonel of the Confederacy: The Life and Times of Henry King Burgwyn Jr.* Chapel Hill: University of North Carolina Press, 1985.
Davis, Burke. *Jeb Stuart: The Last Cavalier.* New York: Rinehart, 1957.
Davis, Jefferson. Collection. MSS 22. Manuscripts, Archives and Rare Books Library, Emory University. Cited as Jefferson Davis Collection.
Davis, Jefferson. *The Rise and Fall of the Confederate Government.* 2 vols. New York: Thomas Yoseloff, 1958.
Davis, Nicholas A. *Chaplain Davis and Hood's Texas Brigade: Being an Expanded Edition of the Reverend Nicholas A. Davis' The Campaign from Texas to Maryland with the Battle of Fredericksburg.* Edited by Donald E. Everett. Baton Rouge: Louisiana State University Press, 1999.

Davis, Robert S., Jr. "'Every Crossroads and Farm': Confederate General Henry DeLamar Clayton's Civil War Maps of Northwestern Georgia." *Georgia Historical Quarterly* 82/1 (Spring 1998): 151–67.

Davis, Stephen. "Atlanta Campaign: Hood Fights Desperately. The Battles for Atlanta. Actions from July 10 to September 2, 1864." *Blue & Gray* 6/6 (August 1989): 8–18, 20–22, 25–26, 28–30, 32–34, 36, 38–39, 45–53, 59–62.

———. *Atlanta Will Fall: Sherman, Joe Johnston and the Yankee Heavy Battalions.* Wilmington, DE: Scholarly Resources, 2001.

———. "An Avalanche of Wounded: Atlanta's Confederate Hospitals and the Challenge of Chickamauga, Fall 1963." *Atlanta Medicine* 78/3 (2005): 5–11.

———. "The Death and Burials of General Richard Brooke Garnett." *Gettysburg Magazine* 5 (July 1991): 107–16.

———. "The Death of Bishop Polk." *Blue & Gray* 6/5 (June 1989): 13–14.

———. "Gen. Sherman's Blunder Led to McPherson's Death." *Civil War News* 42/7 (August 2016): 18–19.

———. "A Georgia Firebrand: Major General W. H. T. Walker, C. S. A." *Georgia Historical Quarterly* 64/4 (Winter 1979): 447–60.

———. "How Many Civilians Died in Sherman's Bombardment of Atlanta?" *Atlanta History* 45/4 (2003): 4–23.

———. "Issues Affecting Jefferson Davis' Decision to Fire Joe Johnston." Paper presented at the Eighteenth Annual Deep Delta Civil War Symposium, Southeastern Louisiana University, June 11, 2004.

———. "John Bell Hood." In Bruce S. Allardice and Lawrence Lee Hewitt, *Kentuckians in Gray: Confederate Generals and Field Officers of the Bluegrass State.* Lexington: University Press of Kentucky, 2008. 152–58.

———. "The Literature of Confederate Victory: Forays in Imaginary History." *Lincoln Herald* 90/2 (Summer 1988): 44–50.

———. "No Hope of Success." *Civil War Times* 57/2 (April 2018): 34–41.

———. "A Reappraisal of the Generalship of John Bell Hood in the Battles for Atlanta." In Savas and Woodbury, *The Campaign for Atlanta*, 49–95.

———. "Sherman in North Georgia: The Battle of Resaca." *Blue & Gray* 31/4 (Summer 2015): 6–9, 17–18, 20–24, 26, 42–50.

———. "'A Very Barbarous Mode of Carrying on War': Sherman's Artillery Bombardment of Atlanta, July 20–August 24, 1864." *Georgia Historical Quarterly* 79/1 (Spring 1995): 57–90.

———. *What the Yankees Did to Us: Sherman's Bombardment and Wrecking of Atlanta.* Macon: Mercer University Press, 2012.

Davis, Steve. "John Bell Hood's 'Addictions' in Civil War Literature." *Blue & Gray* 16/1 (Fall 1998): 28–31.

———. "'That Extraordinary Document': W. H. T. Walker and Patrick Cleburne's Emancipation Proposal." *Civil War Times Illustrated* 16/8 (December 1977): 14–20.

[Davis, Steve, ed.] "The Cost of Kolb's Farm: A Confederate Medical Report." *Blue & Gray* 6/5 (June 1989): 24.

Davis, William C. *Breckinridge: Statesman, Soldier, Symbol.* Baton Rouge: Louisiana State University Press, 1974.

———. *The Cause Lost: Myths and Realities of the Confederacy.* Lawrence: University Press of Kansas, 1996.

———, ed. *The Confederate General.* 6 vols. Harrisburg, PA: National Historical Society, 1991.
———. *The Deep Waters of the Proud.* New York: Doubleday & Company, 1982.
———. *The Generals of the Civil War.* New York: Quadrillion Media, 1993.
———. *Jefferson Davis—The Man and His Hour: A Biography.* New York: HarperCollins Publishers, 1991.
———. *The Orphan Brigade: The Kentucky Confederates Who Couldn't Go Home.* Garden City, NY: Doubleday & Company, 1980.
Davison, Eddy W., and Daniel Foxx. *Nathan Bedford Forrest: In Search of an Enigma.* Gretna, LA: Pelican Publishing Company, 2007.
Dawes, E. C. "The Confederate Strength in the Atlanta Campaign." In Johnson and Buel, *Battles and Leaders of the Civil War*, 4:281–83.
Dawson, Joseph G. "Philip Henry Sheridan." In *Biographical Dictionary of the Union: Northern Leaders of the Civil War*, edited by John T. Hubbell and James W. Geary, eds., 471–72. Westport, CT: Greenwood Press, 1995.
Dean, Jeff. "The Battle of Pickett's Mill." *Blue & Gray* 6/4 (April 1989): 28–37.
Dean, Jeffrey S. *The Hell-Hole in Georgia: Sherman vs. Johnston May 22–June 2, 1864.* Westminster MD: Heritage Books, 2006.
[De Fontaine, Felix G.] *Marginalia; or, Gleanings from an Army Note-Book by "Personne."* Columbia, SC: Steam Power-Press of F. G. de Fontaine, 1864.
DeGress, Francis. "Report of the Battle of Atlanta." Edited by E. Chris Evans. *Blue & Gray* 11/4 (April 1994): 28–30.
Denkler, Kirk, ed. *Voices of the Civil War: Atlanta.* Alexandria, VA: Time-Life Books, 1996.
Dickey, Tom. "Johnston's River Line." *Brown's Guide to Georgia* (March–April 1975): 24–29.
Dodd, William E. *Jefferson Davis.* Philadelphia: George W. Jacobs & Company, 1907.
Dodge, Grenville M. "The Battle of Atlanta." In Kerksis, comp., *The Atlanta Papers*, 487–503.
———. *The Battle of Atlanta and Other Campaigns.* Denver: Sage Books, 1965. Originally published 1910 by Monarch Printing Co. (Council Bluffs, IA).
———. "The Late J. M. Schofield." *Confederate Veteran* 15/10 (October 1907): 460–61.
Dodson, W. C., ed. *Campaigns of Wheeler and His Cavalry 1862–1865.* Jackson, TN: Guild Bindery Press, n.d. [1899].
Dolzall, Gary W. "Opportunity Lost." *America's Civil War* 17/2 (May 2004): 24–30.
Donnell, James. *Atlanta 1864: Sherman Marches South.* Oxford, UK: Osprey Publishing, 2016.
Douglas, H. T. "Hood's Messmates at Yorktown." *Confederate Veteran* 19/2 (February 1911): 78–79.
Dowdey, Clifford. *Lee.* Boston: Little, Brown and Company, 1965.
———. *The Seven Days: The Emergence of Lee.* Boston: Little, Brown and Company, 1964.
———, and Louis H. Manarin, eds. *The Wartime Papers of R. E. Lee.* Boston: Little, Brown and Company, 1961.
Dozer, Jesse. L. "Marching with Sherman through Georgia and the Carolinas: Civil War Diary of Jesse L. Dozer." Edited by Wilfred W. Black. *Georgia Historical Quarterly* 52/3 (September 1968): 308–336.
Drake, Edwin L., ed. *The Annals of the Army of Tennessee and Early Tennessee History* 1/1 (April 1878): 1–13.

———. *The Annals of the Army of Tennessee and Early Western History.* Jackson, TN: Guild Bindery Press, 1998. Originally published 1878 by A. D. Haynes (Nashville).

DuBose, Beverly M., Jr. "Georgia's Civil War Centennial Activities." *Atlanta Historical Bulletin* 10/39 (September 1965): 53–69.

DuBose, John Witherspoon. *General Joseph Wheeler and the Army of Tennessee.* New York: Neale Publishing Co., 1912.

Dunkelman, Mark H. "'The Worst Sight I Ever Saw': The 154th New York Infantry at the Battle of Peachtree Creek." *North & South* 9/2 (May 2006): 64–79.

Dunn, Mathew A. "Notes and Documents: Mathew Andrew Dunn Letters." Edited by Weymouth T. Jordan. *Journal of Mississippi History* 1/1 (January 1939): 110–27.

Dwight, Henry O. "How We Fight at Atlanta." In Commager, *The Blue and the Gray*, 939–43.

Dyer, John P. *"Fightin' Joe" Wheeler.* Baton Rouge: Louisiana State University Press, 1941.

———. *The Gallant Hood.* Indianapolis: Bobbs-Merrill Company, 1950.

Eastwood, Bruce S. "Confederate Medical Problems in the Atlanta Campaign." *Georgia Historical Quarterly* 47/3 (September 1963): 276–92.

Eaton, Clement. *Jefferson Davis.* New York: Free Press, 1977.

Ecelbarger, Gary. "An Ambitious Goal Indeed: An Evaluation of General John B. Hood's Plan and Generalship during the Battle of Atlanta, July 22, 1864." *North & South* 13/2 (July 2011): 21–31.

———. *The Day Dixie Died: The Battle of Atlanta.* New York: Thomas Dunne Books, 2010.

———. "A Human Hurricane on Horseback: General John A. Logan and the Atlanta Campaign." *North & South* 7/7 (November 2004): 70–82.

———. *Slaughter at the Chapel: The Battle of Ezra Church, 1864.* Norman: University of Oklahoma Press, 2016.

Eckenrode, H. J., and Bryan Conrad. *James Longstreet: Lee's War Horse.* Chapel Hill: University of North Carolina Press, 1936.

Edge, Andrew J. Correspondence. #346. Microfilm reel 1. Manuscripts, Archives and Rare Books Library, Emory University. Cited as Edge Correspondence.

Eggleston, Edmund T. "Excerpts from the Civil War Diary of E. T. Eggleston." Edited by Edward Noyes. *Tennessee Historical Quarterly* 17/4 (December 1958): 336–58.

———. *To Succeed or Perish: The Diaries of Sergeant Edmund Trent Eggleston, 1st Mississippi Light Artillery Regiment, CSA.* Knoxville: University of Tennessee Press, 2015.

Eggleston, George Cary. *The History of the Confederate War: Its Causes and its Conduct—A Narrative and Critical History.* 2 vols. New York: Negro Universities Press, 1970.

Einolf, Christopher J. *George Thomas: Virginian for the Union.* Norman: University of Oklahoma Press, 2007.

Eisenschiml, Otto, and Ralph Newman. *The American Iliad: The Epic Story of The Civil War as Narrated by Eyewitnesses and Contemporaries.* Secaucus, NJ: Blue & Grey Press, 1985.

Eiserman, Rick. "'We Have Had a Picture Taken.'" *Civil War Times Illustrated* 49/4 (August 2011): 44–47.

Elan, Mark. "The Road to Atlanta: The Role of Geography in Command and Decision Making During the Atlanta Campaign." Ph.D. dissertation, Florida State University, 1996.

Eleazar, W. D. "Fight at Dead Angle, in Georgia." *Confederate Veteran* 14/7 (July1906): 312.

Eliot, Ellsworth, Jr. *West Point in the Confederacy.* New York: G. A. Baker & Co., 1941.

Elliott, Sam Davis. *John C. Brown of Tennessee: Rebel, Redeemer, and Railroader.* Knoxville: University of Tennessee Press, 2017.

———. *Soldier of Tennessee: General Alexander P. Stewart and the Civil War in the West.* Baton Rouge: Louisiana State University Press, 1999.

Ellis, Barbara, and Stephen J. Dick. "'Who was "Shadow"?' The Computer Knows: Applying Grammar-Program Statistics in Content Analysis to Solve Mysteries about Authorship." *Journalism & Mass Communication Quarterly* 73/4 (Winter 1996): 947–62.

Evans, Clement A. ed. *Confederate Military History.* 12 vols. New York: Thomas Yoseloff, 1962.

Evans, David. "The Atlanta Campaign." *Civil War Times Illustrated* 28/4 (Summer 1989): 13–16, 18, 20–24, 28–30, 32–34, 36–38, 40–61.

———. "The Fight for the Wagons." *Civil War Times Illustrated* 26/10 (February 1988): 16–22.

———. *Sherman's Horsemen: Union Cavalry Operations in the Atlanta Campaign.* Bloomington: Indiana University Press, 1996.

———. "Wool, Women, and War." *Civil War Times Illustrated* 26/5 (September 1987): 38–42.

Ferguson, John H. *On to Atlanta: The Civil War Diaries of John Hill Ferguson Illinois Tenth Regiment of Volunteers.* Edited by Janet Correll Ellison. Lincoln: University of Nebraska Press, 2001.

"Fighting Words." *America's Civil War* 23/4 (September 2010): 30–37.

Fleming, James R. *The Confederate Ninth Tennessee Infantry.* Gretna, LA: Pelican Publishing Company, 2006.

Fletcher, William A. *Rebel Private Front and Rear.* Edited by Bell Irvin Wiley. Austin: University of Texas Press, 1954.

Fordney, Ben Fuller. *George Stoneman: A Biography of the Union General.* Jefferson, NC: McFarland, 2008.

Forstchen, William R. "Lee's Victory at Gettysburg…and Then What?" In Thomsen and Greenberg, *Alternate Gettysburgs,* 324–35.

Foster, Samuel T. *One of Cleburne's Command: The Civil War Reminiscences and Diary of Capt. Samuel T. Foster, Granbury's Texas Brigade, CSA.* Edited by Norman D. Brown. Austin: University of Texas Press, 1980.

Fowler, John D. *Mountaineers in Gray: The Nineteenth Tennessee Volunteer Infantry Regiment, C.S.A.* Knoxville: University of Tennessee Press, 2004.

Frazer, C. W. "Fifth Confederate." In Lindsley, *The Military Annals of Tennessee,* 1:146–53.

Freeman, Douglas Southall, ed. *Lee's Dispatches: Unpublished Letters of General Robert E. Lee, C. S. A. to Jefferson Davis and the War Department of the Confederate States of America.* New York: G. P. Putnam's Sons, 1915.

———. *Lee's Lieutenants: A Study in Command.* 3 vols. New York: Charles Scribner's Sons, 1942–1944.

———. *R. E. Lee: A Biography.* 4 vols. New York: Charles Scribner's Sons, 1934–1935.

Freeman, Douglas Southall, and Grady McWhiney, eds. *Lee's Dispatches: Unpublished Letters of General Robert E. Lee, C. S. A. to Jefferson Davis and the War Department of the Confederate States of America 1862–1865.* New York: G. P. Putnam's Sons, 1957.

Freemon, Frank R. *Microbes and Minie Balls: An Annotated Bibliography of Civil War Medicine.* Rutherford, NJ: Fairleigh Dickinson University Press, 1993.

Fremantle, Arthur J. L. *The Fremantle Diary.* Edited by Walter Lord. Boston: Little, Brown and Company, 1954.

French, S. G. "Lt. General Leonidas Polk at Cassville." *Southern Historical Society Papers* 22 (January–December 1894): 1–9.

French, Samuel G. "Kennesaw Mountain." *Southern Historical Society Papers* 9, nos. 10–12 (October–December 1881): 505–511.

———. *Two Wars: An Autobiography of Gen. Samuel G. French.* Huntington, WV: Blue Acorn Press, 1999. Originally published 1901 by Confederate Veteran (Nashville).

French, Steve. "'Hurry Was the Order of the Day': Imboden and the Wagon Train of Wounded." *North & South* 2/6 (August 1999): 35–42.

Frey, David. *Failure to Pursue: How the Escape of Defeated Forces Prolonged the Civil War.* Jefferson, NC: McFarland, 2016.

[Frobel, Bushrod W.] "Field and Camp; or, the recorded facts of an eye-witness to numerous important events and occurrences, both in field and camp, during the late four years' war. By an Officer." *Scott's Monthly Magazine* 3/1 (January 1867): 1–7; 3/4 (April 1867): 269–75; 3/5 (May 1867): 325–34.

[———.] "The Georgia Campaign; or, a South-side View of Sherman's March to the Sea. By the author of 'Field and Camp.'" *Scott's Monthly Magazine* 6/4 (October 1868): 694–701; 6/5 (November 1868): 756–61; 5/6 (December 1868): 840–46.

Fryman, Robert J. "Fortifying the Landscape: An Archeological Study of Military Engineering and the Atlanta Campaign." In *Archaeological Perspectives on the American Civil War*, edited by Clarence R. Geier and Stephen R. Potter, 43–55. Gainesville: University Press of Florida, 2000.

Fullenkamp, Leonard, Stephen Bowman, and Jay Luvaas, eds. *Guide to the Vicksburg Campaign.* Lawrence: University Press of Kansas, 1998.

Fuller, John W. "A Terrible Day: The Battle of Atlanta, July 22, 1864." In Cozzens, *Battles and Leaders of the Civil War*, 5:546–59.

Fullerton, Dan C. *Armies in Gray: The Organizational History of the Confederate States Army in the Civil War.* Baton Rouge: Louisiana State University Press, 2017.

Funeral Services at the Burial of the Right Rev. Leonidas Polk D. D. Together with the Sermon delivered in St. Paul's Church, Augusta, GA., on June 29, 1864: Being a Feast of St. Peter the Apostle. Columbia, SC: Evans & Cogswell, 1864. There is a copy at Manuscript, Archives, and Rare Book Library, Emory University.

Furqueron, James. R. "The Bull of the Woods: Longstreet and the Confederate Left at Chickamauga." In *James Longstreet: The Man, the Soldier, the Controversy*, edited by R. L. DiNardo and Albert A. Nofi, 99–164. Conshohocken, PA: Combined Publishing, 1998.

Futch, Ovid L. *History of Andersonville Prison.* Gainesville: University of Florida Press, 1968.

Gallagher, Gary W. "The Battle for Joe Johnston." *Civil War* 71 (December 1998): 6–7.

———. *The Confederate War.* Cambridge, MA: Harvard University Press, 1997.

———. *Lee & His Army in Confederate History.* Chapel Hill: University of North Carolina Press, 2001.

———. *Lee and His Generals in War and Memory.* Baton Rouge: Louisiana State University Press, 1998.

———, Stephen D. Engle, Robert K. Krick, and Joseph T. Glatthaar. *The American Civil War: This Mighty Scourge of War.* Oxford, England: Osprey Publishing, 2003.

Galloway, Tammy Harden, ed. *Dear Old Roswell: The Civil War Letters of the King Family of Roswell, Georgia.* Macon: Mercer University Press, 2003.

Garrett, Franklin M. *Atlanta and Environs: A Chronicle of Its People and Events.* 2 vols. Athens: University of Georgia Press, 1969.

———. "Civilian Life in Atlanta." *Civil War Times Illustrated* 6/4 (July 1964): 30–33.

[Garrett, Franklin M. and Wilbur G. Kurtz]. "Key to Map." *Atlanta Historical Bulletin* 2/8 (September 1934): 5–29.

Garrison, Webb. *Atlanta and the War.* Nashville: Rutledge Hill Press, 1995.

Gay, Mary A. H. *Life in Dixie During the War.* Atlanta: Charles P. Byrd, 1897.

"General Joseph E. Johnston's Campaign in Georgia." *Southern Historical Society Papers* 21 (January–December 1893): 314–21.

"General Polk's Death: General Joseph E. Johnston Describes How He Was Killed." *Southern Historical Society Papers* 18 (January–December 1890): 380–81.

"General Sherman's Advance—View of Atlanta from the Signal Station—[Sketched by Theodore R. Davis]." *Harper's Weekly,* August 8, 1864, 501–502.

Gibbons, Robert. "Life at the Crossroads of the Confederacy: Atlanta, 1861–1865." *Atlanta Historical Journal* 23/2 (Summer 1979): 11–72.

Giles, Val C. "The Flag of the First Texas, A. N. Virginia." *Confederate Veteran* 15/9 (September 1907): 417.

———. "The Tom Green Rifles." *Confederate Veteran* 26/1 (January 1918): 20–23.

Gillis Family Papers, 1859–1865. #300. Microfilm reel 1. Manuscript, Archives and Rare Books Library, Emory University.

Gordon, Armistead C. *Jefferson Davis.* New York: C. Scribner's Sons, 1918.

Gordon, John B. *Reminiscences of the Civil War.* New York: Charles Scribner's Sons, 1904.

Gordon, Leslie J. "Courting Nationalism: The Wartime Letters of Bobbie Mitchell and Nettie Fondren." In Gordon and Inscoe, *Inside the Confederate Nation*, 188–208.

———. *General George E. Pickett in Life & Legend.* Chapel Hill: University of North Carolina Press, 1998.

———, and John C. Inscoe, eds. *Inside the Confederate Nation: Essays in Honor of Emory M. Thomas.* Baton Rouge: Louisiana State University Press, 2005.

Gordon Family Papers. No. 318. Georgia Historical Society, Savannah, GA.

Gorgas, Josiah. *The Journals of Josiah Gorgas, 1857–1878.* Edited by Sarah Woolfolk Wiggins. Tuscaloosa: University of Alabama Press, 1995.

Gottfried, Bradley M. *Roads to Gettysburg: Lee's Invasion of the North, 1863.* Shippensburg, PA: White Mane Books, 2001.

Gottschalk, Phil. *In Deadly Earnest: The Missouri Brigade CSA.* Columbia: Missouri River Press, 1991.

Govan, Gilbert E., and James W. Livingood. *A Different Valor: The Story of General Joseph E. Johnston, C. S. A.* Westport, CT: Greenwood Press, 1973.

Grant, Lemuel Pratt. Papers. MSS 100. Atlanta History Center. Cited as Lemuel Grant Papers.

Grant, Ulysses S. "Preparing for the Campaigns of '64." In Johnson and Buel, *Battles and Leaders of the Civil War*, 4:97–117.

Green, John W. *Johnny Green of the Orphan Brigade: The Journal of a Confederate Soldier.* Edited by A. D. Kirwan. Lexington: University Press of Kentucky, 1956.

Greene, A. Wilson. "Stoneman's Raid." In *Chancellorsville: The Battle and Its Aftermath*, edited by Gary W. Gallagher, 65–106. Chapel Hill: University of North Carolina Press, 1996.

———. "On the Edge of Disaster: Lee and McClellan Meet Atop South Mountain." *Civil War Times Illustrated* 24/7 (November 1986): 18–23, 44–46, 49–50.

Grimsley, Mark, and Brooks D. Simpson. *Gettysburg: A Battlefield Guide.* Lincoln: University of Nebraska Press, 1999.

Gruber, Drew. "'Texas need not feel ashamed.'" *Civil War Times Illustrated* 52/3 (June 2013): 52–57.

Guelzo, Allen C. *Fateful Lightning: A New History of the Civil War & Reconstruction.* New York: Oxford University Press, 2012.

———. *Gettysburg: The Last Invasion.* New York: Alfred A. Knopf, 2013.

Guthrey, D. M. "Wheeler's Cavalry around Atlanta." *Confederate Veteran* 13/6 (June 1905): 267.

Hagan, John W. *Confederate Letters of John W. Hagan.* Edited by Bell Irvin Wiley. Athens: University of Georgia Press, 1954.

———. "The Confederate Letters of John W. Hagan, Part II." Edited by Bell I. Wiley. *Georgia Historical Quarterly* 38/2 (June 1954): 268–90.

Hale, Douglas. *The Third Texas Cavalry in the Civil War.* Norman: University of Oklahoma Press, 1993.

Hallock, Judith Lee. *Braxton Bragg and Confederate Defeat.* Vol. 2. Tuscaloosa: University of Alabama Press, 1991.

Hamby, William R. "Fourth Texas in Battle of Gaines's Mill." *Confederate Veteran* 14/4 (April 1906): 183–85.

———. "Hood's Texas Brigade at Sharpsburg." *Confederate Veteran* 16/1 (January 1908): 19–20.

Hanson, Harry. *The Civil War.* New York: New American Library, 1961.

Hardee, William J. Papers. Alabama Department of Archives and History, Montgomery, AL.

Hardee, W. J. Letters. Photocopies and typescripts from Alabama Department of Archives and History, Bell Irvin Wiley Papers, MSS 521, box 173, folder 44. Manuscripts, Archives, and Rare Book Library, Emory University.

Harris, William C. *William Woods Holden: Firebrand of North Carolina Politics.* Baton Rouge: Louisiana State University Press, 1987.

Harrison, Mrs. Burton [Constance Cary]. *Recollections Grave and Gay.* New York: C. Scribner's Sons, 1911.

Harsh, Joseph L. *Confederate Tide Rising: Robert E. Lee and the Making of Southern Strategy, 1861–1862.* Kent, OH: Kent State University Press, 1998.

———. *Sounding the Shallows: A Confederate Companion for the Maryland Campaign of 1862.* Kent, OH: Kent State University Press, 2000.

Hartwig, D. Scott. "Lee's Shrinking Army." *America's Civil War* 29/2 (May 2016): 14, 16.

Hassler, William Woods. *A. P. Hill, Lee's Forgotten General.* Chapel Hill: University of North Carolina Press, 1984.

Hatfield, Eugene Adair. "Battle of Jonesboro." In *A History of Clayton County, Georgia 1821–1983*, edited by Alice Copeland Kilgore, Edith Hanes Smith, and Francis Partridge Tuck, 65–70. College Park GA: Ancestors Unlimited, 1983.

Hattaway, Herman. *General Stephen D. Lee.* Jackson: University Press of Mississippi. 1976.

———. "The General Whom the President Elevated Too High." In *Jefferson Davis' Generals*, edited by Gabor S. Boritt, 84–103, 187–88. New York: Oxford University Press, 1999.
———. "Stephen Dill Lee." In Davis, *The Confederate General*, 4:58–63.
———. "Stephen Dill Lee—A Profile." *Civil War Times Illustrated* 8/5 (August 1969): 17–24.
———, and Archer Jones. *How the North Won: A Military History of the Civil War.* Urbana: University of Illinois Press, 1983.
———, and Richard E. Beringer. *Jefferson Davis, Confederate President.* Lawrence: University Press of Kansas, 2002.
Hay, Thomas Robson. "The Atlanta Campaign" (Part One). *Georgia Historical Quarterly* 7/1 (March 1923): 19–43.
———. "The Atlanta Campaign" (Part Two). *Georgia Historical Quarterly* 7/2 (June 1923): 99–118.
———. "Davis, Bragg and Johnston in the Atlanta Campaign." *Georgia Historical Quarterly* 8/1 (March 1924): 38–48.
———. "The Davis-Hood-Johnston Controversy of 1864." *Mississippi Valley Historical Review* 11 (June 1924–March 1925): 54–84.
———, ed. *Cleburne and His Command by Capt. Irving A. Buck and Pat Cleburne Stonewall Jackson of the West by Thomas Robson Hay.* Jackson, TN: McCowat-Mercer Press, 1959.
Haynie, Paul D. "'I'll Be Damned If I Ever Surrender': How Colonel Thomas J. Harrison Saved McCook's Raid." *North & South* 1/6 (June 1998): 57–63.
Hazen, W. B. *A Narrative of Military Service.* Boston: Ticknor and Company, 1885.
Hebert, Walter H. *Fighting Joe Hooker.* Indianapolis: Bobbs-Merrill Company, 1944.
Henderson, Don. E. "General Hood's Brigade." *Southern Historical Society Papers* 35 (1907): 185–200.
Hennessy, John J. *Return to Bull Run: The Campaign and Battle of Second Manassas.* New York: Simon & Schuster, 1993.
———. "The Second Battle of Manassas: Lee Suppresses the 'Miscreant' Pope." *Blue & Gray* 9/6 (August 1992): 11–34, 46–58.
———. "Thunder at Chantilly." *North & South* 3/3 (March 2000): 47–60.
Herberich, John L. *Masters of the Field: The Fourth United States Cavalry in the Civil War.* Atglen, PA: Schiffer Publishing, 2015.
Hess, Earl J. *The Battle of Ezra Church and the Struggle for Atlanta.* Chapel Hill: University of North Carolina Press, 2015.
———. *The Battle of Peach Tree Creek: Hood's First Effort to Save Atlanta.* Chapel Hill: University of North Carolina Press, 2017.
———. *The Civil War in the West: Victory and Defeat from the Appalachians to the Mississippi.* Chapel Hill: University of North Carolina Press, 2012.
———. "Civilians at War: The Georgia Militia in the Atlanta Campaign." *Georgia Historical Quarterly* 66/3 (Fall 1982): 332–45.
———. *Kennesaw Mountain: Sherman, Johnston and the Atlanta Campaign.* Chapel Hill: University of North Carolina Press, 2013.
———. "On the War Path." *Hallowed Ground* 15/2 (Summer 2014): 18–29.
———, Richard W. Hatcher III, William Garrett Piston, and William T. Shea. *Wilson's Creek, Pea Ridge, and Prairie Grove: A Battlefield Guide with a Section on the Wire Road.* Lincoln: University of Nebraska Press, 2006.

Hewett, James B., Noah Andre Trudeau, and Bryce Suderow, eds. *Supplement to the Official Records of the Union and Confederate Armies.* 100 volumes. Wilmington NC: Broadfoot Publishing Company, 1994–2001.

Hewitt, Lawrence Lee, and Thomas E. Schott, eds. *Confederate Generals in the Western Theater: Essays on America's Civil War*, 4 vols to date. Knoxville: University of Tennessee Press, 2018.

Hewitt, Lawrence L. "James Patton Anderson." In Davis, *The Confederate General*, 1:22–25.

Hibbard, James, and Albert Castel. "Kilpatrick's Jonesboro Raid, August 18–22, 1864." *Atlanta Historical Journal* 29/2 (Summer 1985): 31–46.

Higdon, J. C. "Hindman's Reply to Hood." *Confederate Veteran* 8/2 (February 1900): 69.

Hill, Benjamin H., Jr. *Senator Benjamin H. Hill of Georgia: His Life, Speeches and Writings.* Atlanta: T. H. P. Bloodworth, 1893.

Hill, Louise Biles. *Joseph E. Brown and the Confederacy.* Chapel Hill: University of North Carolina Press, 1939.

Hinde, Paul. "Govan's heroic charge; Battle of Atlanta." *Inn Dixie Magazine (*July 1936): 5–6, 17.

———. "The great night march of Hardee's Corps." *Inn Dixie Magazine (*May 1937): 5–6, 17, 19–20.

Hirshson, Stanley P. *The White Tecumseh: A Biography of William T. Sherman.* New York: John Wiley & Sons, 1997.

Hitt, Michael C. *Charged with Treason: Ordeal of 400 Mill Workers during Military Operations in Roswell, Georgia, 1864–1865.* Monroe, NY: Library Research Associates, 1992.

Hoehling, A. A. *Last Train from Atlanta.* New York: Thomas Yoseloff, 1958.

Hoffsommer, Robert D. "The Aftermath of Gettysburg." *Civil War Times Illustrated* 2/4 (July 1963): 35–38.

Holsworth, Jerry W. "Uncommon Valor: Hood's Texas Brigade in the Maryland Campaign." *Blue & Gray* 13/6 (Summer 1996): 6–20, 50–55.

Hood, Ida Richardson. "In Memory of General J. B. Hood." *Southern Historical Society Papers* 32 (January–December 1904): 151–56.

Hood, J. B. *Advance and Retreat: Personal Experiences in the United States and Confederate States Armies.* New Orleans: Pub. for the Hood Orphan Memorial Fund, 1880. Cited as *A&R.*

———. Letter to James Chesnut, March 7, 1864. GLC02703, Gilder Lehrman Collection. Gilder Lehrman Institute of American History, New York, NY.

Hood, John Bell. Compiled Military Service File, NATF 86. National Archives and Records Administration, Washington, DC. Cited as Hood Service Record.

———. Telegrams Book I. Hood Personal Papers. Photocopy in the possession of Stephen M. Hood, Huntington, WV.

———. Telegram Book III. Hood Papers, National Archives. Photocopy courtesy of Stephen M. Hood.

Hood, Stephen M. "A Shot in the Arm." *America's Civil War* 26/4 (September 2013): 42–43.

———. "John Bell Hood: Pain & Prescriptions." *Civil War Times Illustrated* 54/2 (April 2015): 30–37.

———. *John Bell Hood: The Rise, Fall, and Resurrection of a Confederate General.* El Dorado Hills, CA: Savas Beatie, 2013.

———. *The Lost Papers of Confederate General John Bell Hood.* El Dorado Hills, CA: Savas Beatie, 2015.

"Hood to Hardee." July 21, 1864. Dubose Family Collection. MSS 1020. Box 18, folder 3. Atlanta History Center. Copy courtesy of Melanie Stephan.

"Hood's Headquarters Near Nashville." *Confederate Veteran* 11/8 (August 1894): 231.

Horn, Stanley F. *The Army of Tennessee: A Military History.* Norman: University of Oklahoma Press, 1953.

Hotchkiss, Jedediah. *Make Me a Map of the Valley: The Civil War Journal of Stonewall Jackson's Topographer.* Edited by Archie McDonald. Dallas: Southern Methodist University Press, 1973.

"How General Polk Was Killed." *Confederate Veteran* 31/5 (May 1923): 166.

Howard, Oliver O. *Autobiography of Oliver Otis Howard.* New York: Baker & Taylor Company, 1908.

———. "The Struggle for Atlanta." In Johnson and Buel, *Battles and Leaders of the Civil War,* 4:293–325.

Hubbell, John T., ed. *Conflict and Command.* Kent, OH: Kent State University Press, 2012.

———, and James W. Geary, eds. *Biographical Dictionary of the Union: Northern Leaders of the Civil War.* Westport, CT: Greenwood Press, 1995.

Hubner, Charles W. "Some Recollections of Atlanta During 1864." *Atlanta Historical Bulletin* 1/2 (January 1928): 5–7.

Hudson, Leonne M. *The Odyssey of a Southerner: The Life and Times of Gustavus Woodson Smith.* Macon: Mercer University Press, 1998.

Hughes, Nathaniel Cheairs, Jr. *The Pride of the Confederate Artillery: The Washington Artillery in the Army of Tennessee.* Baton Rouge: Louisiana State University Press, 1997.

———. *General William J. Hardee: Old Reliable.* Baton Rouge: Louisiana State University Press, 1965.

———, and Gordon J. Whitney. *Jefferson Davis in Blue: The Life of Sherman's Relentless Warrior.* Baton Rouge: Louisiana State University Press, 2002.

Hughes, Robert M. *General Johnston.* New York: D. Appleton and Company, 1893.

Hull, A. L. "Maj. Gen. Martin L. Smith." *Confederate Veteran* 6/11 (November 1898): 530.

Hull, Mark M. "Concerning the Emancipation of the Slaves." In Joslyn, *A Meteor Shining Brightly,* 143–77.

Hunter, J. T. "At Yorktown in 1862 and What Followed." *Confederate Veteran* 26/2 (February 1918): 66–68. Continued in 26/3 (March 1918): 112–15.

———. "Lieut. Gen. John B. Hood." *Confederate Veteran* 24/6 (June 1916): 257–59.

———. "A Thirty Days' Scout on the Potomac in '62." *Confederate Veteran* 24/10 (October 1916): 457–59.

"Incidents of Gen. McPherson's Death." *Confederate Veteran* 11/3 (March 1903): 118–19.

Iobst, Richard W. *Civil War Macon.* Macon: Mercer University Press, 1999.

Irvin, Richard A. Letter to William D. Irvin, May 22, 1864. Typescript courtesy of Gary Waddey, Nashville, TN.

Jackman, John S. *Diary of a Confederate Soldier: John S. Jackman of the Orphan Brigade.* Edited by William C. Davis. Columbia: University of South Carolina Press, 1990.

Jacobs, Lee, comp. *The Gray Riders: Stories from the Confederate Cavalry.* Shippensburg, PA: Burd Street Press, 1999.

James, Alfred P. "General Joseph E. Johnston, Storm Center of the Confederate Army." *Mississippi Valley Historical Review* 14 (June 1927–March 1928): 342–59.

James, Robert. "Stonewall's Surprise at Ox Hill." *America's Civil War* 7/6 (January1995): 54–61.

Jamieson, Perry D. *Death in September: The Antietam Campaign.* Fort Worth, TX: Ryan Place Publishers, 1995.

Jenkins, Robert D. *The Battle of Peach Tree Creek: Hood's First Sortie, 20 July 1864.* Macon: Mercer University Press, 2013.

———. "Dalton: The Opening of the Georgia Campaign." *Blue & Gray* 32/1 (December 2015): 6–9, 19–27, 41–50.

———. *To the Gates of Atlanta: From Kennesaw Mountain to Peach Tree Creek, 1–19 July 1864.* Macon: Mercer University Press, 2015.

Johnson, Bradley T., ed. *A Memoir of the Life and Public Service of Joseph E. Johnston.* Baltimore: R. H. Woodward and Company, 1891.

Johnson, David Alan. *Decided on the Battlefield: Grant, Sherman, Lincoln and the Election of 1864.* Amherst, NY: Prometheus Books, 2012.

Johnson, Robert Underwood, and Clarence Clough Buel, eds. *Battles and Leaders of the Civil War.* 4 vols. New York: Thomas Yoseloff, 1956.

Johnston, J. S. "A Reminiscence of Sharpsburg." *Southern Historical Society Papers* 8/10–12 (October–December 1880): 526–29.

Johnston, J. Stoddard. "General J. B. Hood and Chickamauga." *Confederate Veteran* 13/12 (December 1905): 552.

Johnston, Joseph E. "The Dalton-Atlanta Operations." In McClure, *Annals of the War*, 330–41.

———. *Narrative of Military Operations, Directed, During the Late War Between the States.* New York: D. Appleton and Company, 1874.

———. "Opposing Sherman's Advance to Atlanta." In Johnson and Buel, *Battles and Leaders of the Civil War*, 4:260–77.

Jones, J. William. "The Morale of the Confederate Armies." In Evans, *Confederate Military History*, 12:119–93.

Jones, James P. "The Battle of Atlanta and McPherson's Successor." *Civil War History* 7/4 (December 1961): 393–405.

Jones, James Pickett. *"Black Jack": John A. Logan and Southern Illinois in the Civil War Era.* Carbondale: Southern Illinois University Press, 1995. Originally published 1967 by Florida State University Press (Tallahassee).

Jones, John B. *A Rebel War Clerk's Diary.* Edited by Earl Schenck Miers. New York: Sagamore Press, 1958.

Jones, John Beauchamp. *A Rebel War Clerk's Diary at the Confederate States Capital.* Edited by James I. Robertson, Jr. 2 vols. Lawrence: University Press of Kansas, 2015.

Jones, John Beauchamp. *A Rebel War Clerk's Diary at the Confederate States Capital.* Edited by Howard Swiggett. 2 vols. New York: Old Hickory Bookshop, 1935.

Jones, Joseph. "The Medical History of the Confederate States Army and Navy." *Southern Historical Society Papers* 20 (January–December 1892): 109–66.

———. "Roster of the Medical Officers of the Army of Tennessee." *Southern Historical Society Papers* 22 (1894): 165–280.

Jones, Wharton Stewart. "'Glory Enough for All.'" *Confederate Veteran* 38/6 (June 1930): 215–16.

Joslyn, Mauriel Phillips, ed. *A Meteor Shining Brightly: Essays on Maj. Gen. Patrick R. Cleburne.* Milledgeville, GA: Terrell House Publishing, 1997.

Julian, Allen P. "Atlanta's Defenses." *Civil War Times Illustrated* 3/4 (July 1964): 23–24.

Kean, Robert G. H. *Inside the Confederate Government: The Diary of Robert Garlick Hill Kean.* Edited by Edward Younger. New York: Oxford University Press, 1957.

Keegan, John. *The American Civil War: A Military History.* New York: Alfred A. Knopf, 2009.

Kegel, James A. *North with Lee and Jackson: The Lost Story of Gettysburg.* Mechanicsburg, PA: Stackpole Books, 1996.

Keller, Allen. "On the Road to Atlanta: Johnston versus Sherman." *Civil War Times Illustrated* 1/8 (December 1962): 18–22, 32–35.

Kelly, Dennis. "Back in the Saddle: The War Record of William Bate." *Civil War Times Illustrated* 27/8 (December 1862): 118; (December 1988): 27–33.

———. "Epilogue: 'Gen'l J. Is Not the Man We Thought Him.'" *Blue & Gray* 6/5 (June 1989): 59–60.

———. *Kennesaw Mountain and the Atlanta Campaign: A Tour Guide.* Marietta, GA: Kennesaw Mountain Historical Association, 1990.

———. "Mountains to Pass, a River to Cross: The Battle of Kennesaw Mountain and Related Actions from June 10 to July 9, 1864." *Blue & Gray* 6/5 (June 1989): 8–12, 17–28, 30, 46, 48, 50, 52–60.

———. "Second Manassas: The Battle and Campaign." *Civil War Times Illustrated* 22/3 (May 1983): 9–44.

Kennett, Lee. *Marching through Georgia: The Story of Soldiers and Civilians during Sherman's Campaign.* New York: HarperCollins, 1995.

Kerksis, Sydney C. "Action at Gilgal Church June 15–16, 1864." In Kerksis, comp., *The Atlanta Papers*, 831–67.

———, comp. *The Atlanta Papers.* Dayton, OH: Press of Morningside Bookshop, 1980.

Key, William. *The Battle of Atlanta and the Georgia Campaign.* Atlanta: Peachtree Publishers Limited, 1981.

Kime, Marlin G. "Sherman's Gordian Knot: Logistical Problems in the Atlanta Campaign." *Georgia Historical Quarterly* 70/1 (Spring 1986): 102–110.

King, Alvy L. *Louis T. Wigfall: Southern Fire-Eater.* Baton Rouge: Louisiana State University Press, 1970.

Kirkpatrick, Robert Lang. "The Fifth Ohio Infantry at Resaca." In Kerksis, comp., *The Atlanta Papers*, 353–63.

Korda, Michael. *Clouds of Glory: The Life and Legend of Robert E. Lee.* New York: HarperCollins 2014.

Krick, R. E. L. "The Battle of Gaines's Mill." *Civil War* Issue 51 (June 1995): 61–63.

———. "The Men Who Carried This Position Were Soldiers Indeed: The Decisive Charge of Whiting's Division at Gaines's Mill." In *The Richmond Campaign of 1862: The Peninsula and the Seven Days*, edited by Gary W. Gallagher, 181–216. Chapel Hill: University of North Carolina Press, 2000.

Krick, Robert K. "Edward Johnson." In Davis, *The Confederate General*, 3:186–87.

———. "If Longstreet…Says So, It Is Most Likely Not True." In *Three Days at Gettysburg: Essays on Confederate and Union Leadership*, edited by Gary W. Gallagher, 147–68. Kent, OH: Kent State University Press, 1999.

Krumenaker, Lawrence. *Walking the Line: Rediscovering and Touring the Civil War Defenses on Modern Atlanta's Landscape.* Marietta, GA: Hermograph Press, 2014.

Kurtz, Annie Pye, ed. *Atlanta and the Old South: Paintings and Drawings by Wilbur G. Kurtz Artist and Historian.* Atlanta: American Lithograph Co., 1969.

Kurtz, Wilbur G. "At the Dexter Niles House." *Atlanta Constitution Sunday Magazine,* September 29, 1930, pg. 5–6, 20.

———. "Atlanta in the Summer of 1864." *Inn Dixie Magazine* (January 1936): 5–6, 10, 12, 17–19, 21.

———. *The Battle of Atlanta July 22, 1864.* Atlanta Cyclorama, 1954.

———. "The Broken Line and the DeGress Battery." *Atlanta Constitution Magazine,* February 8, 1931, pg. 4–5, 20.

———. Papers. Mss 130. 88 document cases, 43 boxes, and 8 files. Includes notebooks and historical marker texts. Atlanta History Center, Atlanta, GA. Cited as Kurtz Collection.

———. "Data on the capture and removal of the four brass twelve-pounder Napoleon guns of Capt. Max Van Den Corput's Cherokee Battery—Confederate—at Resaca, Ga. by the Second and Third Divisions of the Federal Twentieth Corps, May 15, 1864." Unpublished, handwritten manuscript. MSS 130. Box 26, folder 29. Kurtz Collection.

———. "The Death of Major General W. H. T. Walker, July 22, 1864." *Civil War History* 6/2 (June 1960): 174–79.

———. "General Joseph E. Johnston's Review of Sherman's *Memoirs*, published in Drake's *Annals of the Army of Tennessee*, Nashville, April 1878." Unpublished manuscript (July 20, 1960). Box 81, folder 3. Kurtz Collection.

———. "How the Cyclorama Was Painted." *American Heritage* 7/2 (February 1956): 45.

———. "The Jonesboro Campaign." Typescript of address delivered before the Atlanta Civil War Round Table, June 18, 1957. Box 51, folder 10, Kurtz Collection.

[———]. "Map of Atlanta as of 1938 Showing the Field and Fortified Lines of the Confederate Forces, Together with Those of the Federal Armies—Also the Fields of the Three Major Engagements, during the Summer of 1864." Atlanta Chamber of Commerce, 1938. FF 148, T2, folder 2, Kurtz Collection.

———. "Sherman versus Hood in the Fighting at Atlanta." *Civil War Times Illustrated* 3/4 (July 1964): 8–22.

———. "Some Unofficial Glimpses of General John B. Hood. Read at a joint meeting of Southern Historical Association—the Atlanta Historical Society and the Atlanta Civil War Roundtable. Biltmore Hotel Atlanta Sat. November 11—1950—Kurtz." Box 51, folder 13, Kurtz Collection.

———. "Why Was Snake Creek Gap Left Unguarded?" Unpublished typescript, c. 1935. Box 49, folder 14, Kurtz Collection.

LaFantasie, Glenn W. "The Gentleman General." *Civil War Monitor* 2/4 (Winter 2012): 26–37.

———. *Gettysburg Requiem: The Life and Lost Causes of Confederate Colonel William C. Oates.* New York: Oxford University Press, 2006.

"Lafayette McLaws on Gettysburg." *North & South* Issue 1 (November 1997): 38–40.

Laine, J. Gary, and Morris M. Penny. *Law's Alabama Brigade in the War Between the Union and the Confederacy.* Shippensburg, PA: White Mane Publishing Co., 1996.

Landingham, Irenus Watson. Papers. RG 552. Auburn University Libraries Special Collections and Archives, Auburn, AL. Cited as Landingham Papers.

Lane, Mills, ed. *Times That Prove People's Principles: Civil War in Georgia.* Savannah: Beehive Press, 1993.

Large, George R., and Joe A. Swisher. *Battle of Antietam: The Official History by the Antietam Battlefield Board.* Shippensburg, PA: Burd Street Press, 1998.

Lash, Jeffrey N. *Destroyer of the Iron Horse: General Joseph E. Johnston and Confederate Rail Transport, 1861–1865.* Kent, OH: Kent State University Press, 1991.

———. "A Yankee Bridge Builder for the Confederacy, 1862–1865." *Journal of Confederate History* 6 (1991): 106–36.

Law, E. M. "The Struggle for Round Top.'" In Johnson and Buel, *Battles and Leaders of the Civil War*, 3:318–30.

Lawson, Lewis A. *Wheeler's Last Raid.* Greenwood, FL: Penkevill Publishing Company, 1986.

Lay, Henry Champlin. Papers 1842–85. #418. Southern Historical Collection, Wilson Library, University of North Carolina at Chapel Hill. Cited as Lay Papers.

Lea, H. J. "In the Battle of New Hope Church." *Confederate Veteran* 31/2 (February 1923): 61–62.

"Leading Confederates on Gettysburg: Letter from General John B. Hood." *Southern Historical Society Papers* 4/4 (October 1877): 145.

Lee, Dan. *Thomas J. Wood: A Biography of the Union General of the Civil War.* Jefferson, NC: McFarland, 2012.

Lemon, Mark. *Feed Them the Steel! Being, The Wartime Recollections of Captain James Lile Lemon, Co. A, 18th Georgia Infantry C. S. A.* Privately printed, 2013.

"Letter Written in War Times." *Confederate Veteran* 2/5 (May 1894): 151.

Lewis, Lloyd. *Sherman Fighting Prophet.* New York: Harcourt, Brace & Co., 1932.

Liddell Hart, B. H. Introduction to William T. Sherman, *From Atlanta to the Sea.* London: Folio Society, 1961.

———. *Sherman: Soldier, Realist, American.* New York: Dodd, Mead & Company, 1929.

———. *Strategy.* New York: Frederick A. Praeger, Publishers, 1965.

Linderman, Gerald F. *Embattled Courage: The Experience of Combat in the American Civil War.* New York: Free Press, 1987.

Lindsley, John Berrien, ed. *The Military Annals of Tennessee.* 2 vols. Nashville: J. M. Lindsley & Co., 1886.

Little, Robert D. "Hardee and the Atlanta Campaign." *Georgia Historical Quarterly* 29/1 (March 1945): 1–22.

[Littlefield, J. H.]. "Hood 'Feeling the Enemy.'" In Johnson and Buel, *Battles and Leaders of the Civil War*, 2:276.

Lively, Mathew W. *Calamity at Chancellorsville: The Wounding and Death of Confederate General Stonewall Jackson.* El Dorado Hills, CA: Savas Beatie, 2013.

Livermore, Thomas L. *Numbers and Losses in the Civil War in America, 1861–1865.* New York: Kraus Reprint Co., 1969.

Long, E. B., with Barbara Long. *The Civil War Day by Day: An Almanac 1861–1865.* New York: Doubleday & Company, 1971.

Longacre, Edward G. *Cavalry of the Heartland: The Mounted Forces of the Army of Tennessee.* Yardley, PA: Westholme Publishing, 2009.

———. *Gentleman and Soldier: A Biography of Wade Hampton III.* Nashville: Rutledge Hill Press, 2003.

———. *Mounted Raids of the Civil War.* Lincoln: University of Nebraska Press, 1994.

———. *A Soldier to the Last: Maj. Gen. Joseph Wheeler in Blue and Gray.* Washington: Potomac Books, 2007.

———. *Worthy Opponents: General William T. Sherman U. S. A. General Joseph E. Johnston C. S. A.* Nashville: Rutledge Hill Press, 2006.

Longstreet, James. *From Manassas to Appomattox: Memoirs of the Civil War in America.* Bloomington: Indiana University Press, 1960.

Loop, Charles. "'I Was There.'" Edited by Frank Crawford. *Civil War* 9/1 (January–February 1991): 60.

Losson, Christopher. "The Defender of Cheatham Hill." *Blue & Gray* 6/5 (June 1989): 51–52.

———. *Tennessee's Forgotten Warriors: Frank Cheatham and His Confederate Division.* Knoxville: University of Tennessee Press, 1989.

Lowrey, M. P. "An Autobiography." *Southern Historical Society Papers* 16 (January–December 1888): 365–76.

Luvaas, Jay. "An Appraisal of Joseph E. Johnston." *Civil War Times Illustrated* 4/9 (January 1966): 5–7, 28–32.

Luvaas, Jay, and Harold W. Nelson, eds. *Guide to the Atlanta Campaign: Rocky Face Ridge to Kennesaw Mountain.* Lawrence: University Press of Kansas, 2008.

"M." "The Battle of Dead Angle on the Kennesaw Line, near Marietta, Georgia." *Southern Bivouac* 3/2 (October 1884): 71–74.

Mackall, William W. *A Son's Recollections of His Father.* New York: E. P. Dutton & Co., 1930.

Mackowski, Chris, and Kristopher D. White. *That Furious Struggle: Chancellorsville and the High Tide of the Confederacy, May 1–4, 1863.* El Dorado Hills, CA: Savas Beatie, 2014.

"Maj. J. H. Littlefield." *Confederate Veteran* 10/1 (January 1902): 33.

Major, Duncan K., and Roger S. Fitch. *Supply of Sherman's Army During the Atlanta Campaign.* Ft. Leavenworth, KS: Army Service Schools Press, 1911.

Manigault, Arthur M. *A Carolinian Goes to War: The Civil War Narrative of Arthur Middleton Manigault, Brigadier General, C. S. A.* Edited by R. Lockwood Tower. Columbia: University of South Carolina Press, 1983.

Manning, Wade H. "Surgeons of the Confederacy: John Thomson Darby." *Confederate Veteran* 35/4 (April 1927): 141–42.

"Map Showing Battle Fields of July 20, 22 & 28, 1864. And Other Points of Historical Interest in and around Atlanta," in *Battles of Atlanta.*

Marchant, Peter. Civil War Correspondence. MSS 657f. Atlanta History Center. Cited as Marchant Correspondence.

Marshall, Park. *A Life of William B. Bate.* Edited by Ross Massey. Nashville: Tenth Amendment Publishing, 2014.

Martin, David G. *The Second Bull Run Campaign July-August 1862.* Conshohocken, PA: Combined Publishing, 1997.

Martin, Samuel. "Kill-Cavalry." *Civil War Times Illustrated* 38/7 (February 2000): 23–30, 56–59.

[Martin, Thomas H.] "The Story of The Battle of Peachtree Creek." In Atlanta Business Men's League, *The Proposed Atlanta National Military Park: A Brief Outline of the Project*, 13–26. Atlanta, n.d. (c. 1900).

Mason, A. P. Circular pertaining to construction and fortification of redoubts used by the Army of Tennessee. Atlanta, Georgia, July 7, 1864. GLC05987.43. The Gilder Lehrman Institute of American History, New York, NY.

Maryniak, Benedict R. "Fighting Tom Sweeny, the Irish Bulldog." *Civil War Courier* (August 1997): 53–55.

———. "Private Charles F. Sanscrainte, Company B, 15th Michigan, and His Medal of Honor." *Civil War Courier* (November 1997): 54–56.

Mathews, Byron H., Jr. *The McCook-Stoneman Raid.* Philadelphia: Dorrance and Company, 1976.

Mathis, Ray, ed. *In the Land of the Living: Wartime Letters by Confederates from the Chattahoochee Valley of Alabama and Georgia.* Macon: Mercer University Press, 2015.

Matter, William D. *If It Takes All Summer: The Battle of Spotsylvania.* Chapel Hill: University of North Carolina Press, 1988.

[McCabe, James D.] *The Grayjackets: And How They Lived, Fought and Died for Dixie.* Dahlonega, GA: Confederate Reprint Co., 2009.

McCarley, J. Britt. *The Atlanta Campaign: A Civil War Driving Tour of Atlanta-Area Battlefields.* Atlanta: Cherokee Publishing Company, 1989.

———. *The Atlanta and Savannah Campaigns.* Washington: Center of Military History, 2014.

———. "'The Great Question of the Campaign Was One of Supplies': A Reinterpretation of Sherman's Generalship during the 1864 March to Atlanta in Light of the Logistic Strategy." In *Beyond Combat: Essays in Military History in Honor of Russell F. Weigley*, edited by Edward G. Longacre and Theodore J. Zeman, 26–64. Philadelphia: American Philosophical Society, 2007.

McClure, Alexander K., ed. *The Annals of the War Written by Leading Participants North and South.* Dayton, OH: Morningside House, 1988.

McDonough, James L. *Schofield: Union General in the Civil War and Reconstruction.* Tallahassee: Florida State University Press, 1972.

———. *William Tecumseh Sherman—In the Service of My Country: A Life.* New York: W. W. Norton & Company 2016.

McDonough, James Lee, and James Pickett Jones. *War So Terrible: Sherman and Atlanta.* New York: W. W. Norton & Company, 1987.

McElroy, Robert. *Jefferson Davis: The Unreal and the Real.* New York: Kraus Reprint Co., 1969.

McGuire, J. P. "Thirty-second Tennessee Infantry." In Lindsley, *The Military Annals of Tennessee*, 1:469–82.

McInvale, Morton R. *The Battle of Pickett's Mill: "Foredoomed to Oblivion."* Atlanta: Georgia Dept. of Natural Resources, 1977.

McKay, John E. "Atlanta Campaign." In *Encyclopedia of the American Civil War*, edited by David S. Heidler and Jeanne T. Heidler, 128–46. New York: W. W. Norton & Company, 2000.

McKinney, Francis F. *Education in Violence: The Life of George H. Thomas and the History of the Army of the Cumberland.* Chicago: Americana House, 1991.

McKittrick, Samuel. "A Confederate Officer's Letters on Sherman's March to Atlanta." Edited by Donald W. Lewis. *Georgia Historical Quarterly* 51/4 (December 1967): 491–94.

McLaws, Lafayette. *A Soldier's General: The Civil War Letters of Major General Lafayette McLaws.* Edited by John C. Oeffinger. Chapel Hill: University of North Carolina Press, 2002.

———. "Gettysburg." *Southern Historical Society Papers* 7/2 (February 1879): 64–90.

McMurry, Richard M. "The Affair at Kolb's Farm." *Civil War Times Illustrated* 7/8 (December 1968): 20–27.

———. *Atlanta 1864: Last Chance for the Confederacy.* Lincoln: University of Nebraska Press, 2000.
———. "The Atlanta Campaign." In *The South Besieged,* vol. 5 of *The Image of War: 1861–1865*, edited by William C. Davis, 240–302. 6 vols. Garden City, NY: Doubleday & Company, 1983.
———. "The Atlanta Campaign: A New Look." *Civil War History* 22/1 (March 1976): 5–15.
———. "Atlanta Campaign: Rocky Face to the Dallas Line, the Battles of May 1864." *Blue & Gray* 6/4 (April 1989): 10–18, 20–23, 46, 49, 54–57, 60, 62.
———. "The Atlanta Campaign: December 23, 1863 to July 18, 1864." Ph.D. dissertation, Emory University, 1967.
———. "The Battle of Kennesaw Mountain and the Georgia Campaign of 1864." Master's thesis, Emory University, 1964.
———. "Cassville." *Civil War Times Illustrated* 10/8 (December 1971): 4–9.
———. "Confederate Morale in the Atlanta Campaign of 1864." *Georgia Historical Quarterly* 54/2 (Summer 1970): 226–43.
———. "The *Enemy* at Richmond: Joseph E. Johnston and the Confederate Government." *Civil War History* 27/1 (March 1981): 5–31.
———. "The 'Hell Hole': New Hope Church." *Civil War Times Illustrated* 11/10 (February 1973): 32–43.
———. "John Bell Hood." In Davis, *The Confederate General*, 3:120–27.
———. *John Bell Hood and the War for Southern Independence.* Lexington: University Press of Kentucky, 1982.
———. "Kennesaw Mountain." *Civil War Times Illustrated* 8/9 (January 1970): 19–33.
———. "The Mackall Journal and Its Antecedents." Typescript courtesy of author. From the original in the Joseph E. Johnston Papers, Earl Gregg Swem Library, College of William and Mary.
———. "The Mackall Journal and Its Antecedents." *Civil War History* 20/4 (December 1974): 311–28.
———. "More on 'Raw Courage.'" *Civil War Times Illustrated* 14/6 (October 1975): 36–38.
———. "The *Official Records* of the Rebellion." In *American Historians, 1866–1912*, edited by Clyde N. Wilson, 367–72. Detroit: Gale Research Company, 1986.
———. "Patton Anderson: Major General, C. S. A." *Blue & Gray* 1/2 (October–November 1983): 10–17.
———. "A Policy So Disastrous: Joseph E. Johnston's Atlanta Campaign." In Savas and Woodbury, *The Campaign for Atlanta*, 223–48.
———. "Resaca: 'A Heap of Hard Fiten.'" *Civil War Times Illustrated* 9/7 (November 1970): 4–12.
———. "Riding through Alabama, Part I." *Civil War Times Illustrated* 20/5 (August 1981): 12–17.
———. "Riding through Alabama, Part II." *Civil War Times Illustrated* 20/6 (October 1981): 36–41.
———. "Rise to Glory: A Speculative Essay on the Early Career of John Bell Hood." In *Rank and File: Civil War Essays in Honor of Bell Irvin Wiley*, edited by James I. Robertson and Richard M. McMurry, 39–54. San Rafael, CA: Presidio Press, 1976.
———. *The Road Past Kennesaw: The Atlanta Campaign of 1864.* Washington: National Park Service, 1972. Reprinted in *Battlefields of the Civil War* (New York: Arno, 1979).

McNeely, Patricia G. "Felix Gregory de Fontaine: Chronicling the Horrors of War." In *Knights of the Quill: Confederate Correspondents and their Civil War Reporting*, edited by Patricia G. McNeely, Debra Reddin van Tuyll, and Henry H. Schulte, 45–94. West Lafayette, IN: Purdue University Press, 2010.

McNeill, William J. "A Survey of Confederate Soldier Morale during Sherman's Campaign through Georgia and the Carolinas." *Georgia Historical Quarterly* 55/1 (Spring 1971): 1–25.

McPherson, James M. *Battle Cry of Freedom: The Civil War Era.* New York: Oxford University Press, 1988.

———. *Crossroads of Freedom: Antietam.* New York: Oxford University Press, 2002.

McWhiney, Grady. "A Bishop as General." In Hewitt, *Confederate Generals in the Western Theater*, 1:1–11.

———. "Jefferson Davis and Confederate Military Leadership." In *Leadership During the Civil War: The 1989 Deep Delta Civil War Symposium*, edited by Roman J. Heleniak and Lawrence L. Hewitt, 17–37. Shippensburg, PA: White Mane Publishing Company, 1992.

Merrill, William E. "Block-houses for Railroad Defense in the Department of the Cumberland." In vol. 3 of *Sketches of War History 1861–1865 Papers Prepared for the Ohio Commandery of the Military Order of the Loyal Legion of the United States 1880–1890*, edited by Robert Hunter, 389–421. 6 vols. Cincinnati: Robert Clarke & Co., 1890.

Miles, Jim. "The Campaign for Atlanta." *Civil War* 9/1 (January–February 1991): 12–25, 54–60.

———. *Fields of Glory: A History and Tour Guide of the Atlanta Campaign.* Nashville: Rutledge Hill Press, 1989.

Miller, Brian Craig. *John Bell Hood and the Fight for Civil War Memory.* Knoxville: University of Tennessee Press, 2010.

Milner, W. J. "Lieut. Gen. William Joseph Hardee." *Confederate Veteran* 22/8 (August 1914): 360–64.

———. "The Protest Sent to President Davis." *Confederate Veteran* 22/11 (November 1914): 507.

Minutes of the Board of Directors of the Press Association, Embracing the Quarterly Reports of the Superintendent, October and January. Fourth Session—Atlanta, October 14, 1863. Fifth Session—Augusta, Jan'y 14, 1864. Printed by Order of the Board. Atlanta: Franklin Steam Publishing House, 1864.

Mitchell, Adele H., ed. *The Letters of Major General James E. B. Stuart.* N.p., 1990. Copy at Emory University, Atlanta, Georgia.

Mitchell, Joseph B. *Military Leaders in the Civil War.* New York: G. P. Putnam's Sons, 1972.

Mitchell, Stephens. "Colonel L. P. Grant and the Defenses of Atlanta." *Atlanta Historical Bulletin* 1/1 (February 1932): 32–35.

———. "The Old Ferries and the Ferry Roads." *Atlanta Historical Bulletin* 2/1 (June 1933): 33–41.

Moore, Robert A. *A Life for the Confederacy: As Recorded in the Pocket Diaries of Pvt. Robert Moore Co. G 17th Mississippi Regiment Confederate Guards Holly Springs, Mississippi.* Edited by James W. Silver. Jackson, TN: McCowat-Mercer Press, 1959.

Morgan, Jack. *Through American and Irish Wars: The Life and Times of General Thomas W. Sweeny 1820–1902.* Dublin: Irish Academic Press, 2006.

Morris, Roy Jr. "Richmond's Fate in the Balance." *America's Civil War* 1/1 (May 1988): 35–41.

———, and Phil Noblitt. "The History of a Failure." *Civil War Times Illustrated* 27/5 (September 1988): 36–43.

Mosser, Jeffrey. "I Shall Make Him Remember This Insult." *Civil War Times Illustrated* 32/1 (March/April 1993): 40, 49, 52–57, 60–62.

Mruck, Armin E. "The Role of Railroads in the Atlanta Campaign." *Civil War History* 7/3 (September 1961): 264–71.

Muhlenfield, Elisabeth. *Mary Boykin Chesnut: A Biography.* Baton Rouge: Louisiana State University Press, 1981.

Murfin, James V. *The Gleam of Bayonets: The Battle of Antietam and the Maryland Campaign of 1862.* New York: Thomas Yoseloff, 1965.

Murphree, Joel. "Autobiography and Civil War Letters of Joel Murphree of Troy, Alabama." Edited by H. E. Steryx. *Alabama Historical Quarterly* 19/1 (Spring 1957): 179–208.

Murray, Williamson, and Wayne Wei-Siang Hsieh. *A Savage War: A Military History of the Civil War.* Princeton: Princeton University Press, 2016.

Naisawald, L. VanLoan. *Grape and Canister: The Story of the Field Artillery of the Army of the Potomac, 1861–1865.* Mechanicsburg, PA: Stackpole Books, 1999.

———. "A Nasty Little Battle in the Rain—Chantilly." *Civil War Times Illustrated* 3/3 (June 1964): 11–16.

Neal, Andrew Jackson. Papers. MSS 218. Manuscripts, Archives and Rare Books Library, Emory University, Atlanta, GA. Cited as Neal Papers.

Neal, Diane, and Thomas W. Kremm. *The Lion of the South: General Thomas C. Hindman.* Macon: Mercer University Press, 1993.

Neff, Robert O. *Tennessee's Battered Brigadier (The Life of General Joseph B. Palmer).* N.p., 1988.

Neill, Edward D. *Glimpses of the Nation's Struggle: Papers Read Before the Minnesota Commandery of the Military Order of the Loyal Legion of the United States, 1892–1897.* Volume (Series) 4. Wilmington, NC: Broadfoot Publishing Company, 1992.

Nelson, Larry E. "Independence or Fight." *Civil War Times Illustrated* 15/3 (June 1976): 10–14.

Neul, Robert C. "Battle Most Desperate and Bloody." *America's Civil War* 7/6 (January 1995): 30–36.

Nevins, Allan. *The Organized War 1863–1864.* New York: Charles Scribner's Sons, 1971.

Newton, Steven H. "'Formidable Only in Flight?': Casualties, Attrition, and Morale in Georgia." *North & South* 3/4 (April 2000): 43–56.

———. "Joseph Johnston and Snake Creek Gap." *North & South* 4/3 (March 2001): 56–68.

———. *Lost for the Cause: The Confederate Army in 1864.* Mason City, IA: Savas Publishing Company, 2000.

———. "What Talent Might Have Achieved: The Battle of Resaca." *North & South* 13/5 (January 2012): 27–36.

———. "Why Wouldn't Joe Johnston Fight?" *North & South* 5/6 (September 2002): 44–55.

Nichols, James I. *Confederate Engineers.* Tuscaloosa: University of Alabama Press, 1957.

Niesen, William C. "'The Consequences of Grandeur': A Union Soldier Writes of the Atlanta Campaign." *Atlanta History* 33/3 (Fall 1989): 5–19.

Nisbet, James Cooper. *4 Years on the Firing Line.* Edited by Bell Irvin Wiley. Jackson, TN: McCowat-Mercer Press, 1963.

Noblitt, Phil. "A Flank Unturned at Chickamauga." *America's Civil War* 9/5 (November 1996): 34–40, 96.

———. "The 'Crime at Pickett's Mill.' *America's Civil War* 8/6 (January 1996): 38–44, 95.

Norton, Herman. "Revivalism in the Confederate Armies." *Civil War History* 6/4 (December 1960): 410–24.

Nugent, William L. *My Dear Nellie: The Civil War Letters of William L. Nugent to Eleanor Smith Nugent.* Edited by William M. Cash and Lucy Somerville Howorth. Jackson: University Press of Mississippi, 1977.

Nye, Wilbur S. "Cavalry Operations Around Atlanta." *Civil War Times Illustrated* 3/4 (July 1964): 46–50.

O'Connell, David. *The Art & Life of Atlanta Artist Wilbur G. Kurtz: Inspired by Southern History.* Charleston, SC: History Press, 2013.

O'Connell, Robert L. *Fierce Patriot: The Tangled Lives of William Tecumseh Sherman.* New York: Random House, 2014.

———. "Sherman Closes in." *Civil War Times Illustrated* 53/3 (June 2014): 43–47.

O'Connor, Richard. *Hood: Cavalier General.* New York: Prentice-Hall, 1949.

———. *Thomas: Rock of Chickamauga.* New York: Prentice-Hall, 1948.

"The Opposing Forces at Chickamauga, Ga. September 19th–20th, 1863: The Confederate Army." In Johnson and Buel, *Battles and Leaders of the Civil War*, 3:673–76.

"The Opposing Forces at Gettysburg, Pa." In Johnson and Buel, *Battles and Leaders of the Civil War*, 3:434–39.

"The Opposing Forces in the Maryland Campaign." In Johnson and Buel, *Battles and Leaders of the Civil War*, 2:598–603.

O'Reilly, Francis Augustin. *The Fredericksburg Campaign: Winter War on the Rappahannock.* Baton Rouge: Louisiana State University Press, 2004.

Orendorff, Henry H. *We Are Sherman's Men: The Civil War Letters of Henry Orendorff.* Edited by William M. Anderson. Macomb: Western Illinois University Press, 1986.

Osborn, George C. "The Atlanta Campaign, 1864." *Georgia Historical Quarterly* 34/4 (December 1950): 271–87.

Owen, U. G. "Letters of a Confederate Surgeon in the Army of Tennessee to His Wife." Edited by Enoch L. Mitchell. *Tennessee Historical Quarterly* 5/2 (June 1946): 142–81.

Palumbo, Frank A. *George Henry Thomas: The Dependable General.* Dayton, OH: Morningside House, 1983.

Parker, David B. *Alias Bill Arp: Charles Henry Smith and the South's "Goodly Heritage."* Athens: University of Georgia Press, 1991.

Parker, Eddy R. *Touched by Fire: Letters from Company D 5th Texas Infantry Hood's Brigade Army of Northern Virginia 1862–1865.* Hillsboro, TX: Hill College Press, 2000.

———. *General Leonidas Polk, C. S. A.: The Fighting Bishop.* Baton Rouge: Louisiana State University Press, 1962.

———. *Joseph E. Brown of Georgia.* Baton Rouge: Louisiana State University Press, 1977.

Patchan, Scott C. *Second Manassas: Longstreet's Attack and the Struggle for Chinn Ridge.* Washington: Potomac Books, 2011.

———. "Second Manassas: The Fighting on August 30, 1862." *Blue & Gray* 29/3 (2012): 6–26, 44–50.

Patrick, Rembert W. *Jefferson Davis and His Cabinet.* Baton Rouge: Louisiana State University Press, 1944.

Patterson, Gerard A. "John Bell Hood." *Civil War Times Illustrated* 9/10 (February 1971): 12–21.

Peacock, Jane Bonner, ed. "A Wartime Story: The Davidson Letters, 1862–1865." *Atlanta Historical Bulletin* 19/1 (Spring 1975): 8–121.

Pearce, Haywood J., Jr. *Benjamin H. Hill: Secession and Reconstruction.* Chicago: University of Chicago Press, 1928.

Peddy, George W. *Saddle Bag and Spinning Wheel: Being the Civil War Letters of George W. Peddy, MD Surgeon, 56th Georgia Volunteer Regiment, C. S. A.* Edited by George W. Cuttino. Macon: Mercer University Press, 1981.

Perry, Leslie J. "Davis and Johnston, Light Thrown on a Quarrel among Confederate Leaders. A Question of Rank." *Southern Historical Society Papers* 20 (January–December 1892): 95–108.

Peter, Lawrence J., and Raymond Hull. *The Peter Principle: Why Things Always Go Wrong.* New York: William Morrow, 1998.

Peterson, Lawrence K. *Confederate Combat Commander: The Remarkable Life of Brigadier General Alfred Jefferson Vaughan, Jr.* Knoxville: University of Tennessee Press, 2013.

Petite, Mary Deborah. *"The Women Will Howl!: The Union Capture of Roswell and New Manchester, Georgia, and the Forced Relocation of Mill Workers.* Jefferson, NC: McFarland, 2008.

Pfanz, Harry W. *Gettysburg: The First Day.* Chapel Hill: University of North Carolina Press, 2001.

———. *Gettysburg: The Second Day.* Chapel Hill: University of North Carolina Press, 1987.

Phillips, Jason. *Diehard Rebels: The Confederate Culture of Invincibility.* Athens: University of Georgia Press, 2007.

Polhemus, Marlan H. *The Francis DeGress Story: Civil War Hero, Merchant and Inventor Captain Battery H, First Illinois Light Artillery.* Privately printed, 2013.

Pollard, E. A. *Southern History of the War.* 2 vols. New York: Charles B. Richardson, 1866.

Polley, J. B. "Another 'Charming Nellie' Letter." *Confederate Veteran* 4/4 (April 1896): 118–19.

———. "Brave Texans in the Virginia Army." *Confederate Veteran* 4/5 (May 1896): 158–59.

———. "Crossing Over into Maryland." *Confederate Veteran* 4/8 (August 1896): 275–76.

———. "Hood's Texans in Pennsylvania." *Confederate Veteran* 4/11 (November 1896): 377–79.

———. *Hood's Texas Brigade Its Marches Its Battles Its Achievements.* New York: Neale Publishing Company, 1910.

———. "Second Manassas." *Confederate Veteran* 4/7 (July 1896): 218–20.

———. "To Charming Nellie—No So Funny." *Confederate Veteran* 4/2 (February 1896): 44–45.

Pollock, Daniel A. "The Battle of Atlanta: History and Remembrance." *Southern Spaces.* Emory Center for Digital Scholarship (May 30, 2014). http://southern-spaces.org/2014/battle-atlanta-history-and-remembrance.

Poole, John Randolph. *Cracker Cavaliers: The 2nd Georgia Cavalry under Wheeler and Forrest.* Macon: Mercer University Press, 2014.

Porter, James D. *Tennessee.* Vol. 8 of Evans, *Confederate Military History.*

Powell, David A. *The Chickamauga Campaign—Glory or the Grave: The Breakthrough, the Union Collapse, and the Defense of Horseshoe Ridge, September 20, 1863.* El Dorado Hills, CA: Savas Beatie, 2016.

———. *The Chickamauga Campaign—A Mad Irregular Battle: From the Crossing of the Tennessee River through the Second Day, August 22–September 19, 1863.* El Dorado Hills, CA: Savas Beatie, 2014.

Powell, Robert M. "With Hood at Gettysburg." In Cozzens, *Battles and Leaders of the Civil War*, 6:236–53.

Priest, John Michael. *Antietam: The Soldiers' Battle.* Shippensburg, PA: White Mane Publishing Company, 1989.

———. *Before Antietam: The Battle for South Mountain.* Shippensburg, PA: White Mane Publishing Company, 1992.

Purdue, Howell and Elizabeth Purdue. *Pat Cleburne Confederate General.* Gaithersburg, MD: Olde Soldier Books, 1987.

Purifoy, John. "The Lost Opportunity at Gettysburg." *Confederate Veteran* 31/6 (June 1923): 214–18.

Quint, Ryan T. "Jackson's Flank Attack Reconsidered." In *That Furious Struggle: Chancellorsville and the High Tide of the Confederacy, May 1–4, 1863*, by Chris Mackowski and Kristopher D. White, 143–45. El Dorado Hills, CA: Savas Beatie, 2014.

Quintard, Charles Todd. *Doctor Quintard: Chaplain C. S. A. and Second Bishop of Tennessee Being His Story of the War (1861–1865).* Edited by Arthur Howard Noll. Sewanee TN: The University Press, 1905.

———. *Doctor Quintard, Chaplain C. S. A. and Second Bishop of Tennessee: The Memoir and Civil War Diary of Charles Todd Quintard.* Edited by Sam Davis Elliott. Baton Rouge: Louisiana State University Press, 2003.

Raab, James W. *Deliverance from Evil: General Francis Asbury Shoup, C.S.A.—A Biography.* West Conshohocken, PA: Infinity Publishing, 2012.

———. *Lloyd Tilghman and Francis Asbury Shoup: Two Forgotten Confederate Generals.* Murfreesboro, TN: Southern Heritage Press, 2001.

———. *W. W. Loring: Florida's Forgotten General.* Manhattan, KS: Sunflower University Press, 1996.

Rable, George C. *Fredericksburg! Fredericksburg!* Chapel Hill: University of North Carolina Press, 2002.

Rafuse, Ethan S. "Always 'Fighting Joe': Joseph Hooker and the Campaign in North Georgia, May–July 1864." In *Corps Commanders in Blue: Union Major Generals in the Civil War*, edited by Ethan S. Rafuse, 221–60. Baton Rouge: Louisiana State University Press, 2014.

———. "Generals We Love to Hate." *America's Civil War* 24/6 (January 2012): 30–37.

"Raw Courage." *Civil War Times Illustrated* 12/4 (July 1974): 46.

Read, Ira Beman. "The Campaign from Chattanooga to Atlanta As Seen by a Federal Soldier." Edited by Richard B. Harwell. *Georgia Historical Quarterly* 25/3 (September 1941): 262–78.

Reagan, John H. *Memoirs with Special Reference to Secession and the Civil War.* New York: Neale Publishing Company, 1906.

Records Relating to Ordnance Stores Powder Factory, 1862–1865; Record of Government Stores Received and Issued at Powder Factory 1862–1865. Microfilm UF235 R425, University of Georgia.

"Return of a Confederate Flag." *Confederate Veteran* 22/7 (July 1914): 302.

Reynolds, Daniel H. *Worthy of the Cause for Which They Fight: The Civil War Diary of Brigadier General Daniel Harris Reynolds, 1861–1865.* Edited by Robert Patrick Bender. Fayetteville: University of Arkansas Press, 2011.

Richards, Samuel P. *Sam Richards's Civil War Diary: A Chronicle of the Atlanta Home Front.* Edited by Wendy Hamand Venet. Athens: University of Georgia Press, 2009.

Richardson, Eldon B. *Kolb's Farm: Rehearsal for Atlanta's Doom.* Privately printed, 1979.

Ridley, B. L. "The Battle of Resaca." *Confederate Veteran* 5/1 (January 1897): 36–38.

Ridley, Bromfield L. *Battles and Sketches of the Army of Tennessee.* Mexico, MO: Missouri Printing & Publishing Co., 1906.

Risley, Ford. "Wartime News over Southern Wires: The Confederate Press Association." In *Words at War: The Civil War and American Journalism*, edited by David B. Sachsman, S. Kittrell Rushing, and Roy Morris, Jr., 149–64. West Lafayette, IN: Purdue University Press, 2008.

Robbins, Peggy. "Hood vs. Sherman: A Duel of Words." *Civil War Times Illustrated* 17/4 (July 1978): 22–29.

Roberts, Frank Stovall. "Review of the Army of Tennessee at Dalton, GA." *Confederate Veteran* 26/4 (April 1918): 150.

———. *General A. P. Hill: The Story of a Confederate Warrior.* New York: Random House, 1987.

———. *Stonewall Jackson: The Man, the Soldier, the Legend.* New York: Macmillan Publishing USA, 1997.

Robertson, James I. Jr. *Civil War Virginia: Battleground for a Nation.* Charlottesville: University Press of Virginia, 1991.

Robertson, Jerome B. *Touched with Valor: Civil War Papers and Casualty Reports of Hood's Texas Brigade.* Edited by Harold B. Simpson. Hillsboro, TX: Hill Junior College Press, 1964.

Robinson, Leigh. "General Joseph E. Johnston." *Southern Historical Society Papers* 19 (1891): 337–70.

Robinson, Samuel. "Battle of Kennesaw Mountain." In *The Annals of the Army of Tennessee and Early Western History*, edited by Edwin L. Drake, 109–17. Jackson, TN: Guild Bindery Press, 1998.

———. "First Tennessee Infantry." In Lindsley, *The Military Annals of Tennessee*, 1:164.

Rollins, Richard, ed. *The Returned Battle Flags.* Redondo Beach, CA: Rank and File Publications, 1995.

Roper, Peter W. *Jedediah Hotchkiss: Rebel Mapmaker and Virginia Businessman.* Shippensburg, PA: White Mane Publishing Company, 1992.

Rose, Victor M. *Ross' Texas Brigade.* Kennesaw, GA: Continental Book Company, 1960.

Ross, D. Reid. "Battle of Kolb's Farm: Its Relevance to Sherman's March on Atlanta." *Atlanta History* 32/1 (Spring 1988): 17–33.

Roth, David E. "Battles on the Dallas-New Hope-Pickett's Mill Line." *Blue & Gray* 6/4 (April 1989): 59.

Rowell, John W. "McCook's Raid." *Civil War Times Illustrated* 13/4 (July 1974): 5–9, 42–48.

Rowland, Dunbar, ed. *Jefferson Davis Constitutionalist: His Letters, Papers and Speeches.* 10 vols. Jackson: Printed for the Mississippi Department of Archives and History, 1923.

Roy, T. B. "General Hardee and the Military Operations Around Atlanta." *Southern Historical Society Papers* 8/8–9 (August–September 1880): 337–87.

Rozier, John, ed. *The Granite Farm Letters: The Civil War Correspondence of Edgeworth and Sallie Bird.* Athens: University of Georgia Press, 1988.

Sanders, Stuart W. "The Bishop's Nephew." *Civil War Times Illustrated* 40/1 (March 2001): 24–30, 56, 58–60.

"Saucer Drawing" of the Cyclorama by Theodore Davis, 1887. MSS OS, Box 5.58, Kurtz Collection.

Savas, Theodore P., and David A. Woodbury, eds., *The Campaign for Atlanta & Sherman's March to the Sea: Essays on the American Civil War.* Campbell, CA: Savas Woodbury Publishers, 1994.

Scaife, William R. "Affair at Cassville." Privately printed, 1991.

———. *The Campaign for Atlanta.* Privately published, 1985; 2nd ed. rev., Saline, MI: McNaughton & Gunn, 1993.

———. "The Chattahoochee River Line: An American Maginot." *North & South* 1 (November 1997): 42–58.

———. "Sherman's Great Cavalry Raid." *North & South* 2 (January 1998): 42–46, 51–54.

———, and William D. Erquitt. *The Chattahoochee River Line: An American Maginot.* Atlanta: W. R. Scaife, 1992.

———, and William Harris Bragg. *Joe Brown's Pets: The Georgia Militia 1861–1865.* Macon: Mercer University Press, 2004.

Scales, John R. *Sherman Invades Georgia: Planning the North Georgia Campaign Using a Modern Perspective.* Annapolis: Naval Institute Press, 2006.

Schafer, Elizabeth D. "Jaded Mules, Twisted Rails and Razed Depots." *Civil War* 9/1 (January–February 1991): 40–46.

Schaff, Morris. *Jefferson Davis: His Life and Personality.* Boston: John W. Luce and Company, 1922.

Schmutz, John F. *"The Bloody Fifth": The 5th Texas Infantry Regiment, Hood's Texas Brigade, Army of Northern Virginia.* El Dorado Hills, CA: Savas Beatie, 2016.

Schott, Thomas E. "Lieutenant General William J. Hardee, the Historians, and the Atlanta Campaign." In Hewitt, *Confederate Generals in the Western Theater*, 2:159–85.

Scott, Carole E. "Total War Comes to New Manchester." *Blue & Gray* 12/2 (December 1994): 22–26.

Scribner, B. F. *How Soldiers Were Made; or The War as I Saw It.* Chicago: Donohue & Henneberry, 1887.

Scruggs, Carroll Proctor, ed. *Georgia Historical Markers.* Valdosta, GA: Bay Tree Grove Publishers, 1973.

Sears, Stephen W. "The Bloodiest Day: The Battle of Antietam." *Civil War Times Illustrated* 26/2 (April 1967): 2–47.

———. *Chancellorsville.* New York: Houghton Mifflin Company, 1996.

———. "Fire on the Mountain: The Battle of South Mountain, September 14, 1862." *Blue & Gray* 4/3 (January 1987): 4–22.

———. *George B. McClellan: The Young Napoleon.* New York: Ticknor & Fields, 1988.

———. *Gettysburg.* Boston: Houghton Mifflin Company, 2003.

———. *Landscape Turned Red: The Battle of Antietam.* New Haven: Ticknor & Fields, 1983.

———. "McClellan and the Peace Plank of 1864: A Reappraisal." *Civil War History* 36/1 (March 1990): 57–64.

———. *To the Gates of Richmond: The Peninsula Campaign.* New York: Ticknor & Fields, 1992.

Secrist, Philip L. *Battle of Resaca: Atlanta Campaign 1864.* Macon: Mercer University Press, 1998.
———. "Jefferson Davis and the Atlanta Campaign: A Study in Confederate Command." *Atlanta Historical Bulletin* 17/2 (Fall–Winter 1972): 9–19.
———. "Prelude to the Atlanta Campaign: The Davis-Bragg-Johnston Controversy." *Atlanta Historical Bulletin* 17/1 (Spring–Summer 1972): 9–20.
———. "Resaca: For Sherman a Moment of Truth." *Atlanta Historical Journal* 22/1 (Spring 1978): 9–41.
———. "The Role of Cavalry in the Atlanta Campaign, 1864." *Georgia Historical Quarterly* 56/4 (Winter 1972): 510–28.
———. "Scenes of Awful Carnage." *Civil War Times Illustrated* 10/3 (June1971): 4–9, 45–48.
———. *Sherman's 1864 Trail of Battle to Atlanta.* Macon: Mercer University Press, 2006.
Sehlinger, Peter J. *Kentucky's Last Cavalier: General William Preston, 1816–1887.* Lexington: Kentucky Historical Society, 2004.
Seitz, Don C. *Braxton Bragg: General of the Confederacy.* Columbia, SC: The State Company, 1924.
Sheppard, Jonathan C. *By the Noble Daring of Her Sons: The Florida Brigade in the Army of Tennessee.* Tuscaloosa: University of Alabama Press, 2012.
Sherman, William T. "The Grand Strategy of the Last Year of the War." In Johnson and Buel, *Battles and Leaders of the Civil War*, 4:247–59.
———. *Home Letters of General Sherman.* Edited by M. A. DeWolfe. New York: C. Scribner's Sons, 1909.
———. *Memoirs of General William T. Sherman. By Himself.* 2 vols. New York: D. Appleton & Co., 1875.
Shoup, F. A. "Dalton Campaign—Works at Chattahoochee River—Interesting History." *Confederate Veteran* 3/9 (September 1895): 262–65.
Silverman, Jason H., Samuel N. Thomas, Jr., and Beverly D. Evans IV. *Shanks: The Life and Wars of General Nathan George Evans, C. S. A.* New York: Da Capo Press, 2002.
Simpson, Brooks D., and Jean D. Berlin, eds. *Sherman's Civil War: Selected Correspondence of William T. Sherman, 1860–1865.* Chapel Hill: University of North Carolina Press, 1999.
Simpson, Harold B. *Cry Comanche: The 2nd U. S. Cavalry in Texas, 1855–1861.* Hillsboro, TX: Hill Junior College Press, 1988.
———. *Hood's Texas Brigade: A Compendium.* Fort Worth, TX: Landmark, 1999.
———. *Hood's Texas Brigade: Lee's Grenadier Guard.* Waco, TX: Texian Press, 1970.
———. *Hood's Texas Brigade in Reunion and Memory.* Hillsboro, TX: Hill Junior College Press, 1974.
Simpson, John A. *S. A. Cunningham and the Confederate Heritage.* Athens: University of Georgia Press, 1994.
Skoch, George. "A Test of Rebel Rails." *Civil War Times Illustrated* 25/8 (December 1986): 12–18.
[Smith, Charles H.] "Bill Arp in the Piny Woods. Poetik Meditations among the Tooms." *Atlanta Southern Confederacy,* June 21, 1864.
[———]. *Bill Arp's Peace Papers.* New York: G. W. Carleton, 1873.
Smith, Gustavus W. "The Georgia Militia about Atlanta." In Johnson and Buel, *Battles and Leaders of the Civil War*, 4:331–35.

Sneden, Robert K. "Atlanta. Georgia and Its Rebel Defences." Map. *America's Civil War* 23/4 (September 2010): 31.
Snider, David and William Brooksher. "A Ride down the Sequatchie Valley." *Civil War Times Illustrated* 22/1 (March 1983): 32–39. Reprinted in Brooksher and Snider, *Glory at a Gallop*, 173–84.
Sorrel, G. Moxley. *Recollections of a Confederate Staff Officer.* Dayton, OH: Press of Morningside Bookshop, 1978.
"Southern Newspaper Accounts of These Battles." In Dodson, *Campaigns of Wheeler and His Cavalry 1862–1865*, 213.
Speer, Lonnie R. *Portals to Hell: Military Prisons of the Civil War.* Mechanicsburg, PA: Stackpole Books, 1997.
Spruill, Matt, ed. *Guide to the Battle of Chickamauga.* Lawrence, KS: University Press of Kansas, 1993.
Stackpole, Edward J. *Drama on the Rappahannock: The Fredericksburg Campaign.* Harrisburg, PA: Stackpole Company, 1957.
———. "Showdown at Sharpsburg." *Civil War Times Illustrated* 1/5 (August 1962): 6–10, 42–50.
———. *They Met at Gettysburg.* Harrisburg, PA: Stackpole Company, 1956.
Standard, Diffee William. *Columbus, Georgia, in the Confederacy: The Social and Industrial Life of the Chattahoochee River Port.* New York: William-Frederick Press, 1964.
Starr, Stephen Z. *The Union Cavalry in the Civil War.* 3 vols. Baton Rouge: Louisiana State University Press, 1979–1985.
Steiner, Paul E. *Medical-Military Portraits of Union and Confederate Generals.* Philadelphia: Whitmore Publishing Co., 1968.
Stephenson, Philip D. *The Civil War Memoir of Philip Daingerfield Stephenson, D.D.* Edited by Nathaniel Cheairs Hughes, Jr. Conway: University of Central Arkansas Press, 1995.
Stewart, Alex. P. "The Army of Tennessee: A Sketch." In Lindsley, *The Military Annals of Tennessee,* 1:55–111.
———. "'A Critical Narrative.'" *Confederate Veteran* 16/9 (September 1908): 462–64.
Stickles, Arndt M. *Simon Bolivar Buckner: Borderland Knight.* Chapel Hill: University of North Carolina Press, 1940.
Stinson, Dwight E., Jr. "The Battles of South Mountain." *Civil War Times Illustrated* 1/5 (August 1962): 14–17.
Stokesbury, James L. *A Short History of the Civil War.* New York: William Morrow & Company, 1995.
Stone, Henry. "From the Oostenaula to the Chattahoochee." In Kerksis, comp., *The Atlanta Papers*, 65–97.
———. "Strategy of the Campaign." In Kerksis, comp., *The Atlanta Papers*, 133–62.
———. "Kennesaw Mountain." *Southern Bivouac* 1/4 (December 1882): 135–40.
Storrs, John W. *The "Twentieth Connecticut": A Regimental History.* Anconia, CT: Press of the Naugatuck Valley Sentinel, 1886.
Stout, Samuel Hollingsworth. Papers. MSS 274. Manuscripts, Archives and Rare Books Library, Emory University, Atlanta, GA. Cited as Stout Papers.
Strayer, Larry M., and Richard A. Baumgartner, eds. *Echoes of Battle: The Atlanta Campaign.* Huntington, WV: Blue Acorn Press, 1991.
Strode, Hudson. *Jefferson Davis: Tragic Hero 1864–1889.* New York: Harcourt, Brace & World, 1964.

Strong, William E. "The Death of General James B. McPherson." In Kerksis, comp., *The Atlanta Papers*, 505–39.

Stuart, Jeb. Papers. 1862–1864. Virginia Historical Society. Richmond, VA. Cited as Stuart Papers.

Sutherland, Daniel E. *The Emergence of Total War.* Fort Worth, TX: Ryan Place Publishers, 1996.

"The Swiftsure." *Kennesaw Gazette* 2/9 (May 1, 1887): 1.

Sword, Wiley. *Embrace an Angry Wind—The Confederacy's Last Hurrah: Spring Hill, Franklin and Nashville.* New York: HarperCollins, 1992.

———. *Southern Invincibility: A History of the Confederate Heart.* New York: St. Martin's Press, 1999.

Symonds, Craig L. "Johnston's Toughest Fight." *Military History Quarterly* 16/2 (Winter 2004): 56–61.

———. *Joseph E. Johnston: A Civil War Biography.* New York: W. W. Norton & Company, 1992.

———. *Stonewall of the West: Patrick Cleburne and the Civil War.* Lawrence: University Press of Kansas, 1997.

Talcott, T. M. R. "General Lee's Strategy at the Battle of Chancellorsville." *Southern Historical Society Papers* 34 (January–December 1908): 1–27.

Tankersley, Allen P. *John B. Gordon: A Study in Gallantry.* Atlanta: Whitehall Press, 1955.

Tate, Allen. *Jefferson Davis: His Rise and Fall—A Biographical Narrative.* New York: Minton, Balch & Company, 1929.

Taylor, Thomas T. "A Buckeye Warrior at Atlanta's Gates." *Civil War Times Illustrated* 39/5 (October 2000): 24, 26, 75–79, 81.

Taylor, W. H. "The Campaign in Pennsylvania." In McClure, *Annals of the War.*

Taylor, Walter H. *Four Years with General Lee.* Edited by James I. Robertson. Bloomington: Indiana University Press, 1962.

"The Texans at Sharpsburg." *Confederate Veteran* 22/12 (December 1914): 555–57.

Thomas, Edward J. "A Raid with Joe Wheeler." *Confederate Veteran* 30/5 (May 1922): 169–70.

Thomas, Emory M. *Bold Dragoon: The Life of J. E. B. Stuart.* New York: Harper & Row, 1986.

———. *The Confederate State of Richmond.* Baton Rouge: Louisiana State University Press, 1998.

———. "The Peninsula Campaign: Part I." *Civil War Times Illustrated* 17/10 (February 1979): 4–9, 40–45.

———. "The Peninsula Campaign: Part IV." *Civil War Times Illustrated* 18/3 (June 1979): 10–17.

———. *Robert E. Lee: A Biography.* New York: W.W. Norton & Company, 1995.

Thomas, Wilbur. *General George H. Thomas: The Indomitable Warrior.* New York: Exposition Press, 1964.

Thompson, Phillip F. "Second Manassas: Enfilading Thunder." *Civil War Magazine* 46 (August 1994): 49–51.

Thomsen, Briand, and Martin H. Greenburg, eds. *Alternate Gettysburgs.* New York: Berkley Books, 2002.

Towles, Louis P. "Dalton and the Rebirth of the Army of Tennessee." In *South Carolina in the Civil War and Reconstruction Eras: Essays from the Proceedings of the South Carolina*

Historical Association, edited by Michael Brem Bonner and Fritz Hamer, 79–92. Columbia: University of South Carolina Press, 2016.

Townsend, Mary Bobbitt. *Yankee Warhorse: A Biography of Major General Peter Osterhaus.* Columbia: University of Missouri Press, 2010.

Trammell, Jack. "'Life is Better Than Limb.'" *America's Civil War* 21/6 (January 2009): 46–51.

Trask, W. L. War Journal. Typescript at Kennesaw Mountain National Battlefield Park (courtesy of Dennis Kelly). Microfilm copy of handwritten original at Manuscripts, Archives, and Rare Books Library, Emory University, Atlanta, GA.

Trudeau, Noah Andre. *Gettysburg: A Testing of Courage.* New York: HarperCollins, 2002.

Tucker, Glenn. *Chickamauga: Bloody Battle in the West.* Dayton, OH: Press of Morningside Bookshop, 1972.

Tucker, Spencer C. *Brigadier General John D. Imboden: Confederate Commander in the Shenandoah.* Lexington: University Press of Kentucky, 2003.

Tuthill, Richard S. "An Artilleryman's Recollections of the Battle of Atlanta." In Kerksis, comp., *The Atlanta Papers*, 431–49.

"Two Letters." *Confederate Veteran* 26/9 (September 1918): 395–96.

U. S. War Department. *The War of the Rebellion: A Compilation of the Official Records of the Union and Confederate Armies.* 128 volumes. Washington: Government Printing Office, 1880–1901. Cited as *OR*.

Upson, Theodore P. *With Sherman to the Sea: The Civil War Letters Diaries & Reminiscences of Theodore P. Upson.* Edited by Oscar Osburn Winther. Baton Rouge: Louisiana State University Press, 1943.

Ural, Susannah J. "Fighting 'Too Fast.'" *Civil War Times* 56/6 (December 2017): 34–41.

———. *Hood's Texas Brigade: The Soldiers and Families of the Confederacy's Most Celebrated Unit.* Baton Rouge: Louisiana State University Press, 2017.

———, and Rick Eiserman. "The Winter That Made the Texas Brigade." *Civil War Times Illustrated* 49/4 (August 2011): 36–41.

Vandiver, Frank E. "General Hood as Logistician." *Military Affairs* 16/1 (Spring 1952): 1–11.

———. Introduction to *Narrative of Military Operations* by Joseph E. Johnston. Indianapolis: Indiana University Press, 1960.

———. *Ploughshares into Swords: Josiah Gorgas and Confederate Ordnance.* College Station: Texas A&M University Press, 1994.

———, ed. "Proceedings of the Second Confederate Congress First Session." *Southern Historical Society Papers* 51 (1958): 1–264.

Van Horne, Thomas B. *The Life of Major-General George H. Thomas.* New York: C. Scribner's Sons, 1882.

Vaughan, Turner. "Diary of Turner Vaughan Co. 'C' 4th Alabama Regiment, C. S. A. Commenced March 4th, 1863 and Ending February 12, 1864." *Alabama Historical Quarterly* 18/4 (Winter 1956): 573–601.

Venet, Wendy Hamand. *A Changing Wind: Commerce and Conflict in Civil War Atlanta.* New Haven: Yale University Press, 2014.

———. "From Gate City to Gotham: Sam Richards Chronicles the Civil War in Atlanta." In *Breaking the Heartland: The Civil War in Georgia*, edited by John D. Fowler and David B. Parker, 150–69. Macon: Mercer University Press, 2011.

Vermilya, Daniel J. *The Battle of Kennesaw Mountain.* Charleston: History Press, 2014.

"Vincents Subdivision Map of the City of Atlanta." 1853. Reproduced with key by H. B. Baylor. Atlanta, 1897. Atlanta History Center.

Walker, C. Irvine. *Great Things Are Expected of Us: The Letters of Colonel C. Irvine Walker, 10th South Carolina Infantry, C. S. A.* Edited by William Lee White and Charles Denny Runion. Knoxville: University of Tennessee Press, 2009.

Walker, Mary Hubner. *Charles W. Hubner: Poet Laureate of the South.* Atlanta: Cherokee Publishing Company, 1976.

Walker, Scott. *Hell's Broke Loose in Georgia: Survival in a Civil War Regiment.* Athens: University of Georgia Press, 2005.

Walsh, George. *"Damage Them All You Can": Robert E. Lee's Army of Northern Virginia.* New York: Forge, 2002.

Warner, Ezra J. *Generals in Gray: Lives of the Confederate Commanders.* Baton Rouge: Louisiana State University Press, 1959.

Waters, Zack C. "Death Was Feasting in Our Midst: Major General William B. Bate and the Battle of Dallas, Georgia." Privately printed, n.d.

———, ed. "Lines of Battle: The Partial Reports of Confederate Maj. Gen. William B. Bate." In Savas and Woodbury, *The Campaign for Atlanta*, 197–219.

Watkins, James W. Papers. MSS 413. Manuscripts, Archives and Rare Books Library, Emory University, Atlanta, GA. Cited as Watkins Papers.

Weiss, Timothy F. "'I lead you to battle': Joseph E. Johnston and the Controversy at Cassville." *Georgia Historical Quarterly* 91/4 (Winter 2007): 424–52.

Welch, Richard F. "Gettysburg Finale." *America's Civil War* 6/3 (July 1993): 50–57.

———. "'Old Rosy' and the Fatal Order." *America's Civil War* 1/2 (July 1988): 46–47.

Wellman, Manly Wade. *Giant in Gray: A Biography of Wade Hampton of South Carolina.* New York: Charles Scribner's Sons, 1949.

Welsh, Bill. "Firing the Gap." *America's Civil War* 6/6 (January 1994): 39–44.

Welsh, Jack D. *Medical Histories of Confederate Generals.* Kent, OH: Kent State University Press, 1995.

———. *Medical Histories of Union Generals.* Kent, OH: Kent State University Press, 1996.

Wert, Jeffry D. *Cavalryman of the Lost Cause: A Biography of J. E. B. Stuart.* New York: Simon & Schuster, 2008.

———. *General James Longstreet: The Confederacy's Most Controversial General—A Biography.* New York: Simon & Shuster, 1993.

———. "Little Round Top and Devil's Den." *Civil War Times Illustrated* 27/4 (Summer 1988): 42–49.

———. "Regiment." In *Historical Times Illustrated Encyclopedia of the Civil War,* ed. Patricia L. Faust, 621. New York: Harper & Row, 1986.

———. "Return to the Killing Ground." *America's Civil War* 4/2 (July 1991): 18–25.

Whaley, Elizabeth H. *Forgotten Hero: General James B. McPherson.* New York: Exposition Press, 1955.

"When Georgia Howled: Sherman on the March." Directed by Bruce Burkhardt. Georgia Public Broadcasting, 2015.

White, Cheryl H. *Confederate General Leonidas Polk: Louisiana's Fighting Bishop.* Charleston: History Press, 2013.

White, William Lee. *Bushwhacking on a Grand Scale: The Battle of Chickamauga, September 18–20, 1863.* El Dorado, CA: Savas Beatie, 2013.

Wiley, Bell I. "Billy Yank and Johnny Reb in the Campaign for Atlanta." *Civil War Times Illustrated* 3/4 (July 1964): 18–22. Reprinted as "Billy Yank and Johnny Reb on the Road to Atlanta," *Civil War Times Illustrated* 42/5 (December 2003): 67–74, 76–78, 80, 82.

———. *Embattled Confederates.* New York: Harper & Row, 1964.

———. "A Story of 3 Southern Officers." *Civil War Times Illustrated* 3/1 (April 1964): 6–9, 28–24.

Williams, Alpheus S. *From the Cannon's Mouth: The Civil War Letters of General Alpheus S. Williams.* Edited by Milo M. Quaife. Detroit: Wayne State University Press, 1959.

Williams, David. *Georgia's Civil War: Conflict on the Home Front.* Macon: Mercer University Press, 2017.

Williams, Edward B. *Hood's Texas Brigade in the Civil War.* Jefferson, NC: McFarland, 2012.

Williams, Hiram S. *This War So Horrible: The Civil War Diary of Hiram Smith Williams.* Edited by Lewis N. Wynne and Robert A. Taylor. Tuscaloosa: University of Alabama Press, 1993.

Williams, T. Harry. "The Military Leadership of North and South." In *Why the North Won the Civil War,* edited by David Donald, 33–54. Baton Rouge: Louisiana State University Press, 1960.

Wills, Brian Steel. *A Battle from the Start: The Life of Nathan Bedford Forrest.* New York: HarperCollins, 1992.

———. *George Henry Thomas: As True as Steel.* Lawrence: University Press of Kansas, 2012.

Wills, Charles W. *Army Life of an Illinois Soldier, including a day by day record of Sherman's March to the Sea; letters and diary of the late Charles W. Wills.* Mary E. Kellogg, comp. Carbondale: Southern Illinois University Press, 1996.

Wilson, Jerome P. Papers. MSS 176. Manuscripts, Archives and Rare Books Library, Emory University. Cited as Wilson Papers.

Wingfield, Marshall. *General A. P. Stewart: His Life and Letters.* Memphis: Western Tennessee Historical Society, 1954.

Winkler, Mrs. A. V. *The Confederate Capital and Hood's Texas Brigade.* Baltimore: Butternut & Blue Press, 1991.

Winkler, Mrs. C. M. [Angelina V. S.]. *The Life and Character of Gen. John B. Hood. Written at request, and Published by authority of Hood's Texas Brigade Association, and read before the Association by the authoress, June 27, 1885.* Austin: Draughon & Lambert, Printers, n. d.

Woodward, C. Vann, ed. *Mary Chesnut's Civil War.* New Haven, CT: Yale University Press, 1981.

Woodworth, Steven E. *Chickamauga: A Battlefield Guide with a Section on Chattanooga.* Lincoln: University of Nebraska Press, 1999.

———. *Davis and Lee at War.* Lawrence: University Press of Kansas, 1995.

———. *Decision in the Heartland: The Civil War in the West.* Lincoln: University of Nebraska Press, 2011.

———. *A Deep Steady Thunder: The Battle of Chickamauga.* Fort Worth: Ryan Place Publishers, 1996.

———. "How to Lose a City." *Civil War* 9/1 (January–February 1991): 28–33.

———. *Jefferson Davis and His Generals: The Failure of Confederate Command in the West.* Lawrence: University Press of Kansas, 1990.

———. *No Band of Brothers: Problems in the Rebel High Command.* Columbia: University of Missouri Press, 1999.

———. *Nothing but Victory: The Army of the Tennessee, 1861–1865.* New York: Alfred A. Knopf, 2005.

———. *Sherman: A Biography.* New York: Palgrave Macmillan, 2009.

———. "Soldier with a Blunted Sword: Braxton Bragg and His Lieutenants in the Chickamauga Campaign." In Hewitt, *Confederate Generals in the Western Theater*, 1:123–33

———. *While God Is Marching On: The Religious World of Civil War Soldiers.* Lawrence: University Press of Kansas, 2001.

Worsham, W J. *The Old Nineteenth Tennessee Regiment, C.S.A.* Knoxville: Press of Paragon Printing Company, 1902.

Wright, Mrs. D. Giraud [Louise Wigfall]. *A Southern Girl in '61: The War-Time Memories of a Confederate Senator's Daughter.* New York: Doubleday, Page & Company, 1905.

Wright, Marcus J. *General Officers of the Confederate Army.* Mattituck, NY: J. M. Carroll & Company, 1983.

Wyatt, J. N. "Dalton-Atlanta Campaign." *Confederate Veteran* 5/10 (October 1897): 519–21.

Wynne, Ben. *A Hard Trip: A History of the 15th Mississippi Infantry, CSA.* Macon: Mercer University Press, 2010.

York, R. W. "General Hood's Release from Arrest—An Incident of the Battle of Boonsboro." *Our Living and Our Dead* 2/4 (June 1875): 420–23.

Zaworski, Robert E. *The General and the House on a Hill.* Privately printed, 2011.

Zinn, John G. *The Mutinous Regiment: The Thirty-third New Jersey in the Civil War.* Jefferson, NC: McFarland, 2005.

Index